D1376610

THE NATURALIZED ANIMALS OF BRITAIN AND IRELAND

MANDARIN DUCK DRAKE

THE NATURALIZED ANIMALS OF BRITAIN AND IRELAND

CHRISTOPHER LEVER

First published in 2009 by New Holland Publishers (UK) Ltd
London · Cape Town · Sydney · Auckland
www.newhollandpublishers.com

Garfield House, 86–88 Edgware Road, London W2 2EA, UK
80 McKenzie Street, Cape Town 8001, South Africa
Unit 1, 66 Gibbes Street, Chatswood, New South Wales, Australia 2067
218 Lake Road, Northcote, Auckland, New Zealand

1 3 5 7 9 10 8 6 4 2

ISBN 978 1 84773 454 9

Senior Editor: Krystyna Mayer
Design: Peter Crump
Cartography: Stephen Dew
Production: Melanie Dowland
Publisher: Simon Papps
Editorial Direction: Rosemary Wilkinson

Reproduction by Pica Digital (Pte) Ltd, Singapore
Printed and bound in Singapore by Tien Wah Press

CONTENTS

PREFACE

THIS BOOK, WHICH UPDATES, REVISES AND EXPANDS *The Naturalized Animals of the British Isles* (Lever 1977), describes when, where, why, how and by whom the various alien vertebrate animals now living in the wild in Britain and Ireland were introduced, how they subsequently became naturalized, their present status and distribution, and their ecological and/or socio-economic impact, if any, on the British and Irish environment and economy.

The naturalized vertebrate fauna of Britain and Ireland has altered materially since 1977. Some species have died out, new ones have become established, and the status and distribution of others have changed materially, and in some cases dramatically. The impact of some species, which 30 years ago was considered to be negligible or about which little was known, is now realized to be profound.

The criteria for the inclusion of a species in this book are that it should have been imported from its natural range either deliberately or accidentally by humans, and that in most cases it should currently be established in the wild in self-maintaining and self-perpetuating populations independent of humans. Also included are 'feral domesticated species' (some of which, such as the Reindeer in Scotland, are strictly speaking 'acclimatized' rather than 'naturalized'), 'reintroduced species' (formerly native species that, having become extinct, have been successfully reintroduced), 'proposed reintroductions' and 'ephemeral species' (successfully naturalized species that have since died out). Occasionally, as in the case of, for example, the European Rabbit, when an animal has been extinct for thousands of years, its (re)introduction is treated as an introduction.

Each species account is a monograph on that particular species. In general, only species that have become successfully naturalized are discussed. The inclusion of those that failed to become established would have been of doubtful value unless the precise reasons for each failure were known, and would inevitably have resulted in many omissions. Also not included are:

1. Species such as the European Hedgehog, which are native to Britain and Ireland but have been successfully translocated within a constituent country to places (e.g. from the Scottish mainland to the Hebridean islands) where they do not occur naturally.
2. Species like the European Grayling, which are native to one or more of the constituent countries (England and Wales) and have been translocated to another constituent country (Scotland) where they are not indigenous.
3. Natural colonists such as the Collared Dove.

Terminology

The definitions of some of the terms most frequently associated with the introduction of animals and plants are as follows.

Acclimatized A species living in the wild outside its natural range, which is supported by humans and dependent on them.

Adventive An introduced species that has yet to become naturalized or acclimatized.

Alien A non-native (or introduced) species.

Domesticated 'To dwell in a house. … the inmate of a house. … to be at home with. … to be of the household of.' *Oxford English Dictionary* (*OED*).

Feral 'Animals that have lapsed into a wild form from a domesticated condition' (*OED*). It is thus incorrect to refer to any animal (e.g. American Mink) that has escaped or been released from captivity as 'feral'. To regard the American Mink as a 'domesticated' species is preposterous and wrong.

Introduction The release of a species into a country in which it is not known to have occurred naturally within historic times.

Invasive 'An entrance or an incursion' (*OED*). It is thus again incorrect to describe an invasive species only as one that has caused harm. 'Invasive' surgery is surgery that enters the body, not surgery that causes damage! Thus an 'invasive' species is correctly simply one that has 'invaded' – that is, entered – a country.

Native A species that is a member of the natural biotic community.

Naturalized A species established in the wild in self-maintaining and self-perpetuating numbers without the support of, and independent of, humans.

Reintroduction The release by humans of a species into a country where it was formerly indigenous but where it has since died out.

Translocation The movement by humans of a species from a country where it occurs, either as a native or an alien, to another part of the same country.

British County Names

In 1974 the names and boundaries of several counties in England and Wales, and in 1975 those of some in Scotland, were changed. As most of the events described here antedate these changes, the old names and boundaries have in general been adhered to, with the new names being used for recent events.

References

References are given in the text when the author(s) quoted are the principal (or sole) source(s). Other references are placed at the end of each species account.

Nomenclature

For mammals this follows Harris and Yalden 2008; for birds Dudley et al. 2006; for fish Maitland 2004 and Davies et al. 2004; and for reptiles and amphibians Beebee and Griffiths 2000.

Distribution Maps

Maps for mammal distribution are adapted from Harris and Yalden (2008) by permission of Derek Yalden and © the Mammal Society, and those for fish from Maitland (2004) by permission of Peter Maitland and © the Freshwater Biological Association. Maps for birds were specially produced for this book by the British Trust for Ornithology. The author compiled those for reptiles, amphibians, and feral and reintroduced species. The maps for reintroduced species show reintroduction sites only. The maps for fish include hydrometric areas (river catchments having one or more outlets to the sea or a tidal estuary). For convenience, some hydrometric areas include several smaller catchments with separate tidal outlets.

Status

The status of some species, whether as native, alien, reintroduced or feral, is in some instances hard to determine; in these cases an arbitrary decision about the heading under which to include it has been made.

INTRODUCTION

THE NATIVE WILDLIFE OF BRITAIN AND IRELAND is meagre compared to that of continental Europe. Natural colonization of Britain and Ireland by terrestrial animals was only feasible when they were joined to continental Europe. Gupta et al. (2007) speculate that (c 9500 BC) a megaflood breached the rock dam at the Dover Strait, instigating the catastrophic drainage of an immense post-glacial lake in the southern North Sea basin and the formation of what is now the English Channel (see also Smith 1985; Stringer 2007). Since then, additions to the British fauna have been forced – with or without the assistance of humans – to face the barrier of the sea.

The situation in the Orkneys, Shetlands, Outer Hebrides and Ireland is, as Yalden and Kitchener (2008) point out, more complicated. Because of the depth of the North Channel (between Ireland and south-western Scotland), it is unlikely that Ireland was joined to Scotland at any time in the post-glacial period (Devoy 1986; Yalden 1982, 1999). This suggests that the native land animals of Ireland immigrated during the warmer conditions of the Windermere Interstadial and managed to survive the colder conditions of the Younger Dryas (c 11,000–10,000 BP). Because of the even greater depth of the North Minch between western Scotland and the Outer Hebrides, and the Pentland Firth between north-eastern Scotland and the Orkneys (and Shetlands), they were clearly never joined to the Scottish mainland, and their present terrestrial fauna (with the exception of the Eurasian Otter) has been introduced by humans (Corbet 1961).

Apart from species introduced accidentally, such as, for example, the Orkney Vole, which was probably imported by Neolithic settlers concealed in fodder for their domestic stock, and those like the Grey Squirrel, which was released simply out of curiosity as to the outcome, alien animals have been introduced to Britain and Ireland for three main reasons: economic, ornamental and sporting.

Species in the first category can be divided into those introduced primarily as a source of food, such as the Common Pheasant, and those imported for their pelts, such as the American Mink. The European Rabbit filled an unenviable dual role, being valued for its flesh and for its fur. Animals introduced for ornamental purposes include principally deer, the so-called 'ornamental' pheasants and waterfowl such as the beautiful Mandarin Duck. The Rainbow Trout is a sporting fish par excellence, highly regarded by anglers as a doughty fighter and a fine fish for the table. A minor reason in Britain and Ireland (though a major one elsewhere) for introducing an alien species is as a form of biological control; thus the Chinese Grass Carp has been released to clear waters of unwanted macrophytic vegetation. Alien species introduced to control a pest species may, having eradicated the target species, begin to prey on non-target species, as in the case of the Cane Toad in Australia (see Lever 2001).

A number of factors influence the success or failure of an introduced species in becoming naturalized in an alien environment, including:

- A suitable habitat and climate.

- The existence of a vacant ecological niche.
- An adequate supply of acceptable food.
- An absence of predators to which the introduced species may be unaccustomed in its native range.
- Fecundity.
- A sufficiently large founder stock.
- In the case of birds, the absence or abandonment of the instinct to migrate.

Being a generalist rather than an obligate feeder is, in the case of herbivores, a useful attribute in an invasive species, as also is having non-specific habitat requirements. Cassey (2002) found strong evidence that increased habitat generalism, a lack of migratory tendency and sexual monochromatism together explain a significant variation in the successful naturalization of introduced land birds. When the above criteria are fulfilled, an introduced animal will initially increase fairly rapidly to the maximum population that the colonized area will support. There may then be some contraction in both range and numbers, until a level is eventually reached at which both population and distribution become fully stabilized.

Various far-reaching and unpredictable consequences may attend the naturalization of exotic species. The outcome of an introduction depends on the species itself as well as the native ecosystem; generalists are normally more successful than specialists, and among mammals herbivores are usually more successful than carnivores. Few alien species have the same characteristics and ecological requirements as native species, and this can adversely affect entire communities. The adoption of new foods by an exotic species can seriously damage native ecosystems, and the introduction of an alien species into the range of a native near-congener may give rise to hybridization, almost invariably to the detriment of the latter. Introductions of exotic species can also sometimes lead to speciation. Most introduced organisms will not be successful in becoming established (see the 'Tens Rule' opposite), the most important reasons for failure being abiotic ones (e.g. inclement climatic conditions) and biotic ones (e.g. generalist predators and parasites, pathogens and diseases). Introduced species are most likely to become established when competition from natives is minimal.

The establishment of species outside their natural geographical range is an important driver of changes in global biodiversity. It is thus important to understand why some species are more successful than others. Cassey et al. (2005) found that variables describing characteristics specific to individual introduction events such as propagule pressure/introduction effort are the most important predictors of success (see also Lockwood 1999).

Copp et al. (2007) examined the potential relationship between naturalized populations of freshwater fish and the degree of propagule pressure and the diversity of fish importations. The principal pathways of introduction were importations for ornamental purposes (e.g. aquaria and garden ponds) and sport angling. There were no apparent cases of introductions in ballast water as has occurred, for example, in the Great Lakes of North America.

Invasive alien species are, after habitat destruction, the most important cause of loss of biodiversity through the extinction or reduction of native species. The Millennium Ecosystem Assessment (Alcamo 2003) identified exotic species as one of the six main problems causing 'ecosystems to go into a sudden decline that is irreversible on a human timescale', while as long ago as 1999 the Global Diversity Forum referred to alien species as 'the only form of pollution which spontaneously self-replicates'. This threat from introduced species is increasing with the continued growth of global tourism, and trade and climate change.

In order to properly assess the potential threat of invasive species in Britain and Ireland, a comprehensive checklist of species and of their likely biological impact is required, but strangely does not exist. Monitoring the biota is, as Hill et al. (2005) point out, one of the key requirements for dealing with introduced species agreed under the Convention on Biological Diversity. Many biologists regard non-native species as beneath their notice – indeed, Fitter (1959) records how 'Terry Gompertz, in her pioneer field study of the feral pigeons of London [*Bird Study* 4: 2–13 (1957)], relates her astonishment … at the contemptuous attitude of many ornithologists towards her subject.' This prejudice, which in some circles continues to this day, has led to gross under-recording of some naturalized species. Even such a distinguished ornithologist as Colin Bibby records (*British Birds* 93: 2–3, 2000) that he was 'appalled to see a photograph of Egyptian Geese at a National Trust property where they were welcomed as rare visitors'. Although we are, of course, right to control harmful exotic species, we should not condemn all aliens simply because they are foreign, and we should be prepared to tolerate, even if we feel we cannot welcome, harmless additions to the British and Irish fauna.

Many invasive species pose little threat to the natural environment. A 'Tens Rule' propounded by Williamson and Fitter (1996) states that as a rule of thumb only one in ten of introduced species appears in the wild, only one in ten of those becomes naturalized, and only one in ten of those, i.e. 0.1 per cent of the species arriving, becomes a pest. This low success rate, as Welch et al. (2000) point out, reflects the general unsuitability of most alien species and the lack of ecological niches for colonization.

When, however, non-native species do become established, they can materially transform ecosystems and put at risk native species. All terrestrial and marine natural and semi-natural habitats can be seriously affected. Alien species can damage economic interests, including agriculture, horticulture and silviculture, and can even threaten public health (DEFRA 2003, 2007). They may affect native species through predation, competition (for food and shelter), displacement, community structure, ecosystem functions, genetic integrity through hybridization, and population viability through the introduction of parasites and diseases.

It is becoming increasingly clear that evolutionary processes play an important role during the naturalization of invasive species. Genetic drift during the process of establishment, succeeded by strong selection due to alteration of biotic circumstances, is the precursor of rapid evolutionary changes in naturalized populations. Empirical evidence increasingly suggests that hybridization between alien and indigenous species, coupled with genetic variability following multiple introductions, can contribute to the evolution of new species. Recent research also suggests that genetic bottlenecks may be less of a genetic paradox than was previously believed (Hänfling 2007).

Introduced populations of mammals and birds may exhibit different genetic characteristics to those in native populations. Such introduced populations are frequently formed by the movement of a small founder stock. This results in founder effects and subsequent genetic drift, often resulting in greater differences in allozyme patterns between introduced populations than between naturally occurring populations. In many instances a high percentage of alleles is lost within a few generations of introduction. In some circumstances the mean level of heterozygosity is also greatly diminished (see page 301). Although a reduction in the number of alleles in a populations should in theory cause a reduced potential to track environmental changes, in fact there is little actual evidence for this. Similarly, there is little evidence of any serious inbreeding problems in naturalized populations (see page 301) (Sjoberg 1996).

McNeely (2003) has summarized the human dimensions relating to invasive species. Some species can and do invade new habitats and constantly expand their range, thus posing a threat to indigenous species, human health, and other social and economic interests. Since it seems to be widely considered that in general most invasive species are 'bad', the issue can unite stakeholders who might otherwise be in opposition, such as farmers and conservationists. Drawing attention to the human dimension can shift the focus from the animal itself to the actions by humans that enabled it to become established and spread, or to its management or control, and suggests that concentrating only on the invader may provide only symptomatic relief. A permanent solution demands knowledge of the economic or other motivations that resulted in the species' introduction, though this may sometimes be hard to obtain.

The complicated relationship between globalization and invasion pathways is, as McNeely points out, possibly the most important human aspect of invasive species, and should be a priority for policy makers. An important implication is that concern about invasive species needs to be expressed relative to the resources of the global economic community, which translates into monetary terms. Thus many of those concerned about the threats posed by invasive species have justifiably relied on economics to argue and support their cause.

If the threats posed by introduced species are to be successfully addressed, there is a strong case for conservation non-governmental organizations (NGOs) to become more actively involved. They may have avoided doing so in the past for fear of alienating some of their supporters by advocating the eradication of some animals (the RSPB is said to have lost members through its campaign to exterminate the Ruddy Duck) in order to preserve others. If the support of conservation NGOs and the general public is to be gained, our knowledge of the biology, ecology and human dimensions of introduced species needs to be improved; without it any new arrival can result in unexpected and unpredictable consequences.

The key factor affecting introduced species in Britain during the past half century has been the enactment of the Wildlife and Countryside Act 1981 and the Wildlife (Northern Ireland) Order 1985. Hitherto, those wishing to release exotic species into the wild had what amounted to virtually a free hand. Section 14 Part I of the Act lays down that:

> (1) Subject to the provisos of this Part, if any person releases or allows to escape into the wild any animal which (a) is of a kind which is not ordinarily resident in and is not a regular visitor to Great Britain in a wild state; or (b) is included in Part I Schedule 9, he shall be guilty of an offence.

Schedule 9 lists a total of 42 introduced or reintroduced species (10 mammals, 17 birds, 3 reptiles, 6 amphibians and 6 fish), some of which after a lapse of nearly 30 years have either died out naturally or, as in the case of the Coypu, have been deliberately eradicated by man. On the other hand, since 1981 new species have escaped from captivity and have become naturalized; others, such as the Red Kite, have been successfully reintroduced to England, Scotland and Northern Ireland, and the White-tailed Eagle to Scotland. Corresponding legislation in the Republic of Ireland is the Wildlife Act 1976 and the European Communities (Natural Habitats) Regulations 1997. Section 52 of the former, as amended by the Wildlife (Amendment) Act 2000, makes it an offence to release or allow to escape into the wild any exotic (i.e. non-native) species, or to attempt to establish it in the wild other than in accordance with a licence issued under the Act.

The UK government has international obligations to address the problem of introduced species, principally under the provisions of the Convention on Biological Diversity (CBD), the Bern Convention on Conservation of European Wildlife and Habitats and the EU Habitats and Species Directive. In fulfilment of these obligations, *The Review of Non-Native Species Policy*, published by the Department for Environment, Food and Rural Affairs (DEFRA) in 2003, makes a number of recommendations for controlling invasive animals and plants. Britain is required under the CBD to 'control or eradicate those alien species which threaten ecosystems, habitats or species'. In May 2008, DEFRA launched the Invasive Non-Native Species Framework Strategy for Great Britain; although the highest priority is given to 'prevention' and 'early detection', control or eradication of already established exotics must also be considered (Macdonald and Burnham 2008). This is clearly so, but in Britain only two naturalized mammals have so far been deliberately eradicated – the Muskrat (1937) and the Coypu (1987). A contentious attempt is currently under way to eliminate the Ruddy Duck, and a similar policy is under consideration for the Rose-ringed Parakeet. It is generally acknowledged that stakeholder support is fundamental to the success of any conservation project, and this is especially so when something as sensitive as the control or eradication of a species is contemplated – in particular if it is aesthetically attractive. Consequently, the control of invasive species has come to be recognized as not simply a scientific issue but also a social and political one.

A factor that will have a major influence on the future of both alien and native species in Britain is a change in climate caused by global warming as a result of greenhouse gas emissions resulting from the burning of fossil fuels. Although some introduced (and native) species will benefit from global warming, others will suffer. According to *The Climatic Atlas of European Birds* (2008) and the Intergovernmental Panel on Climate Change, for many species a rise in temperature of only 2°C (very much less than may occur) will cause often devastating reductions in range, habitats and populations. Species that need a colder climate will be driven north from England to Scotland, while some of those already there will disappear. The UK Climate Change Bill originally proposed a reduction in CO_2 emissions of at least 60 per cent by 2050. This target was based on a report by the Royal Commission on Environmental Pollution in 2000. Since then, developments in the science of climate change have shown that this target was far too low to avoid the most serious consequences. WWF-UK, the Tyndall Centre for Climate Change Research, the UN Human Development Report 2007/2008 and leading scientists have all made it clear that the government must accept the latest scientific advice, which clearly indicates that Britain should aim to reduce its CO_2 emissions, relative to those in 1990, by at least 80 per cent by 2050. In October 2008, the government announced that this advice had been accepted and that it would be incorporated in the Climate Change Bill.

References

Alcamo 2003; Blackburn & Duncan 2001; Carlton & Ruiz 2000; Cassey 2001, 2002; Cassey et al. 2005; Corbet 1961; d'Ayala 2003; DEFRA 2003, 2007; Devoy 1986; Genovesi & Shine 2003; Gupta et al. 2007; Hänfling 2007; Henderson 2007; Hill et al. 2005; Holdgate 1986; Lever 1977, 1996, 2005b, in press; Lockwood 1999; McNeely 2003; Mitchell-Jones et al., in Harris & Yalden 2008; Sjoberg 1996; Smith 1985; Sol et al. 2002; Stokes et al. 2004; Stringer 2007; Williamson 1993; Williamson & Fitter 1996; Yalden 1982, 1999; Yalden & Kitchener, in Harris & Yalden 2008.

TABLES

ACKNOWLEDGEMENTS

I owe a special debt of gratitude to the staff of the various libraries in which I carried out my research, especially Paul Cooper, Alison Harding, Kamila Reekie, John Rose and Angela Thresher in the Natural History Museum in London and Tring, where I was given free photocopying facilities, and Gina Douglas and Lynda Brooks in the Linnean Society of London. Other libraries whose staff was most helpful were the Alexander Library of the Edward Grey Institute of Field Ornithology, Oxford; the British Trust for Ornithology, Thetford; the Science Reference Library, London; the Zoological Society of London; and the Imperial College of Science, Technology and Medicine, London.

I am particularly grateful to the British Trust for Ornithology for producing the bird distribution maps. The Director, Andy Clements, kindly authorized this project, which was organized by Mark Rehfisch. Stuart Newson created the maps from data provided by Mark Grantham, Andy Musgrove, David Noble and Graham Austin in conjunction with the WWT, RSPB, JNCC and volunteer counters. Olivia Crowe kindly gave permission for the use of data from BirdWatch Ireland.

I am also most grateful to Peter Maitland and the Freshwater Biological Association for kindly allowing me to adapt the relevant maps from Maitland (2004), and to Derek Yalden and the Mammal Society for permission to adapt the relevant maps from Harris and Yalden (2008).

Without the generous cooperation of a host of correspondents this book would have been very much the poorer. My grateful thanks are due to: Graham Appleton; John Baker; Mike Blair; Keith Bowey; Michelle Bromley; Clarissa Bryan; Duncan Cameron; Ian Carter; Paul Castle; George Christie; Barry Clarke; Damian Clarke; Michaela Clarke; Tres Connaghan; Rachel Coombes; John Cranfield; Oliver Crimmen; Andrew Cunningham; Dylan Davenport; Gareth Davies; Lord Davies; Brian Etheridge; Andy Evans; Jim Foster; Ian Francis; Trent Garner; Martin Gaywood; Tony Gent; David Gibbons; Richard Gibson; Martin Goulding; John Hall; Liz Halliwell; Tristan Hatton-Ellis; Linda van Gucci; Andrew Harby; Nigel Hardwick; Kate Hawkins; Iain Henderson; Nigel Hewlett; Phil Hickley; Mark Holling; Derek Holman; Jules Howard; Liz Howe; Nick Jackson; Martin Kalaher; Steve Langham; Tom Langton; Jenny Lennon; Peter Litherland; Rae Lyster; Peter Maitland; Colin McCarthy; Allan Mee; Pat Morris; Alan Morton; Jim Murphy, M.P.; Peter Newbery; Malcolm Ogilvie; Duncan Orr-Ewing; Lorcan O'Toole; Nigel Phillips; Robert Prys-Jones; James Robinson; Mark Rehfisch; Neil Reid; Roddy Rugman; Richard Saunders; Alison Shaw; Carole Showell; Craig Shuttleworth; Peter Steel; Rob Strachan; Robert Straughan; Ron Summers; Darren Tansley; Angela Taylor; Huw Thomas; Rob Thomas; Roger Trout; Sir Humphrey Wakefield, Bt; Robin Ward; David Waters; Ruth Waters; Carrie Watt; Ian Wellby; Charles Williams; Sir Richard Williams-Bulkeley, Bt; Charles Wilson; Ray Woods; Simon Wotton; Derek Yalden.

For her help and cooperation in acting as midwife to this book I extend my thanks to Krystyna Mayer of New Holland.

Finally, I once again express my grateful thanks to Pat Berry for her patience in deciphering my chaotic and well-nigh unreadable manuscript.

As for my previous books, the working material for this book has been deposited in the library of the Natural History Museum in London.

Christopher Lever,

Winkfield, Berkshire, 2009

NATURALIZED SPECIES

HUMANS ARE INVETERATE AND INCORRIGIBLE MEDDLERS, never content to leave anything as they find it but always seeking to alter and – as they see it – to improve. In no fields is this truer than in those of the animal and plant kingdoms.

One of the ways in which humans have sought to modify the natural environment is by the introduction of animals and plants. They have done this so successfully and so widely that internationally (though not yet fortunately in Britain and Ireland) naturalized (successfully introduced) species have become, after loss of habitat, the principal cause of biological declines and extinctions. Fortunately, the enactment of the Wildlife and Countryside Act 1981 in Britain and similar legislation in the Republic of Ireland have, as mentioned in the introduction, ensured that no further alien species may be released or negligently allowed to escape into the wild in Britain or Ireland.

The species described in the following section – 19 mammals, 21 birds, 6 reptiles, 13 amphibians and 18 fish – were all introduced to Britain and Ireland between Neolithic times and the late 20th century, and are collectively found naturally on every continent except Antarctica. The vast majority have become established in breeding populations, and can thus properly be termed 'naturalized'. A very few, such as the Chinese Grass Carp, European Pond Terrapin and Red-eared Slider, do not breed here because the climate and environmental conditions are inimical to reproduction. This situation could well change should global warming continue as predicted. In any case, such species already exist in sufficiently large, albeit non-reproducing populations to justify inclusion.

Other species that would probably benefit from global warming include those southern European reptiles and amphibians at present established in breeding populations only tenuously in southern England. Such species could well become more firmly established in that area, and might even begin to expand their range northwards. On the other hand, milder climatic conditions might sound the death knell for such species as the (reintroduced) Western Capercaille, which would be driven further north in Scotland until it had nowhere else to go.

Although some naturalized species in Britain and Ireland have become significant ecological and/or economic pests (e.g. the Grey Squirrel, American Mink and Greater Canada Goose) and others (e.g. the American Bullfrog, Topmouth Gudgeon and Sunbleak) have the potential to do so if they ever become widely established, none has so far had the same impact in Britain and Ireland as some species (e.g. the Cane Toad and Small Indian Mongoose) have had in other parts of the world, such as Australia and the Caribbean. Only one species, the Little Owl, has by its destruction of injurious insects become actively beneficial to humans, although considerable aesthetic pleasure can be derived from such beautiful species as the Golden and Lady Amherst's Pheasants and Mandarin Duck.

MAMMALS

MACROPODIDAE (KANGAROOS & WALLABIES)
RED-NECKED WALLABY *Macropus rufogriseus*
(BENNETT'S WALLABY)
PARMA WALLABY *Macropus parma*

Natural Distribution The Australian mainland race of the Red-necked Wallaby, *M. r. banksianus*, occurs from E Queensland S through E New South Wales and S Victoria to SE South Australia. British populations are of the nominate form, *M. r. rufogriseus*, of Tasmania. The Parma Wallaby is restricted to SE Australia.
Naturalized Distribution England; Isle of Man; Scotland.

RED-NECKED WALLABY WITH JOEY IN POUCH, WHIPSNADE, BEDFORDSHIRE

England

In the 1850s, several wallabies (of apparently unrecorded species) escaped from the collection of J.H. Gurney into the woods surrounding Northrepps Hall in north Norfolk; what became of them seems to be unknown.

In the early 1860s, the then Governor of Tasmania reported to the Acclimatisation Society of the United Kingdom, which had been founded earlier in the same year (see Lever 1992), that: 'The only Tasmanian quadruped of much economic value is the brush kangaroo [*M. r. rufogriseus*], very superior leather being made from the skins. Like most of our indigenous marsupial quadrupeds, this animal is fast disappearing before the creatures introduced from Europe.'

The society's founder, Frank Buckland (1861: 22), considered that Bennett's Wallaby[1] 'is extremely hardy, and much the best calculated for acclimatisation in an English park ... with very little attention it would rapidly increase in any of the Midland or southern counties. ... [It] will breed freely in England ... [and] it may be hoped that they may be found useful as food, or at least as a new kind of game.' On 2 April 1865, the society received a consignment of 'three Bennett's kangaroos from Australia'. One, a female, was sent to join the Duke of Marlborough's four wallabies (of unknown species) at Blenheim Palace in Oxfordshire.

PARMA WALLABY

From the 1890s to around 1910, a small population of *M. r. rufogriseus* lived in the wild on Herm in the Channel Islands. There were rumours of wallabies living wild in the Pennine Hills in northern England in the early 20th century, when they were also acclimatized on Lord Rothschild's estate at Tring in Hertfordshire. In the late 1920s, Martin Harman introduced several wallabies to the grounds of Millcombe House on Lundy Island in the Bristol Channel.

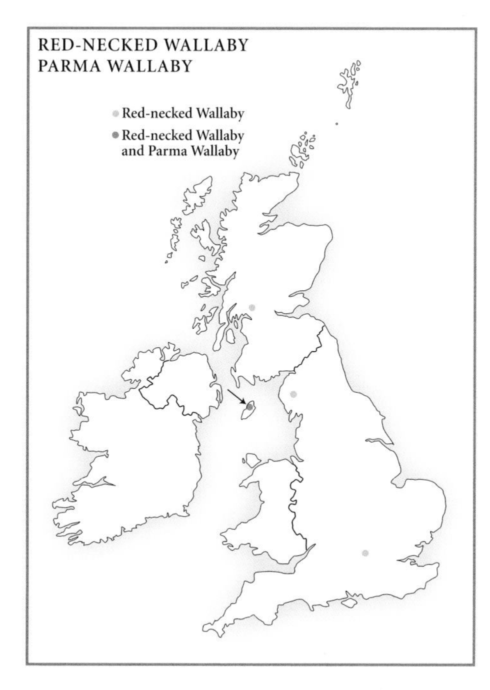

In the mid-1930s, Captain Henry Courtney Brocklehurst (a relation of T.V. Brocklehurst who in 1876 was responsible for the introduction of the Grey Squirrel to neighbouring Cheshire) established a small private collection of animals in the grounds of Roaches House near Leek in Staffordshire.[2] On 11 June 1936, he obtained a pair of Red-necked Wallabies from Whipsnade Zoo; a single joey was born in March 1937, two more in March of the following year, and a fourth in May 1939. These five animals escaped from their enclosure in 1939 or 1940 when, because of the outbreak of the Second World War, it was no longer possible to provide them with winter fodder or to maintain the fence of their enclosure. They were the founder stock of the wild population that became established in the Peak District National Park of Derbyshire and Staffordshire, where they lived near Hoo Moor in the former county, and north-west of Leek, 16 km to the south, in the latter. In 1940–70, Red-necked Wallabies were seen in the wild on several occasions in these two areas, up to 29 km north or east of Roaches House. Some individuals were extremely wary of humans, hopping away at the earliest opportunity, while others allowed a cautious approach to within a few metres (personal observation).

The harsh winter of 1946–7 probably checked the increase in the population in the Peak District. By the early 1960s, however, the numbers had recovered to 40–50, and one observer reported seeing seven animals bolted from cover by terriers – one after the other. The exceptionally hard winter of 1962–3 again caused serious casualties among the animals, though a breeding nucleus managed to survive.

In 1967–70, D.W. Yalden and G.R. Hosey of Manchester University made an intensive study of the Peak District Red-necked Wallabies, and in the latter year they estimated the total population to number around a dozen. The animals appeared to be divided into two discrete subpopulations, which seemed to have little if any contact with each other. By 1971–2, the total population had increased to 14–20, and numbers continued to rise until 1977. In August 1978, when the population had fallen back to its 1971–2 level, a young male from Riber Castle Zoo was irresponsibly released in conjunction with Granada Television. In October 1978, Yalden counted a minimum of 18 Red-necked Wallabies in the Peak District National Park.

In 1977, the 459 ha of land occupied by the Red-necked Wallabies in the park was sold for sheep farming, and overstocked with 2,000 ewes. Competition with the sheep for food and shelter, together with the severe winter of 1978–9, reduced the population still further. Eventually the Peak Park Planning Board acquired some 400 ha of the park, where after the harsh winter of 1981–2 Yalden estimated the population to number about 15, including 2–3 joeys in pouch. In 1970–90, the population varied between about 10 and 20.

Since the turn of the century, the population has been declining, and after surviving for 60 years now appears to be dying out – only a single individual was seen in 2006. The animals' principal controlling factors have been human and canine disturbance, road and rail traffic accidents, predation by Red Foxes and exceptionally severe winters with heavy falls of snow.

A second, but less well-studied, colony of Red-necked Wallabies that became established in 1940 in the Weald of Sussex is believed to have been derived from the collection formed in 1908 by Sir Edmund Loder, Bt at Leonardslee Park, Lower Beeding, near Horsham. In the early 1940s there were many reports of an established breeding colony in the Ashdown Forest/St Leonard's Forest area. No animals were seen in 1965–8, which prompted Sir Giles Loder, Sir Edmund's grandson, to claim, rather rashly, in *The Sussex Mammal Report* of the latter year, that 'conditions now make it impossible for abnormal [sic] animals to survive in a feral state even if the natural habitat is suitable'. Nevertheless, in 1969–70 several reports were received of Red-necked Wallabies north and north-east of Leonardslee; the expansion of Crawley and Horsham prevented any further extension of the animals' range to the north and west respectively. A small number survived in this location until at least 1972 (possibly considerably later), but this population has since died out.

On the Isle of Man[3] anecdotal reports suggest that Red-necked Wallabies first escaped from the Ballaugh Curraghs Wildlife Park (near a designated 193-ha Ramsar site in the north-west of the island) in 1965, 1976 or the early to mid-1980s. Today, a population estimated to number over 100 (plus a small number of Parma Wallabies) lives in the Curraghs,[4] sheltering in the dense willow carr[5] that has invaded the wetland, composed of modified bog and scrub. Although the Ballaugh Curragh area, between Ballaugh and Sulby to the east (a distance of c 3.5 km), is the animals' stronghold, individuals have been seen some distance away, even in open fields. The lack of suitable habitat near Ballaugh Curraghs, however, may preclude further spread of the population in this area (Harby 2008).

The fact that sightings of Parma Wallabies in the wild have been near the wildlife park's wallaby enclosure suggests that the species is a relatively recent escapee. This, of course, is denied by the park authorities, which claim that no wallabies have escaped during the last 20 years. If this is true, given the species' known lifespan it implies that Parma Wallabies in the Ballaugh Curraghs have been breeding in the wild.

Around 1949, seven Red-necked Wallabies escaped from a private collection near East Grinstead in East Sussex, where at least one survived for a year or so. At Whipsnade Wild Animal Park in Bedfordshire there is an acclimatized population in a 200-ha enclosure that in 1979 peaked at about 900, but after the hard winter of 1978–9 collapsed to 400–600. The RSPCA told Newling (2006) of a colony in Cumbria, and there are reports of animals – allegedly escapees from the Alfred McAlpine estate near Fawley – at large in the Henley-on-Thames area of Oxfordshire (Nigel Phillips, personal communication 2008). Baker and Hills (in Harris and Yalden 2008: 786) reported that 'This is among the most frequently reported of exotics (Baker 1990), with reports widely scattered across the country, sometimes in numbers that suggest other undocumented colonies.'

Scotland

Fitter (1959) records that several 'kangaroos', which were probably wallabies though of unknown species, were turned down by the Marquess of Bute at Mount Stuart near Rothesay on the Isle of Bute in about 1912, where at least one survived for some three years and used to lead the line of beaters at pheasant drives.

In about 1975, two pairs of Red-necked Wallabies from Whipsnade were released, by the Countess of Arran,[6] on Inchconnachan in Loch Lomond, Stirlingshire. Wallabies are good swimmers, and by 1979 the first of several individuals had crossed to the mainland, where in 1982 one was seen in Balloch Park at the southern end of the loch. By 1993 the population on the island had multiplied to 28 and by 1999 to 43.

The success of the animals on Inchconnachan has been due largely to the absence of human disturbance and predators and the relatively mild winters. However, this success may prove their downfall, since the current level of browsing poses a threat to the Blaeberry-dominated ground vegetation and its dependent native fauna.

Characteristics

The Red-necked Wallaby is a 'scrub' wallaby, tending to hide during the day in dense vegetation and woodland, from which it emerges at dusk to feed. The Peak District population lived in birch and pine scrub and on adjoining heather moorland, where the animals fed principally on Common Heather, but also on Blaeberries and grasses, and to a lesser extent on pine, the young shoots of Common Bracken, oaks, Rowan and birches. On Inchconnachan they live in open birch/oak/pine woodland, where they feed on a mixture of Common Heather, Blaeberries and grasses. In winter, which is the most stressful time, their diet includes approximately 35 per cent each of Common Heather and Blaeberries and 13 per cent of grasses.

Since it is the Tasmanian nominate form, sometimes known as Bennett's Wallaby, that is established in Britain, and as Tasmania lies at around 42°S and has mountains reaching an altitude of over 1,600 m, the Red-necked Wallaby is reasonably well adapted to the British climate. As, however, females usually give birth to only a single joey, replacement of losses is slow.

Although Red-necked Wallabies survived in the Peak District National Park for over 60 years and in Sussex for some 30 years, they never spread very far. The main reasons for this are probably twofold:

- The suitability of their habitats with an abundant supply of food, which, together with their small populations and basically sedentary nature, made dispersal unnecessary.
- Their vulnerability when attempting to cross roads and railway lines, which accounted for the deaths of at least 11 animals in the Peak District.

The Red-necked Wallabies' dispersal from Inchconnachan to the mainland was probably caused by the comparatively rapid rise in the population and corresponding decline in the amount of available forage.

Impact

The over-browsing by the animals that has occurred on Inchconnachan is an indication of what could happen in the unlikely event that they became widely established elsewhere in Britain. On the Isle of Man, concern has been expressed about the potential impact of a large herbivore on the biodiversity of the Ballaugh Curraghs Ramsar site. So far there is no evidence of this occurring, but the situation requires monitoring. On the positive side, browsing by the animals has created open glades in scrub that may provide roosting and hunting sites for the thriving population of Hen Harriers and potential nesting sites for Corncrakes and Northern Lapwings.

The long-term survival prospects for the animals on the Isle of Man look good, due to the island's relatively milder winters compared with those on the mainland and to an absence of predation by Red Foxes, which have yet to occur in the Ballaugh Curragh area (Harby 2008).

References

Anon. 1965; Baker 1988, 1990; Baker & Hills, in Harris & Yalden 2008; Dubbledam 2008; Edwards 1962; Fitter 1959; Harby 2008; Lever 1977, 1980a, 1980b, 1985, 1992; Long 2003; Mallon 1970; Mitchell 1983, 2001; Neale 1990; Newling 2006; Page & Tittensor 1969; Smith 2006a; Welch et al. 2001; Yalden 1988, 1999, in Corbet & Southern 1977, in Corbet & Hill 1991; Yalden & Hosey 1971.

Notes

1. Named after a former Secretary of the Zoological Society of London.
2. Not at Swythamely Park near Macclesfield (the home of his brother Sir Philip Brocklehurst, Bt) as stated by Fitter (1959: 53).
3. Although strictly a British Crown Dependency, the Isle of Man is here treated as a part of England.
4. Curragh = Manx for acidic boggy ground (Harby 2008).
5. Wet woodland, typically dominated by Common Alder or willows.
6. Long (2003: 42) gives as his source 'Mitchell 1983', but omits this reference from his bibliography.

SCIURIDAE (SQUIRRELS, MARMOTS & ALLIES)

GREY SQUIRREL *Sciurus carolinensis*

Natural Distribution N America, from the Gulf of Mexico N to the Great Lakes and the St Lawrence River, W to the prairies. The form introduced to Britain is believed to be *S. c. leucotus* (NE USA and SE Canada).
Naturalized Distribution England; Scotland; Wales; Ireland.

England

The earliest recorded introduction of the Grey Squirrel to Britain was in 1876, when T.V. Brocklehurst (a relation of Sir Henry Courtney Brocklehurst, Bt, who in the mid-1930s introduced Red-necked Wallabies to neighbouring Staffordshire) released a pair that had been imported from North America, at Henbury Park, near Macclesfield in Cheshire, where the animals became established and bred. At Dunham Park, 24 km north-west of Henbury, Grey Squirrels may have been introduced before 1876, as they are known to have already been there at the time of the first introduction in 1910; they could equally have dispersed to Dunham from Henbury. A pair shot at Highfields in Nottinghamshire in 1884, however, was almost certainly derived from a separate introduction. An apparently abortive attempt to naturalize Grey Squirrels in Norfolk was made at about this time by J.H. Gurney at Northrepps Hall near Cromer, and the species is believed to have been introduced in Kent as early as 1850.

GREY SQUIRREL

In 1889, G.S. Page of New Jersey, USA, imported a small stock of Grey Squirrels from the United States, five of which were liberated in Bushy Park, Middlesex (Greater London), where they failed to gain a footing and where no more were seen until the mid-20th century. Ten of Page's Grey Squirrels were in 1890 released at Woburn Abbey in Bedfordshire by the Duke of Bedford, where they rapidly increased.

In 1902, 100 Grey Squirrels were released in Richmond Park, Surrey, at Kingston Hill, and at around the same time others were freed in Rougemont Gardens, Exeter. Between 1902 and 1929 (the year of the last recorded introduction), a veritable wave of introductions took place in many localities, including Regent's Park, London; Berkshire; Northamptonshire; Oxfordshire; Staffordshire; Devon; Warwickshire; Nottinghamshire; Suffolk and Hampshire (Bournemouth). From some of these centres, Grey Squirrels spread into Gloucestershire and eastern Wiltshire. The stock for most of the introductions came either directly from the United States or from Woburn. Unfortunately, in several instances no records were kept of which new points of release received animals from Woburn and which acquired new blood from America. (A single female released at Farnham Royal in Buckinghamshire 1909 came from South Africa, where the species was introduced at Groote Schuur in the late 19th/early 20th century by Cecil Rhodes, almost certainly using stock from England.) In 1938, it was declared illegal in future to import Grey Squirrels into Britain or to keep them in captivity.

It is difficult to estimate the relative importance of each new introduction because of the considerable numbers involved, and because extensions of range from different points of release often overlapped. In some places new sub-populations merely served to augment already existing ones. By 1930, Grey Squirrels were entrenched throughout much of south-eastern England from Kent to Southampton Water in Hampshire, and thence north-westwards to Northamptonshire and Warwickshire. Further north, the animals' distribution was discontinuous, the sub-populations in Cheshire, Denbighshire and western Yorkshire remaining discrete; in the case of the last-named county, the Pennine Hills seem to have been a formidable barrier.

In 1930–45, the species became established in the Midland counties and in most of southern England apart from Cornwall. No Grey Squirrels were reported from East Anglia, probably because of the unsuitability of the habitat – largely fens, marshland and coniferous woodland favoured by native Red Squirrels, whereas Grey Squirrels prefer deciduous trees. In 1930 there was an isolated record from Alnwick, Northumberland.

By the outbreak of the Second World War in 1939, Grey Squirrels covered a large area of southern and northern England from Kent and Sussex to Cheshire. By its end in 1945, the species had further extended its range northwards and north-eastwards, and had become established in the North and East Ridings of Yorkshire from Middlesborough to Hull. There were, however, still few records from East Anglia and Cumberland and Westmorland (Cumbria) in north-west England, and the species appeared to be declining in Dorset, western Somerset and Devon.

In 1945–56, Grey Squirrels increased in numbers in Cheshire and Yorkshire, moved further east into southern Essex and staged a revival in Dorset, Devon and Somerset. They appeared for the first time in Westmorland in 1944–5, Suffolk (1947–52 – their earliest appearance in East Anglia), southern Lancashire (1950–2) and Sheffield and Cornwall (1955), and also spread further into Derbyshire, Nottinghamshire, Co. Durham, Shropshire, Staffordshire, Worcestershire, Lincolnshire and eastern Cambridgeshire. In the early 1950s, Grey Squirrels crossed the valley of the River Lea near London into Epping Forest.

In 1953–5, a bounty of 1s. (10p) and free cartridges were offered by the Forestry Commission and the Ministry of Agriculture, Fisheries and Food for the destruction of Grey Squirrels. Although the bounty was later doubled and by 1958 a total of £100,000 had been paid out, the population showed no signs of decline and the bounty scheme was discontinued.

By the early 1960s, it was easier to say where Grey Squirrels did not occur than where they did; they had occupied previously vacant areas of Devon, Wiltshire, eastern Cornwall, the southern West Riding of Yorkshire, and parts of Norfolk, Essex, eastern Suffolk, Westmorland, Cumberland and Northumberland. By the middle of the following decade they had spread to parts of western Cornwall, and into the valleys on the eastern side of the Peak District. There were still, however, northern and western regions comprising parts of Cumberland and Westmorland and northern Lancashire where Grey Squirrels were infrequent or unknown, and in the east parts of Norfolk, Suffolk and northern Essex remained uncolonized. Elsewhere, distribution was fairly general throughout the country, although the most heavily overrun areas remained the south, the south-east, the southern Midlands and the southern Welsh Marches.

By the mid-1980s, Grey Squirrels had managed to cross the Pennines from Yorkshire into Lancashire and the southern Lake District National Park (Cumbria), and had spread westwards to extreme western Cornwall. Today, only the Isle of Wight, Brownsea and Furzey Islands in Poole Harbour, Dorset, and a few other offshore islands, eastern Lincolnshire, extreme southern Lancashire, much of Cumbria, and parts of North Yorkshire and Northumberland remain uncolonized.

In the 1880s, a single semi-melanistic Grey Squirrel imported from the United States is said to have escaped from a collection in East Anglia – probably Bedfordshire. Genetic tests have revealed that all of the increasing population of 'black' squirrels around Letchworth in Hertfordshire and elsewhere is descended from this individual. There may currently be up to 25,000 'black' squirrels, which are more aggressive and territorial than normally coloured individuals, in eastern England.

Scotland

In 1892, a pair of Grey Squirrels, imported from the United States by G.S. Page of New Jersey, was released at Finnart on the Dunbartonshire shore of Loch Long. From there the species spread north to Arrochar and Tarbet by 1903, east to Luss (1904), Inverbeg (1906), south-west to Garelochhead and Rosneath, and as far south as Helensburgh (all Argyllshire), Alexandria and Culdross (West Dunbartonshire) by 1912, an overall distance of over 30 km. By 1915, Grey Squirrels had penetrated to the eastern side of Loch Lomond, Stirlingshire, where they established themselves at Drymen. An importation to the Edinburgh Zoo in 1913 resulted in the establishment by the end of the decade of a number of escapees and their progeny outside the zoo grounds at Corstorphine.

Thus, within a quarter of a century, Grey Squirrels had succeeded in becoming established in Scotland over an area of nearly 800 sq km. Before 1929, when introductions ceased, Grey Squirrels were also released in Ayrshire and at North Queensferry, Fife. Isolated records came from Galloway in 1937, Loch Shiel, Inverness-shire (1939) and Selkirkshire (1944). In 1945–56, Grey Squirrels spread east through Fife and northwards into Perthshire (Perth and Kinross). By the 1970s, the species was fairly common in parts of Argyllshire, Clackmannanshire, Dunbartonshire (Argyll and Bute), Falkirk, Fife, Kinross-shire (Perth and Kinross), East Lothian and Midlothian, and the valley of the River Tay in Perthshire. Subsequently, Grey Squirrels spread mainly north-east into

Angus and Aberdeenshire, south-east into Peeblesshire and the Borders, and south-west into Southern Lanarkshire and Ayrshire. Their expansion of range in Scotland seems to have been slower than in England and Wales, but by the 1990s they were spreading into the southern Uplands, and in 2008 an individual was reported near Inverness.

Wales

The first recorded occurrence of the Grey Squirrel in Britain was at Llandisilio Hall, Denbighshire, in October 1828. In a letter to the editor of the *Cambrian Quarterly Magazine* of 1830, the writer said that for some time Grey Squirrels had occurred at Llanfair Caereinion, Llan Eurvyl and Cwm Llwynog (Fox's Dingle) in Montgomeryshire (Powys). The origin of these animals appears to be unrecorded.

In 1903, Grey Squirrels were released in north Wales at Wrexham in Denbighshire, where they were widely established by the outbreak of the Second World War. In 1945–56, they spread to south Wales, though whether from Denbighshire or western England is uncertain – probably a combination of both. In 1922, an introduction of Grey Squirrels was made in Aberdare, Glamorganshire (Rhondda). They appeared in Cardiganshire (Ceredigion) in 1950 and in Merionethshire (Gwynedd) in 1951, and spread further into Brecknockshire, Radnorshire and Montgomeryshire (Powys). By the mid-1980s, virtually the whole of the country had been overrun.

For information on the Grey Squirrel on the Isle of Anglesey, I am indebted to Craig Shuttleworth (personal communication 2008).

The first individual on the island was seen on 10 March 1966 on the Bodorgan estate in the south-west. By 1998, when an eradication programme was initiated, Grey Squirrels were virtually ubiquitous on the island in both coniferous and deciduous woodland. The native Red Squirrel survived only in the 245-ha Pentraeth Forest on the east coast above Red Wharf Bay, where it became infected with the parapoxvirus (squirrelpoxvirus – SQPV), probably contracted from Grey Squirrels, in which it is benign (see page 32 under Impact). By 2003, Grey Squirrels had been eradicated from Pentraeth Forest, and by 2008 they had been all but eliminated from Newborough in the extreme south of the island. In 1998, the pre-breeding population on Anglesey was estimated to number around 3,000; the current population is considered to be less than 100. The eradication programme is scheduled to conclude in 2010, when Grey Squirrels will hopefully have been eliminated on Anglesey.

Since Grey Squirrels probably gained access to Anglesey via the road and/or rail bridges across the Menai Strait, sonic devices may be used to attempt to prevent recolonization, and Grey Squirrels on the mainland may be culled to provide a cordon sanitaire between Anglesey and Gwynedd.

Ireland

The following history of the Grey Squirrel in Ireland is derived largely from Ó Teangana et al. (2000).

Grey Squirrels were first introduced to Ireland from England in 1911 or 1913, when half-a-dozen pairs were successfully released at Castle Forbes in Co. Longford. Others were freed, apparently unsuccessfully, at Ballymahon, Co. Longford, and perhaps in Co. Dublin in 1928. By 1923, Grey Squirrels had spread 16 km from Castle Forbes, and by the outbreak of the Second World War they were continuing to expand their range in a northerly direction. They had reached Westmeath by 1927; Wicklow, Leitrim and Roscommon by at least 1938; Fermanagh (1945); Cavan (from 1946); Armagh (1953) and Monaghan (since at least 1960). Although it is not as well documented, a similar expansion of range occurred to the south and east. The River Shannon has been an effective barrier to the species' spread into Connaught and Clare, although a single individual was recorded in Galway in 1979.

By the late 1960s/early 1970s, Grey Squirrels occurred principally in northern Leinster and southern Ulster; by the end of the latter decade they had extended their range into southern Leinster, northern Munster and mid-Ulster. Since 1979, they have become ubiquitous in the midland counties in the Irish Republic, and have spread to southern Tipperary and Wexford, and expansion to the east coast is almost complete, including recently to urban parks around Dublin. Grey Squirrels do not yet occur in eastern Wicklow, but are common in the west of the county. Although, as mentioned above, the River Shannon provides a barrier to immigration to the west of Ireland, north of the river's source Grey Squirrels are now spreading westwards through Co. Leitrim. Sympatric populations of Red and Grey Squirrels occur in 15 counties in the Irish Republic, and although the latter now seems to be replacing the former in those counties in which the alien has occurred longest (i.e. Cavan, Longford, Monaghan, Meath and Westmeath), there is anecdotal evidence of a decline of Grey Squirrels elsewhere and a resurgence of Red Squirrels.

In Northern Ireland, the Grey Squirrel is more widely distributed than the Red Squirrel, occurring in every county apart from Antrim as far east as eastern Co. Down and north almost to the coast. Its presence in far western Co. Tyrone suggests the likelihood of a sub-population further west in Co. Donegal.

By the turn of the century, Grey Squirrels in Ireland had spread nearly 163 km from Castle Forbes to Co. Wexford in the south-west, at an average speed of 1.94 km per annum. The rate of expansion has varied considerably by region, and has been related to the presence or absence of such geographical features as large loughs, rivers and mountain ranges. Although the animals have spread eastwards unimpeded, and are now present along the coast from Dublin to Down, expansion into eastern Wicklow is impeded by the Wicklow Mountains. To the south, Grey Squirrels have spread unhindered into Tipperary, Kilkenny and Carlow. As the Shannon has hindered the species' spread to the west, so Lower Lough Erne has impeded expansion in a northerly direction, and Lough Neagh and the Lower Bann River have hindered any spread to the north-east. In all cases, however, Grey Squirrels have either outflanked these barriers or, as in the case of the Lower Bann, have simply crossed bridges.

Due to an absence of research studies, the relationships between Grey and Red Squirrels in Ireland are unknown, although anecdotal evidence suggests that the latter may be declining in some localities in the face of the former.

Table 1 Principal sites where Grey Squirrels *Sciurus carolinensis* were introduced in Britain

Date	Locality	Number	Introducer	Source
ENGLAND				
1876	Henbury Park, Macclesfield, Cheshire	4	T.V. Brocklehurst	USA
1889	Bushy Park, Middlesex	5	G.S. Page	USA
1890	Woburn Abbey, Bedfordshire	10	Duke of Bedford	USA (via G.S. Page)
?	Benenden, Kent	?	?	?
?	Nuneham, Oxfordshire	?	?	Woburn
1902	Kingston Hill, Richmond, Surrey	100	?	USA
c 1902	Rougemont Gardens, Exeter, Devon	?	?	?
1903–4	Lyme Park, Cheshire	25	?	?
1905–7	London Zoo, Regent's Park, London	91	Zoological Society of London	Woburn
1906	Scampston Hall, Malton, Yorkshire	36	?	Woburn
1908	Cliveden, Buckinghamshire	?	?	Viscount Astor
1908	Kew, Surrey	4	Royal Botanic Gardens	Woburn
1908	Farnham Royal, Buckinghamshire	4	?	USA
1909	Farnham Royal, Buckinghamshire	1	?	South Africa
1910	Frimley, Surrey	8	?	USA
1910	Dunham Park, Cheshire	2	?	?
1910	Stonewall Park, Edenbridge, Kent	?	?	Lieut-Col E.G.B. Meade Waldo
1911–12	Bramhall, Cheshire	5	?	Woburn
1912	Birmingham	?	?	?
1913	Bedale, Yorkshire	?	?	?
1914	Bingley, Yorkshire	14	?	London Zoo
1914–15	Darlington, Co. Durham	?	?	?
1915	Exeter, Devon	4	?	?
1918	Stanwick, Northamptonshire	2	?	?
1919	Bournemouth, Hampshire	6	?	London Zoo
1921	Hebden Bridge, Yorkshire	8	?	?
1929	Bestwood and Hartsholme, Nottinghamshire	?	?	?
1929	Needwood Forest, Staffordshire	2	?	Bournemouth, Hampshire
SCOTLAND				
1892	Finnart, Loch Long, Dunbartonshire	3	?	G.S. Page (from Canada)
1913	Edinburgh Zoo, Corstorphine	?	Royal Zoological Society of Scotland	Canada
1919	Pittencrieff Park, Dunfermline, Fife	?	?	Canada
?	Ellington Castle, Ayrshire	?	?	?

Table 1 *Continued*

Date	Locality	Number	Introducer	Source
WALES				
1903	Rossett, Wrexham, Denbighshire	5	?	Woburn
1922	Aberdare, Glamorganshire	?	?	London Zoo
IRELAND				
1911 or 1913	Castle Forbes, Co. Longford	6 pairs	Earl of Granard	Woburn
1928	Co. Dublin	?	?	?
1928	Ballymahon, Co. Longford	?	?	?

Sources Adapted from Shorten 1954; Lever 1977. The introduction in Bushy Park, Middlesex, in 1889 is the only one in England that is known to have been unsuccessful. Long (2003: 146–7) misspells a number of place names, e.g. Cum Llyndog for Cwm Llynog; Fidnnart (Finnart); Garelockhead (Garelochhead); Rougemdont (Rougemont); Corstophine (Corstorphine); Rosett (Rossett).

Characteristics

In Britain, Grey Squirrels occur in a wide variety of habitats, being most abundant in mature broadleaved woodlands of oak, Common Beech or Sweet Chestnut, with Common Hazel coppice. They are also found in mixed deciduous/coniferous forests, and occasionally in mature conifer woodland, although densities tend to be higher where broadleaved trees occur nearby, and in hedgerows, parks, gardens and urban localities with mature trees. One of the principal reasons for the Grey Squirrel's success in Britain has been its willingness to accept fragmented habitats.

The young are born in arboreal nests ('dreys') in spring and summer. Grey Squirrels are granivorous herbivores, feeding mainly on tree seeds, especially acorns, beechmast and chestnuts, some of which they cache in autumn, and a wide variety of plant material; they also take birds' eggs and nestlings. Their principal predators are Stoats, Red Foxes, domestic cats and dogs, and some raptors and owls. Other natural controlling factors include shortage of food and severe winter weather. The total British pre-breeding population is estimated to be c 2.52 million (2 million in England, 200,000 in Scotland, 320,000 in Wales).

Impact

Gurnell and Mayle (2003), from whom much of the following account is derived, have summarized the ecological and economic impact of the Grey Squirrel in Britain.

Grey Squirrels cause timber damage by stripping the bark of commercially valuable trees, and also harm those planted for landscaping, and for recreational or conservation purposes. Such damage can occur to the base (e.g. root buttresses of mature trees such as Common Beeches), stem (e.g. thinly barked trees such as Common Beeches, European Hornbeams and larches) or crown of more thickly barked trees such as oaks and pines. The degree and severity of damage can vary considerably from year to year and from place to place, and usually occurs in May–August. In the first half of the 20th century, thinly barked broadleaved trees such as Common Beeches and Sycamores appeared to be the most vulnerable. Since then, however, a wider range of species, including oaks and coniferous trees such as the Scots Pine, Lodgepole Pine and Norway Spruce, has incurred damage. The removal of bark can cause loss of shape, timber quality degradation, loss of vigour and, in the case of ring-barking, tree mortality.

Although Grey Squirrels do consume some of the bark's soft inner cambium layer, tree damage does not seem to be directly related to hunger or thirst, to an attempt to obtain nest material, to a gnawing reflex to wear down their teeth or to marking territorial boundaries. The degree of damage appears to be directly correlated with the density of the Grey Squirrel population, especially juveniles, living near or moving into affected areas. A higher number of animals leads to a higher number of interactions, which in turn leads to an increase in displacement behaviour that is redirected into bark stripping – a sort of 'nervous reaction'.

As well as causing damage to trees, Grey Squirrels may have other unwanted effects on woodland communities. They may, for example, cause an alteration in the composition of the canopy by damaging some species more seriously than others, or by preventing natural tree regeneration by their consumption of seeds. This can cause a decline in the planting of such vulnerable trees as Common Beeches and Sycamores, which in turn will affect the associated invertebrate fauna. The removal of unripe seeds of, for example, Hazel and fruits could adversely affect other seedeaters such as the declining Hazel Dormouse. Grey (and Red) Squirrels take birds' eggs and chicks, although the ecological impact of this has yet to be determined.

Research has shown that the native Red Squirrel has almost everywhere retreated as the Grey Squirrel has expanded its range; the reasons for this, however, remain unclear. Fragmentation of habitat and loss of woodland may be contributory factors (although both species are well able to exist in a fragmented environment), as also may be disease such as the parapoxvirus (mentioned under Wales), of which some 70 per cent of Grey Squirrels are thought to be carriers in Britain: this virus has little if any effect on Grey Squirrels but is invariably fatal to Red Squirrels. Thus the parapoxvirus disease may cause the local eradication of Red Squirrels and contribute to the speed and degree of the replacement of Reds by Greys.

There is little evidence of any food or interference competition between Grey and Red Squirrel adults, despite the fact that the former are considerably larger and heavier than the latter. On the other hand, the pressure of Grey Squirrels does seem to have an inhibiting effect on the growth rate of young Red Squirrels, and reduces the recruitment of young of the latter into the population. This may be because of the dominance hierarchies that operate in both species, whereby the young are subordinate to adults. Also, female Red Squirrel reproduction success tends to be lower in the summer in the presence of Grey Squirrels (this problem does not occur in spring), and in winter and spring Greys steal the seed-food caches of Reds. The abundance of tree seeds (nuts) is crucially important to the survival of both species in winter, and has a major bearing on their population density and demography. That each species favours different types of woodland is of particular relevance. Grey Squirrels prefer broadleaved woodlands, although they can survive well (albeit at lower densities) in coniferous forests – preferably spruce or fir. Red Squirrels favour conifer trees, although they, too, are well able to live in pure deciduous or mixed deciduous/coniferous woodland, without in their case any lowering of population density.

Grey Squirrels can also impinge directly on humans. They cause damage to horicultural, agricultural and pomological crops (especially bulbs, corms and newly sown seeds), and take food intended for garden and game-birds and domestic fowl. In house lofts they damage thatched or shingle roofs, and gnaw through plastic and lead water pipes, telephone cables and electric wiring – a number of domestic fires have been attributed to the activities of Grey Squirrels.

The International Union for Conservation of Nature and Natural Resources (IUCN) advocates eradication (complete removal), control (reduction in numbers) and containment (limitation of spread) as the most effective ways of alleviating the impact of non-native species. Even if the eradication of Grey Squirrels were possible, many people who find the animals attractive, especially those living where Red Squirrels do not occur, would oppose it. On islands, however, where Grey Squirrels have not yet invaded (e.g. the Isle of Wight, and Brownsea and Furzey Islands in Poole Harbour, Dorset), every effort should be made to prevent them from gaining access. The same policy might be adopted in such important conservation areas as the ancient Caledonian Pine forests of the Scottish Highlands. On those islands where Grey Squirrels are present, attempts should be made, as on Anglesey, to remove them.

Methods of management include live-trapping (labour intensive but effective); poisoning under the Grey Squirrel (Warfarin) Order 1973 (effective but also kills non-target species), and illegal where Red Squirrels and Pine Martens occur (its total withdrawal as a means of control could have a disastrous effect on woodland and forestry preservation); shooting and/or drey-destruction (ineffective on a large scale); and lethal trapping (only moderately effective, impacts on non-target species and opposed by many people on welfare grounds). Continued research into the development of an immunocontraceptive for Grey Squirrels and of an effective vaccine for parapoxvirus are both of high priority – indeed, together with habitat management (see below), this may be the way forwards. The cost/benefit ratio of managing Grey Squirrels also needs to be taken into consideration, especially given that in many cases such management will have to be long term.

Although it seems clear that Red and Grey Squirrels are unable to live sympatrically in a given area, they may be able to survive allopatrically in the same area by habitat separation. Since the numbers of both species are dependent on habitat composition, it may be possible to manage large areas (e.g. >2000 ha) of coniferous forest in favour of native Red Squirrels and to the disadvantage of alien Grey Squirrels.

References

Anderson & Hughes 1995; Baker 1990; Barrington 1880; Benham 1953; Bremner & Park 2007; Bryce 1997; Clinging & Whitely 1980; Colhquhoun 1942; Crichton 1974; Deane 1964; DEFRA 2003; Fairly 1984; Fitter 1959; Forestry Commission 1953; Freethy 1983; Gurnell, in Corbett & Harris 1991, 1996, 1999; Gurnell & Mayle 2003; Gurnell & Pepper 1993; Gurnell & Steele 2002; Gurnell et al., in Harris & Yalden 2008; Harvie-Brown 1880–1; Henderson 1947; Jackson 1961; Johnston 1937; Kenward 1983, 1989; Kenward & Hodder 1998; Kenward & Parish 1986; Kenward et al.1998; Lever 1969, 1977, 1979, 1980c, 1985, in press; Lister & Lewis 2008; Lloyd 1962, 1983; Long 2003; Lowe 1993; MacRae 2006; MacKinnon 1978; Malvern 2008; Ministry of Agriculture, Fisheries & Food 1962; Middleton 1930, 1931, 1932, 1935; Middleton & Parson 1937; Moffat 1938; Moore 1997; Ní Lamhna 1979; Okubo et al. 1989; Ó Teangana et al. 1989; Reynolds 1985; Ruttledge 1924; Sheail 1999, 2003; Shorten 1946, 1948, 1951, 1953, 1954, 1957, 1962, 1964; Shorten & Courtier 1955; Smith 2006c; Staines 1986; Stelfox 1927; Taylor 1966; Thompson & Peace 1962; Tibbets 2007; Tittensor, in Corbet & Southern 1977; Usher 1992; Usher et al. 1992; Vizoso 1967; Watt 1923; Welch et al. 2001; Yalden 1999.

GLIRIDAE (DORMICE)

EDIBLE DORMOUSE *Glis glis*

(FAT DORMOUSE; SQUIRREL-TAILED DORMOUSE; GREY DORMOUSE)

Natural Distribution S and E Europe (apart from most of Iberia) E to Iran and the Caspian Sea in Asia.
Naturalized Distribution England.

EDIBLE DORMOUSE EATING HAZELNUT

According to Flower (1929: 769), 'On or about February 4th,[1] 1902, some individuals of *Glis glis*[2] from the Continent of Europe (probably Germany or Switzerland) were turned loose at Tring Park [Hertfordshire] by Walter (later Lord) Rothschild.' (According to Jones-Walters and Corbet, 1991, who were informed by the grandson of the original importer, the source was Hungary.)

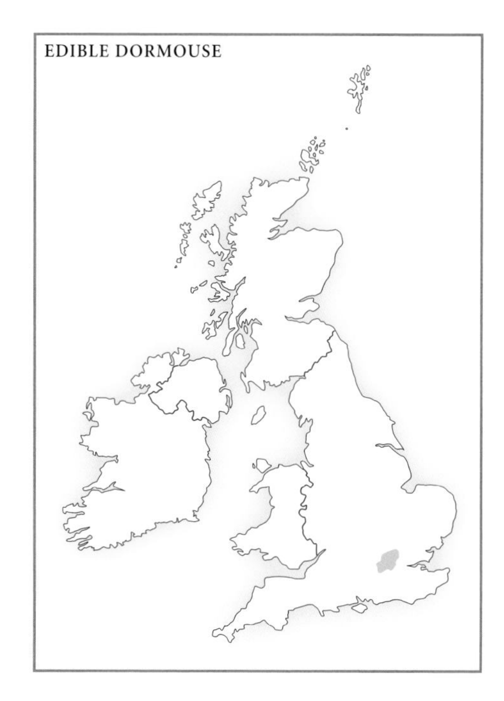

Soon after the animals' release at Tring, reports were being received of damage to crops and calls were made for their removal. It was not, however, until the mid-1920s that Edible Dormice began to be seen in any numbers in and around Tring Park, and intrusion into houses was soon reported. In 1926, Pendley Manor near Tring, where no fewer than 39 animals were caught, was overrun, and others were reported from Aldbury and Wigginton in Hertfordshire, and Drayton Beauchamp and Aston Clinton in Buckinghamshire.

In the 1930s, numerous reports of Edible Dormice were received, including from Whipsnade in Bedfordshire in 1931; Wendover, Buckinghamshire (1933); and in scrub and on nearby downs between Lion Pit and Holly Findle, Hertfordshire (1935–6); in the latter year 75 were caught. By 1938, reports were being received of Edible Dormice as far afield as Worcestershire and Wiltshire, with unconfirmed sightings in Berkshire, Gloucestershire, Hampshire, Northamptonshire, Oxfordshire and Surrey. Dormice were reported from Ludlow, Shropshire in 1941–2 and Coventry in 1945. All these records, if correctly identified (Edible Dormice are not dissimilar to young Grey Squirrels) almost certainly arose from deliberate transportation by humans. In 1941 the animals appeared at Great Pednor and in 1945 in Amersham, both in Buckinghamshire, and in 1946 in Incents House at Berkhamsted School, Hertfordshire.

Early in 1951, a census was taken of Edible Dormice in England, which resulted in the discovery of no less than 24 new locations, some of which were in hitherto unoccupied districts such as Ashley Green, Cholesbury, Hyde Heath, Great Missenden and Pitstone in Buckinghamshire, and Ashridge Park, Little Gaddesden, Ringshall and Rossway in parts of neighbouring Hertfordshire.

Edible Dormice can thus be seen to have colonized only a limited number of places in the first three decades after their introduction, the main expansion of range having taken place almost entirely after the fourth decade. By the mid-1970s, they occupied a triangle of around 260 sq km in the Chiltern Hills formed by Beaconsfield, Aylesbury and Luton. Isolated records from outside this area came from Ibstone and Bledlow Ridge on the Buckinghamshire/Oxfordshire border and Potters Bar, Hertfordshire.

By the 1990s, Edible Dormice had increased their range to around 600 sq km between Aylesbury, High Wycombe and Whipsnade. An enquiry conducted by P.A. Morris in 1995 added a significant number of 1 km squares (e.g. Windsor, Berkshire; Sandy Bedfordshire; Eastleigh, Hampshire) to the known distribution of 1980, implying some increase in both numbers and range. Most of the 1995 records came from Buckinghamshire (68 per cent) and Hertfordshire (30 per cent), with a single report from Stevenage, Hertfordshire, some 30 km east of Tring.

According to Morris (1997, 2003), Edible Dormice may already be established in the New Forest in Hampshire (a particularly favourable habitat), to which they are known to have been deliberately translocated at an unrecorded date by humans, and/or near Oxford, some 45 km west of Tring.

Why the Edible Dormouse has spread so little during its century in England, compared to what it is capable of in its native Eurasian range, is an unresolved mystery. Although to the west lies the open farmland of the Vale of Aylesbury, an unfavourable habitat for a largely arboreal species, an apparently suitable environment northwards into the Chiltern Hills and south into Oxfordshire also remains uncolonized. Expansion has probably been hindered by increasingly fragmented woodlands and man-made barriers such as major roads and urban areas. Climatic and environmental factors, of which we are as yet unaware, may also be implicated. Nevertheless, and despite the creature's failure to spread far in over 100 years, Morris (1997: 7) considers 'It is only a matter of time before this animal becomes commonplace in other parts of England, together with its associated problems.' It is noteworthy that where Edible Dormice do occur they are usually abundant.

Characteristics

In its native range the Edible Dormouse favours Common Beech woodland, which it uses for both food and dens, especially when associated naturally with conifers. Such mixed woodlands do not occur in Britain, except where Common Beech saplings are protected by shelter plantings of conifers. This combination is found in the Chiltern Hills, particularly around the site of the animals' introduction, and as Morris (2003) points out, if Walter Rothschild had lived elsewhere, where woodlands are dominated by other tree species, Edible Dormice might not have become so successfully established in England.

The young of Edible Dormice (normally at least seven) are born in nests that may be based on old ones of birds, especially Common Woodpigeons, or in Grey Squirrel dreys or nest boxes. In autumn Edible Dormice enter into hibernation, which may last for as long as seven[3] months, during which they can lose up to 50 per cent of their body weight. Hibernation, which may be either communal or solitary, takes place underground (e.g. among the roots of rotted tree-stumps or in holes) or under the floors of houses. Edible Dormice are mainly arboreal and nocturnal feeders (lying up in nests during the day), eating a wide range of tree fruits (especially beechmast, hazelnuts and acorns), berries (particularly those of Bramble and Elder), buds, apples, bark (especially of fruit trees and willows), insects, carrion, fungi and occasionally birds' eggs or nestlings. They gnaw the bark of conifers and that of some deciduous trees, either for the inner cambium layer or for exuded sap. In England their main predators are Tawny Owls, Stoats, Weasels and domestic cats. The total English population is estimated to number at least 10,000.

The Edible Dormouse was a favourite food of the Romans, by whom it was reared in oak and beech groves and fed on currants and chestnuts.

Impact

Morris (1997, 2003, 2008) has summarized the impact of Edible Dormice in England.

Although as early as the 1920s, Edible Dormice were said to be proving a pest of growing crops and in houses, it was not for a further 40 years that they had become numerous enough to cause significant damage in commercial forests, where the trees most affected are Norway Spruces and larches. Damage tends to occur mainly in June and July, when short strips of bark are gnawed to gain access to the cambium layer and/or for sap. Sometimes ring-barking occurs several metres above ground level, when the tree crown will die, and in subsequent gales the bole is liable to snap at the point of damage. Minor harm occurs in trees such as Scots Pines and birches, which may result in fungal infection. Fruit trees are also sometimes damaged. Trees are attacked more in some years than in others – possibly because of an enhanced flow of sap or because of a dearth of other foods as, for example, when Common Beeches fail to flower profusely. In plantations damage is likely to be exacerbated because trees are all of the same age and are thus a potential source of food at the same time.

Where they gain access to the roof spaces of houses, Edible Dormice make a great deal of noise at night, disturbing the sleeping inhabitants. They gnaw timbers and electric wiring (thus creating a fire hazard), eat stored fruit (especially plums and apples), nest in airing cupboards where they also urinate and defecate (thus posing a potentially significant threat to human health), and drown in household and garden water-storage cisterns and lavatory bowls. In 1943–61, almost 600 Edible Dormice were caught in homes in the Amersham area alone, including more than 60 in a single house.

Although the Edible Dormouse's flea can carry typhus, and in central Europe its tick has been implicated in the transmission of Lyme disease (a form of arthritis caused by bacteria), no significant disease has yet been associated with the English population.

In spite of considerable research, led by P.A. Morris, a number of questions, apart from that of slowness of spread referred to above, regarding the species in England remain unanswered. Why does it enter houses? Why does breeding fail in some years – is it linked to Common Beech flower and seed production? Why are trees attacked in some years but not in others? With only a single litter being produced per annum, and with a complete breeding failure in an average of two years in five, how has the population become so large in such a relatively short space of time? The answers to these and other intriguing questions at present remain an enigma.

References

Baker 1990; Bieber 1998;[4] Bowen 2003; Carrington 1950; Corbet, in Corbet & Southern 1977; Cowdy 1966; Fitter 1959; Flower 1929; Harris et al. 1995; Jackson 1994; Jones-Walters & Corbet, in Corbet & Harris 1991; Leutscher 1954; Lever 1969, 1977, 1980a, 1985, 1994; Lloyd 1947; Long 2003; Morris 1997, in Poland, in Harris & Yalden 2008; Morris & Hoodless 1992; Morris & Temple 1998; Platt & Rowe 1964; Street 1955; Thompson 1953; Thompson & Platt 1964; Vesey-Fitzgerald 1936, 1938; Vevers 1947; Yalden 1999.

Notes

1. As Morris (1997) points out, an unlikely time to release a hibernating species.
2. The Edible Dormouse is one of the few species whose scientific name is sometimes used in the vernacular.
3. Hence the German name '*Der Siebenschläfer*' ('The Seven Sleeper'). In its early days in England, the Edible Dormouse was also known as the Chinchilla (to which it has a passing resemblance) or the Spanish Rat (although it only occurs in the extreme north of Spain).
4. Bieber (1998) refers to the species as *Myoxus glis*. This, as Morris (1997) points out, is because the name *Glis* has been challenged and the substitution of the next oldest generic name (*Myoxus*) has been advocated. This (see Morris, in Harris and Yalden 2008) has not yet been implemented.

CRICETIDAE (VOLES, LEMMINGS, HAMSTERS & ALLIES)

BANK VOLE *Myodes glareolus*

Natural Distribution From N of the Arctic Circle S through Scandinavia and Britain to the Pyrenees and Italy, W to C Siberia. Distribution is discontinuous in the E part of its range.
Naturalized Distribution Ireland.

BANK VOLE

Ireland

Although the Bank Vole occurs naturally throughout Britain, including on many offshore islands (to some of which it may have been translocated from the mainland), the Irish population, discovered in 1964, was possibly formed by animals accidentally introduced from Germany in the 1920s. Genetic studies confirm that the population was founded from a small, introduced stock. The species has so far been recorded in Counties Waterford, Tipperary, Limerick, Galway, Clare, Cork and Kerry in the south-west, where it is spreading at an estimated rate of c 3 km a year.

Characteristics
Bank Voles are most common in mature mixed and deciduous woodlands, but also occur in grassland, hedgerows and fens, and along road verges. Thick ground cover is an essential requirement. They are mainly herbivorous.

Impact
In continental Europe, Bank Voles are a minor pest in commercial forestry plantations, where they eat seeds, seedlings and bark saplings, and gnaw roots; bark eating has been recorded in Britain and Ireland.

References
Ryan 1996; Shore & Hare, in Harris & Yalden 2008; Sleeman 1997; Smal & Fairley 1984.

CRICETIDAE (VOLES, LEMMINGS, HAMSTERS & ALLIES)

ORKNEY AND GUERNSEY VOLES *Microtus arvalis*

Natural Distribution The Common Vole is widespread in Eurasia from NC Spain through the Caucasus to C Siberia. The nominate form occurs in lowland W Europe.
Naturalized Distribution England (Guernsey, Channel Islands); Scotland (Orkney Islands).
In Britain and Ireland, the Common Vole is restricted to the island of Guernsey in the Channel Islands (where it is known as the Guernsey Vole), and to eight of the Orkney Islands (the Orkney Vole).

England (Guernsey, Channel Islands)
'Guernsey populations may have been introduced or may be a relict population from the end of the Pleistocene when the island was connected to continental Europe; both scenarios consistent with close relationship to adjacent continental populations indicated by epigenetic polymorphisms (Berry and Rose 1975)' (Gorman and Reynolds, in Harris and Yalden 2008: 108).

On Guernsey, *M. a. sarnius* (the local form) occurs more frequently in damp meadows than in drier hedgebanks.

Scotland (Orkney Islands)
Mammal bones found in Orkney that were associated with other material radiocarbon-dated to around 4000 BP were identified by Corbet (1975) as those of *M. arvalis.* The species was introduced to Orkney by Neolithic man, probably between about 5700 BP

(the date of the earliest known settlement) and 5400 BP (the date of the earliest strata containing *arvalis* remains). Radiocarbon dates of *arvalis* bones from Westray were 4800± 120 and 3590± 80 years BP (OxA-1081, 1080). Berry and Rose (1975) argued, from epigenetic characteristics, that the animals came from south-eastern Europe, but this was disputed by Corbet (1986), who thought an origin in the Low Countries was more likely. However, molecular studies confirm that the animals in the Orkneys are of the same phylogenetic lineage as those from France and Spain (Haynes et al. 2003), and south-western Europe is therefore a more probable source. This implies an early movement and trading links of Neolithic people.

In the Orkney Islands, the species currently occurs on Mainland, Eday (to which it was deliberately translocated from Westray in 1986–7), Westray, Sanday, Burray, Hunda, South Ronaldsay and Rousay. Neolithic remains from Holm of Westray could indicate transportation by raptors or an extinct population. The species could have occurred on Shapinsay until 1906, though this may be a case of misidentification; it has certainly not been present on the island since 1943. The total Orkney Islands summer population is estimated to exceed four million.

Characteristics

In the Orkney Islands, the species is found in a wide range of habitats, including deciduous and coniferous woodland, wet heath and marshes, heather moorland, plateau blanket bog, old peat cuttings, and waste ground associated with cliff tops, ditches, fence lines, road verges and gardens, but seldom on arable land, silage fields and short-cut pastures. It is a herbivore, feeding on the leaves, stems, roots and other parts of a wide variety of grasses and dicotyledons.

In the Orkneys no fewer than five subspecies have been named, of which perhaps two are valid: *M. a. orcadensis* (Mainland, Rousay and South Ronaldshay) and *M. a. sandayensis* (Sanday and Westray). All Orkney Voles (wholly melanistic individuals occur

ORKNEY VOLE, WESTRAY, ORKNEY ISLANDS

from time to time) are distinctive, being much larger morphologically than most European individuals, though they are believed to be smaller than in the Neolithic. The species' principal predators in the Orkney Islands are Hen Harriers, Short-eared Owls, Common Kestrels and domestic cats.

Impact

In the Orkneys, the species enters hay and oat stacks and grain stores, but it remains at ground level and little damage is caused. No impact has been recorded on Guernsey.

References

Beirne 1947, 1952; Berry 1985; Berry & Rose 1975; Bishop & Delany 1963; Booth & Booth 1994; Corbet 1961, 1969, 1974, 1983; Corbet & Walters, in Corbet & Southern 1977; Gorman, in Corbet & Harris 1991; Gorman & Reynolds, in Harris & Yalden 2008; Haynes 2003; Hewson 1951; Lever 1977, 1985; Long 2003; Turner 1965; Yalden 1999.

MURIDAE (RATS, MICE, GERBILS & ALLIES)

HOUSE MOUSE *Mus domesticus*[1]

Natural Distribution From W and S Europe and N Africa E through the Middle East to Iran. As a result of natural and artificial dispersal, now one of the most widespread of all non-human mammals.
Naturalized Distribution England; Scotland; Wales; Ireland.

England; Scotland; Wales; Ireland

Corbet (1974: 187–8) records that:

> The house mouse was probably the first mammal to be introduced to Britain by human agency. None of the very early historical references to mice allows clear identification of species and information on its date of introduction can therefore only come from subfossil finds. Many of these are difficult to interpret because of the possibility of intrusion through burrows into older strata, but one recent record, hitherto unpublished, seems to confirm its presence in the pre-Roman Iron Age. This is based on the rostrum[2] of two animals and one mandible,[3] identified by the author … from a site at Gussage All Saints, Dorset. They were in a sealed layer with other small mammals and with no sign of subsequent disturbance.

Other contemporaneous records of House Mice in Britain include finds at Maiden Castle in Dorset and at Danebury in Hampshire. House Mice form a minor but fairly significant component of the small mammal fauna of many pre-Roman sites throughout England.

In *A Testimonie of Antiquitie*, Aelfric 'Grammaticus' (fl. 1006), at one time a monk at Winchester and subsequently abbot of Cerne and Ensham, mentions an animal he refers to as '*raturus raet*'. Aelfric's book, however, is a translation from Provençal authors, for whom *raturus* meant the House Mouse. By the late medieval period there were, according to Armitage (1985: 67), 'sizeable populations of House Mice in towns everywhere' throughout much of Britain.

HOUSE MOUSE RAIDING GRAIN STORE

As Easteal (1981: 94) points out: 'The colonization of a new area by a species may be a major event in the evolution of that species and can result in the formation of new species. This can occur if the colonizing event itself causes isolation between different populations, which then diverge genetically as the result of micro-evolutionary processes, or if the colonizing event itself, in cases where it involves very few individuals, brings about a radical genetic change in the founding population.'

HOUSE MOUSE

The tendency of isolated offshore islands to allow the development of different forms of one species displaying new genetic characteristics is typified by the now-extinct St Kilda House Mouse on the island of Hirta in the St Kilda group in the Outer Hebrides. Although this animal, which was quite distinct morphologically from the mainland form, did not become known to science until 1899, it must have been living on St Kilda for many hundreds of years. Berry (1970) showed that the St Kilda House Mouse more closely resembled the House Mice of the Faeroe and Shetland Islands than those of mainland Britain. The likelihood, therefore, is that the Norsemen accidentally brought the animals to the Faeroes, Shetlands and St Kilda from Scandinavia over 1,000 years ago in cargoes of hay imported to feed their domestic stock. The species died out on St Kilda, partly due to predation by feral cats and competition from Long-tailed Field (Wood) Mice, after the last human inhabitants abandoned St Kilda in 1930.

Today, House Mice are widely distributed in England, Scotland, Wales and Ireland, including the majority of small, inhabited offshore islands.

Characteristics

The House Mouse is a largely commensal species, though the urban population is markedly more so than the rural. The former lives in close association with humans in a wide variety of buildings and even in cold stores and coalmines. The latter occurs in farm buildings, granaries, poultry houses, piggeries, rubbish dumps, hedgerows and formerly corn ricks, and to a lesser extent in open arable fields, where it is largely replaced by the Wood Mouse. House Mice are omnivorous, in the country favouring cereal grains, insects, seeds and plant material. Their main controlling factor is cold weather, and even a short spell can account for the deaths of large numbers. Predation by rats, mustelids, domestic cats and Barn Owls is not a statistically important controlling factor.

Impact

Wherever it occurs, the House Mouse causes much damage to stored foods both directly through consumption and indirectly through contamination with urine and faeces. Gnawing by House Mice of electric cables and insulation material in buildings can pose a serious fire hazard.

House Mice in Britain and Ireland harbour a flea, *Nosopsyllus fasciatus*, and a blood-sucking louse, *Polyplax serrata*, which can transmit a number of dangerous diseases to humans. Cases of rat-bite fever (caused by the bacteria *Spirillum minor* or *Streptobacillus moniliformis*), tularaemia (caused by the bacterium *Francisella tularensis*), murine typhus (caused by *Rickettsia mooseri*, transmitted by fleas) and scrub typhus have all been associated with House Mice, as has rickettsial pox, leptospirosis, favus (caused by the fungus *Achorion quickeanum*), lymphocytric choriomeningitis, *Salmonella* bacteria, and tapeworms *Hymenolepis nana* and *H. diminuta* (Berry 1991).

Most methods of control by man involve the use of poison, though the animals' inherited resistance to anticoagulants such as Warfarin has seriously reduced the effectiveness of this widely used means of removal.

References

Amori & Clout 2002; Armour-Chelu 1991; Baker 1990; Berry 1970, in Corbet & Harris 1991; Berry et al., in Harris & Yalden 2008; Bishop & Delany 1963; Chitty & Southern 1954; Corbet 1974; Coy 1984; Crowcroft 1966; Easteal 1981; Fairley 1971; Fitter 1959; Key et al. 1996; Kitchener 2001; Lever 1969, 1977, 1980c, 1994; Long 2003; Morris 1986; Rowe, in Corbet & Southern 1964; Schwarz & Schwarz 1943; Sutcliffe & Kowalski 1976; Thaler et al. 1981; Yalden 1977, 1986, 1999.

Notes

1. It was formerly believed that all House Mice in Europe were of one species, then known as *Mus musculus*. In the 1970s, however, it was found that House Mice in Denmark, and thence eastwards across Europe, divide into two separate sub-populations; the north-eastern one retains the Linnaean designation *M. musculus*, the south-western population being renamed *M. domesticus* (first described by Rutty 1772) in Thaler et al. 1981.
2. A beak-like projection.
3. In mammals (and fish) the lower jawbone.

MURIDAE (RATS, MICE, GERBILS & ALLIES)

COMMON RAT *Rattus norvegicus*

(NORWAY RAT; BROWN RAT; SEWER RAT)

Natural Distribution Probably originally confined to the steppes of C Asia. Now, as a result of natural and anthropogenic dispersal, virtually cosmopolitan apart from polar regions.
Naturalized Distribution England; Scotland; Wales; Ireland.

The Common Rat appears to have reached Europe via Russia early in the 18th century. The German-born naturalist Peter Simon Pallas records that vast hordes dispersed westwards from Asia following a severe earthquake in 1727, soon spreading through Eastern Europe to the shores of the Baltic Sea. They had already arrived in Copenhagen, probably in Russian ships from Asia, in 1716, and had reached Paris and parts of eastern Prussia by 1750, either from Copenhagen or from the east. They were first recorded on the Norwegian mainland (where by 1776 they were said to be more common than the Ship Rat) in 1762; on the Faeroe Islands – which they reputedly reached in the wreckage of the *King of Prussia* – in 1768; in Brunswick, Germany and Greenland around 1780; and in Sweden in 1790. Common Rats reached Spain and Italy in the mid- to late 18th century, and Switzerland in about 1809. They first arrived in North America, presumably on ships from England, in about 1775.

COMMON RAT ON FARM

England

An old legend records that the first Common Rats to reach England arrived from Europe in the ship that landed the future King William III of Orange at Torbay in 1688. Another story, propagated by opponents of the House of Hanover such as Jacobites, Tories and Roman Catholics, tells that the animals first reached England in the ship that brought George I from Germany in 1714, earning them the soubriquet of 'Hanoverian Rats'. Charles Waterton, the delightfully eccentric Roman Catholic 'Squire' of Walton Hall in Yorkshire and an inquisitive naturalist, fostered this idea, which he inherited from his father. In his *Essays* (1838) he describes *norvegicus* as 'a little grey-coloured short-legged animal ... known to naturalists as the Hanoverian rat'. His father, he continues, was 'always positive that it actually came over in the same ship which conveyed the new dynasty to our shores'.

According to Thomas Pennant, in his *British Zoology* (1766) and *History of British Quadrupeds* (1781), the first Common Rats reached England around 1728–9, probably in ships from Russian ports but certainly not in Norwegian timber vessels, as was at one time erroneously believed since, as mentioned above, they did not arrive in Norway until 1762. In the latter work, Pennant noted that by 1776 'the Norway-rat has greatly lessened [the Ship Rat's] numbers, and in many places almost extirpated them', while Pennant's friend, Gilbert White, commented in 1777 that 'the Norway rats destroy all the indigenous [*sic*] ones', i.e. the Ship Rat.

Scotland

In his *Mammalia Scotica* (published in 1808 but compiled in 1764–74), the naturalist John Walker wrote:

> *Mus fossor* ... the Norway Rat ... First brought, as they say, to Scotland in ships from Norway ... Wheresoever it pitches its abode, it pitches out the Black Ratten utterly ... the Black Ratten ... and the Norway Rat were previously entirely unknown in Annan [southern Dumfries and Galloway] ... however about twenty years agone [i.e. c 1744–54] the Norway Rat was cast on shore from ships driven to the mouth of the River Annan, and now is scattered through almost the whole region of Annan.

The *New Statistical Account of Scotland* for the parish of Newlands in Peebles relates that:

> The brown or Russian or Norwegian rat ... a good many years ago invaded Tweeddale, to the total extirmination [*sic*] of the former black rat inhabitants. Their first appearance was at Selkirk, about the year 1776 or 1777 [where the townspeople feared that their burrows would undermine the buildings], and passing from Selkirk, they were next heard of in the mill at Traquair; from thence ... they appeared in the mills of Peebles; then entering by Lyne Water, they arrived at Flemington Mill, in this parish ... about the year 1791 or 1792.

In the Scottish Highlands, the Common Rat was first observed by the Rev. G. Gordon, in Morayshire in 1814. As late as 1855, the species was being described as 'recently introduced' in some Scottish localities.

Wales

On the Isle of Anglesey, Common Rats were so numerous in 1762 that they are said to have eaten ears of standing corn while it was being reaped. They gained access to Puffin Island (Priestholm) off the east coast of Anglesey when they swam ashore from a wrecked Prussian vessel in about 1816.

Ireland

According to the physician John Rutty, in his *Essay Towards the Natural History of the County of Dublin* (1772), Common Rats (allegedly the size of cats and rabbits!) 'first began to infest these parts about the year 1722'. If true, this predates their arrival in England by 6–7 years.

Summary of Status

Today, Common Rats are found throughout Britain and Ireland except perhaps in some of the more exposed montane regions of the Scottish uplands north and south of the Great Glen and on some of the smaller offshore islands.

Characteristics

Common Rats are typically associated with farms, rubbish tips, sewers, mills and warehouses, but are also found in hedgerows surrounding cereal and root crops such as sugar beet, turnips and swedes throughout the year. They prefer places with dense groundcover near water. They also occur independently of humans in many coastal localities, particularly salt marshes. Common Rats are omnivorous, and although they favour starch- and protein-rich foods such as cereals, they will consume almost anything edible they find, and in urban habitats even 'inedible' substances such as candles and soap.

Impact

As one of the most widely distributed vertebrate pest species in Britain and Ireland (and, indeed, the world), the Common Rat is responsible for extensive damage, especially to stored food products such as grain, both directly through consumption, and indirectly through contamination with faeces, urine and hair, though it is said that the economic damage caused to the bags containing the grain considerably exceeds the losses due to consumption or contamination of their contents. In rural habitats, Common Rats are also, albeit to a lesser extent, a pest of growing crops. Where they live in close proximity to humans, they cause significant damage to buildings and pose a fire hazard by gnawing through electric cables and insulating material.

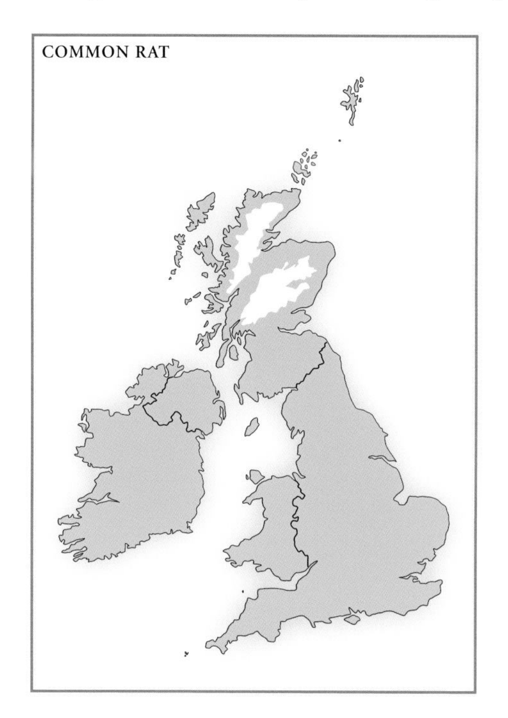

Among the most significant impacts of Common Rats (and, to a lesser extent only because of their comparative rarity, of Ship Rats) in Britain and Ireland, is the havoc they can wreak among colonially nesting seabirds on offshore islands. To take but three examples, on Puffin Island (Priestholm) off the east coast of Anglesey (see under Wales), they almost wiped out the population of Atlantic Puffins (and European Rabbits), and the avifauna of Ailsa Craig and Little Cumbrae in the Firth of Clyde has also been severely affected, as has the population of Manx

Shearwaters on Lundy Island in the Bristol Channel, from which both Common and Ship Rats were removed in 2005.

As Moors et al. (1992) point out, rats do not affect birds solely as predators. In their consumption of the inflorescence of numerous plants and seedlings they can compete directly with some bird species for food, and also for shelter and nest sites.

In Britain and Ireland, leptospirosis is probably the most common disease transmissible to humans by Common Rats, which excrete the *Leptospira* organism in their urine, though few carry the *ichterohaemorrhagiae* serovar that is believed to be the causative agent for Weil's disease (infectious jaundice) in humans. Other rat-borne parasites in Britain that can cause diseases in humans include *Cryptosporidium parvum* (cryptosporidiosis), *Toxoplasma gondii* (toxoplasmosis), *Coxiella burnetii* (Q fever) and hantavirus (which causes various febrile haemorrhagic diseases, often associated with kidney damage or failure). Heavy infestations with *Ixodes* ticks, which sometimes occur in British Common Rats, are of concern because such ticks carry the spirochaete responsible for causing Lyme disease (a form of arthritis) in humans.

The mechanism whereby the Common Rat has largely replaced the Ship Rat in Britain is, like the replacement of the native Red Squirrel by the alien Grey Squirrel, little understood, though the larger size, heavier weight and more aggressive nature of the Common Rat are likely to have been contributory factors. It is interesting to note, as Yalden (1999) points out, that this replacement of Ship by Common Rats has not been mirrored in southern Europe, where the former remains common and the latter is scarce and is chiefly confined to seaports, which were formerly the preserve in Britain of the Ship Rat. In New Zealand, also, although the Common Rat was the earlier arrival it is the Ship Rat that is the dominant species. Yalden (1999: 184) postulates that 'the longer tail and larger ears of the Black [Ship] Rat must give it a better ability to lose heat in warm climates, but a worse ability to withstand cooler ones'. Though this could account for the Ship Rat's success in southern Europe and New Zealand, it begs the question of how it managed to survive for so long in more temperate Britain.

Although young Common Rats are vulnerable to the majority of both terrestrial and avian predators, the aggressive nature of adults renders them largely immune to most non-human predators. Most human methods of control involve the use of anticoagulant poisons, to which some populations have developed resistance, and the continued use of anticoagulants may result in more widespread resistant populations. However, irrespective of their resistance to anticoagulants, most Common Rats survive limited bait consumption because improved farming methods, which provide an abundant supply of food, result in baits being largely ignored. Rats are notoriously shy of new objects, and are likely largely to avoid bait containers, traps and other control equipment. Different control methods (e.g. non-lethal ones such as denying access to food and cover, proofing of buildings, the erection of barriers around food sources and repellents) provide at best only short-term control.

References

Bramwell 1990; Barett-Hamilton & Hinton 1910–21; Barnett 1951; Bishop & Delany 1963; Chitty & Shorten 1946; Chitty & Southern 1954; Corbet 1974; Fitter 1959; Hinton 1931; Lever, 1969, 1977, 1980c, 1985, 1994; Long 2003; Matheson 1931, 1962; Moors et al. 1992; Quy & Macdonald, in Harris & Yalden 2008; Taylor, in Corbet & Southern 1977; Taylor et al., in Corbet & Harris 1991; Twigg 1975; Yalden 1999.

MURIDAE (RATS, MICE, GERBILS & ALLIES)

SHIP RAT *Rattus rattus*

(BLACK RAT; ROOF RAT; HOUSE RAT; ALEXANDRINE RAT)

Natural Distribution Originally probably restricted to SE Asia. Now, as a result of natural and anthropogenic dispersal, virtually cosmopolitan.
Naturalized Distribution England; Scotland; Wales; Ireland.

SHIP RAT

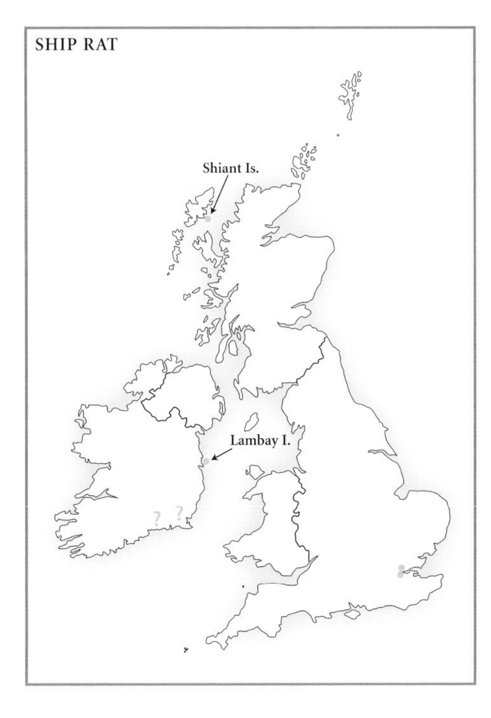

The Ship Rat was probably one of the first mammals to associate with early humans. It may have spread westwards from India to Egypt in the 4th century BC, and thence along trade routes into Europe. The tradition that it first reached Western Europe in the 12th century in the baggage of returning Crusaders is now regarded as a myth. Its spread by humans was probably augmented by natural dispersal west through the steppes corridors. It first reached North America in the early 17th century, and by the early 20th century most of the world had been overrun.

England

Well-stratified remains of Ship Rats have been recorded in Roman sites in London, York and Wroxeter (near Shrewsbury), probably dating from the 3rd–5th century AD. Since, however, there are no records from Anglo-Saxon times, it seems possible that the species died out around the end of the Roman period, and was introduced again during the 9th or 10th century from Scandinavia. Interestingly, as Twigg (1992) points out, the House Mouse was recorded throughout the whole of this period. Further specimens of Ship Rats have been found in archaeological sites in the City of London dated to the 3rd and 4th centuries AD, as well as in several other widely distributed sites elsewhere dated to the 3rd century.

In 1297, a number were caught at a village named Weston, where they were sold for a farthing (¼d.) each, and more are recorded as being trapped in Oxford in 1335 and again in 1363. In the *Vision of Piers Plowman* (c 1362–92) occurs the following disparaging couplet: 'Had ye rattes youre wille/Ye couthe nouyt reule yourseule.' The same work goes on to refer to a 'ratoner' (ratcatcher) and to a 'route of ratones and smale mys mo than a thousande with hem'. In the 'Pardoner's Tale' from the *Canterbury Tales* (c 1387), Geoffrey Chaucer writes of a harassed householder who:

> Forth he goeth, no longer wold he tary,
> Into the town unto a Pothecary,
> And praied him that he him wolde sell
> Som poison, that he might his ratouns quell.

In his *Polychronicon* (before 1364), the Benedictine chronicler Ranulf Higden describes Ship Rats as *nocentissimos* ('most harmful'), and a rat trap is listed as an item of expenditure in the accounts for 1469 of the churchwardens of St Michael's, Cornhill, in the City of London.

Summing up the situation in the late Middle Ages, Armitage (1985: 67) wrote that there were 'sizeable populations of Black [Ship] Rats … in towns everywhere, with the rat infestation a particular problem in the more densely settled urban areas'.

The fortunes of the Ship Rat seem to have declined rapidly following the introduction of the Common Rat to England in 1728–9. In *The Universal Directory for Taking Alive and Destroying Rats and All Other Kinds of Four-footed and Winged Vermin in a Method Hitherto Unattempted: Calculated for the Use of the Gentleman, the Farmer and the Warrener* (1768), Robert Smith, who was ratcatcher to Princess Amelia, daughter of George III, described how:

> the black ones do not burrow and run into shores and sewers as the others do but chiefly lie in the ceilings and wainscots in houses, and in outhouses they lie under the ridge tiles, and behind the rafters, and run about the side-plates: but their numbers are greatly diminished to what they were formerly, not many of them being now left, for the Norway rats always drive them out and kill them wherever they can come at them; as a proof of which I was once

> exercising my employment at a gentleman's house, and when the night came that I appointed to catch, I set all my traps going as usual, and in the lower part of the house in the cellars I caught the Norway rats, but in the upper part of the house I took nothing but black rats. I then put them together in a great cage to keep them alive until the morning, that the gentleman might see them, when the Norway rats killed the black rats immediately and devoured them in my presence.

In his *British Zoology* (1776), Thomas Pennant considered that 'the Norway-rat has also greatly lessened their [Ship Rat] numbers, and in many places almost extirpated them'. He goes on to say that 'among other officers his *British* majesty has a *ratcatcher*, distinguished by a particular dress, scarlet embroidered with yellow worsted in which are the figures of mice destroying wheat-sheaves'. Thomas Swaine, a ratcatcher who worked in 15 counties, mentions in *The Vermin Catcher* (1783) that he found Ship Rats only at High Wycombe in Buckinghamshire and in Middlesex.

The charmingly eccentric naturalist Charles Waterton of Walton Hall in Yorkshire believed that the Ship Rat would soon be eradicated by the Common Rat. He only ever saw 'one solitary specimen' of the former, which in his *Essays* (1838) he addressed as: 'Poor injured Briton! Hard, indeed, has been the fate of thy family! In another generation at fartherest it will probably sink down to the dust forever.' Waterton's words were to prove, if somewhat premature, prophetic.

By 1845, the Ship Rat was uncommon enough in London for a ratcatcher of St Giles – condescendingly described in the journal *Notes and Queries* for 1854 as 'an intelligent man and not a bad naturalist for his station in life' – to be able to charge collectors 3 guineas (£3.30) each for specimens from his cage outside the National Gallery in Trafalgar Square. In 1850, Ship Rats were discovered in a number of old granaries in London, and as late as 1875 a house in Cornhill in the City was found to be overrun. In the Channel Islands, they were said still to be 'common on Alderney and Herm' in 1862.

In the 19th and early 20th centuries, the English population was augmented on a number of occasions by the survivors of wrecked ships from foreign ports. In 1866, a fruit-carrying vessel was driven ashore in a gale at Seascale in Cumberland (Cumbria), where a number of Ship Rats escaped. After the Italian grain-carrier *Espagnol* was wrecked in Acton Cove near Marazion in Cornwall in 1875, 'the whole of the surrounding district was swarming with these little rats'. More escaped from an Austrian-registered barque wrecked near St Mary on the Isle of Man in 1883, and from a Greek vessel that was stranded on Lundy Island in the Bristol Channel in 1928.

By 1890, although it was still not uncommon in parts of Cheshire, the Ship Rat was thought to be more or less extinct in the rest of mainland Britain. In fact, Ship Rats were probably never even near extinction.[1] Almost certainly the two brown forms, *R. r. frugivorus* and *R. r. alexandrinus*, were often mistaken for Common Rats, although the corollary – that the black form of the Common Rat was frequently misidentified as the Ship Rat – is also probably true. In 1925–38, Matheson (1939) collected records of Ship Rats killed in the ports of Bristol, Weymouth, Southampton, Plymouth, Grimsby, Middlesborough and Liverpool, and lesser numbers elsewhere, including London. Fisher (1950) records that during the Second World War the Ship Rat was common not only, as would be expected, in the docks and quays of the port of London, but also in clubs, restaurants, theatres and even residential properties in the West End, where it was the most common rat in Oxford Street. At the outbreak of war in 1939, some 90 per cent of rats in the City of London were Ship Rats.

After his survey of the Ship Rat in Britain in 1951–6, Bentley (1959: 301) wrote: 'There is little likelihood of completely eliminating the species from Britain. It can be expected to persist for a considerable time yet in the commercial quarters adjacent to the docks in London, Bristol and Liverpool (and perhaps Edinburgh and Belfast) where conditions for its continued existence are most favourable.'

After a second survey, Bentley found that on Lundy Island in the Bristol Channel the Ship Rat had existed for many years (probably since 1928), and that in the Channel Islands it was still present on Sark and probably on Alderney, but was not recorded on Herm. In 1956–61, Bentley (1964) found that the number of infested sites had fallen from 48 to 28, a decline of some 42 per cent. Away from London, the Ship Rat was believed to be established only in the Channel Islands, Lundy Island, Bristol, Avonmouth, Sharpness, Gloucester, Salford, Wallasey, Liverpool and Bootle, with a decrease in numbers (though not in distribution) in Sharpness. The ports of Hull, Portsmouth and Southampton each reported five or less infestations, and a single individual was caught in Middlesborough. Sunderland, South Shields, Ellesmere Port and Eston (near Middlesborough) each reported a single infestation.

In 1961, Bentley (1964) found Ship Rats at Walthamstow (one only), at Martley on the River Severn in Worcestershire, at Congleton in Cheshire (a solitary individual), and perhaps at Hardington Mandeville near Yeovil, Higham Ferrers, Northamptonshire, and Cranbrook in Kent – from all of which the species had been absent in 1956.

In London in 1961, Ship Rats were established in Bethnal Green, Battersea, Southwark, Stepney, Bermondsey, Finsbury, Holborn, the City, Poplar and West Ham. In all but the first three of these localities the population had decreased, and in Poplar the animals were confined to a single wharf. In Southwark, on the other hand, Ship Rats had spread further away from the river. A single infestation was recorded in Erith, which in 1956 had been rat free. More than five colonies were reported from the City of Westminster, and five or less from Greenwich and Shoreditch.

'It is clear,' wrote Bentley (1964: 372), 'the status of the ship rat in Britain has further diminished and the species is on the way to losing its permanent foothold in the United Kingdom ... A further reduction in status can be expected over the next few years.' Bentley's opinion was prophetic, for within a decade Ship Rats only survived in any numbers on a few offshore islands such as Lundy (c 500) in the Channel Islands, and in the docks, warehouses and quays of some major ports such as London and Liverpool. The species has not been recorded in Manchester since 1978, Bristol since 1980 and Liverpool since 1983. In 1984–91, it was only reported in England at Southwark in London and at Avonmouth, Bristol. Most mainland reports today are of small, transient groups accidentally introduced in merchandise. Ship Rats are now believed to occur sporadically along both banks of the Thames in London, and there is a long-standing sub-population in Tilbury. Whether the species survives on Alderney and Sark in the Channel Islands is unknown, but the populations of Ship (and Common) Rats were allegedly eradicated in 2005 because of their predation on the eggs of colonially nesting seabirds.

Scotland

The earliest mention, albeit a negative one, of Ship Rats in Scotland dates from 1578, when John Leslie (or Lesley), Bishop of Ross, wrote in his *History of Scotland* of 'a wonder, the rattoune lyves not in Buquhane [Buchan] ... in this cuntrey no Rattoune is bred, or brought in from any other place, there may lyve'. Even by 1630 it was said that: 'There is not a ratt in Sutherland, and if they doe come thither in shipps from other parts (which often happeneth) they die presentlie, as soon as they doe smell the aire of that cuntrey. But they are in Catteynes [Caithness], the next adjacent province, divyded onlie be a little strye or brook from Sutherland.'

The same tradition appears to have been maintained in neighbouring Ross-shire, for when in 1656 an ex-Cromwellian trooper, Richard Franck, found himself in that county, he wrote: 'The inhabitants will flatter you with an absurd opinion that the earth in Ross hath an antipathy against rats … to the best of my observation I never saw a rat: nor do I remember of any one that was with me ever did.' The parishes of Dunnet in Caithness (Highland), Annan in Dumfriesshire (Dumfries and Galloway) and Liddersdale, Inverness-shire (Highland) were also reputedly free of Ship Rats, and their soil was much in demand for flooring granaries in the belief that this would make them rat-proof.

One of the earliest positive references to Ship Rats in Scotland was made by Martin Martin in *A Description of the Western Islands of Scotland* (1703), where in the Outer Hebrides he saw 'a great many Rats in the Village Rowdil [on Harris], which have become very troublesome to the Natives, and destroyed all their Corn, Milk, Butter, Cheese, etc. They could not extirpate these Vermin for some time by all their endeavours.'

In 1813, the Ship Rat was the only species of rat known in the town of Forfar (Angus), and it was described as 'not rare' throughout Scotland. In 1830, Ship Rats were said still to be common in rural Aberdeenshire (they survived at Cairnton of Kemnay until 1855) and in Morayshire until the 1840s. In 1838, it was said that in the bordering county of Banffshire 'in Keith, which is at a greater distance from the coast, it [the Ship Rat] is not very uncommon'. In Edinburgh, Ship Rats were found still to be inhabiting 'the garrets of the high houses in the old city' until 1834.

On Scottish offshore islands, Ship Rats were still infesting the Orkneys in considerable numbers in 1808, where on South Ronaldshay they were said to be fairly abundant in 1813. In the Hebrides, they still occurred on Benbecula until the 1880s, and in the Shetland Islands they survived until 1904.[2]

Sixty years later, Bentley (1964) recorded their presence only in Lerwick in the Shetland Islands, on the Clyde, in Aberdeen, Dundee, Edinburgh, Leith, Glasgow and Faslane (Dunbartonshire), with a decrease in numbers (though not of distribution) in all the above (60 per cent in Dundee) apart from Edinburgh, Lerwick and on the Clyde. None have occurred in Glasgow or Aberdeen since 1983. They also survived on Westray in the Orkney Islands, and on Garbh Eilean and Eilean Mhuire in the Shiant Islands off Harris in the Inner Hebrides, where the population in the 1990s was estimated to number around 500, but is now believed to be up to 1,000. Ship Rats are believed to have arrived in the Shiant Islands in 1900 after a shipwreck on the nearby Galtachean Island. The Shianats are the only known British islands where Ship Rats occur without the presence of Common Rats (Angus and Hopkins 1996; Key et al. 1998).

Wales

In his *Itinerarium Cambriae* (probably written in the late 12th century), Giraldus De Barri, commonly named Cambrensis (?1146–?1220), wrote of *murium majorum, qui vulgariter Rati dicuntur* ('the larger species of mice, commonly called Rats'), which were said to have attacked a man and to have eaten a woman and her child at Merrion near the coast of southern Pembrokeshire.

Information on the subsequent history of the Ship Rat in Wales seems to be hard to come by. It is known that by 1890 it was not uncommon in parts of the north, and

Matheson (1939) found accounts of the species killed in the ports of Swansea and Cardiff. In the port of Cardiff, Bentley (1964) received reports of only a small number of overrun sites, including Milford Haven, Pembrokeshire, where the species had not occurred in 1956.

Ireland

What may be the earliest reference to the Ship Rat in Ireland is that made by the Welsh topographer Giraldus De Barri, commonly called Cambrensis (?1146–?1220), a native of Pembrokeshire, who described how Bishop St Yvorus of Bergery (Bergerin) (d AD 500) reputedly cursed '*mures majores qui vulgariter Rati vocantur*' ('the larger species of mice, commonly known as Rats') and expelled them from the Ferneginan region of Leinster because they had gnawed his books. Similarly, St Colombcille (AD 521–97) allegedly drove *mures* from a number of Irish islands, including Tory Island off the coast of Co. Donegal and Inishmurry, Co. Sligo.

In his *Topographia Hibernica* (compiled in 1183–7), Cambrensis wrote: 'There is another thing remarkable in this island [variously identified as either Aran in Galway Bay off Co. Clare or Inishglora or Caher off the coast of Co. Mayo] – although mures swarm in vast numbers in other parts of Ireland, here not a single one is found. No *mus* is bred here nor does it live if introduced.'

The earliest appearance of Ship Rats (or possibly House Mice) in Irish iconography is in the *Book of Kells*,[3] a late 8th- or early 9th-century AD illuminated manuscript which depicts, on the monogram page and plate 107, a representation of two domestic cats and four animals that are either Ship Rats or House Mice – most probably the former. In a Register of about 1377 of Archbishop Milo Sweetman, there is mention of agricultural damage caused by Ship Rats in his See in Co. Armagh. In his *Description of Ireland* (1617), the traveller Fynes Moryson quotes the following derogatory doggerel about the state of Ireland in the early 17th century:

> For four vile beasts Ireland hath no fence,
> Their bodies lice, their houses rats possess.
> Most wicked priests govern their conscience,
> And ravening wolves[4] do waste their fields no less.

After the arrival in Ireland of the Common Rat in the 1720s, the Ship Rat began to suffer the same decline there as in Britain. A single individual was found in Co. Cork in 1842, and unconfirmed sightings were reported in Counties Kerry, Armagh, Dublin and Antrim. In 1876, a litter was found in Leviston in Co. Kildare, and as late as 1911 a colony was discovered in a warehouse in Dungarvon in Co. Waterford. In 1935, a solitary individual was observed on Lambay Island north of Dublin.

Bentley (1964) recorded infestations of Ship Rats in 1961 only in Belfast and Londonderry, but the species may also occur in Counties Wexford and Waterford. On Lambay Island, off Dublin, Ship and Common Rats coexist.

Summary of Status

Twigg (1992), from whom the following account is derived, has summarized the status of the Ship Rat in Britain and Ireland in 1961–89. (For the sake of continuity, England, Scotland, Wales and Ireland are all discussed collectively.)

In 1961–70, Ship Rats were recorded from only five land sites, including Lundy Island, where they appeared to be declining, and in 1966 on a ship at Grimsby from where there have been no subsequent records. They were also present on Westray, Orkneys in 1960, where the last recorded individual was killed at Skaill eight years later.

In 1971–80, Ship Rats were recorded at 16 sites, where control operations were implemented in docks and warehouses in Cardiff and the Vale of Glamorgan; Spalding, Lincolnshire; Glasgow (1971); Cumbria (1974); Birkenhead (1976); Belfast (1978) and Manchester (1971 and 1980). By the end of the decade, they survived in these sites only in Birkenhead. They were reported in Great Yarmouth in 1979 and in Ipswich, Suffolk in 1982, but have not been seen there since. In 1971–80, they occurred in Cardiff docks and in the latter year in Glasgow, but no control measures were undertaken in either city; they had apparently disappeared from both these places by 1989. They still survived on Lundy Island, and a single dead specimen was found at Chippenham in Wiltshire in 1974.

Possibly the most surprising feature of this, and indeed of subsequent decades, is, as Twigg (1992) points out, the decline of the Ship Rat in London; in 1971–80 there were only two records – from Barking and Dagenham (1979) and in the same year from Harrow. The last Ship Rat was reported in Islington in 1976.

In 1981–4, Ship Rats were observed in private houses in Edinburgh and in the city's docks at Leith. A small number were killed in the Avonmouth docks near Bristol in 1982, when an infestation also occurred in buildings at Bristol University. In 1983, they were noted on Lundy Island and on several occasions on Sark in the Channel Islands, where they appeared to be 'fairly common'. In the same year, some were found in a cargo of wool at Tilbury in London. In 1984, they were recorded in the port at Lowestoft and on a ship at Felixstowe (both in Suffolk), where they were eradicated.

In 1985–9 (the final year of Twigg's survey), there was an increase in the number of records. In both 1988 and 1989, Ship Rats were reported in Dundee and Edinburgh respectively, and from time to time they were found on ships on Teeside and West Hartlepool. They were not often found on land, although in 1986 two were discovered in Gateshead in a cargo of bananas from a ship that had docked in Wales. During this time, they were recorded at Goole and Selby in Yorkshire and in Liverpool city centre. Single individuals were caught at Avonmouth in 1985, at Paddock Wood, and at Wood Walton Hill, Cambridgeshire in 1989.

In 1985–6, they occurred in a warehouse at West Byfleet in Surrey, from where they had disappeared by the end of the decade. They were also seen in grain silos at Tilbury, where they was still present in 1989, and in 1987 in a flour-mill, grain silos and buildings belonging to the Port of London Authority.

In Ireland, a small number of Ship Rats in a piggery in Londonderry in 1987 was eradicated with Warfarin. In London, there were only two records in 1985–9 – one of which was of a solitary individual in Hammersmith in 1985, with further probable sightings in 1986 or 1987.

On Lundy in 1985–6, the population was apparently 'stable and moderate', and in 1989 the species' presence was confirmed in the harbour of St Anne on Alderney in the Channel Islands.

The 1985–9 records of the Port of London Authority show that the animals were reported on 23 sea-going vessels at Tilbury, but on no river or coastal boats. Three ships examined by the Falmouth and Truro Port Health Authority in 1989 contained Ship Rats, as did four craft inspected by the Hull and Goole Port Health Authority in 1985–9.

The contraction in range of the Ship Rat in Britain that has occurred since the 1950s was a crucial period in the species' history, and brought it to a position from which it has never recovered.

In 1961–70 there were, with the exception of the sub-population on Lundy Island, no records south of the Mersey. In 1971–80, however, the situation changed dramatically, largely due to the species' presence in the London area and in parts of East Anglia and the Bristol Channel.

In 1981–4, there was another fall in the number of sites where the Ship Rat occurred, with only two records north of Lowestoft. These shifts demonstrate the ephemeral nature of non-insular sub-populations, indicating short-lived introductions of small numbers that are either easily controlled or fail to survive in the absence of larger populations into which they can become assimilated.

The period 1985–9 witnessed a renewed increase in the number of Ship Rat sites, with a total of 16 land records, including those on Lundy and Alderney. Some of the records for this period referred only to the unconfirmed sightings of single individuals. This raised the possibility of mistaken identification of the melanistic phase of the Common Rat, and also suggested the absence of large populations. The total of 23 sites in the years 1981–9 was higher than that for the two previous decades, but there was no evidence of a corresponding increase in the total population.

Twigg (1992) thought that the potential for the introduction of Ship Rats still existed, because of the number of occasions on which they were being found on ships in ports, where although the numbers are usually small no fewer than 75 were discovered on a ship in Hull in 1978.

Before the Second World War, Matheson (1939) considered that the Ship Rat was capable of maintaining a considerable presence in British seaports without much further recruitment from abroad. As a result of the massive post-war building programme and the increasing use of steel and concrete in the construction industry, however, there was a decline in habitats suitable for Ship Rats, and also in sea-borne cargoes, which were mainly replaced by air-freighting; once in Britain, cargoes tended to continue their onwards journey to inland locations by road or rail rather than by canal boats.

The Ship Rat's contraction of range in various localities in mainland Britain may have been partially due to its propensity for living in buildings, where it is more vulnerable to control measures. Its near-disappearance has also often coincided with the demolition of quayside flour- and seed-mills, where it was formerly hard to control. Other contributory factors are likely to include:

- The introduction of measures to reduce infestation on coastal vessels.
- Improved methods of dockside building to deny the animals shelter.
- More effective means of control.
- Improved hygiene.
- More secure methods of storing goods on ships and on shore.

Port records, however, indicate that Ship Rats do continue to arrive in Britain by ship, and it is likely that both global warming (resulting in more suitable climatic conditions that might enable them to survive away from buildings) and the construction of the Channel Tunnel (giving an alternative means of access) will enable the species to re-establish more than its current tentative toehold in Britain.

In Britain the Ship Rat is today mainly confined to dockside warehouses and food-processing plants, though it has also been reported in supermarkets, restaurants and large shops. Its preferred habitat is in cavity walls, behind wall panelling and in ceilings. On the Shiant Islands (and formerly on Lundy Island) it lives among rocks and on cliff faces, and in the Channel Islands it is said to be arboreal.

Characteristics

The Ship Rat is omnivorous – although more vegetarian than the Common Rat – and is particularly partial to fruit. Most agricultural crops are consumed – especially cereals.

In Britain, the domestic cat is probably the Ship Rat's principal natural predator. Although in confined spaces Common Rats will kill Ship Rats, elsewhere the two species live sympatrically (as they formerly did on Lundy), and it seems doubtful if, in the wild, the Common Rat is ever a significant predator of the Ship Rat. Replacement of the latter by the former in temperate regions is said probably to be due to the Ship Rat's lower tolerance of severe weather and inability to forage well in adverse conditions; this begs the question as to how the Ship Rat has survived for so long in temperate Britain, where the current population is estimated to number around 1,300.

Impact

The widely held belief that the Ship Rat was responsible for the transmission in Europe in the 14th century of the 'Black Death' or bubonic plague has been challenged on a number of grounds, not least because bubonic plague, caused by the organism *Pateurella (Yersinia) pestis*, does not spread well in a cool and temperate climate (Twigg 1984). Although both Ship and Common Rats can and do carry bubonic plague, and in the tropics the former in particular may be an important vector, a high percentage of both species succumb to the disease, so that even in tropical regions neither acts as a really satisfactory long-term reservoir. Ship Rats appear to carry many of the same parasites and diseases as Common Rats.

The Ship Rat has developed a remarkable ability to live sympatrically with humans, and since it is much better able to live in buildings than the Common Rat it is generally the more important pest in urban localities, where damage and fouling of stored food was formerly widespread. In common with Common Rats and House Mice, Ship Rats pose a safety hazard by gnawing through insulation material and electricity cables. Ship Rats are controlled by humans in much the same way as the other two murine species.

Wherever they occur with colonially nesting seabirds, as on the Shiant Islands, Ship Rats are a threat to such birds as Atlantic Puffins, Guillemots, Razorbills, Kittiwakes and gulls, and even to other ground-nesting species like the Great Skua.

References

Armitage 1985; Armitage et al. 1984; Bayliss 1980; Bentley 1959, 1964; Bishop & Delany 1963; Chitty & Southern 1954; Corbet 1974; Davies 1987; Deane 1952; Dobney & Harwood 1998; Fisher 1950; Fitter 1959; Hinton 1931; Key et al. 1998; Lever 1969, 1977, 1980c, 1985, 1994; Long 2003; Matheson 1931, 1939, 1958; Matthews 1952; Moffat 1938; O'Connor 1991, 1992; Perrin & Gurnell 1971; Rackham 1979; Smith 1993; Taylor, in Corbet & Southern 1977, in Corbet & Harris 1991; Twigg et al., in Harris & Yalden 2008; Walsh 1988; Welch et al. 2001; Yalden 1999.

Notes

1. Long (2003: 191) claims that Ship Rats in England 'became widespread again in the 1930s', whereas they had been widely distributed for many years.
2. Long (2003) says that the Ship's Rat only reached the Shetland Islands in 1904, whereas it survived there up to that date.
3. In Trinity College, Dublin.
4. Wolves died out in Ireland in 1692–1786.

LEPORIDAE (RABBITS & HARES)

EUROPEAN RABBIT *Oryctolagus cuniculus*[1]

Natural Distribution Present in the Iberian Peninsula and S France, where presumed to be endemic since at least the Middle Pleistocene, c 300,000 BP.
Naturalized Distribution England; Scotland; Wales; Ireland.

England

Rabbit remains have been found at the interglacial sites of post-Cromerian Boxgrove and Hoxonian Swanscombe, Kent, in the Middle Pleistocene, but not at subsequent Neolithic, Iron Age, Roman or Anglo-Saxon sites, suggesting that, as in the case of the Ship Rat, the species died out in the wild. However, in his *Rerum Rusticarum*, compiled in 54 BC, Marcus Terrentius Varro records that the Romans imported domesticated rabbits to Britain, where they were reared in captivity in *leporaria*. Rabbit embryos (*laurices*) were a delicacy of Roman gourmets, especially during times of fasting. John Whitaker, in *The History of Manchester* (1771) appears to be the first British writer to refer to the importation of domestic rabbits to Britain by the Romans.

There is, however, as Fitter (1959) points out, no evidence that any Roman-introduced European Rabbits escaped into the Romano-British countryside. There is no Celtic or

EUROPEAN RABBITS

Anglo-Saxon name for the European Rabbit, and no mention of the animal in that most comprehensive survey of English land resources, the *Domesday Book* (1086).

By the 15th century, however, the Latin *cuniculus* was being translated as *conynge* or *conninge*, later becoming coney; a cunicularia or cuningera was a coneygarth or warren. Until the 18th century the name rabbit, derived from the medieval French *rabette*, was used only for juveniles, which were an esteemed culinary delicacy.

Veale (1957) and Sheail (1971), from whom the following account is derived, have traced the early history of European Rabbits in Britain. The former author is responsible for discovering what may be the earliest British reference to the species:

> In 1176 there were rabbits in the Scilly Isles, where Richard de Wyka granted to the abbey of Tavistock his title *de cuniculus* 'which for some time I had unlawfully withheld, believing that the tithes were not payable on things of this sort'.[2] At some time between 1183 and 1219 the tenant of Lundy Island [in the Bristol Channel] was entitled to take fifty rabbits a year from certain *chovis* (coves?) on the island.[3]
>
> Evidence also survives as to the existence of rabbits in the early thirteenth century on the Isle of Wight, where in 1225 there was a *custod' cuniculorum* in the manor of Bowcombe, Carisbrooke[4]

Thus Veale concludes that:

> The rabbit became established in the late twelfth century on the small islands off the English coast; that in the middle years of the thirteenth century coneygarths were being more widely set up on the mainland, but that even late in the century rabbits were to be found only on certain estates. By the early fourteenth century, although owners of warrens still valued them highly and frequently haled poachers before the law, rabbits seem to have been more numerous, and the earliest trace of what was later to become a profitable export trade in their skins can be found in the export of 200 skins from Hull in 1305.

The reason for the installation of European Rabbits on offshore islands was presumably to safeguard them from human and non-human predators, and because the animals found the often lighter soil easier for burrowing. On the mainland, warrens were usually on royal or private estates, where they were frequently artificially enhanced by humans to make them more acceptable.

When in 1199 King John, as Earl of Moreton (Mortain in Normandy), gave permission for his tenants outside the forest of Dartmoor in Devon, to take (Brown) Hares and other animals, no mention was made of European Rabbits.[5] This suggests that even as late as the end of the 12th century, they were still scarce on the English mainland.

These dates tend to suggest that European Rabbits were not introduced to Britain by the Normans (r 1066–1154), as has usually been suggested, but between the reigns of the succeeding early Plantagenet kings Henry II (1154–89) or Richard I (1189–99), by whom they may have been imported to England on their return through continental Europe (which they reached in the Middle Ages) from the Crusades[6] (Lever 1977; Sykes 2007).

The earliest European Rabbit remains discovered on the English mainland appear to date from the late 12th/early 13th centuries, though the species' burrowing habits, with the risk of intrusion, can make accurate dates difficult to determine. They were found in a midden at Rayleigh Castle in Essex,[7] which was, wrote Veale (1957: 88):

> In royal hands from 1163–1215. ... It seems probable that the castle itself fell into disrepair some time during the first quarter of the thirteenth century, and it was no longer occupied

> after about 1220. … Possibly the rabbits once eaten there had come from the islands just off the Essex coast, such as Foulness or … Wallasey, which were manors in the Honour of Rayleigh.

Sheail (1971) refers to European Rabbit bones found in the Buttermarket in Ipswich, which are also believed to date from about the 12th century.

In 1221, 6,000 European Rabbit skins – which may have been English but could also possibly have been imported – were referred to in a Devon plea. The earliest certain mention of home-grown animals on the English mainland occurs in 1235, when Henry III made a gift of *decem cuninos vivos* from the royal park at Guildford in Surrey. Six years later he ordered hay to be conveyed from his *cuningera* at Guildford, and in the following year he gave instructions for the collection of 40 or 50 European Rabbits *secundum quod invenerint prefatam fertilem* from the same source.

During succeeding decades the number of royal and other *cuningera* or coneygarths seems to have multiplied, and European Rabbits began increasingly to be conveyed alive around the country. Veale found that for Henry III's Christmas of 1240:

> 100 [European Rabbits] were to be supplied from the lands of the bishopric of Winchester, 200 from those of the earl of Warenne, and 200 by the King's escheator. In 1241 the sheriffs of Hampshire, Sussex, Surrey and Kent were to produce 100, 50, 100 and 500 respectively. In 1243 180 rabbits were required from the estates of the bishop of Winchester, 100 coming from the Isle of Wight, and 300 from those of the archbishop of Canterbury; 300 were to come from the lands of the Bishop of Chichester in 1244, and 200 in 1245. Similar orders were going to the sheriffs of Essex, Hertfordshire and Middlesex in 1248, and to those of Buckinghamshire and Bedfordshire in 1249. The coneygarth belonging to the manor of Kempston, Bedfordshire, held by the earl of Chester, is referred to as early as 1254.
>
> In 1241, the keepers of the bishopric of Winchester were ordered to take 100 rabbits within the bishopric … alive to Sugwas, the manor of the bishop of Hereford. In the same year the keepers of the lands of the bishopric of London supplied the King's uncle, Peter of Savoy, with eighty live rabbits from Clacton, Essex, for his warren at Cheshunt, and by 1244 the King himself had begun to stock his park at Windsor. The sheriff of Surrey sent some rabbits from Guildford; the keepers of the bishopric of Chichester and the earl of Derby produced others … the earl of Aumale sent some to the royal park at Nottingham at the same time … from Lincolnshire … or the Holderness estates.

Veale suggested that perhaps some French ecclesiastic, such as Peter des Roches (d 1238), Bishop of Winchester, may have educated the English palate to a taste for European Rabbit flesh. During the late 13th and 14th centuries, they began to appear with increasing frequency on the menus of banquets and feasts. For his Christmas dinner of 1251, Henry III and his court consumed 450, and they were included on a menu in Canterbury in 1309, at the coronation feast for Henry IV in 1399, and at the enthronements of the Archbishops of Canterbury in 1443 and York in 1465; at the last, for George Nevill 'Chancelour of Englande', no fewer than 4,000 'conyes' were consumed. In his *Booke of Nurture* (1460), John Russell, English bishop and chancellor, gives an early example of instructions for cooking them – 'the cony, ley hym on the bak in the disch, if he have grece' – and the technical term for carving one – 'unlace that cony'.

During the many movements of European Rabbits around the countryside, many must have managed to escape and become established in the wild. Even by the mid-13th century, the animals were becoming recognized as a pest, since as early as 1254–7 the burgesses of Dunster in Somerset were complaining about this destructiveness. Sheail (1971) mentions a tax return for Sussex of 1340 that contains complaints from the inhabitants of West Wittering in Sussex that animals owned by the Bishop of Chichester had destroyed their wheat. In Ovingdean in the same county, 40 ha of arable land were ravaged by European Rabbits belonging to the aptly named Earl of Warenne. Serious crop damage in the Breckland area of Suffolk was caused by European Rabbits throughout the medieval period, and the manor of Methwold was said to have fallen into disuse by 1522 because the land was in close proximity to a warren.

In a Pipe Roll of 1245,[8] a European Rabbit (or possibly a Brown Hare) is depicted being attacked by an unidentified species of raptor. Sheail (1971) refers to another early representation of a European Rabbit in English iconography – in this case a pair depicted in a 14th-century painting at Longthorpe Tower, Peterborough.[9]

The earliest documented case of trespass involving the species occurred in 1268, when Richard, Earl of Cornwall and King of Almain, complained that his coneygarth at Isleworth in Middlesex had been plundered by gangs of poachers. A serious case of trespassing took place during the reign (1377–1400) of Richard II, when an unspecified number of poachers were excommunicated for stealing some 10,000 European Rabbits from the manor of North Curry in Somerset. In 1290, warrens (*cunicularia*) are mentioned in *Fleta*, a book compiled by an anonymous author. In the contemporaneous *Britton* – the earliest summary of the law of England written in Norman French probably by either John le Breton, Bishop of Hereford (d 1275) or Judge Henry de Bracton (d 1268) – mention is made of *De veneysoun et de pessoun et de conyis*. In 1389, 'conynges', together with warrens, 'connigeries' and ferrets for hunting European Rabbits, first appear in the State Book of Richard II.[10]

From the time of European Rabbits' introduction until the late 14th century, their meat and skin must, from the prices obtained for them and from their inclusion in royal and ecclesiastical banquets, have been regarded as luxuries. These high prices also suggest that the animals must have been only locally distributed. At Farnham in Surrey, for example, they sold for an average of 2½d. each (a huge price) in 1253–1376. In 1270 in Cambridgeshire, they fetched 5d. each, and even as late as 1395 animals for a Determination Feast at Merton College, Oxford, cost 6–8d. a couple plus the cost of transportation from Bushey in Hertfordshire of ½d. each: this implies that European Rabbits were rare, or even unknown, in Oxfordshire at the time. An Inquisition on Lundy Island in the Bristol Channel in 1274 revealed that around 2,000 European Rabbits were killed each year; they fetched 5s. 6d. 'each 100 skins because the flesh is not sold'.[11] In 1272, the capture of European Rabbits with hawks, dogs and ferrets at Waleton (where they were sold for 2½d. each) is mentioned in A. Rogers's *History of Agriculture and Prices in*

England, which also refers to prices of between 10 for 1s. and 26 for 1s. 11d. obtained in Oxford in 1310–13. For the induction banquet of Ralph de Borne, Abbot of St Austin's Abbey, Canterbury, in 1309, animals were purchased at a cost of 6d. each.

By the early 16th century, one dozen 'rabbet ronners' were being sold for 2s., and by 1530 in Yorkshire they were fetching 5d. per couple. Twenty years later, the price had risen again to 4d. each, and in November 1610 Thomas Cocks of Canterbury noted in his account book that he paid 'for a rabett 9d.' By the end of the 18th century, it had been calculated that the value of a skin in proportion to that of the carcass was greater than that of an ox or a sheep.

Scotland

During the reign (1214–49) of Alexander II, the royal warrens were protected from poachers by statute, and in 1264 in the succeeding reign (1249–86) of Alexander II, the Keeper of the King's Warren at Crail in Fife received an annual salary of 16s. 8d. for his services. Other, later 13th-century royal warrens were established at Perth and near Cramond, Edinburgh. In the first year of the reign (1329–71) of David II, the King's Chamberlain paid 8s. to four men for catching European Rabbits on the Isle of May in the Firth of Forth. The same monarch granted one William Herwart, keeper of his warrens in Fife, 'charter in life-rent' of the office of 'Keeper of the King's Muir' in Crail and of its 'cuningare' at a salary of 40s. per annum. As early as 1377, European Rabbits are mentioned as an article of commerce in Berwickshire,[12] and a warren is referred to on the linksland[13] of Aberdeen in 1424, when an Exchequer Roll records that a duty of 12d. per 100 was levied on exported 'cunning' skins. In 1457, the killing of the animals when snow was on the ground was made the subject of 'dittay',[14] with a fine on conviction of £10.

The rent book of 1474 for Cupar Abbey in Fife indicates that the monks employed a 'warander of kunynyare', and in the following year one Gilbert Ra (or Rae) undertook to protect the Abbey's 'conygar from all harm … and put it to all profit within his power'. The Earl of Orkney's rent book for 1497–1503 records that European Rabbits ('114 cunnings' and '1274 cunningis skinnis') formed part of his annual rent received in kind.

European Rabbits were clearly sufficiently numerous and well established to be regarded as a pest in at least one district in Scotland by 1511, when 'Schir Robert Egew, Chaiplan to My Lord Sinclair', complained that 'There will be too many cunnings within two years – they have riddled all the banks of the Links right well', and there are more complaints about their destructiveness during the uprising of 1549 led by Robert Ket. At this time, one animal fetched 'two shillings unto the Feast of Fastens Eve [Shrove Tuesday] next to come, and from thenceforth at twelve pennies'.

In 1551, because of 'the great and exhorbitant dearth', young European Rabbits ('rabbetis') were protected by statute for a period of three years, except from falconry by the nobility. A charter of 1583, during the reign of James VI, granted to the Provost and Baillies of Aberdeen, refers to a *cunicularum de Abirdene* near the 'Gallowhills' of that city.

On Scottish offshore islands, European Rabbits are known to have occurred on Little Cumbrae between the Isle of Bute and Renfrewshire as early as 1453, and in the Orkney Islands (on Orkney, Lamb Holm and Sanday) by 1529. In his *Description of the Western Isles of Scotland called Hybrides* (1549), Donald Monro, known as 'High Dean of the Isles', refers to them on Mull; on 'Inche Kenzie' (Inchkenneth), off the west coast of Mull, which was said to be 'full of conyngis about the shoiris'; and on the unidentified 'Sigrain-moir-Magoinein, that is to say the Cuninges Isle, wherein there are many cuninges'. Several Islands in the Firth of Forth (e.g. Cramond, Inchcolm, Inchmickery, Craigleith, May) were apparently 'verie full of conies', and the animals were also reported 'on a little isle, with a chapel on it, called Cavay' (Cava) in the Orkneys. The German traveller and writer Von Wedel saw European Rabbits on the Bass Rock when he visited North Berwick in 1584. Great Cumbrae and Ailsa Craig on the west coast were overrun by 1612, and by 1677 they had spread as far west as Lewis and North and South Uist in the Outer Hebrides. Seven years later, warrens had become established on Burray Island south of the Mull of Kintyre. European Rabbits were introduced to Gigha Island off the Mull of Kintyre's west coast in 1763. In 1840–90, a number of introductions were made to the Island of Raasay off the east coast of Skye, and further releases were made on North and South Uist. In 1865, they became established on Seaforth Island in Loch Seaforth on the Isle of Lewis, and two years later some were released by lobster fishermen on Lunga in the Treshnish Isles west of Mull, presumably as a source of food. In the 1870s, some pet animals were freed on Foula in the Shetland Islands, where they soon became established; others were released on Hascosay in the same archipelago in 1900.

Not all introductions to Scotland were successful, especially in the Shetlands; these include South Havra (where they were killed by purposely imported cats); Trondra (where they were introduced on several occasions but never survived for more than a year); Holm of Melby (where four released for sporting purposes increased rapidly and survived an importation of cats, but died out around 1930 from unknown causes); and Oxna (where they were allegedly drowned in a flood).

On the Highland mainland, European Rabbits were released in several localities from the mid-18th century onwards. In 1750 they appeared in Morayshire, when they were found on the golf-links at Lossiemouth, having allegedly escaped from a vessel from Europe moored in Branderburgh harbour. They were introduced to Caithness sometime between 1743 (when they were unknown) and 1793 (when they were said to be abundant). They first appeared in Wester Ross in 1850. Further south, they were first seen in Gullane, East Lothian in 1807, in Dumfriesshire (Dumfries and Galloway) in 1815 and in Kintyre in 1845.

Wales

Matheson (1941) traced the early history of the European Rabbit in Wales, where the earliest written evidence for its presence dates from 1282, when Richard le Forester was paid 3s. 6d. for catching them and keeping ferrets for the King at Rhuddlan Castle in Flintshire.[15] In 1284, the commote of Estimaner in Merionethshire (Gwynedd) paid 8s. for the maintenance of a *haracium* (warren).[16] In a Pipe Roll of 1325–6, European Rabbits are referred to on the islands of Schalmey, 'Schokolm' (Skokholm) and Middleholm, and in 1386–88 a total of 6,120 were obtained from these three islands. They also doubtless occurred on other islands off the south Welsh coast, such as Ramsey, Skomer and Caldey. In 1376, there was a warren at Castle Kerdyf (Cardiff), and in 1492 another on Flat Holm, a small island in the Bristol Channel south of Cardiff. Matheson quotes from an inventory of warrens in Glamorgan in 1578 listed by the Elizabethan historian Rice Merrick: Llandaff, Barry Yland, Mynidd Glew, Wyke, Wenny, Morgan and Britton Fery – all on or near the coast of the Bristol Channel.[17]

In 1517, the Prior of Pill on the mainland in Pembrokeshire granted a 40-year lease of property including a warren to Morris Butler, reserving for himself and the monastery the right to hunt within the warren thrice yearly.[18] In *Leland's Itinerary*, written around 1540, the English antiquarian John Leland (or Leyland) refers to numerous 'conies' on St Tudwall's Island, St Dwynwen (Llanddwyn Island), Anglesey, and on Hilbre Island at the mouth of the River Dee. During his tour in Wales in about 1773, Thomas Pennant observed European Rabbits on Puffin Island (Priestholm) – where they were eradicated by introduced Common Rats in the 19th century – and the Skerries, both off the coast of Anglesey. In his *Tour in Wales* (1810), Pennant also remarked on the 'vast and profitable warren' noted 'for the delicacy of the rabbits, by reason of their feeding on the maritime plants', owned by Sir Pyers Mostyn at Talacre on the Flintshire coast.

Ireland

The topographer Giraldus De Barri, commonly called Cambrensis, when discussing the Irish Hare in his *Topographia Hibernica* (compiled in 1183–7), wrote: 'There are hares ... closely resembling rabbits [*cuniculi]*', although as Fitter (1959) points out, he may have seen European Rabbits when he visited Paris in 1167.

Veale (1957) discovered the earliest reference in Britain and Ireland to rights *in warennis cunigariis*, which occurs in a charter granting lands in Connaught to Hugh de Lacy in 1204. An early reference to the species in Ireland is contained in a late 13th- or early 14th-century poem entitled 'Da choinin a Dhúmha duinn' ('Two conies from Dumho Duinn'). As the first documentary evidence of them in Ireland dates from shortly after their arrival in England, it seems likely that they were introduced there either directly from Europe by the later Normans or early Plantagenets, or subsequently from England.

In 1282, 20 skins from Balisax (Ballysax) in Co. Kildare were sold for 1s. 4d., where five years later 100 'great coneys' from the same source fetched 13s. 4d. In 1324, the profits made by hunting in the *cunicularium* at Rosslare in Co. Wexford formed part of the tax return made on his land submitted by Aymer de Valence. During the 14th and 15th centuries, a considerable export trade in European Rabbit skins built up in parts of Ireland, as described in *The Libel of English Policy* (c 1430). In his *Description of Ireland* (1617), the traveller Fynes Moryson refers to 'Felles of kydde and conies grete plente' met with during his travels.

When he visited the manor of Ollort (Oulart) at Ferns in Co. Wexford in July 1635, Sir William Brereton, Bt saw European Rabbits on the banks of the River Slaney, and at a park near Wexford there was an 'abundance of rabbits, whereof here there are too many, so as they pester the ground'. In Northern Ireland there were extensive warrens by the 17th century. In his *Tour in Ireland* (1780), the agriculturist and author Arthur Young said that when he was in the country in the late 1770s, the animals were common in many localities. As recently as 1906, they were misguidedly released on Clare Island in Clew Bay, Co. Mayo, and in 1911 on Rathlin Island off the north coast of Co. Antrim.

Naturalization of the European Rabbit

Although the history of the European Rabbit in Britain and Ireland is now fairly well documented, it is still uncertain as to exactly when the animals first began to escape from their warrens to form naturalized populations in the wild. In the early days, they were largely restricted on the mainland – especially in Scotland and Wales – to the sandy linksland between the foreshore and the hinterland, and to a number of offshore islands. The reasons for this are probably two-fold:

- They take more naturally to burrowing in soft, sandy soil.
- The larger terrestrial predators tend not to flourish on small islands, which usually lack the necessary amount of prey species and shelter.

It seems likely that European Rabbits began to escape from their warrens in any numbers in the mid-13th century, when they and other livestock were being transported from one part of the country to another. Fitter (1959) suggests that the 'Black Death' of 1349, with a resulting reduction in the country's labour force, may have been a contributory factor. He mentions the Bohemian traveller Schascek, who in 1465 saw 'rabbits and hares without number' in Clarendon Park, Wiltshire, though these may still have been in an enclosed warren. They were clearly not at this period regarded as animals worthy of the chase, for the anonymous 15th-century *Master of the Game* scathingly records that 'of conies I do not speak, for no man hunteth them unless it be fur hunters, and they hunt them with ferrets and with long small [mesh] nets'. Even by the 18th century, Gilbert White in Letter VII (undated, but 1767) to Thomas Pennant in his *Natural History of Selborne* (1789) wrote: 'but these [European Rabbits] being inconvenient to the huntsman, on account of their burrows … they ['deer-stealers'] permitted the country people to destroy them all'.

By 1551, the Swiss naturalist Conrad Gesner was able to write of *copia ingens cunicularum* in lowland England, and that 'there are few countries wherein coneys do not breed, but the most plenty of all is in England'. Even here, however, there were many locations where European Rabbits were extremely scarce or unknown until at least the late 18th century.

The second half of the 18th century witnessed a phenomenal explosion in the British European Rabbit population. Ritchie (1922) attributed this firstly to the advances made in agricultural practices, which provided the animals with both an increase in favourable habitat and an additional supply of food, especially in winter when it was most needed; and secondly to the rise of game preservation, which led to the wholesale destruction of avian and mammalian predators.

Ecology

Cowan and Hartley (in Harris and Yalden 2008) have summarized the ecology and impact of European Rabbits in Britain and Ireland.

The favourite habitat of European Rabbits is short grassland, either natural as on dry heathland or machair,[19] or closely cropped pastureland with nearby shelter. On well-drained land with adequate protection, they may occur up to the tree-line. They are never abundant in extensive conifer plantations. The extent of their burrows depends on the type of soil. In loose soil (e.g. sand) they are normally excavated in the vicinity of such supporting structures as tree roots or shrubs; burrows constructed on chalk land tend to be more extensive and to have more interconnected entrances.

European Rabbits eat a wide variety of herbage, especially grasses, of which they prefer the young and succulent shoots and leaves of more nutritious types such as *Festuca*. Winter wheat is preferred to maize, and dicotyledons in mixed arable areas. When food is short, they select those parts of plants that contain the highest nitrogen content.

Although Red Foxes, Stoats, Weasels, Polecats, feral domestic cats, Eurasian Badgers, Common Buzzards and in northern Scotland Wildcats are among the European Rabbit's principal predators (see Impact on Animals, page 66), their most important controlling factor has been myxomatosis.

The myxoma virus arrived in England (Kent) from France in the autumn of 1953, from where by the following year it had spread to Ireland. The virus was isolated from the South American Forest Rabbit, in which it is non-lethal. During the next two years myxomatosis, which is species-specific, spread throughout Britain via the European Rabbit

Flea, resulting in more than 99 per cent mortality in the wild European Rabbit population. The virulence of the disease, which is now endemic, has since declined, and the genetic resistance of the species to the virus is increasing. In most populations, however, outbreaks still occur biannually, usually in spring and late summer/early autumn, when the highest mortality rate is usually no more than around 30 per cent, and is declining.

From an estimated population of around 100 million European Rabbits before myxomatosis, the number fell to about 1 million in 1959 and was only 20 million in 1986; a decade later it had recovered to only some 37.5 million.

In 1970, 59 per cent of agricultural holdings were overrun with European Rabbits, compared to 94 per cent before the arrival of myxomatosis. Post-1970 surveys have shown an increase in both numbers and distribution of the animals, although a review in Scotland in 2000 found no overall change in their numbers on cultivated land since the early 1990s. Today, they are most widely distributed in eastern and south-eastern England and Wales, and most numerous in eastern and south-eastern England and Scotland, and on the coastal Scottish Highlands and islands. They do not occur on the islands of Rum or Tiree, on some of the smaller Scottish islands such as Gunna off Coll in the Inner Hebrides, Sanday in the Orkney Islands and most of the Treshnish Islands (see under Scotland, page 61), or on the Isles of Scilly.

In 1992, Rabbit haemorrhagic disease (RHD) entered the British and Irish domestic population, spreading to the wild population two years later. Like myxomatosis, RHD is species-specific to *O. cuniculus*, and is extremely contagious: the disease peaks in late autumn/early winter, and in those individuals that are susceptible (many are immune) the mortality rate averages around 30 per cent.

In the wild, European Rabbits are potential vectors of VTEC *Escherichia coli* 0157 and have been implicated as a potentially serious source of infection in humans (Bailey 2002). Where they live near domestic cattle excreting *E. coli* 0157, they can also act as carriers of the disease (Scaife 2006).

Methods of control by man include gassing (the most effective, though repeated applications are usually required), shooting, trapping, snaring and the use of ferrets. Non-lethal methods include the use of exclosure fencing and, for forestry, repellents. About £5 million is spent annually on controlling the animals on agricultural holdings.

Economic Impact

The European Rabbit has had a greater negative economic and ecological impact in Britain and Ireland than any other introduced animal.

The species is the cause of considerable economic damage to a wide variety of crops, including cereals, roots, grassland, horticulture, arboriculture, pomology and commercial forestry. The bark of many tree species is eaten, and if ring barked a tree will die, making European Rabbits a major inhibitor of natural tree generation in forestry plantations. Before myxomatosis, they were the principal vertebrate pest of agriculture, causing an estimated £50 million of damage per annum in 1952 at then current prices. An increase in numbers as mortality from myxomatosis has declined has caused increasing problems, and the annual damage to British agriculture is now around £115 million. The damage is most serious where the animals heavily graze seedlings, in particular winter wheat. They damage pastures not only by grazing herbage intended for domestic stock, but also by the alterations in the characteristics of the vegetation caused by their activities. Because they reduce herbage to a height of 3 cm, the lateral buds of plants are so damaged that normal regeneration cannot occur. The important grasses and clovers are thus gradually eradicated, and are replaced by species of little agricultural value. Eventually, when the land is not too dry, a more or less continuous carpet of mosses, which even European Rabbits will not eat, becomes established. In montane habitats, grazing by the animals can eliminate species that are favoured by domestic sheep.

Impact on Flora

The serious impact of European Rabbit grazing on native species of flora was not appreciated until after the arrival of myxomatosis. Their removal resulted in:

- An increase in both the height and variety of plant species, in particular palatable and economically valuable legumes and grasses.
- An increase in the abundance of woody plant seedlings.
- An improvement in plant flowering.
- Faster plant succession due to enhanced seed germination and seed establishment.

A greater variety in plant species became evident shortly after the arrival of myxomatosis, including the plentiful flowering of scarce or less common ones. In the continued absence of the animals, these flowering plants were succeeded by the more dominant grass and woody species, and large areas of chalk downland reverted to scrub, especially Common Hawthorn. Where no grazing occurred, scrub and grassland developed into more open woodland. On acid land, European Rabbits graze principally on such species as *Vicia sativa*, *Trifolium repens*, *Festuca rubra* and *Agrostis capillaris*. In moorland and heathland habitats, grazing by European Rabbits can eliminate *Calluna vulgaris*, *Vaccinum myrtillus*, *Nardus stricta* and *Deschampsia caespitosa*, resulting in succession by grass heathland dominated by *Agrostis* and *Festuca* spp., *Deschampsia flexuosa* and *Carex arenaria*. Under continued heavy grazing, grasses are eventually replaced by moss-dominated vegetation and finally by lichens. These changes in vegetation type can also result in soil erosion, sometimes producing scree, and in the degradation and destabilization of coastal sand dunes.

Impact on Animals

The increase in the height of vegetation after myxomatosis benefitted snails, woodlice, spiders, the Marbled White butterfly, the Lulworth Skipper butterfly, the Six-spot Burnet moth, and small mammals, especially the Short-tailed Vole. Honey Bees benefitted briefly from the profusion of flowers until succession by grasses.

Sand Lizards, on the other hand, lost breeding sites, and the Adonis Blue butterfly suffered a reduction in population, while the Large Blue disappeared entirely by 1979. The Large Blue lays its eggs on Wild Thyme, where egg and larvae survival is best on closely cropped sward where the host ant *Myrmica sabuleti*, which rears the larvae underground, is most abundant. The recovery in the European Rabbit population resulted in the restoration of this optimum habitat, and provided conditions for the Large Blue's successful reintroduction since 1983.

The populations of birds dependent on short-sward grasslands, such as the Woodlark, Northern Lapwing, Red-billed Chough, Stone Curlew and Northern Wheatear and other species, declined after myxomatosis. Succession to taller grasses succeeded by scrub initially favoured Skylarks and Meadow Pipits, then Common Linnets, Yellowhammers and Common Whitethroats.

Birds such as Tawny Owls, Barn Owls, Short-eared Owls and Common Kestrels that are serious predators of small rodents were variously affected by changes in rodent populations following myxomatosis. Some species experienced good breeding years while mice and vole populations increased (see below), but suffered poor breeding seasons in subsequent years when rodent populations declined due to increased predation by other predators such as the Red Fox due to a shortage of European Rabbits. The breeding success of the Red Kite in Wales declined catastrophically after myxomatosis, though it is uncertain whether this was related to a fall in the numbers of European Rabbits, and sheep carrion is known to be of greater importance to Welsh Red Kites. However, in northern Scotland, European Rabbit carcasses form 70 per cent of the diet of the Red Kite. Common Buzzards suffered widespread breeding failure after myxomatosis, and their breeding success remains strongly related to the abundance of European Rabbits. The species can also comprise up to 72 per cent of the diet of the Golden Eagle, but the bird's breeding success does not depend on its abundance. The number of Brown Hares in some localities increased explosively after myxomatosis, mainly due to enhanced survival of leverets as a result of increased cover provided by longer grass.

Wildcats are generalized predators, and when available, as in the eastern Highlands of Scotland, young European Rabbits and diseased adults can comprise over 90 per cent of their diet. The availability of European Rabbits may control Wildcat numbers, which declined after myxomatosis. Foxes, too, exploit prey according to its availability, and are not reliant on the abundance of European Rabbits. Just after the arrival of myxomatosis, Red Foxes increased due to the abundance of affected animals; as the latter began to disappear, Red Foxes switched to the Short-tailed Vole as a major food item. Among mustelids, populations of Stoats declined most after the arrival of myxomatosis, but have since recovered correlatively with those of the European Rabbit. Weasels increased in number as a result of the increase in voles (see above). The impact of myxomatosis on Polecats and Eurasian Badgers is uncertain and is unlikely to have been great, although European Rabbits comprise up to 85 per cent of the former's diet. The availability of the species is known to be important to the breeding success of female American Mink.

References

Andrews et al. 1959; Armour & Thompson 1955; Armstrong 1982; Backhouse & Thompson 1955; Bailey 2002; Baker 1990; Birks 1999; Corbet, G.B. 1986; Corbet, L. 1978; Cowan, in Corbet & Harris 1991; Cowan & Hartley, in Harris & Yalden 2008; Davies & Davis 1973; Davis & Davis 1981; Easterbee 1988; Fenner & Ratcliffe 1965; Henderson 1997; Kirkham 1981; Lever 1969, 1977, 1979, 1980a, 1980c, 1985, in press; Lloyd 1970, in Corbet & Southern 1977; Lloyd & Walton 1969; Lockley 1940, 1964; Long 2003; Macdonald 1999; Matheson 1941; Matthews 1952; Phillips et al. 1952; Ritchie 1922; Ritchie et al. 1954; Scaife 2006; Sheail 1970, 1971; Spencer 1956; Stephens 1952; Sumption & Flowerdew 1985; Sykes 2007; Thomas, A.S. 1956; Thomas, J.A. 1980; Thompson 1954, 1956, 1994; Thompson & Worden 1956; Tittensor 1982; Trout 1986; Veale 1957; Wildman 1998; Yalden 1999.

Notes

1. Strictly speaking, as an escaped domestic animal, the European Rabbit should be included in the section Feral Domestic Species. Since, however, it would seem rather incongruous to discuss it alongside, for example, sheep, cattle and goats, it is included in the main body of the book.
2. Finsberg, H.P.R. 1947. 'Some early Tavistock Charters'. *English Historical Review* 62: 365.
3. Exeter City Archives. Miscellaneous Deeds, D. 614.
4. P.R.O. Exchequer. Foreign Rolls 8 Henry III. E. 364.
5. Rowe, S. 1848. *Perambulation of Dartmoor.*
6. Cowan and Hartley (in Harris and Yalden 2008: 206) state, however, that the European Rabbit was 'Introduced to England and Ireland by [the] Normans'.

7. Francis, E.B. 1913. Raleigh Castle. *Transactions of the Essex Archaeological Society* 12: 184. Hinton, M.A.C. 1912–13. 'On the remains of vertebrate animals found in the middens of Raleigh Castle'. *Essex Naturalist* 17: 16–21.
8. Hampshire Record Office. MS. Eccl. 2, 159287.
9. Clive-Rouse, E. and Baker, A. 1955. 'The wall paintings at Longthorpe Tower, near Peterborough'. *Archaeologia* 96: 1–58.
10. 13 Richard II. I. c. 13.
11. See also: *Boke of Kervinge*, published in the 13th century by Wynkyn de Worde.
12. *Proceedings of the Berwickshire Naturalists' Club*. 1863–8.
13. Sandy ground near the sea covered with coarse grasses.
14. The matter of the charge or the grounds for an indictment against someone for a criminal offence.
15. *The Antiquary*. 1911 (August).
16. *Archaeologia Cambrensis*. 1884.
17. *A Booke of Glamorganshire Antiquities* (ed. J.A. Corbet).
18. Pritchard, *E.M. History of St Dogmael's Abbey*.
19. Low-lying land near the coast formed from wind-blown sand and shell fragments, especially in Scotland.

LEPORIDAE (RABBITS & HARES)

BROWN HARE *Lepus europaeus*

(EUROPEAN HARE; COMMON HARE)

Natural Distribution Much of Europe, mainly N of the Alps, but following the development of agriculture has spread eastwards into Siberia as far as Lake Baikal.
Naturalized Distribution England; Scotland; Wales; Ireland.

England

Yalden (1999: 127) has summarized the status of the Brown Hare in England:

> There is a conspicuous absence of hare bones beyond the early Mesolithic, when the Mountain Hare *L. timidus* seems to have been eliminated from England ... by the spread of forests. ... it is at least arguable that Neolithic and later farming has created the habitat for the Brown Hare. ... Caesar (*De Bello Gallico*, Book V, 12) says of the Ancient Britons 'Hares, fowl and geese they think it unlawful to eat, but rear them for pleasure', which implies that they had been introduced by Iron Age times, but perhaps not long before. There are archaeological records of hares from the Iron Age sites of Danebury Hill Fort (Coy 1984), Winnal Down (Maltby 1985), Maiden Castle (Armour-Chelu 1991) and Blunsdon St Andrew (Coy 1982), as well as from such Roman sites as Portchester Castle (Grant 1976), Exeter (Maltby 1979), Silchester (Maltby 1984), Colchester (Luff 1982), Caerleon (Hamilton-Dyer 1993), Segontium (Noddle 1993), Wroxeter (O'Connor 1987) and Hod Hill (Fraser 1968). These certainly confirm that hares, presumably Brown Hares, were present in southern England in pre-Roman and Roman times. There are some possible earlier records: from Bronze Age sites at Brean Down (Levitan 1990) and Manor Farm, Borwick (Jones et al. 1987). ... Most intriguingly, Turk (1984) suggests that both Brown and Mountain Hare occur together in a Bronze Age site at Hartledale, Derbyshire; if these are really contemporary, they are both the latest native English Mountain Hares, and perhaps the earliest Brown Hares.

Brown Hares today are widespread in England, especially in the south, south-east, Midlands and East Anglia, and have also been introduced successfully to the Isle of Man – traditionally by Sir John Stanley when he was granted the Lordship of the island by Henry IV in 1405 (Fargher 1977). In Derbyshire and on the Isle of Man, where above around 300 m farmland gives way to heather moorland, the Brown Hare is replaced (as in Scotland) by the Mountain Hare.

Scotland

Brown Hares first appeared in the western Highlands in the late 18th century, when the country was being opened up by the construction of a network of roads that facilitated their movement. Among Scottish offshore islands there are few to which Brown Hares have not at some time been introduced. The *Old Statistical Account* records their release on Coll around 1787 and on Lewis before 1797, and it is known that they arrived on Mull in 1814–15, on Islay (before 1816) and in the Shetlands in the early 19th century. Despite frequent reinforcement, they died out on Mull later in the century. There were at least three introductions to the Shetlands, where all the animals were killed by the crofters whose crops they ate.

BROWN HARE

Today, Brown Hares occur regularly in scattered localities on the Scottish mainland, especially in the south and east, and on a number of offshore islands, including the Orkneys (Rousay and Mainland), the Inner Hebrides (Tiree, Skye, Scalpay, Staffin, Coll, Luing, Mull, Islay and Gigha) and islands in the Firth of Clyde (Arran, Bute and sometimes Davaar). Above about 300 m, where agricultural land gives way to heather moorland, Brown Hares in Scotland (as in England) are replaced by Mountain Hares.

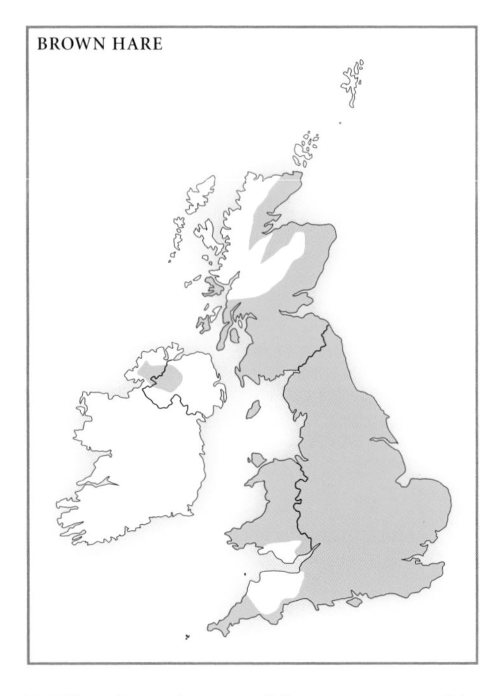

Wales

According to Yalden (1999), the Mountain Hare was seemingly eliminated from Wales during the Mesolithic as a result of the spread of forests. There are possible archaeological records of hares from three Neolithic Welsh sites – Gwaenysgar, Pant-y-saer and Bosclodiad-y-gawres (Caseldine 1990), where bones of hares, presumably Brown Hares, occur with those of domestic ungulates. The species is now established in scattered localities throughout Wales, including the Isle of Anglesey.

Table 2 Early introductions of the Brown Hare *Lepus europaeus* into Ireland

Date	County	Locality	Introduced by
c 1852	Cork	Castlemartyr	Earl of Shannon
c 1858	Cork	Castle Hyde, Fermoy	Mr Hyde
c 1865–6	Wicklow	Powerscourt	Viscount Powerscourt (via Edward Heneage, Lincolnshire, England)
c 1868	Armagh	Brownlow House	Lord Lurgan
c 1868	Down	Copeland Island	David Kerr
1876–7	Tyrone	Strabane	John Herdman
? 1880s	Galway	Salrock, Killary Bay	General Thompson et al.
? 1880s	Fermanagh	Cleenish Island, Lough Erne	Captain Collum
? 1880s	Londonderry	?	Mr Stevenson
c 1892	Donegal	Ballyconnell	Mr Olpert
?	Cork	Trabulgan Coursing Ground	Lord Fermoy

Source Barrett-Hamilton (1898).

Ireland

Brown Hares were introduced for greyhound coursing in the mid- to late 19th century, when Barrett-Hamilton (1898) recorded the introductions listed in Table 2 (opposite).

Although several populations of Brown Hares became established and the species proliferated, by the end of the century most had died out. Since then the species has occurred only in Co. Tyrone and in eastern Co. Donegal (Fairley 2001; Sheppard 2004).[1] During their survey in 2005, Reid and Montgomery (2007) found a substantial population in Co. Londonderry and Co. Tyrone west of Lough Neagh in mid-Ulster, and further sightings were recorded near Baronscourt in the latter county. Neither of these localities had been sampled during previous hare surveys in the late 1990s/early 2000s. The animals at Baronscourt are likely to be descended from an introduction there in 1876, while those in mid-Ulster could be the descendants of some released at an unknown site in Co. Londonderry. During 2005, the mid-Ulster sub-population was estimated by Reid and Montgomery (2007) to number 700–2,000 individuals, indicating firm establishment. The size of the sub-population near Strabane in Co. Tyrone (where 65 animals from Norfolk were released in 1876 and where in 1910 they were still numerous) is unknown, but the total Irish population may number in the low thousands – probably <5 per cent of the country's Irish Hare population. Sheppard (2004) suggested that the species could also occur in eastern Co. Donegal near the Strabane and Baronscourt introduction sites, but the failure of the Hare Survey of Ireland 2006/07 to detect any Brown Hares in the Irish Republic suggests that, if they do occur, their distribution must be restricted.

Characteristics

Brown Hares are most common in strips of uncultivated land in arable regions, where cereals (especially wheat), sugar beet and peas predominate. They favour cultivated areas over non-cultivated ones such as pasture, set-aside and woodland, and since they require permanent cover (woodland, shelter-belts and hedgerows), small fields and mixed farming provide a more suitable habitat than monoculture in 'prairie'-type fields. High densities of livestock, especially sheep, and the presence of humans are inhibiting factors.

Brown Hares eat mainly grasses and herbs (wild varieties are preferred to cultivated ones), arable crops – especially young cereals, but also maize, peas, beans and sugar beet. Grasses comprise up to 90 per cent of their diet in winter, though they are important year-round, and in summer 40–60 per cent of their food consists of herbs.

The principal predator of Brown Hares is the Red Fox, which can curtail the species' population growth and density; in parts of southern England 80–100 per cent of annual production of leverets may be taken by Red Foxes. Other predators in Britain and Ireland include Stoats, American Mink, Polecats, Eurasian Badgers, Northern Goshawks and Golden Eagles. If Eurasian Eagle Owls succeed in establishing more than a toehold in Britain they could, as in continental Europe, become a significant predator.

Impact

Reid and Montgomery (2007) have summarized the impact of Brown Hares, with particular reference to Ireland.

The fact that 53 per cent of the hare population in mid-Ulster in 2005 consisted of Brown Hares suggests that a proportion of the population of indigenous Irish Hares (which are smaller and have shorter ears than the Brown Hare) may have been replaced by the introduced species. While the post-breeding density of hares in mid-Ulster was not materially different from the pre-breeding density, the mean estimates suggest that the density of Brown Hares may have increased (+ 18 per cent) by a similar extent as the decrease (- 19 per cent) in that of the native Irish Hare. This change in the density of the two species in mid-Ulster suggests that competition between Brown and Irish Hares tends to favour the former.

The hypothesis that there is interspecific competition between Brown and Irish Hares is supported by their use of habitat. Even in winter, when the overlap in mutual habitat usage is at a minimum, it is still high. The distribution of hares in late winter, however, the peak of the breeding season, may not be an accurate reflection of habitat-specific food or space utilization, since the distribution of males may be skewed by the presence of oestrous females. Thus an examination of habitat usage and overlap in autumn may provide a better assessment of the potential for competition, since distribution then is unlikely to be influenced by extraneous factors. Moreover, at the end of the growing season the preferred food resources have probably been exhausted, increasing the likelihood for competition. Brown and Irish Hares show different habitat selection preferences, but in autumn they share similar proportions of the principal components of their habitat space, and thus niche overlap is complete. Both species show a preference for improved grassland, but whereas Irish Hares favour neutral grasslands rather than arable areas in autumn, Brown Hares prefer arable areas over other habitats.

The similarity between the habitat choices of both hares and the high degree of usage overlap suggests the potential for competition if resources are limited. Reid and Montgomery (2007: 134) hypothesize that 'the strength of competition between both species may be influenced not only by the availability of habitats but the subtle differences between their habitat choices'. They suggest that 'in areas with greater prevalence of arable horticulture,[2] brown hares may find it easier to establish and push out the native Irish hare, while in predominantly pastoral areas the Irish hare may remain the dominant species'. The relatively small number of Brown Hares introduced to Ireland, and the competition between them and native Irish Hares, could have contributed to the failure of the former to spread far since their introduction. If global warming were to continue, the warmer and drier summers would be likely to increase cereal yields, which in some areas favour the expansion of range and numbers of Brown Hares.

Hybridization between the two species may constitute a threat to the genetic integrity of Irish Hare populations where they occur sympatrically and at the same elevations as Brown Hares. Where suitable conditions exist, as in the Peak District of Cumbria in England and on the Isle of Man, where both Brown and Mountain Hares occur, each settles into altitudinally distinct ranges, with the latter occupying higher elevations.

The establishment of Brown Hares in Ireland could pose a significant threat to the native species. With such a relatively small population and a restricted range it should not be impossible, if desired, to remove the Brown Hare from the list of Irish mammals.

In Britain, Brown Hares can be a significant pest of sugar beet, horticultural crops and plantations of young trees and shrubs. Because of a steady fall in the population (c 817,500 in 1995, Harris et al. 1995) during the last hundred years or so, the Brown Hare is included in the UK Biodiversity Action Plan, which inter alia states that habitat management should be implemented bearing in mind the requirements of Brown Hares, and that legislation on shooting (c 390,000 are shot annually) and selling should be reviewed. The aim is to maintain the species' range, expand the population and double the spring numbers by 2010. This is likely to be best achieved by retaining hedges and headlands that provide Brown Hares with food and cover, coupled with the control of Red Foxes, which are Brown Hares' main predator.

References

Barrett-Hamilton 1898; Caseldine 1990; Corbet 1986; Coy 1984, 1987; Fargher 1977; Fraser 1968; Grant 1976; Hamilton-Dyer 1993; Harris et al. 1995; Hewson, in Corbet & Southern 1977; Hill et al. 1995; Jennings, in Harris & Yalden 2008; Jones et al. 1987; Lever 1985; Levitan 1990; Long 2003; Luff 1982; Maltby 1979, 1984, 1985; Matthews 1952; O'Connor 1987; Reid & Montgomery 2007; Sheppard 2004; Tapper, in Corbet & Harris 1991; Turk 1964; Yalden 1999.

Notes

1. Although Brown Hares were introduced to Co. Fermanagh in the 1880s, they do not occur there now as stated by Long (2003: 124).
2. A 'linguistic error' for 'agriculture' (N. Reid, personal communication 2008).

SORICIDAE (SHREWS)

LESSER WHITE-TOOTHED SHREW *Crocidura suaveolens* (SCILLY SHREW)

Natural Distribution Widely distributed in Europe and Asia, occurring in most of the Palearctic from N Spain to S Korea. Also found in N Africa and on Crete, Corsica and Menorca in the Mediterranean.[1]
Naturalized Distribution England (Isles of Scilly; Channel Islands).

England

The remains of what may have been Lesser White-toothed Shrews have been found in Pre-Ipswichian glaciation deposits at Aveley in east London (Rzebik-Kowalska 1998). It is thus just possible that the species occurs on the Isles of Scilly and in the Channel Islands as a glacial relict. A more likely explanation, however, is that it was accidentally introduced by Iron Age or earlier traders from France or northern Spain at the time when they visited Cornwall in search of tin; otherwise its spasmodic distribution on the islands is hard to explain. Pernetta and Handford (1970) record the species' presence in possible Bronze Age deposits on Nornour in the Isles of Scilly. Today, it occurs on most, if not all, the larger Scilly Isles (certainly on St Mary's, St Martins, Tresco, Bryher, St Agnes, Gugh, Samson and Tean, and possibly formerly on Annet), and on Jersey and Sark in the Channel Islands. The total population in 1995 was estimated at c 14,000.

Characteristics

Lesser White-toothed Shrews are found in a wide variety of habitats affording sufficient shelter, especially tall vegetation, including Common Bracken, and along hedgerows and in woodland. In the Isles of Scilly they occur among boulders and seaweed near the high-tide line on sheltered beaches. On Jersey in the Channel Islands they are found mainly in coastal sand dunes and among boulders in Common Heather and scrub.

The species' main food is arthropods – especially beetles – flies, insect larvae and centipedes, and also gastropods and earthworms. On the coast of the Scilly Isles they feed largely on crustaceans, especially the coastal amphipod *Talitroides dorrieni*, which was introduced in shipments of plants from New Zealand and occurs in large numbers on boulder-covered beaches (Temple and Morris 1999). They also eat millipedes, adult and larval flies (especially *Thoracochaeta zosterae* in rotting seaweed), adult and larval beetles (including *Cercyon littoralis*), spiders and mites (Pernetta 1973).

In the Channel Islands the species' main predators are Barn Owls, Common Kestrels, Stoats and domestic cats, and in the Isles of Scilly Common Kestrels and cats.

Impact

Because of their predation on potential invertebrate pests, Lesser White-toothed Shrews are of benefit to humans.

LESSER WHITE-TOOTHED SHREW MOTHER AND YOUNG

References

Beirne 1947, 1952; Churchfield, in Corbet & Harris 1991; Churchfield & Temple, in Harris & Yalden 2008; Corbet 1961, 1969, 1974; Harris et al. 1995; Hinton 1924; Jenkins, in Corbet & Southern 1977; Lever 1977, 1985, 1994; Long 2003; Pernetta 1973; Pernetta & Handford 1970; Rezbik-Kowalska 1998; Temple & Morris 1997; Yalden 1999.

Note

1. Churchfield and Temple (in Harris and Yalden 2008: 277) state that *C. suaveolens* is 'very widely distributed in Europe, Asia and Africa', whereas their map (Fig. 7.25 on page 278) indicates the species' absence from Africa. In fact it occurs only in North Africa.

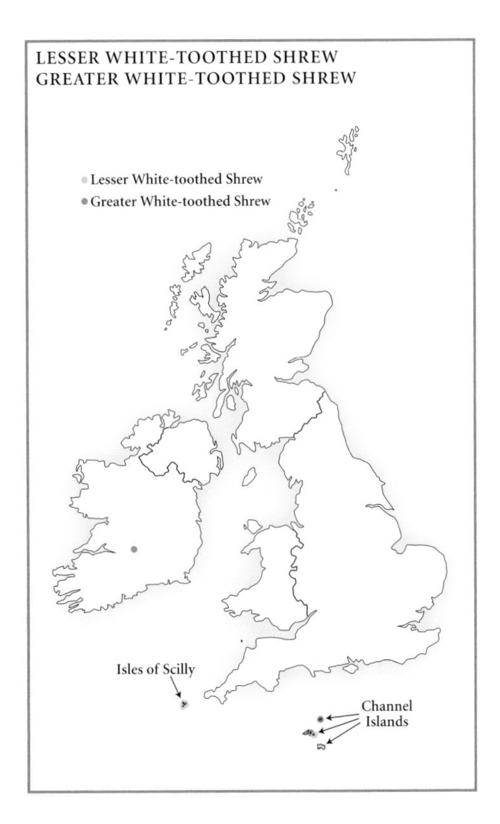

SORICIDAE (SHREWS)

GREATER WHITE-TOOTHED SHREW *Crocidura russula*

Natural Distribution Continental W and SW Europe E to C Germany, and N Africa from Morocco to Tunisia, and on Sardinia and Ibiza in the Mediterranean.
Naturalized Distribution England (Channel Islands); Ireland.

England (Channel Islands)

Fossil remains of uncertain date (but probably Pre-Ipswichian) of *Crocidura* (probably *C. russula*) have been found at several sites in Essex, and in Quaternary deposits in Tornewton Cave in Devon (Rzebik 1968). It is thus just possible that the Greater White-toothed Shrew is a glacial relict in the Channel Islands (where it occurs on Guernsey, Alderney and Herm), but an accidental introduction by humans is more likely. Its irregular distribution is otherwise difficult to explain.

Greater White-toothed Shrews are fairly common in the Channel Islands in woodland, hedgerows, grassland and cultivated areas, especially around human settlements and farm buildings, though much less abundant than in continental Europe. They feed in the Channel Islands on a wide variety of invertebrates; in Europe small amounts of plant material and occasionally lizards and small rodents are also eaten.

In the Channel Islands the species' main potential predators are the same as those of the Lesser White-toothed Shrew.

GREATER WHITE-TOOTHED SHREW EATING INSECT

Ireland

The following account of the discovery of the species in Ireland is derived from Tosh et al. (2008).[1]

During a study of the ecology of the Barn Owl in Ireland, a total of 53 shrew skulls – subsequently identified as those of the Greater White-toothed Shrew – were collected on 30 September 2007 in 10 pellets in a roost in Co. Tipperary. Further field surveys to locate Barn Owl and Common Kestrel roosts, conducted in March 2008 in Counties Clare, Cork, Limerick and Tipperary, revealed the presence of Greater White-toothed Shrew remains in the pellets of these two species in 14–15 locations in Counties Limerick and Tipperary. Later in the same month, five females and two males were caught at four sites in Tipperary. These initial investigations suggest a range within these two southern counties of considerably more than 30 x 60 km.

Since it is extremely unlikely that the species has been present but undiscovered in Ireland as fossil material, it seems almost certain that it has become a member of the Irish mammalian fauna as a result of introduction – probably accidentally – by humans from continental Europe since 2001, the year of completion of the most recent survey of small mammals in Ireland. A combination of a long breeding season, multiple litters of up to 11 young each, early maturation, communal breeding, and the use of buildings and other habitats in close proximity to humans should help to facilitate the species' expansion of range in the Irish Republic.

Impact

The frequency of the presence of Greater White-toothed Shrew remains in raptor pellets in parts of south-central Ireland suggests the potential for the species to become an important food item for birds of prey (and mammalian predators) where it occurs. Its arrival in Ireland, where there is a dearth of small mammals compared to the number in Britain, could prove beneficial to predators such as the declining Barn Owl; in some parts of its range in continental Europe, it is an important part of the Barn Owl's diet.

The species is also potentially of benefit to humans through its consumption of injurious invertebrate pests.

On the other hand, research in Europe suggests that the Greater White-toothed Shrew has the potential to displace the almost ubiquitous and also introduced much smaller Pygmy Shrew, and there may be other, as yet unassessed, negative ecological consequences; research into this is currently continuing.

References

Beirne 1947, 1952; Churchfield, in Corbet & Harris 1991, in Harris & Yalden 2008; Corbet 1961, 1969, 1974; Hinton 1924; Jenkins, in Corbet & Southern 1977; Lever 1977, 1985, 1994; Long 2003; Pernetta 1973; Pernetta & Handford 1970; Rzebik 1968; Temple & Morris 1997; Tosh et al. 2008; Yalden 1999.

Notes

1. I am indebted to Dr Derek Yalden for drawing this reference to my attention.

SORICIDAE (SHREWS)

PYGMY SHREW *Sorex minutus*

Natural Distribution Widespread throughout the Palaearctic, from W Europe (including Britain), except for parts of the Mediterranean region, E through Siberia to Lake Baikal.

Naturalized Distribution Ireland.

Ireland

The occurrence of the Pygmy Shrew in Ireland, where there are no native shrew species, has been explained by the possible existence of a temporary land bridge between Britain and Ireland at the end of the Devensian. The low level of mtDNA variation, however, indicates a very small founder stock, and suggests an introduction by boat from southern Europe (perhaps France, Spain or Andorra) by humans. This would be consistent with the species' presence on, for example, the Orkneys and Outer Hebrides, which were certainly not attached to the British mainland in the late Devensian.

Pygmy Shrews are now almost ubiquitous on the Irish mainland, where they occur in a wide variety of habitats with plenty of ground cover. The most important food items in Ireland are beetles, woodlice, adult flies, insect larvae, spiders and bugs.

Impact

The Pygmy Shrew is generally beneficial to humans because it preys on large numbers of potentially injurious invertebrate pests.

References

Churchfield & Searle, in Harris & Yalden 2008; Mascheretti et al. 2003; Yalden 1981, 1999.

PYGMY SHREW

MUSTELIDAE (WEASELS & ALLIES)

AMERICAN MINK *Mustela vison*

Natural Distribution Most of North America, including the coniferous and deciduous zones and the tundra (but excluding the majority of islands within the Arctic Circle), SW USA and Mexico.
Naturalized Distribution England; Scotland; Wales; Ireland.

England

American Mink were first imported from eastern Canada (*M. v. vison*) and Alaska (*M. v. ingens* and *M. v. melampeplus*) to fur farms in Britain – mostly in England – in 1929, from which escapes were soon being reported from places as far apart as Morecambe Bay in Lancashire and Stratford-on-Avon in Warwickshire. In the 1950s the industry began to expand, and large numbers of American Mink – as many as 700 in a single shipment – were imported from North America and Scandinavia. By 1962, the number of farms had risen to around 700, with an annual pelt production of 160,000. By 1971, the number of farms had fallen to about 240, but because of their larger size the annual pelt harvest had increased to around 300,000. By 1984, there were only 54 farms still operating in Britain.

The first unconfirmed rumours of escaped American Mink breeding in the wild occurred in the mid- to late 1940s at Singleton near Blackpool in Lancashire, where there were a number of large fur farms. One specialized in selling pairs of American Mink to prospective breeders – some of them demobilized soldiers who used their military gratuity to start an income-producing business – and private householders by whom they were inadequately housed and from whom a number of escapes took place. By the 1950s, some local damage by escaped animals was being recorded by poultry farmers, although they did not become established in the wild.

In 1953, a stretch of the River Teign near Chudleigh and Bovey Tracey in Devon was found to be devoid of Common Rats following the establishment of a colony of American Mink, believed to have originated from animals that escaped or were deliberately released from a recently defunct fur farm. In the same year a male was shot on the Teign near Moretonhampstead, having probably followed the river up from Chudleigh and Bovey Tracey. In 1956, a female with young was seen in an old mill leat in the same area, and two more were shot near Moretonhampstead, and another on the Teign. There were several other reports of American Mink seen on the Teign as far upriver as Chagford, and in November 1957 Ian Linn announced at a meeting of the Mammal Society that they were breeding in the wild. Naturalized animals were also reported in the late 1950s at Teignmouth, and on the River Avon in the New Forest, Hampshire. In 1958, a pregnant female was trapped on the River Deben near Woodbridge in north-east Suffolk, and there were reports of American Mink in the wild in Yorkshire.

In the late 1950s/early 1960s, the species began to extend its range, and was reported from South Brent on the Devon Avon and in 1961 from Throof, Bisterne, Somerley, Kingston, Fordingbridge and Bodenham on the River Avon in Hampshire and Wiltshire; on the Rivers Test and Itchen in the former county; and on the River Stour in Dorset, the Avon in Gloucestershire, and the Wharfe and Lune in Lancashire and Yorkshire.

In the mid- to late 1960s, American Mink increased in numbers in Devon, Lancashire, Wiltshire, Hampshire and Dorset, and first appeared in Sussex on the Rivers Cuckmere

AMERICAN MINK

and Ouse (where 24 were killed in 1964–5 and 30 in 1965–6), and in Somerset, Kent, Northamptonshire and Northumberland. In Hertfordshire two were trapped on Wilstone Reservoir near Tring in 1962; one was killed at Wigginton near Tring in 1965, and in the same year another was shot on the River Chess near Latimer. In 1966, a single American Mink was seen on the Thames at Medmenham near Henley-on-Thames, and another was observed with a Tufted Duck carcass on Weston Turville Reservoir in 1967.

By the end of the decade, the species was being reported in the wild on the Rivers Taw, Dart, Exe, Axe, Coly, Culm, Tavy, Mole, Torridge, East and West Webburn and Yeo in Devon, and on much of Dartmoor. By 1967, it had spread west to Cornwall, where it occurred on the Rivers Tiddy, Hayle and Tressillian at St Clements, and one was found dead on Goss Moor. In 1968, naturalized American Mink killed 74 ducks (species unrecorded) on the banks of the River Looe. In 1975, one was seen killing an adult Common Moorhen on Marazion Marsh near Penzance. In 1953–67, a total of 2,888 American Mink were caught in the wild in England; by the mid-1970s, 4,875 had been killed in 41 English and Welsh counties.

By the late 1970s, the species occurred in significant numbers on the Rivers Exe and Teign in Devon; the Avon, Wylye and Beaulieu (Hampshire and Wiltshire); the Stour (Dorset); the Wharfe and Lune (Yorkshire and Lancashire); the Wyre (Lancashire); and the Cuckmere and Ouse in West Sussex.

Today, American Mink occur in some 10–20 per cent of sites surveyed by Jeffries (2003) in southern and central England and in the north-east, but in less than 10 per cent of surveyed sites elsewhere. The English population was estimated by Harris et al. (1995) to number c 46,750.

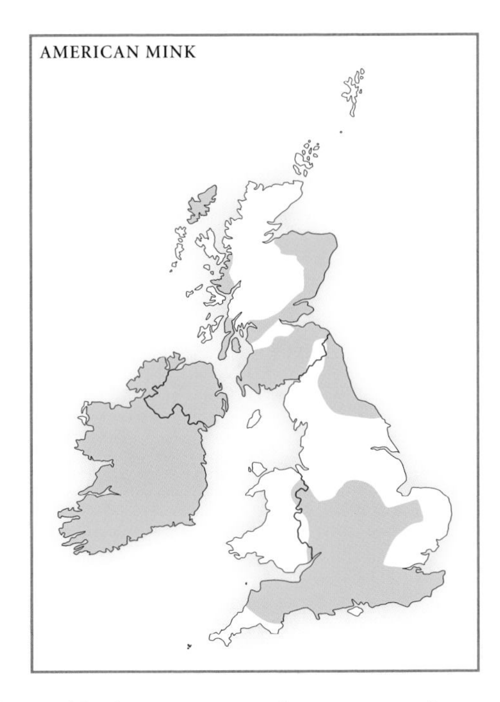
AMERICAN MINK

Scotland

The earliest report of escaped American Mink in Scotland came in the 1930s from a farm at Pawston Hill in Berwickshire. This seems to have been an isolated incident, and it was not until 1962 that animals were observed in any numbers in the wild, and then only on the River Tay in Perthshire, the Dee and Ugie in Aberdeenshire, the Deveron in Banffshire and the Urr in Kirkcudbrightshire (Dumfries and Galloway). In 1953–67, a total of 1,256 American Mink were killed in the wild in Scotland; by the mid-1970s the total had risen to 1,946 from 29 counties.

By the early 1970s, American Mink occurred throughout mainland Scotland, particularly in Ayrshire, Stirlingshire (Falkirk), Banffshire (Aberdeenshire), Midlothian, Roxburghshire (Borders), Kincardineshire (Angus), Fife, Selkirkshire (Borders), Dumfriesshire (Dumfries and Galloway) and Morayshire. By the late 1970s, the species was established on the Rivers Tweed, Gala Water, Teviot and Kale Water in Roxburghshire, Selkirkshire and Berwickshire; the Forth and Teith (Perthshire and Stirlingshire); at Boquhan (Forth), Blane, Earlsburn Reservoir, Balfron (Selkirkshire and Stirlingshire); the Tay, Almond and Earn (Perthshire); the Dee, Deveron and North and South Ugie and Cruden Water (Banffshire and Aberdeenshire); the Spey (Banffshire and Morayshire); the Doon (Ayrshire); the Urr Water, Dalbeattie (Kirkcudbrightshire); and at Dalkeith, Gorebridge and Penucuick, Midlothian.

On Scottish offshore islands, American Mink were first reported on Lewis in the Outer Hebrides in 1969 and on the Isle of Arran in the late 1960s, from where some are known to have swum the 800 m to Holy Island.

Today, wild American Mink appear at the highest density in Morayshire and Aberdeenshire, and on Harris and Lewis in the Outer Hebrides. In 'Argyll and the Western Isles', Welch et al. (2001: 191) record them as 'locally abundant' in 69 localities in Argyll in 1994, compared to only four places in 1965. The pre-breeding population in Scotland was estimated by Harris et al. (1995) to number c 52,250.

Wales

In the late 1950s/early 1960s, American Mink were observed in the wild on the Cilgerran Gorge/Newcastle Emlyn stretch of the River Teifi on the Carmarthenshire/Cardiganshire (Ceredigion) border; in the Fishguard/Letterston area of Pembrokeshire; and on the Rivers East Cleddau[1] and Taf on the Pembrokeshire/Carmarthenshire border. By the mid- to late 1960s, the population in south and west Wales began to expand considerably.

In 1953–67, a total of 195 wild American Mink was taken in Wales. By the late 1970s, they occurred on the River Teifi (Carmarthenshire/Cardiganshire), and on the Cych, Ceri, Dulas and East and West Cleddau and Taf (Pembrokeshire/Carmarthenshire).

Today, wild American Mink are found throughout the Welsh mainland on less than 10 per cent of sites surveyed by Jeffries (2003). The pre-breeding population was estimated by Harris et al. (1995) to number c 9,570.

Summary of Status in England, Scotland and Wales

During the 1970s, the then Ministry of Agriculture, Fisheries and Food ceased its costly and largely ineffective attempt to control the population of naturalized American Mink in Britain, although considerable numbers continued to be culled by private individuals and other interested parties. In 2007 the following regional Wildlife Trusts were involved in American Mink management projects: Anglian (Central and Eastern); North East (Dales); Southern (Kent); South West (Cornwall and Wessex); and Thames (North East, South East and West). Other control projects included those in East Anglia, Midlands (Lower Severn), Kent and the Thames in England; Anglesey, Gwynedd, Conwy and Denbighshire in Wales; and the Cairngorms and Outer Hebrides in Scotland.

In 1977–2004, American Mink continued to expand their range in England (especially in East Anglia and east Yorkshire), Scotland (particularly in the Hebrides) and Wales; since then there appears to have been a general decline in numbers and a contraction of range, especially where the native Eurasian Otter has become established, as in parts of south-west England – mainly Devon and Cornwall. Control of American Mink continues as an important part of Water Vole recovery projects in England, Scotland and Wales, and a concerted attempt is being made to eradicate the former from the islands of Harris, Lewis and North and South Uist in the Outer Hebrides.

Harris et al. (1995) estimated the total British pre-breeding population of American Mink to be c 110,000 – a figure that had fallen to c 40,000 by the end of the decade. The last British mink farm ceased to operate in 2003.

Ireland

In Ireland the earliest fur farms were formed in 1950–3, and by the end of the decade some 40 breeders had become established. Most of the American Mink imported came from Scandinavia, supplemented by stock from North America and Britain.

By the 1970s, the animals were naturalized in some numbers in Co. Tyrone and throughout the watershed of the Rivers Dodder and Liffey in Co. Dublin. Today, they occur throughout the Irish mainland in >10 per cent of the sites surveyed by Jeffries (2003). No estimate has been made of the population in Ireland, but the country is probably less densely populated than Britain because of the presence of a thriving Eurasian Otter population.

Characteristics

American Mink are normally associated with aquatic habitats, preferring eutrophic streams, rivers and lakes with plenty of bankside vegetation such as reed beds and carr.[2] They are less abundant on oligotrophic waters or where waterside cover is absent or sparse. Quite dense populations may occur in undisturbed rocky coastal habitats with a broad shoreline and plenty of nearby cover. American Mink are also sometimes found in estuaries and on rivers and canals near urban areas if sufficient cover – an essential requirement wherever they occur – and prey are available.

Table 3 Early reports of wild American Mink *Mustela vison* in Ireland

Date	Locality	Remarks
1961	Near Omagh, Co. Tyrone	30 escaped; some established on R. Strule; by 1967 had spread to R. Mourne and Glenelly; by 1968 to Victoria Bridge, 24 km west of Omagh
1962	Dunsinea, Co. Dublin	1 killed
1963	Mount Merrion, Co. Dublin	1 killed
1964	R. Dodder, near Tallaght, Co. Dublin	2 seen
1965	Tynan Abbey, R. Tynan, Co. Armagh	1 shot*
1965	R. Connswater (Newtownards Road, E. Belfast), Co. Down	1 near-melanic killed*
1965	Grand Canal, near Stradbally, Co. Leix	1 seen
1966	Carrickfergus, Co. Antrim	1 seen
1966	Deramore Estate, S Belfast, Co. Down	1 captured and escaped
1966	Ballycurry, Co. Wicklow	Several escaped from farm
1967	R. Comeragh (above Drumkeena Bridge), Co. Kerry	1 seen
1967	Castledermot, Co. Kildare	2 escaped from farm
1968	R. Margie, Ballycastle, Co. Antrim	1 shot
1968	Lissanoure, Flushwater, Co. Antrim	1 trapped
1968	Ashford, Co. Wicklow	1 seen
1968	R. Finn (between Stonebridge and Rosslea), Co. Fermanagh	1 seen
1968	Brookeborough, Co. Antrim	1 killed on road
1968	Dublin/Navan road	1 seen
1968	R. Boyne, Oldsbridge, Co. Meath	?
1968	R. Quoile, Finnebruge, Co. Down	1 killed on road*
1968	Valencia Island, Glanleam, Co. Kerry	1 shot
1969	Dunshaughlin (5 km from)	1 melanic female shot
1969	Rathangan, Co. Kildare	1 shot

Source Dean & O'Gorman 1969. *In Ulster Museum.

Mink are opportunistic predators, consuming a wide variety of vertebrate and invertebrate animals. European Rabbits, where numerous, are the most heavily preyed upon mammal (replacing the Muskrat in North America), and can be the most important constituent of American Mink diet. Various small rodents and insectivores are also, though less heavily, predated, and even on occasion Brown Hare adults and leverets. On inland waters ducks, Common Coots and Common Moorhens are the animals to suffer the highest predation, while in coastal habitats gulls – usually juveniles or as carrion – are the most common avian prey. In rocky coastal habitats, especially in the Outer Hebrides, a variety of colonially nesting seabirds may suffer from American Mink predation. Differences in predation on fish species depend on availability. Eels are killed because of their vulnerability, as are Blenny in confined rock-pools. On beaches, Shore Crabs are sometimes eaten. Where they occur, Brown Trout (and their anadromous form the Sea Trout) and Atlantic Salmon are predated, and the native Freshwater Crayfish can be seriously affected. Other foods occasionally eaten by American Mink include amphibians, snakes, beetles, earthworms and molluscs. Male animals, which outside the breeding season are solitary and territorial, tend to prey on larger species than the smaller females. Domestic poultry and game-birds comprise less than 1 per cent of the the species' diet in Britain.

The species' successful colonization of Britain suggests, in the apparent absence of any major shift in its feeding strategy, the exploitation of an ecological niche not fully occupied by the low population of native Eurasian Otters. This hypothesis is confirmed by the recent decline of American Mink in waters reoccupied by returning Eurasian Otters. It is significant that in Ireland, where Eurasian Otter densities have been higher than in Britain, American Mink have become far less densely established.

Impact

The impact of American Mink on native wildlife in Britain has for long been a matter of contention, and indeed confusion, among biologists and laymen alike.

American Mink became established in Britain at a time when the populations of two other (native) mustelids were in serious decline – the amphibious Eurasian Otter and the terrestrial Polecat. The populations of these two animals are now recovering and they are spreading into places already occupied by American Mink. The implications of this re-establishment seem likely to impinge directly on the population dynamics of all three species, and thus indirectly on their collective prey. The relationship between American Mink and Polecats in Britain is unknown, but evidence from Eastern Europe suggests that American Mink have a negative impact on at least female Polecats (Macdonald and Harrington 2003).

Although it was at one time believed that wild American Mink could inhibit the recovery of the Eurasian Otter, recent research has now shown the reverse to be the case and that the latter's populations are expanding at the expense of the American Mink. In the past 20–30 years in south-west England, there has been a noticeable increase in the number of Eurasian Otters and a corresponding decrease in that of American Mink; this suggests that the latter may suffer from competition for food and habitat with the much larger Eurasian Otter. Bonesi and Macdonald (2004a: 9) found that 'mink declines when the density of otters increases. … our findings suggest that the reestablishment of otter populations is likely to lead to a decline of mink that may, in turn, be beneficial to native species [e.g. the Water Vole] threatened by this invasive'. In a subsequent study, Bonesi and Macdonald (2004b: 509) found that 'mink coexist with otters for longer in areas with abundance of habitats hosting terrestrial prey because, by not relying on aquatic prey, mink can segregate from its competitors. … the characteristics of the habitat closer to the riverbank will affect the length of time the two species coexist, because mink are still tied to the water even in the presence of otters'. Thus habitat heterogeneity plays an important part in the coexistence or otherwise of the two species.

A discussion of the impact of American Mink on other species in Britain can be broadly divided between freshwater riverine and still-water systems and the marine coastal and insular environment. In the former, the native Water Vole has undoubtedly been the species most seriously affected by American Mink. Although Water Voles have been in serious and rapid decline for over a century due to habitat loss, degradation and fragmentation acting synergistically, and are now one of our most threatened animals, this decline has accelerated sharply since the 1960s/1970s, when American Mink were spreading and becoming widely established. Being semi-aquatic, they are able to prey on Water Voles both on land and in the water, and females and juveniles are sinuous enough to enter Water Vole burrows. Unless most places occupied by Water Voles are kept free of American Mink, the Water Vole (the much-loved 'Ratty' of Kenneth Grahame's 1908 classic *The Wind in the Willows*) may become extinct in much of Britain. Recognition of this possibility is the inclusion of the Water Vole in the UK Biodiversity Action Plan.

An intriguing complication to the American Mink/Water Vole relationship is that in some circumstances both species seem to be able to coexist satisfactorily. Since the 1970s, they have lived sympatrically at Stodmarsh near Canterbury in Kent, and at Cors Canon (Tregaron Bog) on the Upper River Teifi in Wales (Strachan and Jeffries 1996), where they coexist with Polecats, Eurasian Otters and raptors. Why this should be so remains something of a mystery, but it may be significant that both these localities consist of large expanses of highly structured wetlands comprising dense and luxuriant vegetation or adjacent to stands of willow carr and reed fen.

In many places American Mink have been solely responsible for eradicating Water Voles, whatever the quality of the habitat. As American Mink expand their distribution, Water Vole populations become increasingly isolated and fragmented, and consequently more vulnerable to predation and to their own population fluctuations and environmental changes. 'The evidence ... irrefutably demonstrates,' wrote Macdonald and Strachan (1999: 145), 'that mink have contributed substantially to the decline of the water vole. ... It is also hard to argue against the contention that water voles will be largely eliminated from mainland Britain if action is not taken. ... water voles may already be slipping too fast down the slope to extinction to be rescued.' These findings put into perspective the comments of such apologists as Dunstone[3] (1993: 201), who claimed that 'In Britain there have not been any extinctions *or drastic reductions* [author's italics] of any species of prey that can be directly and unequivocably [*sic*] attributed to mink.'

Aleutian disease, an uncontrollable disorder caused by a parvovirus and common among farmed American Mink, has been identified in the naturalized population in the Thames Valley. Aleutian disease affects Red Foxes, domestic dogs, cats and House Mice, and also other mustelids including Eurasian Otters and Polecats, and could thus threaten the revival of these two species.

Where it occurs in riverine habitats, the native Freshwater Crayfish – already declining in the face of competition from the larger alien Signal Crayfish and crayfish plague – is a favourite prey of American Mink. Although evidence is only circumstantial, the combined effects of competition, predation and disease may well be hastening the native species' decline. The impact of predation on riparian birds is less clear, although such species as Common Coots, Common Moorhens, Little Grebes, Mallards and White-throated Dippers can be locally depleted. Ground-nesting non-riverine waterbirds sometimes affected by American Mink predation include the Tufted Duck, Common Shelduck, Red-breasted Merganser, European Teal, Eurasian Wigeon and Northern Shoveler. There is anecdotal evidence to suggest that in some places (e.g. Duddington Loch, near Edinburgh) their presence has an inhibiting impact on migratory wildfowl.

In a marine environment, colonially ground-nesting seabirds are the animals most at risk from American Mink predation. American Mink are quite capable of swimming for considerable distances across the open sea, and because of their habit of 'surplus killing' (killing more than they are able to eat) are a significant threat to the fauna of offshore islands. In western Scotland (especially in the Western Isles), heavy predation by the species has seriously affected the productivity of nesting seabirds, and thus the conservation status of local populations. Among birds most affected are terns (Common and Arctic Terns) and gulls (Lesser Black-backed, Herring, Common and Black-headed Gulls), some colonies of which have been eradicated. In western Scotland, a decline of 52 per cent has been reported for Black-headed Gulls, 30 per cent for Common Gulls and 36 per cent for Common Terns (Macdonald 2003). Other colonially nesting seabirds potentially at risk include Razorbills and Common Guillemots, burrow-nesting birds such as Common Shelduck, Atlantic Puffins and shearwaters, and terrestrially nesting seabirds such as Common Eiders, Red-breasted Mergansers and Eurasian Oystercatchers.

The American Mink's ability to dive for fish and crabs and to scavenge on the coastline ensures its survival on islands when nesting seabirds are absent, and also enables it to move from island to island. The machair[4] habitat on Benbecula and North and South Uist provides some of the most important breeding sites for wading birds in Europe. The arrival of American Mink from Harris and Lewis, where they have occurred for nearly 40 years, threatens the future of all ground-nesting birds on these islands. The decline of terrestrially breeding birds on the Uists is usually attributed to European Hedgehogs, translocated from the Scottish mainland, or to Common Gulls or Hooded Crows, but American Mink are also likely suspects. Scottish Natural Heritage has been awarded a grant of £1.65 million by the EU to fund their eradication in the Outer Hebrides, starting with Benbecula and North and South Uist. The success, or otherwise, of this project could provide valuable pointers for other eradication programmes elsewhere.

In the 1970s, an exchange of views on the impact of American Mink in Britain was conducted in the *New Scientist* (e.g. Lever 1978a, b; Linn and Chanin 1978a, b), culminating in a letter from Linn and Chanin to that journal (1 March 1979. 81: 709). Given the findings described above, it reads somewhat ironically: 'the mink should be regarded, at worst, as a rather minor pest. … on a national scale the total damage is really quite small. … mink are at worst a limited pest … what cannot be cured must be endured. *And even, dare we suggest, enjoyed?* [author's italics].'

References

Aars et al. 2001; Baker 1990; Bignal 1978; Birks 1986, 1990, 1999; Birks & Dunstone, in Corbet & Harris 1991; Bonesi & Macdonald 2004a, b; Craik 1997, 1998; Cunningham 1987; Cuthbert 1973; Deane & O'Gorman 1969; Dunstone 1986, 1993; Dunstone & Macdonald, in Harris & Yalden 2008; Fairley 1980; Fitter 1959; Gatins 1980; Halliwell & Macdonald 1996; Harris et al.1995; Jones 1976; Lever 1977, 1978a, b, 1980c, 1984a, 1985, 1994, in press; Linn & Chanin 1978a, b; Linn & Stevenson 1980; Long 2003; Macdonald 1999, 2003; Macdonald & Harrington 2003; Macdonald & Strachan 1999; Strachan 2008; Strachan & Jeffries 1996a, b; Strachan et al.1998; Thompson 1962a, b, 1964, 1965, 1967, 1968a, b, 1966, 1969, 1971, in Corbet & Southern 1977; Turk 1968–69; Welch et al. 2001; Yalden 1999.

Notes

1. Long (2003: 284) incorrectly says Cladden.
2. Wet woodland, typically dominated by the Common Alder or willows.
3. Dunstone (1993: 22) rightly says that 'Breeding populations in the wild which are derived from released or escaped domestic animals are termed feral.' However, his references to 'feral' mink, with the implication that the American Mink is a 'domestic' animal, misunderstand the meaning of both 'feral' and 'domestic' and are incorrect.
4. Low-lying land near the coast formed of wind-blown sand and shell fragments. It is found only in western Scotland and parts of Ireland.

CERVIDAE (DEER)

SIKA *Cervus nippon*
(JAPANESE SIKA)

Natural Distribution Formerly from the Ussuri region of Siberia S through Korea and Manchuria to North Vietnam, Japan and Taiwan. Now probably extinct in North Vietnam and China, apart from Manchuria.
Naturalized Distribution England; Scotland; [Wales]; Ireland.

SIKA STAGS AND CALF, DORSET

The taxonomy of the Sika is somewhat confused – up to 13 subspecies have been recognized by some authorities. On the Asian mainland, Sika are now a continuum of intermediate forms, and there is evidence that these mainland forms are themselves ancient hybrids between Sika and Red Deer or Chinese Wapiti. Many of the putative forms are thus ill defined, and the general consensus now is that all Sika are included in two main subspecies: the smaller nominate one, *C. n. nippon* from the Japanese islands, and the larger *C. n. hortulorum* from the mainland and Taiwan. Both races have been introduced to Britain and Ireland, but only the nominate one has become successfully naturalized.[1]

In addition to the naturalized Sika populations in Britain and Ireland, the species is still widely kept in deer parks, where around 1,000 are of the nominate *C. n. nippon* of Japan and 500 *C. n. hortulorum* of Manchuria and Taiwan.

England

Sika were first introduced to England in 1860, when a pair was presented to the Zoological Society of London in Regent's Park. Thereafter a number of private parks in England (and Scotland) were stocked with the deer, many derived from Powerscourt Park in Ireland (see page 92).

Table 4 Principal deer parks in England known to have been stocked with Sika *Cervus nippon*

County	Estate	Date
Buckinghamshire	Waddesdon (from Powerscourt)	c 1874
Dorset	Melbury	c 1880
Hampshire	Hursley, Winchester	c 1885
Essex	Weald, Brentwood	c 1890
Kent	Knole, Sevenoaks	c 1890
Rutland	Exton, Oakham	c 1890
Sussex	Leonardslee, Horsham	1892
Dorset	Brownsea Island, Poole Harbour	1896
Northamptonshire	Whittlebury, Towcester	Late 19th century
Surrey	Park Hatch, Shillinglee	Late 19th century
Westmorland	Rigmaden, Kirkby Lonsdale	Late 19th century
Bedfordshire	Woburn	c 1900
Norfolk	Melton Constable, Fakenham	c 1900
Kent	Surrenden-Dering	Early 20th century
Yorkshire	Park Nook, Bolton-by-Bowland	1904–6
Dorset	Hyde, Wareham	Early 20th century
Northumberland	Hulne, Alnwick	Early 20th century
Nottinghamshire	Rainworth Lodge, Mansfield	Early 20th century
Oxfordshire	Crowsley, Henley-on-Thames (from the Mull of Kintyre)	Early 20th century
Oxfordshire	Great Barrington, Burford	Early 20th century and again c 1928
Oxfordshire	Fawley Court, Henley-on-Thames	Early 20th century
Oxfordshire	Nuneham, Abingdon	Early 20th century
Sussex	West Grinstead, Horsham	Early 20th century
Westmorland	Lowther, Penrith	Before 1914
Shropshire	Weston, Shifnal	1925
Devon	Lundy Island, Bristol Channel (from Surrenden-Dering, Kent)	1927 or 1929
Kent	Mersham Hatch, Ashford	1930s
Buckinghamshire	Langley and Ashridge	Between the two world wars
Sussex	Arundel Castle	Between the two world wars
Sussex	Eridge, Tunbridge Wells	?
Sussex	St Leonard's, Horsham	?
Wiltshire	Rushmore, Tollard Royal, Cranborne Chase	?
Yorkshire	Allerton, Knaresborough	?

Sources Whitehead 1964; Lever 1977, 1985.

Inevitably, escapes (and in some cases deliberate releases) from some of the parks took place, and Sika became established in various localities in England, principally in the south, where they became most common on the heathlands of south-east Dorset between Wareham Woods and Puddletown and on the Bagshot and Reading Beds, and in Hampshire in the south of the New Forest. They appeared to be more successful in establishing on the acid soil of the Tertiary deposits than on such rich or chalky land as that around Melbury and Tollard Royal, Cranborne Chase. A small number became established in Tyneham Wood near the Bagshot Beds on the Wealden, and also between Beaminster and Sherborne. Some of the deer introduced to Brownsea Island in 1896 swam across Poole Harbour – allegedly on the day of their introduction – to the Isle of Purbeck on the mainland, where they became established. When Hyde Park was commandeered by the War Office on the outbreak of war in 1939, a number of the deer escaped to join up with those already on the Purbeck peninsula; these deer were the ancestors of the present population of Sika in south-east Dorset.

In 1904, a pair escaped from the collection of Lord Montagu at Beaulieu, in Hampshire. The animals became established in the nearby Ashen Wood and a year later were joined by a deliberately released pair. These four deer became the ancestors of the present New Forest population. This is based on the Beaulieu/Brockenhurst area and the Beaulieu woods west of the Beaulieu River, and has increased from an estimated 60 head in 1967 to 150–200 today.

In Devon and Somerset, the deer that escaped from Pixton Park (the only known animals to have been imported to England direct from Japan) and became established in the Barle Valley southwards along the valley of the River Exe in the mid-1960s, had died out a decade later. A population of less than 100 occurs on Lundy Island in the Bristol Channel, where Sika were originally introduced in 1927 or 1929; they also still occur on Brownsea Island in Poole Harbour. The most extensive population in England today, in Dorset, numbers something under 2,000 and is now extending westwards into Devon.

In southern Oxfordshire and Buckinghamshire, a small number of Sika became established before the 1970s in the woods of the Chiltern Hills around Stonor, Fawley and Crowsley near Henley-on-Thames, and in the Hambleden valley. Today, they occur on the Oxfordshire/Buckinghamshire border and also on the Oxfordshire/Wiltshire/Gloucestershire borders, and in parts of Northamptonshire and Bedfordshire. Elsewhere in southern England, Sika occasionally seen sporadically between Charing and Challock in Kent are descended from some that escaped from Surrenden-Dering Park at the outbreak of the Second World War, while those around Sevenoaks are the descendants of escapees from Knole Park. Deer reported infrequently in the Leonard's Forest area of Sussex originate from Leonardslee and West Grinstead Parks, while those recorded from time to time in the Shillinglee/Chiddingfold district on the Surrey/West Sussex border are believed to be descended from some that escaped from Park Hatch Park in 1939.

In Yorkshire and Lancashire, Sika that became established in Pendle Forest and Ribblesdale, between Bolton-by-Bowland and Ribchester, were descended from some deliberately released in Park Nook in 1904–6 by Lord Ribblesdale and Captains Peter and Alec Ormerod of Wyresdale Park, Garstang, for hunting by the Ribblesdale Buckhounds.

Most 'Sika' populations in England – especially those in the southern Lake District and Lancashire – are now in fact Sika x Red Deer hybrids (see under Impact, page 94). The only pure-bred Sika populations in England may be those in the New Forest.

Despite their widespread introduction to deer parks in the late 19th and early 20th centuries, Sika in the wild in England today are restricted to only a few localized major populations. Bowland in Lancashire (about 200) and the Lake District of Cumbria are, with the New Forest in Hampshire and the Poole area of south-east Dorset, the species' current main strongholds in England. Harris et al. (1995) estimated the total English population to number around 2,500, but Putman (in Harris and Yalden 2008) believed the figure had perhaps doubled since then.

For the likely spread of Sika south from the Scottish borders into Northumberland, see under Scotland, below.

Scotland

Table 5 Principal estates in Scotland to which Sika *Cervus nippon* are known to have been introduced

County	Estate	Introduced by	Date
Fife	Tulliallan	Daughter of Admiral Lord Keith	c 1870
Ross and Cromarty	Achanalt, Garve; Lochrosque, Achnasheen	Sir Arthur Bignold	c 1887–9
Argyllshire	Carradale, Mull of Kintyre (from Fawley Court, Buckinghamshire)	Austin Mackenzie	c 1893
Sutherland	Rosehall Park	W. Ewing Gilmour	Late 19th century
Inverness-shire	Glenmazeran and Glenkyllachy (from Fawley Court, Buckinghamshire)	William D. Mackenzie	c 1900
Inverness-shire	Aldourie Castle, Inverness (from Rosehall, Sutherland)	Col E.G. Frazer Tytler	c 1900
Peeblesshire	Dawyck Park, Stobo (from Japan)	F.R.S. Balfour	1908
Ross and Cromarty	Coulin, Wester Ross	?	c 1919
Inverness-shire	Loch Morar	?	1920
Caithness	Berriedale (from Welbeck Abbey, Nottinghamshire)	Duke of Portland	c 1920 and c 1930
Ross and Cromarty	Rosehaugh, Black Isle	Douglas Fletcher	Early 20th century
Angus	Kinnaird Castle, Brechin	Earl of Southesk	?

Sources Whitehead 1964; Lever 1977, 1985.

Sika that escaped from Dawyck Park in 1912 became established on the 1,600-ha estate and eventually crossed the border into Dumfriesshire (Dumfries and Galloway). In Angus, escaped deer established themselves in a 40-ha wood outside the Kinnaird Castle deer park, while at Tulliallan Sika crossed the border into the Devilla Woods near Alloa in Clackmannanshire.[2] In Argyllshire, deer that escaped from Carradale colonized Saddell, Glen Lussa and Torrisdale to the south, and Cour, Crossaig and Claonaig near Skipness to the north. Sika also became established in woods around Achanaglachach in South

Knapdale and as far north as Poltalloch north of the Crinan Canal. By the early 1970s, Sika were being reported south of Campbeltown and on the west coast of Kintyre.

Further north, in Inverness-shire, Sika became established in the Great Glen on both sides of Loch Ness and the Caledonian Canal; to the west they colonized Balmacaan and Creag-nan-Eun; to the east escapees from Aldourie Castle spread along the eastern side of Loch Ness through Glencoe, Dell, Fort Augustus and Culachy, south to Aberchalder near Invergarry. Away from Loch Ness, they established themselves on Corriegarth near Gorthleck and at Flichity, Strathnairn.

In Ross and Cromarty, Sika that escaped from Achanalt, Lochrosque, and Coulin colonized the former area and became established around Strathbran between Inchbae and Garbat northwards to Kinlochluichart, Strathvaich, Amat Lodge and Alladale (by 1968) in Strathcarron. In Caithness and Sutherland, the descendants of deer that escaped from Rosehall spread to Achany, Glencassley and Glenrossal, and by 1970 to Berriedale and around Langwell Water.

The only Sika in Scotland known to have been imported directly from Japan were those introduced to Dawyck in Peeblesshire in 1908 (see Table 5, opposite).

Today, Sika in Scotland are 'extremely well established and widespread, with some 14,000 km^2 colonized by c 10,000 individuals. Main populations in Argyll, Inverness-shire, Peebles-shire, Ross and Cromarty and Sutherland, although actively extending their range (Abernethy 1998; Harris et al.1995; Putman, in Harris and Yalden 2008: 590).

This extension of range, which in Scotland averages 0.7–3.7 km per annum, has been summarized by Livingstone et al. (2006). In the north and north-east it seems to be constrained by the Perth/Inverness railway line, although the Glasgow/Oban line has not impeded the expansion of the deer from the Kintyre Peninsula in the north-east, where their rate of spread (aided, as elsewhere, by newly maturing conifer plantations) is the highest in Scotland. In the far north they have been spreading mainly to the north-east and in a westerly direction. In the west it seems likely that the species will soon have reached the coast.

Livingstone et al. (2006) also attempt to predict the future spread of Sika in Scotland. Since they are principally animals of coniferous woodlands, the presence of commercial forests has enabled them to expand their range into new areas. Roads, railway lines, human settlements and non-forested areas appear to have hindered their spread. They also seem to have a preference for elevations of 130–400 m asl. These criteria suggest that in Aberdeenshire it is highly likely that they will continue to spread in an easterly direction through forested areas. Around Stirling, Sika have already surmounted the barrier of the highland railway line and seem likely soon to reach an area with suitable habitat in the south-east. Since the Peebles sub-population has already succeeded in crossing the

M74 to Dumfries and Galloway, it seems probable that it will continue to spread further north-east into Ayrshire. The deer appear already to have begun colonizing the south Borders towards Northumberland, where there are few inhibiting barriers and extensive areas of suitable coniferous forests such as Kielder, Wanehope and Redesdale. They currently occur in at least 36 per cent of the Scottish mainland, and it appears probable that in the course of time they will colonize all suitable mainland habitats.

Harris et al. (1995) estimated the Scottish population of Sika to number around 9,000, but Putman (in Harris and Yalden 2008) believed the total may have doubled since then.

The majority of 'Sika' populations in Scotland are now actually (as in England and Ireland) hybrids with native Red Deer. The only remaining pure-bred Sika in Scotland may be those in Peeblesshire and Morayshire.

[Wales]

Between about 1900 and 1950, a herd of some two-dozen Sika was kept in Vaynol Park near Bangor in Caernarvonshire (Gwynedd). More recently a small number was maintained in Llannerch Park near St Asaph in Denbighshire, but as far as is known no escapes took place from either collection.

Ireland

Sika were first introduced to Ireland in 1860, when a stag and three hinds were acquired, at a price of £8–10 each, from the German animal dealer Carl Jamrach by Viscount Powerscourt, who transferred them to his home, Powerscourt Park near Enniskerry in Co. Wicklow.[3] This park became the most important source of supply for the original stocking of a number of parks elsewhere in Ireland, and also in England and in Scotland (see above).

By 1884, the four Sika introduced to Powerscourt by Lord Powerscourt in 1860 had increased to over a hundred.

The stag and two hinds released at Kenmare in 1865 became established in the woods around Muckross Lake, and by 1935 were said to have increased to 3,000–5,000 head. By 1972, over 500 were established on the 40,500 ha of the Kenmare estate in Co. Kerry,

Table 6 Principal deer parks in Ireland known to have been stocked with Sika *Cervus nippon*

County	Estate	Date
Wicklow	Powerscourt, Enniskerry (from Carl Jamrach)	1860
Kerry	Kenmare, Muckross	c 1865
Fermanagh	Colebrooke	1870
Tyrone	Baronscourt, Newtownstewart	1891–2
Down	Castlewellan	?
Monaghan	Rockcorry	?
Limerick	Glenstal	?

Sources Whitehead 1964; Lever 1977, 1985. All parks were stocked direct from Powerscourt, apart from Baronscourt, which was supplied from Colebrooke, which had itself been stocked by Powerscourt.

including the deer forests of Derrycunnihy and Glena, and in the Bourn Vincent Memorial Park at Muckross, as well as in the state-owned forest of Killarney. From Co. Kerry, deer crossed the border into Co. Cork, where they took up residence in the woods of Glengariff. They also became common in most of the 16,000 ha of woods of the eastern Wicklow Mountains, especially in the Glencree area and around Enniskerry, where by 1972 there were around 300. From there they wandered north-west across the River Dodder to Saggart in Co. Dublin.

The stag and five hinds brought from Powerscourt to Colebrooke in Co. Fermanagh in 1870 by Sir Victor Brooke had increased to about 300 by 1891. From Colebrooke, Sika swam to Inismore Island on Lower Lough Erne, and by 1902 had crossed to Tempo and the woods around Crom Castle. By the 1950s they had arrived at Ballinglen and the Glen of Imaal, some 50 km from Powerscourt. In 1885–97, Sika x Red Deer hybrids were being reported from Colebrooke. By the 1970s there were 150–60 Sika (or hybrids) in this area.

In 1896, several deer escaped from Baronscourt Park in Co. Tyrone and established themselves on Bessy Bell Hill, where by the 1970s they numbered around 250. Sika, presumably from Baronscourt, appeared in Lislap Forest at Gortin in 1962 and at Seskinore, Knockmany Hill and in Glengannce Forest in 1968.

Sika in Ireland were almost wiped out by the exceptionally severe winter of 1962–3. Since then, however, they have staged a remarkable recovery, and within a decade the Irish population was estimated to number some 2,500. Today, 'In S Ireland, major populations in Killarney (believed to be pure Japanese Sika), Limerick, Donegal and Wicklow. In N Ireland, significant populations associated with the old deer park of Colebrooke, Co. Fermanagh, and in Tyrone, probably continuous with populations in Donegal. Smaller populations were established as escapees from deer parks in Monaghan and Down, but the current status of these is unknown.' (Putman, in Harris and Yalden 2008: 590).

Hayden and Harrison (2000) estimated the population of Sika in Ireland at 20,000–25,000. Hybridization with Red Deer in Ireland is widespread, with populations in the Wicklow Mountains (except perhaps on Luggala) being entirely of hybrid status. Only animals of the Muckross peninsula in Killarney are believed to be of pure-bred Sika stock.

Characteristics

Sika are principally associated with dense woodlands and scrubby vegetation on acid soils, the majority of populations occurring in areas of coniferous plantations and heathland. The species is, however, adaptable in its habitat requirements, and can be encountered in a wide variety of other environments, including (e.g. on the Arne peninsula on the Isle of Purbeck in Dorset) estuarine reed beds. It is, however, dependent on trees for shelter and cover. Typically, the deer rest during the day in dense thickets, only emerging at dusk to feed in forest clearings or on nearby open ground. In the New Forest in southern England, they occupy a more varied habitat of acid grasslands, heathland and broadleaved as well as coniferous woodland.

Sika are opportunistic foragers in dense thickets, feeding principally on grasses and Common Heather, but also to a lesser extent on pine needles, Common Gorse and bark. The New Forest populations eat considerable quantities of both deciduous and coniferous browse in winter, and in summer feed largely on grasses and Common Heather, but with a not inconsiderable intake of forbs, deciduous browse, Common Gorse and conifer needles; in autumn acorns, European Holly and fungi are added to their diet.

Impact

Although arguably the most serious threat posed by the species in Britain and Ireland is through its hybridization with native Red Deer, it also has a negative impact on forestry and agriculture.

Damage to commercial forestry through the browsing of both lateral and leading shoots, and in severe weather bark stripping, have been recorded, especially of young Norway Spruce. Mature trees may suffer damage through bole scoring, especially of conifers, which appears to be unique to Sika: this consists of the scarring of tree trunks with deep vertical grooves by the antlers of adult stags, mainly during the autumn rut. Local damage to agricultural crops is sometimes recorded. In areas where they occur at a high density, Sika can also have a significant impact on ground vegetation on open heathland and/or such wetland habitats as salt-marshes and reed beds (e.g. on the Arne peninsula in Dorset), altering species composition and vegetation structure.

It has been known for some time that Sika and Red Deer readily interbreed (Powerscourt 1884). McCurdy (1988) refers to what may be one of the earliest cases of hybridization when he quotes from the Colebrooke (Co. Fermanagh) Deer Book, which records that in 1885 three hybrid calves were born on the estate. The earliest recorded hybrid stag in England may be one that was roused in Barton Wood in the Lune Valley in the southern Lake District by the Ribblesdale Buckhounds in 1940.

Hybridization seems to occur most frequently when colonizing stags move into territories already occupied by established populations of the other species. It has been known for some time that as well as in the southern Lake District in England, deer in the Wicklow Mountains in Ireland are entirely hybrids, and it is now clear that the same problem is arising in Scotland. In Co. Wicklow, where Sika and Red Deer have lived sympatrically for many years, complete genetic and phenotypic introgression seems to have occurred, and the same outcome seems likely in due course in the Kintyre Peninsula in Argyll (and probably elsewhere) in Scotland. Indeed, the gloomy possibility exists that 'Our largest native land mammal, the Red Deer, looks as though it is doomed to be lost into a hybrid that is neither Red Deer nor Sika' (Yalden 1999: 203).

References

Abernethy 1994, 1998; Baker 1990; Buckhurst et al. 1964; Carter 1984; Chadwick 1996; Deane 1972; Delap 1936–7, 1967, 1968; De Nahlik 1959; Diaz 2006; Diaz et al. 2007; Edlin 1952; Fitter 1959; Harrington 1973, 1980, 1982; Harris & Duff 1970; Harris et al. 1995; Hayden & Harrington 2000; Horwood 1973; Horwood & Masters 1970; Larner 1977; Lever 1969, 1977, 1980a, 1985, 1994, in press; Livingstone et al. 2006; Long 2003; Lowe & Gardiner 1975; Lowe, in Corbet & Southern 1977; Matthews 1952; Mayle 2003; McCurdy 1988; Mitchell & Robinson 1971; Moffat 1938; Mooney 1952; Mulloy 1970; Pemberton 2006; Powerscourt 1884; Putman, in Harris & Yalden 2008; Ratcliffe 1987, 1989, in Corbet & Harris 1991; Robinson 1973; Taylor 1939, 1948; Ward 2005a, b; Welch et al. 2001; Whitehead 1964; Willet & Mulloy 1970; Yalden 1999.

Notes

1. Long (2003: 425) incorrectly says that 'Three subspecies occur in Britain: *C. n. nippon*, *C. n. mantchuricus* and *C. n. taisuanus* … which interbreed.'
2. Long (2003: 426) – 'Clackmanannonshire'.
3. At the same time, Lord Powerscourt also released Wapiti, Sambar and Eland.

CERVIDAE (DEER)

REEVES'S[1] MUNTJAC *Muntiacus reevesi*
(CHINESE MUNTJAC; BARKING DEER)

Natural Distribution Densely wooded hilly regions (200–400 m asl) in SE China and Taiwan.[2]
Naturalized Distribution England; [Scotland]; Wales.

Chapman et al. (1994, 1995) unravelled the enigma of the status and early history of muntjac in Britain.

Until relatively recently it was generally believed that two species, Reeves's Muntjac and the Indian Muntjac, were at large in the wild in Britain and were hybridizing. It is now known, however, that although both species have occurred in the wild and will interbreed, the offspring are invariably infertile. Muntjac in the wild in Britain have the skull characteristics and chromosome count of Reeves's Muntjac, and the naturalized population is thus exclusively of that species.

England

There has for long been a widespread belief that muntjac were first imported to Britain by the 11th Duke of Bedford. Research by Chapman et al. (1994, 1995), however, from whom much of the following account is derived, revealed that the first Reeves's Muntjac introduced to England were a pair donated to the Zoological Society of London (ZSL) in 1838, Indian Muntjac having been imported as early as 1829. Although no further

REEVES'S MUNTJAC BUCK WITH ANTLERS IN VELVET

muntjac were apparently acquired by the ZSL until 1862, by 1845 the Earl of Derby had both Indian Muntjac, imported from Java, and Reeves's Muntjac, obtained from an unrecorded source, in his collection at Knowsley Hall, Prescot, Lancashire (Merseyside).

In 1881, the Hon. Lionel Walter (later Lord) Rothschild acquired the first Reeves's Muntjac for his menagerie at Tring Park in Hertfordshire. The records of the collection at Tring were all lost during the Second World War, so subsequent introductions of Reeves's Muntjac are unknown, but it seems probable that escapes or deliberate releases from the park contributed to some of the early records of the species in the wild.

In 1890–1, Lord Herbrand Russell acquired some Indian Muntjac from the ZSL for his collection at Cairnsmore in Dumfriesshire (Dumfries and Galloway). When in 1893 Lord Herbrand succeeded his brother as the 11th Duke of Bedford part of this collection, including three of the Indian Muntjac, was transferred to Woburn Abbey in Bedfordshire, where in the following year it was joined by half-a-dozen Reeves's purchased from the animal dealer William Jamrach.[3] These animals probably came from collections in mainland Europe (where the Jardin des Plantes in Paris was a major supplier of exotic fauna) or directly from China, but it is not impossible that they were acquired from the ZSL (which in the late 19th century was another important source of both species) or from Knowsley. By 1906, a total of 28 Reeves's (and 115 Indian) Muntjac had arrived at Woburn.

Since neither species thrived in captivity, where mortality was high (21 Reeves's and 55 Indian died in the first decade) and the birth rate was low (only 8 Reeves's and 5 Indian fawns were born), in 1901 both species were released into the woods on the estate, at least some of which were outside the park. Other than a brief mention in 1903, nothing more was heard of the Indian Muntjac outside the park at Woburn, and for whatever reason the species apparently died out in the wild, though it may have persisted inside the park until 1930.

In 1928–30, the Duke of Bedford presented the ZSL's collection at Whipsnade Park in Bedfordshire (only 14 km from Woburn) with a pair of Reeves's Muntjac and 12 Indian Muntjac. Subsequent confusion between the species may have led to the belief that only Indian Muntjac survived at Whipsnade, whereas since at least the late 1960s it is in fact only Reeves's Muntjac that have occurred in the park. It seems probable that Whipsnade has contributed to the naturalized population that became established in that part of southern Bedfordshire.

Considerable numbers of Indian Muntjac other than those referred to above were introduced to the four collections previously described; for full details see Champan et al. (1994). Many muntjac of both species were also introduced to several smaller private collections scattered around England, and these were presumably the source of the early records of Reeves's Muntjac in the wild well away from Woburn, Whipsnade and Tring.

Early sightings of muntjac in the wild away from Woburn were few and far between, and the identity of the species involved was confused, although most, if not all, were almost certainly Reeves's Muntjac. By the 1930s, the population was beginning to build up and the number of records was correspondingly starting to increase, though the animals were still only patchily distributed. By 1938, the species' stronghold was Maulden Wood, only 10 km from Woburn. However, this wood is adjacent to Ampthill House, where Sir Anthony Wingfield maintained a collection of exotic species, though whether he had any Reeves's Muntjac is unknown. By the 1950s, Reeves's Muntjac had been reported in several parts of Bedfordshire, although even by the early 1960s they were still sparsely distributed, occurring mainly in the woods around Ampthill and Luton, extending as far as Odell near the Bedfordshire/Northamptonshire border, still only 25 km from Woburn.

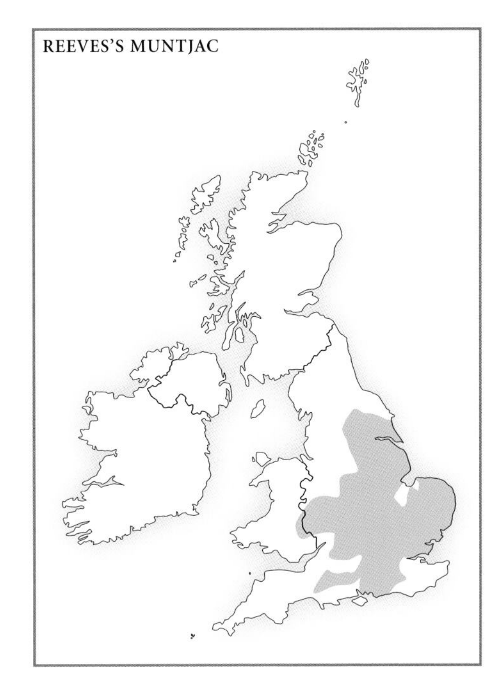

In 1947–9, a minimum of five deliberate releases of Reeves's Muntjac were made from Woburn with the aim of founding new naturalized populations. These were at Bix Bottom, near Henley-on-Thames and at Bicester (both in Oxfordshire), at Elveden on the Norfolk/Suffolk border, near Corby in Northamptonshire and at an unrecorded location in Kent. These liberations were, with the apparent exception of the one in Kent, all successful, and were responsible for the establishment of sizeable populations, and nearly all the suitable woods located in Buckinghamshire, Northamptonshire, Oxfordshire, and the Brecklands of Norfolk and Suffolk hold populations of Reeves's Muntjac. In about 1950, some animals are believed to have been sent from Woburn to Endsleigh, the Duke of Bedford's property in Devon, and in 1967 to Holcombe Park in the same county, where permanent populations apparently failed to materialize.

Away from Woburn, Reeves's Muntjac are recorded as having been released or escaped in Lincolnshire and Greater London (Enfield and West Hampstead). On offshore islands, 3 bucks and 1 doe were in 1977 planted on Steep Holm in the Bristol Channel (Legg 1990), where the population in 1994 was around 20.

From the 1940s until at least the early 1980s, the population and range of Reeves's Muntjac in England increased more rapidly. By the mid-1990s, they had been recorded from nearly half the 10-km squares in England and Wales, and only Cleveland, Durham, Greater Manchester, Merseyside and Tyne and Wear in England were apparently uncolonized. The bulk of the wild population was established in Berkshire, Buckinghamshire, Hertfordshire, Oxfordshire and Warwickshire, with smaller numbers in Cambridgeshire, Essex, Hereford and Worcester, Norfolk, Northamptonshire, Suffolk and Wiltshire. In the majority of counties the animals were only sparsely and locally established. By about 2007, Reeves's Muntjac were 'well established over much of England, especially [the] southern half' (Chapman, in Harris and Yalden 2008: 567).

A factor that contributed to the successful establishment of Reeves's Muntjac in England was the existence of a near-vacant ecological niche in which Fallow Deer were the only deer present. Fallow Deer are predominantly grazers, and thus there is little niche overlap between the two species. Because of the absence of Roe Deer – which, like Reeves's Muntjac, are mainly browsers – in localities initially colonized by Reeves's Muntjac, the latter faced no foraging competition.

The false impression of a rapid rate in the increase in the population and its distribution (the average rate of dispersal is only 1 km a year) has been due, at least in part, to artificial spread by releases or escapes from widely scattered loci, which continue to this day. The diversity of sources of the naturalized population has been confirmed by the discovery of genetic variability between sub-populations determined by examination of mitochondrial DNA. Many of the existing discrete sub-populations appear to have become established through the presence of only a small founder stock. Indeed, since the females breed throughout the year, have a post-partum oestrus (Chapman, in Corbet and Harris 1991) and are pregnant for virtually their entire adult lives, a single doe is theoretically capable of founding a new sub-population on her own. The general pattern seems to be one of slow growth for 10–20 years after colonization, succeeded by a much more rapid rate of increase. Because much of the present range has only been occupied relatively recently, a significant increase in the population is likely to occur in due course. A period of over 40 years without an exceptionally hard winter has undoubtedly made a significant contribution to the recent rate of spread, and the advent of global warming is likely to accelerate the process. The amount of suitable habitat available is increasing year by year, with new woodland plantings in much of rural England, and this should be of benefit to colonizing Reeves's Muntjac.

Table 7 Earliest county records of Reeves's Muntjac *Muntiacus reevesi* in England

Year	County	Locality
1922	Bedfordshire	Wrest Park
1925 (or 1928)	Hertfordshire	Ashridge Park; Welwyn (or Broxbourne)
1928 (and 1938)	Greater London	Finchley
1930	Leicestershire	Beaumanor estate
1933–4	Buckinghamshire/ Northamptonshire border	Salcey Forest; Yardley Chase; Hazelborough Forest
1934	Warwickshire	Ryton Woods, Kineton
c 1937	Northamptonshire	Kelmarsh, near Kettering
c 1938	Wiltshire	Savernake Forest
1941	Staffordshire	Needwood Forest
1941	Essex	Colchester; Easthorpe
1941	Warwickshire	Walton Wood, near Kineton
1948	Derbyshire	Matlock
1940s	Hereford and Worcester	Various
1951	Norfolk	Walsingham
1952*	Suffolk	Lowestoft
1952	Oxfordshire	Stoke Lyne, near Bicester
1956	Berkshire	Twyford
1958	Middlesex	Enfield
1958	Gloucestershire	Withington Woods, Cheltenham
1959	Avon	Hanham Abbots
1959	Huntingdonshire	?
1950s	West Midlands	Various
Late 1950s	Kent	Lullingstone Castle, Dartford
1960	Buckinghamshire	Bow Brickhill
1961	Cambridgeshire	Hayley Wood

Table 7 *Continued*

Year	County	Locality
1962	Dorset	Melbury Osmond
1966	Somerset	Frome
1966	West Sussex	Various
1966	Surrey	Barnes Common
1970+	Hampshire	Basingstoke
1971	Rutland	Rutland Wood
c 1971	Devon	Holcombe Park
1972	Cornwall	Mithian, near Truro
1972	Nottinghamshire	Widmerpool
1973	Humberside	Flamborough Head
1974	Lincolnshire	Dobbins Wood, near Bourne
Early 1980s	Cumbria	High Hay Bridge
Before 1981	Shropshire	Telford
1981	Yorkshire	Near Skipton
1986	East Sussex	Maresfield
1989	Cheshire	Christleton; Kelsall
1989	Northumberland	Near Felton
1992	Lancashire	Near Silverdale; near Hothersall Hall

Sources Compiled from Lever (1977, 1985) and Chapman et al. (1994). *In 1941 or 1942, a possible Reeves's Muntjac was shot at Parham. +Reeves's Muntjac may have occurred in Alice Holt Forest and the New Forest in 1948 and 1949 respectively.

There is no evidence to suggest that the early spread of Reeves's Muntjac in England was in any way explosive. Although they were first released in the woods outside Woburn Park as long ago as 1901, they were still uncommon in Bedfordshire and neighbouring Hertfordshire by the end of the Second World War, and even by the 1960s the population was still low and patchily distributed in the two counties. This may have been due, at least in part, to the exceptionally severe winters of 1939–40, 1941–2, 1947–8, 1956–7 and 1962–3, when deep and long-lying snow cover made foraging difficult or impossible. Elsewhere, well-established populations occur in Northamptonshire and Warwickshire, where the animals appear to have been present locally in viable numbers for around 30 years. Although Chapman et al. (1994) suggest that the animals in these two counties must have become naturalized as a result of deliberate releases, it must be at least possible that the founder stock escaped from local private collections. Outside Northamptonshire and Warwickshire, sporadic records of individual animals are likely to have been a result of other undocumented deliberate releases or accidental escapes.

Records of Reeves's Muntjac in urban and suburban localities are both numerous and widespread. Towns and cities where they have been seen (listed by Chapman et al. 1994) include Aylesbury, Berkhamstead, Birmingham, Brandon, Bury St Edmunds, Camberley, Cambridge, Chelmsford, Colchester, Daventry, Derby, Great Yarmouth, Ipswich, Leamington Spa, Leicester, Leominster, Luton, Northampton, Norwich, Oxford,

Peterborough, Reading, Rugby, St Albans, Solihull, Stratford-on-Avon, Swaffham, Thetford, Ticknall, Watford, Wellingborough, Witham and Wolverhampton. Sightings in inner London include Finchley (as early as 1928 and again in 1938), Barnes Common, West Ham, Bushy Park, Palmers Green, Bloomsbury (Woburn Square), Marylebone, Stratford and Hackney.

In the Greater London suburbs, Reeves's Muntjac became established in gardens, on railway embankments and riverbanks, in woodlands and on golf courses in Harrow, Pinner and Ruislip. Further west they have occurred on the urban edge of Hounslow Heath, while to the north-east they became established on a small urbanized plot of land at Gunnersbury. Other west London sightings include Brent Reservoir, Osterley Park, Ickenham and Wembley.

The greatest potential for the species' future extension of range appears to be Kent and Sussex, and to a lesser extent north into Lincolnshire, Nottinghamshire and southern Yorkshire, with a westwards expansion into Cheshire and northern Shropshire. Future changes in habitat, such as the planting of new woodlands, and milder winters with an absence of prolonged snow cover due to global warming, along with deliberate releases and accidental escapes, may well result in the colonization of other localities.

Harris et al. (1995) estimated the English population at 52,000. Chapman (in Harris and Yalden 2008), who suggested that the population doubles in less than eight years, estimated a figure of around 104,000.

[Scotland]

Possible sightings of Reeves's Muntjac in the wild in Scotland include Argyllshire; the Tweed valley in the Borders Region, including Galashiels, Kelso and Ancrum; Pitlochry, Perthshire; Inverarary and Kilmarnock (both Strathclyde); near Turriff and Inverurie (both Grampian); between Etteridge and Kingussie and near Garve (all Highland Region); Inversnaid (Central); Caerlaverock and west of Dumfries (both Dumfries and Galloway); and Easter Bush (Lothian). In the 1970s a small semi-wild population became temporarily established near Cumbernauld, Strathclyde.

The map for Reeves's Muntjac in Harris and Yalden 2008) indicates 'rare occurrences, vagrants or scarce' in parts of Fife, Perth and Kinross, Clackmannanshire and Stirling, and north of the Great Glen south-west of Inverness. There are, however, no established populations of the animals in Scotland today.

Table 8 Early county sightings of Reeves's Muntjac *Muntiacus reevesi* in Wales

Year	County	Locality
1957 or 1958	Blaenau Gwent	Llanfair Grange
1962	Powys	Llangorse Lake
1981	Pembrokeshire (Dyfed)	Near Narberth
1989	Gwynedd	Betws-y-Coed Forest*
1993	West Glamorgan	Llwchwr Bridge
1992	Clwyd	Mostyn

Source Chapman et al. 1994. *Possibly descended from a deliberate release in 1968.

Wales

Although there are many records of (mostly individual) Reeves's Muntjac in Wales, where Chapman et al. (1994: 159) considered that 'it appears that small well-separated populations are becoming established', Chapman (in Harris and Yalden 2008), whose map indicates 'rare occurrences, vagrants or scarce' in south-western and west central regions and 'scattered but regular occurrence' in parts of the south-east, refers only to 'a few records from Wales'. Harris et al. (1995) estimated the total Welsh population to number c 250; Chapman (in Harris and Yalden 2008), who suggested that the population doubles in <8 years, estimated a population of around 500.

Characteristics

Reeves's Muntjac favour a dense habitat containing a wide variety of vegetation, such as young unthinned plantations, scrub, and overgrown and neglected gardens. They also occur in commercial conifer plantations with some deciduous trees and cover, where they prefer places with a diversity of ground and shrub layers, Brambles and mature nut-bearing trees. It appears, however, that Reeves's Muntjac are fairly catholic in their choice of habitat, and that only the presence of arable land seems an important requirement, with marginal upland localities being largely avoided. Contrary to popular belief, Reeves's Muntjac do not necessarily avoid open ground.

As concentrate selectors rather than bulk feeders, Reeves's Muntjac are primarily browsers (rather than grazers) in deciduous woodlands, feeding on the leaves and flowers of many species of plant and shrub and on young tree shoots. In autumn and winter, fungi and nuts become an important part of their diet.

Ecological Impact

Mayle (2003), from whom much of the following account is derived, has studied the impact of introduced deer on the natural environment.

All deer have some impact on their habitat through their foraging behaviour, and particular concern has been raised about the effect of Reeves's Muntjac (and Fallow Deer) in semi-natural woodland habitats. Heavy grazing or browsing generally reduces the number of tall-growing herbs and ferns (apart from Common Bracken) and increases the number of lower-growing species and grasses. Palatable species such as Bramble are especially heavily browsed.

The most biologically rich woodlands in Britain tend to be small (<50 ha), ancient and semi-natural, and many have been traditionally managed by coppicing, which benefits both the ground flora and its associated invertebrate community. Many of these woodlands are occupied by Reeves's Muntjac and Fallow Deer (and also native Roe Deer), which pose a particular problem. Small coppice coupes[4] attract deer, and the regrowing coppice stools[5] are susceptible to browsing for the first 2–5 years. Fraying[6] of saplings also occurs. Muntjac browse coppice regrowth up to 1 m high and will snap taller stems to reach higher foliage. Browsing by deer also affects shrub and ground flora, especially Bramble, which can be reduced to very small plants and seedlings.

Because Reeves's Muntjac have a small home range that they utilize intensively, they can have a profound impact on the ground flora. Several rare or nationally important species associated with coppiced woodlands are known to be vulnerable to browsing by deer – especially Reeves's Muntjac – including Early Purple, Greater Butterfly and Pyramidal Orchids. Other species to suffer include marsh orchids *Dactylorhiza* spp., Bee Orchids, helleborines Orchidaceae, violets *Viola* spp., Common Bluebells, Primroses,

Common Spotted Orchids, Cuckoo Flowers or Lady's Smock, and Wood Anemones. Studies on these species and others, such as Lords and Ladies and Dogs' Mercury, have shown that significant numbers of inflorescences are destroyed, thus reducing the amount of successful seed-setting. This can have a negative impact on such species' long-term survival, and the removal of vegetation reduces plant size and vigour and the size of inflorescences. In some species, such as the Oxlip, whereas Reeves's Muntjac remove only the inflorescence, Fallow Deer consume many of the leaves too, thus reducing the plant's vigour and the size of its flowers.

The impact of deer on woodland vegetation can also have an effect on woodland fauna. Browsing by Reeves's Muntjac on Honeysuckle removes the preferred low egg-deposition sites of the White Admiral butterfly. Several threatened butterfly species associated with coppice woodland (the Heath Fritillary, High Brown Fritillary, Pearl-bordered Fritillary, Small Pearl-bordered Fritillary, Wood White and Duke of Burgundy Fritillary) rely on food plants such as violets, Primroses and Cowslips, which are a favoured food of the deer. Were they ever to become widely established in Kent, the county with some of the most important English orchid woodlands, their impact on the local flora could be devastating. Similar concern has been expressed were populations of Reeves's Muntjac to increase in the Oxlip woods of parts of East Anglia.

Heavy browsing and grazing by Reeves's Muntjac and Fallow Deer tends to reduce the amount of ground cover provided by Bramble, and thus the abundance and variety of small mammals in woodland settings, including the Wood Mouse and Bank Vole.

Any reduction in the shrub and ground flora understorey of woodlands is likely to affect populations of woodland birds that depend on this layer for nesting, shelter and food. Among the species most affected are Common Blackbirds, Song Thrushes, Dunnocks, Bullfinches and Nightingales, and several warblers (Sylviidae).

Heavy defoliation of Bramble and other shrubs by Reeves's Muntjac during the winter may be detrimental to both native Roe Deer and introduced Chinese Water Deer. Research has shown that Reeves's Muntjac compete advantageously with Roe, Fallow and Chinese Water Deer, and European Rabbits and Brown Hares. High densities of Reeves's Muntjac seem to exclude the other species – especially Roe Deer – in some areas, and Chinese Water Deer are reduced in numbers and confined to wetter areas of reed bed and willow-carr[7] as a result of the limited amount of browse available in winter due to defoliation of Brambles and other shrubs by Reeves's Muntjac.

This catalogue of problems caused by the species in the natural environment in England makes a mockery of Dansie's suggestion (in Corbet and Southern 1977: 451) that the species 'appears to be a harmless introduction'.

Economic Impact

So far there is little evidence that Reeves's Muntjac cause major damage to agricultural crops – their small mouth precludes much harm to root crops. They can, however, occasionally be responsible for serious losses in market gardens, orchards and sometimes vineyards. Their principal economic impact is felt in private and municipal gardens, where their effect – especially on roses – can be severe.

References

Anderson 1986; Anderson & Cham 1987; Chapman, in Corbet & Harris 1991; 1996, 2005, in Harris & Yalden 2008; Chapman & Chapman 1972; Chapman & Harris 1993; Chapman et al. 1993, 1994, 1995; Chapman, in Corbet & Harris 1991; 1996, 2005; Clark 1971; Cooke 1994, 1997,

2004, 2005, 2006a, b, 2006–07; Cooke & Farrell 1995, 2001; Dansie 1966, 1969a, b, 1970, 1971, in Corbet & Southern 1977; 1983; De Nahlik 1959, 1974; Edlin 1952; Fitter 1959; Harris & Duff 1970; Kirby 2001; Leeke 1970; Legg 1990; Lever 1969, 1977, 1979, 1985, 1994, in press; Long 2003; Matthews 1952; Mayle 2003; Perrins & Overall 2001; Pickvance & Chard 1960; Pollard & Cooke 1994; Pritchard 1982; Samstag 1985; Smith-Jones 2004–5; Soper 1969; Taylor 1939, 1948; Warner 1972; Whitehead 1964; Willet & Mulloy 1970; Yalden 1999.

Notes

1. The species is named after John Reeves (1774–1856), who in 1812–31 was employed by the East India Company in Canton, from where he shipped a number of species (e.g. Reeves's Pheasant) to England. In 1817 he was elected a Fellow of the Royal Society and of the Linnaean Society.
2. The nominate form, which is the one introduced to Britain, is little differentiated from *M. r. micurus* of Taiwan.
3. Presumably the son of Carl Jamrach, who in 1860 had provided Sika for Lord Powerscourt in Ireland.
4. An area that is cut in coppice rotation.
5. The permanent base of a coppiced tree.
6. The rubbing of (usually newly grown) antlers by deer to remove the protective covering of 'velvet' (a soft, downy layer of skin).
7. In East Anglia, wooded wet areas with low nutrient levels, dominated by the Common Alder or willows.

CERVIDAE (DEER)

FALLOW DEER *Dama dama*

Natural Distribution Originally believed to have been restricted to Anatolia (Asia Minor) east of the Aegean Sea (Turkey and Iran). Introduced in the early Neolithic to the Mediterranean region, and subsequently widely elsewhere.
Naturalized Distribution England; Scotland; Wales; Ireland.

A large form of the Fallow Deer, *D. d. clactoniana*, lived in Britain during the Hoxonian period, some 250,000 BP. By the Ipswichian period, around 100,000 BP, it had become only slightly larger than present-day Fallow Deer. It died out in Britain during the last (Würm) glaciation, surviving only in southern Europe. Fallow Deer were reintroduced to Britain in the 11th century by the Normans, who released them in forests for hunting. Because it died out in Britain thousands of years ago, the species is treated here as an introduction rather than a reintroduction.

England

Fallow Deer remains appear on sites, especially manors and castles, across England in the Norman period. Specimens from the early medieval manor of Goltho (Jones and Ruben 1987) and Dudley Castle (Thomas 2002) suggest an introduction during the Norman period (1066–1135). Because at that time Fallow Deer did not occur in northern France (which would otherwise have been the logical source of the introduced stock), Anatolia, Greece or Sicily have been suggested as the most probable sources, with the last-named being generally accepted as the most likely. Not only is it known that Fallow Deer occurred on Sicily at that time, but the island was occupied by the Normans as part of the Norman Empire for over a hundred years from at least the mid-11th century.

FALLOW DEER BUCK, KNOLE PARK, KENT

During and after the Norman period, numerous hunting reserves and parks in England (and to a lesser extent in Scotland, Wales and Ireland) were stocked with Fallow Deer – the *Domesday Book* of 1086 records no fewer than 31 such enclosures in southern England alone. By the 13th century, the hunting of Fallow Deer had become a popular sport of the nobility, and vast tracts of woodland in England were enclosed for this purpose. In the county of Essex alone, for example, it is known that in 1200–1390 there were over 130 such enclosures, and Rackham (1990) suggested that by about 1300 there were around 3,200 deer parks in England, comprising some 2 per cent of the country. It seems almost certain that at an early stage Fallow Deer escaped from some of these enclosures and became naturalized in the English countryside.

By the early 17th century, there were still more than 700 parks in England enclosing 'more fallow in a single English county than in the whole of Europe. Every English gentleman of £500 to £1,000 rent by the year hath a park for them, enclosed by pales of wood for 2 to 3 miles [3–5 km] compass.' During the Civil War (1642–9), the fences of many such parks fell into disrepair, enabling the deer to escape and become established in the wild. In the 18th century there was a revival of the fashion of maintaining deer parks, and by the end of the 19th century an estimated 71,000 Fallow Deer were imparked in almost 400 parks. The two world wars of the 20th century saw the break-up of many of the deer parks, and the escape of further large numbers of Fallow Deer into the countryside.

Although Fallow Deer still occur in forests, where they have been naturalized for centuries, most of the present-day population is found in woodland in the vicinity of former parks, especially those abandoned during the last century. The fact that they have spent a long time in captivity (not 'as domesticated animals', Yalden 1999: 156) is confirmed by a very low degree of genetic diversity, which shows that the population has passed through a genetic bottleneck. This is in marked contrast to the pelage colour variations (principally common, menil, black and white, with many other lesser varieties) that characterize British park and naturalized herds, and is, as Yalden points out, exactly what man tends to select, whereas hidden genetic protein polymorphism becomes easily lost.

Today, Fallow Deer are the most widespread of the four introduced species of deer naturalized in England, being found in numerous localities and in the majority of counties, principally in the south, east, south-east, south-west and Midlands. In the following overview of some of their main strongholds, updated and expanded from Chapman and Chapman (1975, 1980) and Lever (1977, 1985), the likely park sources (and dates where known) of sub-populations in individual counties are listed within square brackets.

West Kent Occur from Eynsford and Sevenoaks to at least as far east as Maidstone and Chatham [Knole, Surrenden-Dering, Chilham, Mersham, Hatch, Waldershare].

East Sussex Well established from the Northam/Robertsbridge/Battle area westwards to beyond Ashdown Forest.[1] In West Sussex, occur from St Leonard's Forest near Horsham to the South Downs. North of the Downs, established around Midhurst, and west of the Rother Valley to beyond the Surrey and Hampshire borders [Ashburnham, Britling, Buckhurst, Arundel, Petworth, Parham, Cowdray].[2]

Hampshire Widely established throughout much of Hampshire, especially in the coniferous and mixed plantations of Bramshill Forest in the north-east, around Petersfield, near the coast at Emsworth on Chichester Harbour, in the deciduous and coniferous woodlands of the Forest of Bere in the south-east, in Harewood Forest near Andover, around Hursley and Farley Mount near Winchester, and west of the River Test towards Salisbury, Wiltshire [Aldermaston, Hackwood, Hurstbourne, Hursley, Uppark]. The largest population lives in the south-west in the 27,000-ha New Forest (created by William the Conqueror in 1079), one-third of which comprises deciduous and coniferous plantations, the remainder consisting of natural deciduous woodland – mainly Common Beech, Pedunculate or Common Oak and Sessile Oak with an understorey of European Holly – moorland, grassland, heath and agricultural land. A census of 1670 revealed the presence of 7,593 Fallow Deer in the New Forest. As a result of the Deer Removal Act of 1851, designed to clear forests of deer so that they could be replanted with oaks, the Fallow Deer were all but eliminated, and the population recovered only to around 200 by the end of the century. The New Forest population is now about 1,000.

Dorset Present in the east on the New Forest border, between Lulworth Cove and Wool in the south-east, between Maiden Newton and Bridport in the south-west and in the 2,000 ha of mixed hardwoods and softwoods of Cranbourne Chase in the north-east [Hooke, Hyde House, Melbury, Sherborne and possibly Charlborough].[3]

Devon Occur sporadically from the Dorset border to the Otter Valley.[4] West of the River Exe, considerable numbers established in the woodlands of the Haldon Hills.[5] On the eastern fringe of Dartmoor, found in Teign Valley – in particular between Fingle Bridge[6] and Dunsford – in populations of Douglas Fir, Norway Spruce and Japanese Larch, as well as in deciduous woodlands of birches, Rowan, oaks, Common Hazel and Common Beech, with an understorey mainly of European Holly, Bramble, grasses and Wild Thyme. In south Devon, occur in Plym Forest north-east of Plymouth [Bicton, Blatchford, Cready, Shute, Werrington, Whiddon].[7]

Cornwall Smaller numbers found on Torpoint [Mount Edgecumbe], to the south of Bodmin Moor, near Lostwithiel and Falmouth, and around Truro.

Somerset In north-western Somerset, found on Croydon Hill south of Dunster, and around Birds Hill, Combe Sydenham [Combe, Dunster, Hatch Court, Nettlecombe, Sydenham].

Gloucestershire Widely distributed in suitable woodlands throughout the county, where their stronghold is the 13,000-ha Forest of Dean [Brockhampton, Great Barrington, Lydney, Over, etc].

Wiltshire Savernake Forest is the stronghold in Wiltshire; also found north-west towards Marlborough and east of Salisbury.

Surrey A few have from time to time occurred in the Farnham area [Farnham, Witley].

Berkshire Found in small numbers at least as far east as Maidenhead Thicket, and northwards to the Oxfordshire border [Aldermaston, Hall Place, Hampstead, Littlecote, Welford, Windsor].

Oxfordshire Widely distributed, especially in the Woodstock/Charlbury area, in Bernwood Forest east of Oxford and around Stonor, Henley-on-Thames, and east of Chinnor and Aston Rowant in the Chiltern Hills on the Buckinghamshire border [Cornbury, Ditchley, Glympton, Great Barrington, Nuneham, Shotover, Stonor].

Hertfordshire Large numbers occur in the 1,600 ha of mainly deciduous woodland (principally Common Beech, oaks, Common Ash, Sweet Chestnut and Wild Cherry), common land and scrub (mainly Common Bracken, Common Gorse and birches at Ashridge [Ashridge, etc.]. Also found at Watford, Cuffley, Northaw and Hitchin, and between Nuthampstead and Bishop's Stortford on the Essex border [Ashridge (1928), Hatfield, King's Walden, Woodhall (1916)].

East Anglia Widespread in suitable habitats. In Essex, occur mainly in the north and west, being particularly numerous in Epping Forest on the outskirts of London (where they have been established since the 15th century), Navestockside, Mill Green, Roxwell,

Hatfield Forest, Strethall, Markshall and Clacton-on-Sea [Chelmsford, Dunmow, Easton Lodge, Langleys, Weald].[8] In west Suffolk, their main stronghold is the largely coniferous King's Forest; also found in deciduous woodland in the south-east around Saxham, Whepstead, Lawshall, Long Melford, Hawkedon and Cavendish near Bury St Edmunds, and westwards through Dalham, Ousden and Lidgate near the Cambridgeshire border.[9] Established in the southern half of east Suffolk through Felsham, Chelsworth, Ashbocking, Helmingham, south of Ipswich, and between the Rivers Orwell and Deben to the mainly coniferous Tunstall and Rendlesham Forests [Henham (1914), Ickworth, Livermere Hall, Woolverstone]. In Norfolk, occur principally around King's Lynn and Sandringham in the west. A few are also found between Cromer on the north coast south to Norwich [Blickling, Dudwick, Holkham, Houghton, Sandringham].[10]

Cambridgeshire and **Lincolnshire** Although much of the area comprises flat and treeless expanses of intensively cultivated farmland, Fallow Deer are found in a few localities in both counties. In Cambridgeshire, these include Gamlinghay in the south-west and the Soke of Peterborough in the north-west, especially in the triangle formed by Peterborough, Stamford and Wansford-in-England. In Lincolnshire, they have been long established in the Stamford/Bourne/Kirkley/Underwood/Grantham district; in the 1980s they moved into Linwood and Willingham Forests and to near Wragby and Woodhall Spa west of Wolds [Brocklesbury, Burghley, Grimsthorpe].[11] In 1977, some were released on the 100-ha Read's Island upriver of Barton-upon-Humber, where their present status (if they survive) is uncertain.

Midlands In some Midland counties, established in the wild for several hundred years. In Northamptonshire occur between Oundle and Corby, in the mixed woodland of Rockingham (where they have existed since at least the early 13th century) and in Whittlewood Forest in the south [Fawsley, Milton etc.].

Nottinghamshire Found in Sherwood Forest (where they were established by the 13th century) and the Underwood/Annesley area [Rufford, Thoresby, Welbeck].

Staffordshire The 2,400 ha of mixed woodland, heath and scrub that constitute Cannock Chase between Birmingham and Stoke-on-Trent in Staffordshire have been the home of Fallow Deer since at least 1271; they have been established in Needwood Forest near Cheadle since before 1313, and also occur around Abbots Bromley.

Leicestershire Found in the deciduous woodland of Charnwood Forest [Calke Abbey, Donington Hall, Staunton Harold].

Rutland Live between Pickworth and South Witham, especially in Morkery Wood [Exton, etc.].

Derbyshire Found in small numbers around Spondon, on the outskirts of Derby, and Matlock.

Warwickshire Found around Edge Hill south-east of Stratford-upon-Avon [Ragley Hall, and possibly Alscot, Charlecote, Ettington].

Worcestershire Occur mainly in the woodland, pasture and scrub on Bredon Hill [Hagley Park].

Shropshire The most interesting population lives in the 3,600 ha of mainly coniferous plantations, intermixed with some deciduous woodland, of Mortimer Forest south-west of Ludlow [Garnstone (black variety), Hampton Court, Haye, Ludford, Moccas, Moor], where a long-haired variety has been established for many years. These deer (which are unique to Mortimer) have body hairs of over double the length of that of other populations, with even longer, frequently curly hairs on the tail, inside the ears and on the forehead (Springthorpe et al. 1969). Long hair is dominant over normal-length hair and occurs in various colours (Smith 1980). The hairs grow more rapidly and for a longer period, but this characteristic declines with age (Johnson and Hornby 1980). Elsewhere in Shropshire, normally coated Fallow Deer have lived in the oak and Common Beech woodlands of Wyre Forest[12] since the 13th century [Attingham, Cleobury, Longnor, Loton, Mawley Hall (c 1880), Mortimer].

Herefordshire Occur mainly near Hereford, Leominster and Ross-on-Wye in the south of the county.

Northern England Much smaller numbers are found than in the Midlands and south, but a few small herds are scattered at least as far north as southern Northumberland and in parts of the Lake District of Cumbria. In Yorkshire, Fallow Deer are known north-west of Barnsley;[13] in Stang Forest south of Barnard Castle; in the Swale Valley west of Richmond; near Swinton Park, Masham and West Tanfield; in the Laver Valley; in woodlands in Nidderdale north-west of Harrogate; near Knaresborough east of Harrogate;[14] and in woods south-east of Wakefield [Allerton, Duncombe, Nostell Priory, Studley Royal, and perhaps Castle Howard and Scampston]. In Cumbria, a few Fallow Deer occur in Common Ash and other deciduous woodlands near Arnside near the Lancashire border [Dalemain, Lowther].[15] For many years a small herd of Fallow Deer lived near Alnwick in eastern Northumberland [?Alnwick].

Scotland

By tradition, monks first introduced Fallow Deer to Kildalton on the Isle of Islay around AD 900. They were first referred to on the Scottish mainland in 1283, when a herd became established in the King's Park at Stirling. In 1424, Fallow Deer first came under the protection of Scottish law when it was decreed that those who 'slay deare, that is to say harte, hynde [Red Deer], doe [Fallow], and rae [Roe]' should be fined 40s. and their lairds £10. In about 1539, James V and Mary of Guise 'hunted Fallow Deer on the Lomonds', and they were among the game at the 'tinchel' (deer drive) organized in Glen Tilt on the Braes of Atholl by the Earl of Atholl for Mary Queen of Scots in 1564. Twenty-three years later, legislation was enacted that rendered the 'slayers and schutters of hart, hinde, doe, roe, haires, cunninges and other beasts' liable to the same penalties as those who stole horses or oxen.

By the middle of the 17th century, Fallow Deer were established in a number of forests in central Scotland. The *Wardlaw Chronicle* of 1642 records that 'a gallant, noble convoy, well appointed and envyed by many' went hunting in the Forest of Killin in Perthshire, where they found 'fallow-deer hunting to their mind'.

During the 18th century, there were a number of new introductions of Fallow Deer to Scotland, e.g. to Ross-shire by 1729 and Dumfriesshire in 1780. As in the rest of Britain and Ireland, numerous deer parks were enclosed in Scotland during the 19th century from which, as elsewhere, there were inevitable escapes. The present naturalized population of around 4,000 Fallow Deer is confined to some of the Lowland and Border counties, and to three islands in the Inner Hebrides (see below).

In the Borders, Fallow Deer are known from Kirkudbrightshire, Dumfriesshire (Dumfries and Galloway) and, to a lesser extent, Wigtownshire, to which a small number were translocated in 1971. In Kirkudbrightshire, they are found in the extreme south of the 6,200 ha of mixed deciduous Sessile Oak woodland and softwood plantations of Galloway Forest near Newton Stewart. In Dumfriesshire, the 200 or so Fallow Deer in the woods around Raehills, near Moffat, are descended from some introduced in 1780.

Further north, Fallow Deer occur around the shores of Loch Lomond in Stirlingshire and Dunbartonshire, as well as, at least until recently, on three islands – Inchlonaig, Inchcailloch and Inchmurrin – in the south of the loch.

East of Loch Lomond, straddling the Stirlingshire/Perthshire border, a small population became established in the early 1970s in the 17,000-ha Queen Elizabeth Forest Park, comprising the forests of Rowardennan, Loch Ard and Achray. Their stronghold in the latter county lies to the east of the River Tay between Dunkeld and Ballinluig; smaller numbers have also occurred between Blairgowrie and Dunkeld.

On the west coast, a small number of Fallow Deer are believed to be established at Inverary in Argyllshire and on the Mull of Kintyre.

From time to time Fallow Deer are reported from a number of Highland counties, including Banffshire (Arndilly House and Craigellachie); Inverness-shire (Balmacaan and Loch Ness); Morayshire (between Fochabers and Craigellachie in the Spey Valley); Ross and Cromarty (Achnalt, Garve, Corriemoillie, Kinlochluichart, Lochrosque and Scatwell); and near Dunrobin in Sutherland.

Although Fallow Deer are known to have occurred in the past on no fewer than 11 islands off the west coast of Scotland, they currently survive on only three – Mull (around Loch Ba', Glenforsa and Ben More), Islay (near Kildalton) and Scarba (south of Mull) – in the Inner Hebrides.

Wales

Fallow Deer have been established in Wales for well over 700 years, and by the 16th century they are known to have occurred in the Royal Forest of Snowdon in Caernarvonshire (Gwynedd). In his *Tours in Wales* (1772), the traveller and naturalist Thomas Pennant quotes the following warrant from H. Sydney to John Vaughan, forester of Snowdon, of 14 August 1561: 'These are to require you to delyver to my friend Maurice Wynne, Gent. or to the bringer hereof in his name, one of my fee stags or bucks [Fallow Deer] of this season, due to me out of the queens majestys forest of Snowdon.'

There were also by this date a number of deer parks established in Wales, including one at St Dinothes (St Donat's Castle) in Glamorgan referred to by the antiquarian John Leland (or Leyland) in his *Itinerary in Wales* (1710), and another at Baron Hill near Beaumaris on the Isle of Anglesey where, according to Pennant, Sir Richard Bulkeley, Bt 'kept two parks well stored with … Fallow deer'. George Owen, in his *Description of Pembrokeshire* (1603), says that Fallow Deer were enclosed 'in two small parkes onlye, and not in anye fforest or chase, and the nomber very fewe'.

As in the rest of Britain and Ireland, a number of deer parks were formed in Wales, especially during the 19th century, from which there were the usual escapes, mainly after the two world wars. The present range of wild Fallow Deer in Wales is not extensive nor is the population large.

South-east In Monmouthshire, established around Monmouth, on both sides of the River Wye, especially in Beaulieu and Lady Park Woods and in Redding Enclosure towards Staunton in the east, and at Hendre and White Hill in the west [probably Wyastone Leys and Hendre].
South-west Found in the forests of Brechfa (principally conifers with some oaks) and Caeo (principally oaks, Common Beech and Sycamore) between Llandilo in Carmarthenshire [Golden Grove Park] and Lampeter in Cardiganshire (Ceredigion); small numbers also found in woods south of Carmarthen.
Montgomeryshire (Powys) Established on Breidden Hill, Welshpool and the Long Mountain [Powis Castle (1947), Loton].
Glamorganshire (Neath) Established in Margarn Forest near Port Talbot, where conifer plantations predominate among deciduous woodlands.
Brecknockshire and **Radnorshire (Powys)** Occur in the native hardwood and coniferous plantations of the Wye Valley along the Monmouthshire and Herefordshire borders.
Denbighshire Found in the east to south of Wrexham and between Denbigh and Colwyn Bay on the north coast [Wynnstay Park].
Merionethshire (Gwynedd) A mass escape of 90 deer from Nannau Park near Dollgellau in 1963 resulted in the formation of a number of wild herds,[16] some of which moved northwards to the mainly coniferous plantations of Coed-y-Brenin Forest, while others colonized oak and softwood plantations on the slopes of Dovey Valley.
Caernarvonshire (Gwynedd) Occur near Bangor on the Menai Straits. Across the straits on the Isle of Anglesey, a small number became established on the sand dunes and in the conifer plantations of Newborough Warren on the west coast around Bodorgan near Aberfraw.

Ireland

Although Fallow Deer were almost certainly introduced to Ireland by the Normans in the 11th century for hunting, there is frustratingly little mention of them in the early literature. Fitter (1959: 44) says 'there were some in woods in Munster by the twelfth century', but does not quote his source. Mooney (1952: 14) suggests there was an 'introduction into the Glencree [Co. Wicklow] Royal Park in 1244. One interesting record (Fairley 1984) tells of a gift of 12 Fallow Deer in 1296 from the Royal Forest at Glencree to Eustace le Poer, ancestor of the Powers of Curraghmore. The traveller Fynes Moryson, in his *Description of Ireland* (1617), wrote that 'The Earl of Ormond, in Munster, and the Earl of Kildare, in Leinster, had each of them a small park enclosed for Fallow-Deer ... they have also about Ophalia and Wexford, and in some parts of Munster, some Fallow-Deer scattered in the woods.' This establishes the existence of Fallow Deer in the wild in Ireland by at least the early 17th century, although they almost certainly became naturalized well before that date. In his *Natural History of the County of Dublin* (1772), the physician John Rutty refers to 'the *Cervos platyceros*, the Buck, or Fallow-deer', which was by then quite common in Co. Dublin. During the following century a number of parks containing Fallow Deer were enclosed in Ireland.

In 1920–2, when many estates were illegally taken over by thugs of the Irish Republican Army, a number of deer fences became broken down through lack of maintenance, allowing the deer to escape. In at least one instance – at Charleville Castle in Co. Offaly – the owner deliberately opened his gates to let his deer escape rather than be slaughtered by the murderous Sinn Feiners. As a result, by the mid-1920s herds of naturalized Fallow Deer had become established in nearly every Irish county.

Northern Ireland

Today, Fallow Deer are well established in Co. Antrim, Co. Armagh, Co. Down and Co. Tyrone. The species' stronghold is between Glaslough (in Co. Monaghan in the Irish Republic) and Caledon in Co. Down, near the border with Co. Armagh. In Co. Down, Fallow Deer occur on the Clandeboye estate between Bangor on the north-east coast and Newtownards at the northern end of Strangford Lough. To the north, in Co. Antrim, they are more widely distributed near Ranaldstown north of Lough Neagh [Shane's Castle Park, c 1933; dark variety]. Although there are believed to be few if any Fallow Deer in Co. Londonderry, there may be some around Colebroke, Florence Court and Crom Castle in Co. Fermanagh [Crom Castle, Florence Court, Kiltiernay House].

Republic of Ireland

Co. Waterford Widely distributed throughout much of Co. Waterford except in the extreme east, principally in woodlands south of the River Suir near Kilsheelan and Clonmel, along the River Nier, north of Cappoquin and near Villierstown [Curraghmore Park, c 1909].

Co. Tipperary Widely distributed and have inhabited plantations on the slopes of the Knockmealdown Mountains near Clogheen and Ballyporeen, in the Glen of Atherlow in the Galtry Mountains, and in smaller woods around Bansha, Dundrum, Rear Cross, Newport, Silvermines Mountains and near Kilcooley Abbey in the east [Ballinacourty, Dundrum House, c 1918; Gyrteen, 1890s; Castle Lough, c 1920]. Smaller numbers occur south of Tallow on the eastern border of Co. Cork.

Co. Clare North of the River Shannon, occur in the Bunratty/Tulla region and near Lough Cutra, Dromoland Castle and Raheen parks astride the Clare/Galway border; in eastern Galway also established in woodlands near Derrybrien on the Slieve Aughty Mountains and at Kylebrack and Woodford. In 1970, in Portumna Forest on the northern shore of Lough Derg, a population of no fewer than 370 lived in an area of no more than 560 ha.[17] Further north, also occur in smaller numbers around Ahascragh, on both banks of the River Clonbrock and near Ballygar.

Co. Offaly and Leix Widely distributed: in the former principally near Birr Castle and Clonfinlough and around Tullamore [Bin Castle, Charleville, Kinitty Castle, Knockdin Castle, Thomastoun]; in the latter around Clonaslee, Slieve Bloom Mountains, Camross, Montrath, Emo, Ballyfin, Ballybritas and Stradbally [Emo Park].

Co. Wicklow Established in the Wicklow Mountains since 1244,[18] where on the western forested slopes they range from the Kiltegan region northwards to the Lackan Reservoir, eastwards towards Delgany and south through plantations around Carrick Mountain to the Wicklow coast [Ballycurry, Powerscourt]. Further south, found south of Aughrim and across the border into Co. Wexford.

Lambay Island In 1888, Count Considine released some Fallow Deer on this island 25 km north-east of Dublin, where their descendants still survive. The herd, which is composed mainly of the black variety, is maintained at around 30 and roams at will over the 1,000-ha island.

Co. Westmeath Occur north and south of Mullingar.

Co. Roscommon Herd established near Boyle [Rockingham Park, c 1910].

Co. Mayo Herd occurs on the slopes of Partry Mountain near Lough Mask.

Co. Sligo Found in woodlands between Collooney and Lough Gill and around Slish Wood [Marktree Castle].

Co. Monaghan In northern Co. Monaghan, near Glaslough, and in the south between Rockcorry and Cootehill near the border of Co. Cavan [Hilton Park].

From time to time Fallow Deer are also recorded from other Irish counties, both in the north and in the Irish Republic, but seldom with regularity.

Summary of Status

The Fallow Deer is today the most widespread deer species in Britain and Ireland, where its distribution and population have both increased since the 1970s. Harris et al. (1995) estimated the British population to number around 100,000, of which 95,000 were in England, 4,000 in Scotland and 1,000 in Wales. In Ireland, where some 2,000 a year are legally shot, the extent of the population has not been estimated.

Characteristics

Fallow Deer are usually found in mature woodlands, showing a preference for deciduous or mixed ones with an established understorey; they will, however, also colonize coniferous plantations with some open areas. They are preferential grazers, grasses comprising >60 per cent of their diet between spring and autumn and >20 per cent in winter. Herbs and broadleaf browse form a significant proportion of their forage intake, including Bramble, European Holly, Ivy and Common Heather. Nuts (especially acorns, chestnuts and beechmast) and other fruits such as Crab Apples, Blackberries and Bilberries are readily taken in autumn and winter.

Impact

The Fallow Deer 'is a pest of both agricultural and forest crops in some areas' (Welch et al. 2001: 35). It can cause damage to young plantations and prevent coppice regeneration, and in severe winters bark stripping of mature trees (both deciduous and coniferous) can be a problem. Considerable damage to ground flora, e.g. Oxlip, is reported in some localities, and bucks may inflict serious harm to trees by thrashing them with their antlers to clear them of velvet[19] when they are newly grown in late summer and in aggression before the autumn rut. Fallow Deer occurring on agricultural land can cause some local damage to pastures and crops – especially cereals.

For a more detailed assessment of the impact of Fallow Deer in Britain and Ireland see under Reeves's Muntjac, page 101.

References

Chapman & Chaplin 1967; Chapman, in Corbet & Southern 1977; Chapman & Chapman 1969, 1970, 1975, 1980; Chapman & Putman, in Corbet & Harris 1991; Deane 1972; Delap 1936–7; De Nahlik 1974; Doney & Packer 1998; Eastcott 1971; Fitter 1959; Gill 1992; Harris & Duff 1970; Idle & Mitchell 1968; Johnson & Hornby 1980; Jones & Ruben 1987; Langbein et al., in Harris & Yalden 2008; Lever 1969, 1977, 1980a, 1985, 1994, in press; Long 2003; Marshall 1974; Matthews 1952; Moffat 1938; Mooney 1952; Mulloy 1970; Page 1954; Pemberton 1993; Putman 1994a, b; Putman & Moore 1998; Rackham 1986; Scott 1967; Smith 1980; Springthorpe 1969; Springthorpe & Voysey 1969; Sykes 2004, 2005; Taylor 1939, 1948; Tubbs 1986; Venner 1970; Ward 2005a, b, c; Welch et al. 2001; Whitehead 1964; Willet & Mulloy 1970; Yalden 1999.

Notes

1. Conisbee, L.R. 1977. *Hastings and East Sussex Naturalist* 12: 43–6.
2. Carne, P.H. 1970. *Wild Deer of the West Sussex Downs.* West Sussex Deer Control Society: Petersfield, Hampshire.

3. Carne, P.H. 1967. *Proceedings of the Dorset Natural History & Archaeological Society* 88: 93–101.
4. Carne, P.H. 1967. *Journal of the British Deer Society* 1: 42–4.
5. Carter, D. 1967. *Journal of the British Deer Society* 1: 46.
6. Scott, W.A. 1967. *Journal of the British Deer Society* 1: 45.
7. Penistan, M.J. 1967. *Journal of the British Deer Society* 1: 44–5.
8. Chapman, D.I. 1977. *Essex Naturalist* n.s.1: 1; 3–30.
9. Cranbrook, Earl of and Payn, W.H. 1970. *Transactions of the Suffolk Naturalists' Society* 15: 222.
10 Goldsmith, J.G. (ed.). 1972. *Norfolk Bird and Mammal Report* 23: 60–9.
11. Johnson, M. 1976. *Transactions of the Lincolnshire Naturalists' Union* 19: 8.
12. Hickin, N.E. 1971. *The Natural History of an English Forest: the Wildlife of Wyre.* Arrow Books: London.
13. Robinson, J. 1975. *Deer* 3: 222.
14. Ibid., pp. 320–2.
15. Delap, P. In Hervey, G.A.K. and Barnes, J.A.G. 1970. *Natural History of the Lake District*: 176–93. Warne: London.
16. Vaughan, H.V. and Carne, P.H. 1971. *Deer* 2: 767.
17. Mulloy, F. 1970. *Deer* 2: 502–4.
18. Deane, C.D. 1972. *Deer* 2: 920–6.
19. A protective covering of soft skin on newly grown antlers.

CERVIDAE (DEER)

CHINESE WATER DEER *Hydropotes inermis*

Natural Distribution Formerly reed beds and coarse riverine grassland in E China from the lower Yangtze Kiang basin W to Hupeh and N to Korea. Now probably extinct or greatly reduced in numbers in much of its range, except perhaps in N, C and parts of the S.
Naturalized Distribution England.

England

The following account of Chinese Water Deer in England is derived partly from Chapman (1995).

The first Chinese Water Deer were imported to London Zoo in 1873. Subsequently a single doe, obtained from an unrecorded source, was introduced to Woburn Abbey in Bedfordshire by the Duke of Bedford in April 1896. Later, more were introduced to Woburn where, helped by their fecundity (2–6 fawns are usually born), the deer prospered and became established. By the end of 1913, when a total of 19 had been introduced, 115 young had been born and 6 adults had died, and the population was recorded as 126. During the First World War, when Woburn Abbey became a military hospital, the animal records in the park were abandoned, but it is known that in 1935–6, 180 Chinese Water Deer died at Woburn. Recording was resumed in the 1960s, and a census taken in 1967 showed a total of nearly 300. At least one importation, of 1 buck and 2 does, was made in 1970, and 2 years later there was a large enough population outside the park for 20 to be culled in the surrounding farmland.

In 1929, the Duke of Bedford gave 18 Chinese Water Deer to the Zoological Society of London's collection at nearby Whipsnade Park, followed a year later by a further 14. By late 1933, the Woburn population numbered some 200, but shortly afterwards crashed to only about 60 due, it was believed, to an attack of enteritis associated with a heavy infestation of nematodes. By 1936, the population had recovered to 115, and during the Second World War it was deliberately maintained at a low level. Stadler (1988) estimated the population at about 400–500, ranging over more than 200 ha of grassland. Neonatal losses were high (50 per cent in 1986), probably due to sub-optimal forage.

CHINESE WATER DEER BUCK

The earliest record of a Chinese Water Deer in the wild in England was in Buckinghamshire in 1940. The first known escapes from Woburn took place in the 1940s, although because of confusion with Reeves's Muntjac the earliest confirmed record was not until 1954. During the 1940s, deer from Woburn were sent to several other English collections, including Leckford Abbas near Stockbridge and Farleigh Wallop near Basingstoke (both Hampshire), and from the latter to Cwmllecoediog, Aber Angell, near Machynlleth in Montgomeryshire; Studley Royal near Ripon (Yorkshire); Foxwarren Park, Cobham (Surrey); Walcot Park, Lydbury North, near Bishops Castle, and thence to Hope Court, Hope Bagot, near Ludlow (both Shropshire). From several of these collections deer managed to escape and become, albeit temporarily, established in the wild.

By the 1960s, Chinese Water Deer were well established in parts of the Bedfordshire countryside, from where very small numbers spread to Ashridge in west Hertfordshire and to neighbouring areas of the Chiltern Hills in Buckinghamshire, and to Yardley Chase in Northamptonshire; in none of these locations, however, did the species become established.

Today, the best-known population of naturalized Chinese Water Deer in England is based at Woodwalton National Nature Reserve in Cambridgeshire (formerly Huntingdonshire). The earliest confirmed identifications were made in 1971, and in the following year the population was estimated by Cooke and Farrell (1981) to number 50–75. This population originated from a small number of animals from Woburn released in 1947–52. From Woodwalton, Chinese Water Deer have spread to a number of woods and onto arable land nearby.

The earliest records for Norfolk were two seen in 1958 at Swim Coots and Ling's Mill near Hickling in eastern Norfolk, which were later found to have escaped from a private collection at nearby Stalham. In 1969 another, presumably from the same source, was the victim of a road-traffic accident near Stalham. In 1972 several were seen at Hickling Broad, and it became clear that a population had become established in the extensive reed-bed system of the Norfolk Broads – a habitat very like much of that in the species' native range in the Far East.

Further south, on the Suffolk coast, a Chinese Water Deer (Chapman, 1995, in error says a Reeves's Muntjac) was first sighted at Minsmere in 1989. In southern Norfolk small numbers became established on the northern edge of Thetford Forest, and also on the Suffolk/Cambridgeshire border.

Away from eastern England there have been unconfirmed sightings in the Northlew/Highampton area of Devon, near the Sutton Bingham Reservoir in Somerset, and perhaps in Hatfield Forest, Essex.

Today, Chinese Water Deer are established discontinuously in parts of Bedfordshire (first recorded 1954), Cambridgeshire (1971), Norfolk (1968) and Suffolk (1987). There are relatively few between the principal concentrations of western Bedfordshire, the Fens of Cambridgeshire and the Norfolk Broads, although there has been some expansion of range north-west from the Broads. Harris et al. (1995) estimated the English population at only 650, a figure increased by Ward (2005a) to 1,500 in the wild in 2004, plus at least 600 in captivity.

Characteristics

In the Far East, Chinese Water Deer live near rivers, lakes and coasts with reeds and other tall grasses for cover, and in hilly grassland. In England they inhabit reed beds and woodlands – preferably of mixed vegetation for cover with open areas for foraging – and in lower densities on agricultural land. They are selective feeders on the tender parts of a variety of grasses, sedges, herbs and woody species.

In China the Chinese Water Deer, with a total population of only 10,000–30,000, is under threat from habitat fragmentation, and excessive hunting and poaching for food and for use in traditional Chinese medicine; coastal populations suffer from typhoons and tidal waves. The English captive and wild populations (the only substantial ones outside China) are thus of some conservation significance, and the necessity for an eventual repatriation of English stock to the Far East is not impossible.

Impact

When other forage is scare, Chinese Water Deer will eat root crops such as carrots (especially the green tops), sprouting winter wheat and potatoes left after harvest; on arable land they are often observed to eat weeds rather than crop plants. Partly because of its small population, the fact that it is not an extreme concentrate selector and because it occurs in less valuable habitats, the economic and ecological impact of the species is so far negligible. Having no antlers,[1] only long, curved upper canines, Chinese Water Deer bucks cause no harm to trees by fraying[2] or by bark stripping.

The principal controlling factors of Chinese Water Deer in England are road-traffic accidents, cold and wet winters, shortage of food due to snow cover, and some predation of fawns by Red Foxes. Where they occur sympatrically with the also introduced Reeves's Muntjac, the latter appear to restrict the former to the wetter areas of reeds and willow carr,[3] and heavy defoliation of Bramble and other shrubs by Reeves's Muntjac limits the amount of winter forage that is available for Chinese Water Deer. The same changes in Chinese Water Deer numbers have been observed in China wherever the ranges of the two species overlap.

References

Bedford 1995; Chapman 1995; Clark 1987; Cooke & Farrell 1981, 1987, 1998; in Harris & Yalden 2008; Dansie, in Corbet & Southern 1959; De Nahlik 1959; Edlin 1952; Farrell & Cooke, in Corbet & Harris 1991; Harris & Duff 1970; Lever 1977, 1985, in press; Long 2003; Mayle 2003; Middleton 1937; Page 1954; Stadler 1988; Taylor 1939, 1948; Whitehead 1964; Willet & Mulloy 1970.

Notes

1. The species' scientific name, *Hydropotes inermis*, means 'weaponless water-drinker'.
2. Clearing the soft protective skin from newly grown antlers.
3. Wet woodland, typically dominated by Common Alder or willows.

BIRDS

ANATIDAE (DUCKS, GEESE & SWANS)

BAR-HEADED GOOSE *Anser indicus*

Natural Distribution Montane regions of C Asia, Mongolia and China (4,500–5,000 m asl), wintering in N India and N Myanmar.
Naturalized Distribution England; [Scotland]; [Wales]; [Ireland].

England

In 1991, the total number of free-flying Bar-headed Geese in the wild in Britain was reported by Delany (1992) to be 85 at 27 localities, although the only record of successful breeding was at Stratfield Saye in Hampshire, where a flock of 19 included 9 juveniles in 3 broods. Other records in 1991 were a flock of 11 on Highfield Lake, South Yorkshire, and flocks of 6 at each of Abberton Reservoir, Essex and the Otter Trust, Bungay, Suffolk. In northern England there were reports from Cheshire, Greater Manchester, South Yorkshire and Shropshire, and in eastern England from Bedfordshire, Greater London, Essex and Kent. Further west, Bar-headed Geese were reported in Oxfordshire, Berkshire, Gloucestershire and Avon, and in the south from the Isle of Wight.

In 1996–2001, reports of successful breeding came from Avon, Derbyshire, Nottinghamshire, Greater London, Greater Manchester, Hampshire, Somerset, Surrey, Sussex and West Midlands. In 2000, there was a population of 52 birds at 18 localities in 8 counties, with a single breeding pair at Stratfield Saye. From a population in the early 2000s of more than 100 birds in about 30 locations, at least 5 pairs bred successfully in

BAR-HEADED GOOSE

most years, and their numbers and range in Britain have been slowly increasing. In 2002, up to five pairs occurred at five sites in four counties. In Kent, a flock in the Grove/Stedmarsh area north-east of Canterbury between August and October included at least three juveniles; the species may also have bred near Wingham, east of Canterbury. Two adults and two immatures were present in Chichester Harbour, Sussex, in late August/early September, but it is uncertain where breeding occurred. In Warwickshire, a pair reared a single gosling on Kineton Lake, while in Wiltshire a pair reared four young on Edington Lake. No breeding was recorded in either Hampshire or Derbyshire in 2002, although in recent years breeding had been fairly regular. Reports of singletons and small flocks came from several counties.

In 2003, at least three pairs bred in Chichester Harbour, Sussex, where six adults with up to three juveniles were seen in the autumn; in Warwickshire, where a single pair successfully reared six goslings on Kineton Lake; and in Greater Manchester at Hope Carr Nature Reserve, where a clutch of eggs was destroyed under licence. In 2004, 9 adults with 2 juveniles were seen in Chichester Harbour, and a flock of over 20 birds in Kew Gardens, London, suggested local breeding; Bar-headed Geese were present but did not breed on Kineton Lake. In 2005, a pair fledged a single gosling in Wiltshire, and a further clutch of eggs was legally destroyed in Hope Carr Nature Reserve. No breeding was reported in 2005 in Kew Gardens, Chichester Harbour or Kineton Lake.

[Scotland]

In 1991, six non-breeding Bar-headed Geese were observed on Castle Loch in Dumfries and Galloway, and there are individual records from Orkney, Tayside and Lothian.

[Wales]

Bar-headed Geese have been recorded in Pembrokeshire and Denbighshire.

[Ireland]

Non-breeding birds have been observed on Strangford Lough, Co. Down.

Summary of Status

Although breeding by Bar-headed Geese has been recorded in England annually since 1996, with a maximum of eight pairs in 1999 – all south of Manchester – breeding numbers remain low and there is little indication of continuous breeding over a number of years. A restraint against more successful reproduction may be the lack of flocks large enough to stimulate breeding.

References

Blair et al. 2000; Delany 1992, 1993, 1995; Holling & RBBP 2007; Lever 2005; Mead 2000; Ogilvie & RBBP 1999–2004; Rowell et al. 2000.

ANATIDAE (DUCKS, GEESE & SWANS)

SNOW GOOSE *Anser caerulescens*

Natural Distribution NE Siberia, N Alaska and NW Canada, wintering in S USA, N Mexico and Japan. Also NE Canada and NW Greenland, wintering in NE USA.
Naturalized Distribution England; Scotland; [Wales]; [Ireland].

SNOW GOOSE

England

In 1991, a flock of 32 Snow Geese was established at the Linch Hill Leisure Park in Oxfordshire and another, comprising 22 adults of the larger *A. c. atlanticus* race, at the Wildfowl and Wetlands Trust headquarters at Slimbridge, Gloucestershire. In Norfolk there was a flock of 23 birds of the smaller nominate subspecies on the Babingley River. Blue morphs of the nominate form were reported from Avon, the Isle of Wight, Greater London, Norfolk and Kent. Birds of unspecified race were observed in Leicestershire, Bedfordshire, Cumbria, Dorset, Kent, Norfolk, Oxfordshire and Hampshire. Successful breeding was recorded only on the Babingley River in Norfolk, at Radwell Gravel-pit in Bedfordshire and at Stratfield Saye in Hampshire.

In the spring of 1996, a flock of 50 Snow Geese was seen in Sandringham Park, Norfolk, but no breeding occurred. In 1999 and 2000, they bred successfully in Hampshire. In the latter year they were recorded on Thamesmead Lakes in Greater London (22), at Eversley (13) and Stratfield Saye (9) in Hampshire, on the University of York Lake in North Yorkshire (10) and in Blenheim Park, Oxfordshire. In 2002, two pairs were present at the regular breeding site at Stratfield Saye, but no breeding is believed to have occurred. In 2003, they bred successfully at two sites; at Stratfield Saye a pair raised one gosling, and at Somerley Lake a Snow Goose mated with a Greylag Goose and fledged seven young. No breeding was recorded in 2004, but in 2005 two pairs at Stratfield Saye reared a total of five young. Nevertheless, the population in Hampshire continued to decline – the maximum number at Eversley Gravel-pits being six in January, and a maximum of four were counted at the end of the year.

Scotland

For the last 20 years or so, the largest and apparently self-sustaining breeding population of Snow Geese in Britain has been established in north-western Mull and on the neighbouring island of Coll, Argyll, in the Inner Hebrides. In 1991, when the British population was estimated by Delany (1993) at 182 at 27 sites, 40 occurred on Mull, where 6 were blue morphs of the nominate *A. c. caerulescens.* Six birds of the larger *A. c. atlanticus* form occurred at Tankerness on Orkney. Snow Geese of unspecified race were reported from Dumfries and Galloway.

In 1993, Snow Geese bred successfully at Haunn on Mull, where 14 of the 40 birds seen were juveniles. At least 49 were counted in 1997 on Coll, where the population was judged to be self-sustaining. In 1999 and 2000, they bred successfully on Coll and Mull, where up to 30 were observed in the former year and 30–40 in the latter. In 2002, a flock of 24 adults and 10 goslings was rounded up and ringed on Coll during the birds' annual wing-moult in July, which was believed to be the entire population on the island. This flock commutes between Coll and Mull (a flight of about 12 km) during the year. In July 2003, the flock on Coll consisted of 25 adults and 4 juveniles, indicating successful breeding by at least 1 pair; in 2004, 5 pairs successfully reared 12 young; and in 2005, 6 pairs hatched 22 goslings, of which only 6 fledged.

[Wales]

In 1991, Snow Geese of unspecified race were observed on the Isle of Anglesey, but no breeding attempt has been recorded.

[Ireland]

Snow Geese have been observed on Strangford Lough in Co. Down.

Summary of Status

By the turn of the century, the British population exceeded 250 birds, but is believed to be very under-recorded. About 20 pairs were breeding annually, mostly in Scotland. The British population is derived from escaped birds from avicultural collections.

Impact

Snow Geese can be very aggressive when feeding or breeding, and have hybridized with 11 other Anatidae. They therefore pose a potential local threat to native wildfowl.

References

Blair et al. 2000; Delany 1992, 1993, 1995; Holling & RBBP 2007; Lever 2005; Mead 2000; Ogilvie & RBBP 1999–2004; Rowell et al. 2004; Welch et al. 2001.

ANATIDAE (DUCKS, GEESE & SWANS)

EMPEROR GOOSE *Anser canagicus*

Natural Distribution NW Alaska and NE Siberia. Winters in Aleutian Islands and the Alaska Peninsula, and in Asia S to Commander Island and Kamchatka.
Naturalized Distribution England.

EMPEROR GOOSE

England

In 1991, Delany (1993) found 14 Emperor Geese at 14 localities, and 9 years later Rowell et al. (2004) recorded the same number at 5 sites. In 2000 and 2001, they bred successfully on Walney Island off Barrow-in-Furness, Cumbria. No breeding was reported in 2002, but in 2003 out of a flock of 21, 2 pairs successfully fledged 3 young, and in the following year a single pair raised 2 goslings; in 2005, 2 pairs succeeded in rearing 3 young.

In 2001 breeding by Emperor Geese was also recorded in Surrey.

Reference

Blair et al. 2000; Delany 1993; Holling & RBBP 2007; Ogilvie & RBBP 2004; Rowell et al. 2000.

ANATIDAE (DUCKS, GEESE & SWANS)

GREATER CANADA GOOSE *Branta canadensis*

Natural Distribution Greater Canada Geese breed in the Bering, N Kuril and Aleutian Islands, and in much of mainland N America. They winter in Japan, SW Canada and the USA S to Texas, and in Mexico. The British nominate race breeds in E Canada and winters in the E USA.
Naturalized Distribution England; Scotland; Wales; Ireland.

England

Greater Canada Geese (formerly known as Canada Geese) of the nominate subspecies[1] have been kept in captivity in Britain for well over 300 years. The earliest mention of them in England appears to refer to birds in the collection of Charles II (r 1660–85) in St James's Park in London. Here, on 9 February 1665, John Evelyn saw, according to his *Diary*: 'Numerous flocks of severall sorts of ordinary and extraordinary wild fowle … which for being so neere so great a Citty, and among such a concourse of souldiers and people, is a singular and diverting thing.'

Francis Willughby and John Ray in their *Ornithologia* (1676–8) saw Greater Canada Geese in the royal collection before 1672: 'the name shows the place whence it comes. We saw and described both this and the precedent [the Spur-winged Goose of Africa] among the King's wildfowl in St James's Park.'

Kirby and Sjöberg (1997) assert that introductions were made 'from *c.* 1650', but provide no evidence in support of this claim.

GREATER CANADA GOOSE FAMILY

The earliest record of the species in the wild in Britain seems to be of one that was shot on the Thames at Brentford in Middlesex in 1731. In 1785, John Latham wrote: 'This species is now pretty common in a tame state ... in England. ... they are thought a great ornament to the pieces of water in many gentlemen's seats, where they are very familiar and breed freely.'

In the last decade of the 18th century, the father of Charles Waterton (1782–1865), the pioneer conservationist, kept a flock of 13 at Walton Hall in Yorkshire. 'The fine proportions of this stately foreigner,' wrote Waterton in his *Essays on Natural History* (1838), 'it's voice and flavour of its flesh are strong inducements for us all to hope that erelong it will become a naturalized bird throughout the whole of Great Britain.'

Other 19th-century estates in England that are known to have had Greater Canada Geese include those at Lilford Hall in Northamptonshire and Bicton House near Exeter in Devon. In 1844, a pair nested at Groby Pool in Leicestershire, and William Yarrell in his *History of British Birds* (1845) refers to free-flying birds that were 'obtained in Cambridgeshire, Cornwall, Derbyshire, Devonshire, Hampshire, Oxfordshire and Yorkshire'. In 1885, a pair nested in the wild in Edgbaston Park on the outskirts of Birmingham, Warwickshire, and a second pair bred at Gerandon Pond in Leicestershire, where Alexander Montagu Browne, in *The Vertebrate Animals of Leicestershire and Rutland* (1889), described Greater Canada Geese as 'an introduced species, often found at large, especially in winter, and roaming so far afield as to give rise to the doubt if it may not soon become feral'. In 1892, two pairs of the species bred at Rydal Water in Westmorland (Cumbria), and three years later M.A. Mathew and W.S.M. D'Urban, in their *Birds of Devon*, wrote of Greater Canada Geese as 'breeding freely and wandering at will over the country'.

During the 20th century, Greater Canada Geese were liberated at several places in England, including Radipole Lake near Weymouth in Dorset (before 1932); West Wycombe Park in Buckinghamshire by Sir John Dashwood, Bt in 1933–4; the Sevenoaks/Tunbridge Wells area of Kent by the Wildfowlers Association of Great Britain and Ireland in the 1950s; Frampton in Gloucestershire by the then Wildfowl Trust in 1953; and (from Stapleford Park, Leicestershire, and Swinton Park, Yorkshire) in Staffordshire and on the Serpentine in Hyde Park, London in 1955.

In 1953, N.G. Blurton-Jones for the British Trust for Ornithology, and in 1967–9 M.A. Ogilvie for the Wildfowl Trust, conducted censuses of naturalized Greater Canada Geese in Britain and Ireland. The former calculated the English population at 2,670–3,482 individuals, while the latter estimated it at a minimum of 10,090, distributed in discrete subpopulations in almost every county.

As late as the 1940s, Greater Canada Geese occurred principally only in parks and lakes of large private estates. By the 1970s, however, partially as a result of human translocation, they also became established on reservoirs, flooded gravel-pits, rural meres, urban lakes, marshes and sluggish rivers. By the middle of the following decade, the British population had increased to perhaps 34,000, occurring in England mainly in the south-east, East Anglia (where, however, in Norfolk the numbers fell by some 50 per cent in the 1960s), Derbyshire, the west Midlands, Yorkshire, the Lake District of Cumbria and south Devon. By around 1990, the British population had risen to some 59,500–64,000; in 1989–2000 it increased by a staggering 166 per cent (9.3 per cent p.a.). The latest published report gives a total British population of some 90,000. The increase has occurred principally in places where the population density was low, especially in waterless lowland habitats. Both the population and its distribution continue to expand.

Scotland

The earliest mention of Greater Canada Geese in Scotland refers to their presence in the 19th century on the Earl of Wemyss's estate at Gosford House, Longniddry, in East Lothian, and William Yarrell in his *History of British Birds* (1845) states that free-flying birds 'have been taken [i.e. shot] in Scotland'.

By the outbreak of the Second World War, Greater Canada Geese were well established in the Tay and Forth areas, though scarcer further north. During the war the birds markedly declined when winter feeding had to be abandoned, and by 1953 the population had shrunk to only 150, most of which were in Dumfriesshire.

In their Greater Canada Goose surveys of 1953 and 1967–9, N.G. Blurton-Jones and M.A. Ogilvie estimated the Scottish population at 147–264 and 100 respectively, mostly in Dumfriesshire, Kinross-shire, Midlothian and East Lothian, Perthshire, Morayshire, Inverness-shire and Renfrewshire, and on the island of Colonsay off the west coast. This distribution remains much the same today, although birds are now colonizing upland regions in central Scotland. The main populations are on the North Solway Plain and on Loch Leven.

From time to time, isolated individuals have occurred on islands off the west and north coasts (e.g. Islay, South Uist, the Inner and Outer Hebrides, Orkney and Shetland). These are believed to be natural migrants from eastern North America, possibly via Greenland, where they join up with Greater White-fronted Geese and Barnacle Geese before continuing their journey to Britain and Ireland. These transatlantic vagrants are much smaller than the nominate Greater Canada Goose, and are believed to be Richardson's Canada Goose *B. c. hutchinsii* (Arctic central Canada) and/or the Lesser Canada Goose *B. c. parvipes* (central Alaska and central Canada). A few hybrids have also been observed.

Wales

What may have been the earliest introduction of Greater Canada Geese to Wales was made at Leighton Park, Montgomeryshire in around 1908. In 1955, a number were introduced from Stapleford Park, Leicestershire and Swinton Park, Yorkshire to Pembrokeshire and the Isle of Anglesey, Gwynedd. N.G. Blurton-Jones in 1953 recorded 68–84 in Montgomeryshire, of which 30–50 occurred in Powis Park; in 1967–9 M.A. Ogilvie reported a total of some 660 (20 in Newport, Monmouthshire; 40 in Fowborough, Pembrokeshire; 400 in the Severn Valley near Welshpool; and 200 on Anglesey). Today, Greater Canada Geese remain sparsely distributed in the principality.

Ireland

It is believed that Greater Canada Geese may have been first introduced to Northern Ireland in the early 19th century, from where they later spread south into the Irish Republic. Nevertheless, up to around 1900 very few wildfowl collections in Ireland held Greater Canada Geese, and records were infrequent, on the east coast and almost invariably between January and June. This has led to the suggestions that they may have been the result of escapes from English wildfowl collections.

In 1912–20, a free-flying flock of up to 20 existed in the grounds of Dublin Zoo, but the birds failed to establish any satellite populations away from the zoo. A small number occurred on St Ann's estate, Clontarf, near Dublin until 1914, but thereafter disappeared. In 1935, some were translocated from Baronscourt to Caledon, Co. Tyrone, and at around the same time others were reported on Lough Neagh and in Ward Park, Co.

Down. N.G. Blurton-Jones recorded a population of 47–120 in 1953. In 1967–9, M.A. Ogilvie found a total of 229, 70 of which were on Strangford Lough, Co. Down. Merne (1970) reported the species in Co. Cork (the Lough, c 55); Co. Wexford (Burrow, Rosslare, 4) and North Slob (2); Dublin (St Stephen's Green, 27); Co. Down (Strangford Lough, Castle Ward, Killyleagh, Ringdufferin, Newtonards, over 110); Tollymore, Newcastle, Castlewellan (1 each); Kircublin (2); Co. Antrim (Ballymoney, 8); Co. Tyrone (Caledon, 3); and Co. Fermanagh (Ely Lodge, Enniskillen, 32, and Irvinestown, 6).

By the late 1980s/early 1990s, it was estimated that the Irish population may have numbered 575–700. By the end of the latter decade, most of the population (which is believed to be increasing at a rate of some 3 per cent p.a.) of 500–650 adults occurred on the Woodford River in Co. Cavan and Fermanagh. Today, most of the 23 breeding sites are in Northern Ireland, where over 530 of the total population of 970 occurs on only 8 sites. The rate of increase of the population and its distribution is slower in Ireland than in Britain.

As on the islands off the west and north coasts of Scotland, smaller races have occurred in Ireland, as transatlantic vagrants from North America via Greenland. Only 10 were recorded before 1900 and none between then and 1954. Since that year, however, there have been more frequent records, principally on the North and South Slobs in Co. Wexford, but also in Counties Down, Offaly, Sligo and Longford.

Characteristics

It was not until the 1930s that Greater Canada Geese started to live predominantly in the wild in Britain and Ireland, and within 20 years they began increasingly to come into conflict with farmers due to their predilection for newly sown grass leys, spring and winter wheat and sugar beet, as a result of which the removal of adult birds during the time of moult and of unfledged young to hitherto unoccupied areas was undertaken as an attempted form of control. This redistribution almost invariably resulted in the formation of new sub-populations in previously unoccupied areas, and the donor colonies soon resumed their former numbers, since previous non-breeders now had sufficient space for nesting. This policy of translocations was largely responsible for the widespread increase in numbers and distribution during the 1970s and 1980s.

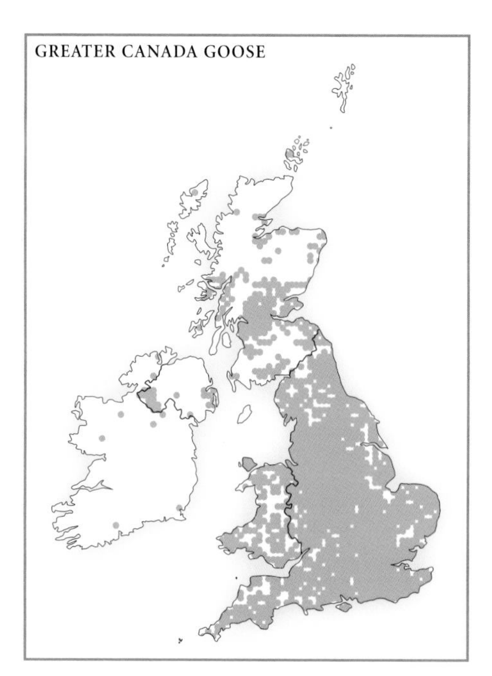

The principal reasons for the successful naturalization of Greater Canada Geese in Britain and Ireland are the existence of a near-vacant ecological niche for a large aquatic bird that breeds on waters in open woodlands and urban and suburban parks (where its only competitor is the Mute Swan), the availability of suitable habitats and the species' abandonment of the instinct to migrate.

The birds have also been helped by the creation of new and favourable habitats, such as reservoirs and gravel-pits; changes in land-management practices, e.g. the 'improvement' of marginal pasturage and an increase in the hectarage of winter wheat, may also have contributed to the population expansion by providing an additional food supply in late winter.

Greater Canada Geese are most abundant in the Midland counties and south-east of England, which partly reflects their slow rate of dispersion from the sites where they occurred in the past, and partly the availability of reservoirs, gravel-pits, lowland lakes and urban waters. They are, however, now increasingly found in moorland regions and on upland water bodies (both natural and artificial), and are also increasing in density in occupied areas.

Adult birds have few predators apart from humans, though American Mink, Eurasian Otters and Mute Swans may sometimes take eggs and goslings. Greater Canada Geese are long-lived and have a high potential breeding rate, so their sedentary nature is a major factor controlling the population growth and dispersal. High densities cause maximum concentrations of birds and of territory size, and restrict the number of pairs able to reproduce through shortage of space. In North America (and in some parts of continental Europe), Greater Canada Geese are highly migratory. In Britain and Ireland, perhaps because of the more temperate climate, they do not migrate, though there may be some local dispersal in autumn and winter. This abandonment of the instinct to migrate in a species that is migratory in its native range is a major factor in its successful naturalization elsewhere.

Dennis (1964) and Walker (1970), however, established that a moult-migration (which is believed to have existed since 1947), then involving over 300 non-breeding adults and immatures, takes place annually in late May/early June between the Yorkshire sub-population and the Beauly Firth in Inverness-shire, mirroring similar migrations in North America. Many of these birds make the return journey to Yorkshire (a round trip of around 840 km) in late August/early September. Some immature birds do not undertake this northwards moult-migration from Yorkshire but remain behind in northern England. Since the 1970s, birds from some north Midland counties (e.g. Derbyshire and Nottinghamshire) have joined the Yorkshire sub-population on its flight north, which could help to explain the increasing numbers at Beauly. Moult-migrations enable non- and failed breeders to take advantage of rich food sources outside their normal range at an energy-expensive period of the annual life cycle.

Impact

The ecological and economic impacts of Greater Canada Geese in Britain and Ireland have been summarized by Watola et al. 1996.

- On agricultural land, the birds are recognized as a significant, if local, pest. They tend to remain on good feeding sites for a long period of time, which can often lead to cumulative damage to pastures and crops, especially those near water. Damage can also be caused by over-grazing such crops as cereals, oilseed rape and sugar beet. Additionally, spring pasturage can be adversely affected by trampling, grazing and fouling.
- In urban localities, in particular south-east England, urban and rural parks, golf courses and other amenity areas can be damaged by fouling, trampling and over-grazing. Footpaths may become slippery and dangerous to the public, and the birds' faeces may contain pathogens that pose a hazard to human health.

- They contribute to the eutrophication of the waters on which they occur, and the increase in nutrients caused by their droppings can promote the formation of algal blooms, the toxins of which are extremely dangerous to humans, domestic stock and fish.

- The birds have been reported to eat the young shoots and submerged rhizomes of the Common Reed Mace and willows, causing riverbank erosion.

- They may compete with native waterfowl for food and nesting and roosting sites. Anecdotal evidence suggests that, by sheer weight of numbers, they may drive away ducks and Mute Swans, and compete for grazing with Eurasian Widgeon. On the other hand, they give these and other species early warning of potential predators and, by uprooting submerged vegetation, make it available to dabbling ducks.

- They have been involved in a number of aerial bird strikes, both in North America and Britain. Their size, weight and tendency to form large flocks make them a special threat to air traffic where they occur near major airfields.[2]

Watola et al. 1996 also summarized the methods of controlling Greater Canada Geese, which broadly fall into three categories: culling adults, reducing reproductive output and behaviour modification. In Britain, the species is protected under the Wildlife and Countryside Act 1981, so that although during the shooting season Greater Canada Geese may be legitimately shot, control outside the shooting season can only be done legally under licence.

Short-term control is usually achieved by shooting the adults in the shooting season or rounding them up during the moult, when they are flightless, in early July. Shooting has the disadvantage of encouraging survivors to move elsewhere, thus spreading the population even more widely into urban areas where shooting is impossible. Culling during the moult, which is the most humane way of lowering the number of adults, and shooting both have the disadvantage of attracting public opposition. Shooting also tends to drive away non-target species.

In the absence of techniques for the chemical or immunological sterilization of Greater Canada Geese, control of reproduction can be achieved either by shooting adults on the nest (an emotive solution likely to be unacceptable to the public) or by the substitution of real eggs with false ones in the nests of early breeders, and the removal without substitution from late breeders. (The removal of eggs from early nesters without the substitution of false ones would encourage renesting elsewhere, while the eggs removed from late breeders are not replaced.) Alternatively, the eggs can be pricked, hard-boiled or coated with liquid paraffin, which kills the embryo, before being returned to the nest. This method has the advantage of being seemingly more acceptable to the public.

Methods of influencing the birds' behaviour include relocation, scaring, exclusion, habitat modification and the use of chemical repellents. Of these techniques, relocation tends to result in the formation of new sub-populations in hitherto uncolonized areas, or the return of the birds to the site of origin. Scaring by an auditory or visual contrivance is generally unsuccessful in the long term, since the birds quickly become used to the stimuli produced, and is also likely to arose the ire of the neighbouring human population. Physical exclusion by fencing or the use of wire netting over water can be effective, but is time consuming and expensive. Since Red Foxes predate many nests not on islands, excluding the birds from islands and thus forcing them to nest elsewhere can be an effective means of control. Habitat modification, such as the planting of unpalatable ground cover, more shrubs to reduce the openness of sites, and tall trees to adversely affect the angle of flight approach, can also help to reduce local sub-populations. These methods may also, however, impinge on non-target species. Among chemical repellents, methyl anthranilite is both non-toxic to humans and biodegradable, and is known to repel geese, but is only effective in the short term. All these methods of reducing the population of Greater Canada Geese in Britain are likely to prove at best a short-term palliative measure rather than a long-term solution.

References

Austin et al., in Wernham et al. 2002; Austin et al. 2007; Berry 1939; Blair et al. 2000; Blurton-Jones 1956; Browne & O'Halloran 1998; Carter, in Gibbons et al. 1993; Delany 1992, 1993; Dennis 1964; Fitter 1959; Gibbons & Avery 2001; Guthrie 1903; Harrison 1988; Holloway 1996; Kear 1965, 1990; Kirby & Sjöberg, in Hagemeijer & Blair 1997; Lever 1977, 1980, 1987, 1994, 2005, in press; Limentani 1975; Long 1981; Mead 2000; Merne 1970; Ogilvie 1977; Park, in Lack 1986; Rehfisch et al. 2002; Royal Society for the Protection of Birds 2006; Salomonsen 1968; Sangster et al. 2005; Sharrock 1976; Walker 1970; Welch et al. 2001; Wood 1964.

Note

1. It is believed that some British and Irish birds may be of the subspecies *B. c. maxima* (south-central Canada). This form was previously known as the Giant Canada Goose.
2. Greater Canada Geese are believed to have been responsible for the forced landing of a commercial aircraft on the Hudson River, New York, in January 2009.

ANATIDAE (DUCKS, GEESE & SWANS)

BLACK SWAN *Cygnus atratus*

Natural Distribution S Australia and Tasmania.
Naturalized Distribution England; Scotland; [Wales]; [Ireland].

England

First imported from Australia around 1791, Black Swans did not breed in the wild until 1851 at Carshalton, Surrey, and at Bicton, Devon, in the following decade. Since 1902, they have nested intermittently on the Thames, mostly in the vicinity of London and under the nominal tutelage of the Worshipful Company of Vintners, by whom they were acquired from Australia in exchange for Mute Swans. By the mid-1980s, Black Swans, presumably from this source, had been recorded from all over the home counties, especially Berkshire, Essex, Hertfordshire and Surrey, and even occasionally from further afield, but were nowhere properly established.

In the 1990s, several-dozen birds that had presumably escaped or been released from avicultural collections were reported in the wild, and in 1996–7 breeding was recorded in Northamptonshire, Essex and Wiltshire. By the end of that decade, a small breeding population had become established in the Broadlands of east Norfolk (where the species had first been observed before 1980), based on Salthouse Broad and the River Bure at Wroxham, and to a lesser extent on the Trinity Broads. In 1989, a pair bred at Walcott, and in the late 1980s at least one pair nested regularly, though not always successfully, at Salthouse. By 1994, there was a minimum of three breeding pairs in the Salthouse/ Wroxham area, and another pair raised five cygnets at Waxham. Since then Black Swans have been slowly expanding their range in Norfolk, where by the end of the decade the population numbered 20–40.

In 1996–2001, reports of successful breeding in England were received from Cleveland, Devon, Essex, Greater Manchester, Northamptonshire, Sussex and Wiltshire, and the birds in Essex, Northamptonshire and Wiltshire were said to be well established. The highest number of breeding pairs, in 2001, was nine.

Black Swans are known to have bred successfully in England annually in 1996–2005. In 2002, 2 pairs nested in 3 places in 2 counties. In Greater Manchester, 1 pair nested unsuccessfully at Arley Hall, and 2 immatures were seen on Elton Reservoir. In Sussex, a pair was reported with a downy cygnet on Benbow Pond in Cowdray Park. In 4 places (Devon, Lothian, north-east Scotland and Orkney), where breeding occurred in 2001, no breeding was reported in the following year.

In 2003, at least 8, and possibly 14 pairs, nested in Devon (where in July a flock of 10 adults included 1 juvenile); Wiltshire (2 pairs, 1 of which hatched a single young); Essex (1 pair raised 4 young); Greater London (where a pair nested in St James's Park); Kent (1 or 2 pairs, but no breeding observed); Sussex (2 pairs at 2 sites hatched a total of 4 young); Cambridgeshire (present at 2 sites, where 1 out of 2 pairs nested); and Greater Manchester (a single pair fledged 1 young).

In 2004, 11 breeding pairs (a record) were confirmed, with perhaps a maximum of 16 pairs, distributed between Gloucestershire (1 pair bred); Wiltshire (4 pairs bred at 4

BLACK SWAN COURTSHIP DISPLAY

sites); Essex (2 pairs bred at 2 sites); Hertfordshire (at 1 site 2 pairs hatched 12 young between them, of which half survived); Kent (birds were present at 10 sites but no breeding was confirmed); Sussex (no breeding recorded); Cambridgeshire (4 pairs at 3 sites, but only 1 pair laid eggs, 4 in total, none of which hatched); Lincolnshire (a single pair fledged 4 young); and Lancashire and North Merseyside (1 pair present but there was no evidence of breeding).

In 2005, 8 pairs are known to have bred, and possibly a further 5. In Wiltshire, 5 pairs (of which 2 bred) occurred at 4 sites; in Hertfordshire, 2 pairs were present at 2 sites but did not breed; in Kent, 1 pair with 2 immatures was seen in July; in Sussex, 2 pairs raised 4 young at 2 sites; in Staffordshire 2 pairs bred; and in Greater Manchester a single pair fledged 4 young.

Scotland

In 1996–2001, reports of successful breeding by Black Swans were received from Lothian, Orkney, and the north-eastern mainland, and the birds in Lothian were said to be well established. However, although the cob from the breeding pair in Lothian in 2001 remained in 2003 and a replacement pen was introduced, no breeding was observed in that year and there have been no subsequent records from the site. Also in 2003, a pair remained at the former breeding site on Orkney, but no breeding occurred. In 2004, in Dumfries and Galloway, a pair was present early in the year and three in September, and breeding was suspected.

[Wales; Ireland]

Black Swans have been recorded in the wild at a number of locations in Wales, mostly in the south, west and north, including the Isle of Anglesey. In Northern Ireland they have occurred on Loughs Neagh and Carlingford and in Co. Armagh, and in the Irish Republic in Co. Cork.

Summary of Status

Many county bird reports refer to non-breeding individual Black Swans present over lengthy periods, and the figures given above, especially for 2005, may underestimate the total population.

References

Allard 1999; Blair et al. 2000; Fitter 1959; Holling & RBBP 2007; Lever 1987, 2005; Mead 2000; Ogilvie & RBBP 1999–2003.

ANATIDAE (DUCKS, GEESE & SWANS)

EGYPTIAN GOOSE *Alopochen aegyptiacus*

Natural Distribution Sub-Saharan Africa. (Formerly also SE Europe.)
Naturalized Distribution England; [Scotland]; [Wales]; [Ireland].

England

Egyptian Geese were commonly kept in captivity in continental Europe in the 17th century (or perhaps even earlier). The species was first seen in England in the collection of Charles II (r 1660–85) in St James's Park in London, where it was referred to by Francis Willughby and John Ray in their *Ornithologia* (1676–8) as 'The Gambo-Goose or Spur-Wing'd Goose [*Plectropterus gambensis* – also of Africa]', although the accompanying illustration (Plate LXXI) clearly shows the Egyptian Goose. In 1785, it was described by John Latham as: 'common at the Cape of Good Hope, from whence numbers have been brought to England: and are now not uncommon in gentlemen's ponds in many parts of this Kingdom'.

In January 1795, an Egyptian Goose was shot at Thatcham in Berkshire, and this may be the earliest record of the species in the wild. In 1803 or 1804, a further six were shot near Buscot in the same area, and in 1808 two were killed at Harleston in Norfolk, the earliest record in the wild in East Anglia. In his *Diary*, the wildfowler Colonel Peter Hawker noted that he: 'Killed 2 [Egyptian Geese] in Norfolk and 3 at Longparish in Hampshire in the winter of 1823, and the next year again during some tremendous gales from the west, a flock of about 80 appeared near the same place when 2 more were killed.'

In April 1830, a flock of five Egyptian Geese was seen on the 'Fern' (Farne) Islands in Northumberland, and until 1880 small numbers occurred regularly in winter on the nearby Fenham Flats, where they were known locally as 'Spanish Geese'. These birds probably came from Gosford House in East Lothian, as did those seen on the River Tweed. Other references to Egyptian Geese in the wild during the first half of the 19th century include Dorset (1836); the Isle of Man (1840); the River Severn near Bridgwater in Somerset (1840); near Leicester (1843); Romney Marsh in Kent (1846); Blenheim Palace and Shelswell, Oxfordshire (1847); Shoreham and Shermanbury, Sussex (1848); Ormsby Broad, Norfolk (1848); and Derwent Water, Cumberland (Cumbria) (1849).

By the middle of the 19th century, a number of free-flying colonies were established on 'gentlemen's ponds', mainly in southern and eastern England. The largest flocks were those at Blickling Hall, Gunton Park, Holkham Hall (Earl of Leicester) and Kimberley Hall (Earl of Kimberley) in Norfolk; at Bicton and Crediton in Devon; and at Woburn Abbey in Bedfordshire (Duke of Bedford). From these and other estates, birds frequently strayed to other parts of the country. In their *Birds of Devon* (1895), M.A. Mathew and W.S.M. D'Urban wrote: 'The Egyptian Goose … [is] frequently kept on ponds in a semi-domesticated state, breeding freely and wandering at will over the country'.

In 1934–7, Egyptian Geese were recorded at Hamper Mill in the Colne Valley, Hertfordshire, at Reading in Berkshire in 1935, and on Connaught Water in Epping Forest, Essex in 1936. In the following decade a single bird lived on Ruislip Reservoir in Middlesex in 1941–2.

In 1954–6, they bred (or were at least seen) on Hillington Lake, Stradsett Lane and Beeston Hall, all near King's Lynn, Norfolk. In the same county others were present in

Gunton Park, Fustyweed Gravel-pits and Salhouse Broad. By the end of the decade, the largest full-winged flocks were those at Holkham and Beeston Halls, with smaller numbers at Woodbastick near Salhouse (all Norfolk) and Woburn Abbey, Bedfordshire. By the 1960s, the post-breeding population was estimated at 300–400.

By the 1970s, Egyptian Geese occurred mainly in Norfolk in the Holkham/Beeston area and in the valley of the River Bure. In 1971–5, sub-populations or individuals were recorded at Barton, Beeston Hall and Beeston St Lawrence, Blickling, Cley, Didlington, Felbrigg, Gunton, Hillington, Holkham, Horeton, Houghton, Lenwade, Lexham, Lyng, Narborough, Narford, Salhouse, Sparham, Stradsett, Swanton Morley, West Acre, West Newton and Wroxham. There had also been an expansion of range to the Brecklands of Suffolk. In September 1973, a total of 98 was counted at Holkham, and in the following February 141 at Beeston Hall. In Suffolk in the spring of 1973, a pair was present on Fritton Lake near the coast, and others were seen at Sotterley. In the following year a pair

EGYPTIAN GOOSE WITH YOUNG

Table 9 Breeding and non-breeding records of Egyptian Geese *Alopochen aegyptiacus* in England, 2003–5

Year	County	Sites	Pairs	Breeding Results
2003	Hampshire	1	1	9 young fledged
2003	Buckinghamshire	2	2	Did not breed
2003	Essex	?	16	3 pairs bred
2003	Greater London	2	4	23 young fledged
2003	Hertfordshire	2	2	10 young fledged
2003	Kent	1	1	Did not breed
2003	Surrey	?	14	3 pairs bred; 1 pair probably bred; 10 pairs possibly bred
2003	Lincolnshire	?	5	3 pairs bred; 1 pair probably bred; 1 pair possibly bred
2003	Norfolk	43	?	Bred at all sites
2003	Suffolk	19	30	8 pairs bred; 1 pair probably bred; 21 pairs possibly bred
2003	Leicestershire	1	11	54 young fledged
2003	Warwickshire	1	1	Possibly bred
2004	Hampshire	1	1	9 young fledged
2004	Buckinghamshire	1	1	7 young fledged
2004	Essex	3	5	3 pairs bred; 2 pairs possibly bred
2004	Greater London	?	?	At least 4 pairs bred
2004	Hertfordshire	1	1	Possibly bred
2004	Kent	1	1	Did not breed (but autumn counts suggest local breeding)
2004	Surrey	?	10	2 pairs bred; 8 pairs possibly bred
2004	Lincolnshire	1	2	1 pair fledged 3 young; 1 pair probably bred
2004	Norfolk	38	?	Bred at all sites
2004	Suffolk	At least 28	?	15 pairs fledged at least 20 young; 10 pairs possibly bred
2004	Leicestershire	1	?	At least 9 pairs fledged at least 14 young
2005	Hampshire	3	3	25 young fledged
2005	Devon	1	1	6 young fledged
2005	Berkshire	?	14	All bred; now occur widely in county
2005	Buckinghamshire	2	2	9 young fledged
2005	Hertfordshire	1	1	3 young hatched, but not all survived
2005	Norfolk	30	?	Breeding pairs on all sites
2005	Suffolk	?	13	All bred and 7 young reported, but believed to be under-recorded
2005	Leicestershire	1	At least 11	At least 33 young hatched, of which at least 26 fledged
2005	Nottinghamshire	3	7	?

Source Adapted from Holling & RBBP 2007.

bred near Leiston. Away from East Anglia, a pair raised three goslings on Effingham Pond in Surrey.

By the mid-1980s, the population had risen to around 500, of which some 80 per cent occurred in Norfolk, and by the turn of the decade to 750–900, of which the vast majority were in Norfolk. By the late 1990s, the English breeding population was believed to number around 380–400 pairs, and by the turn of the century there was thought to be a total of approximately 950 adults in England, including some 300 breeding pairs.

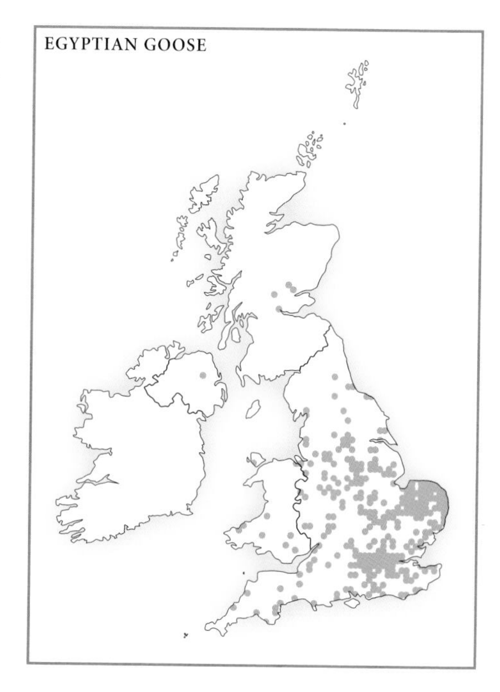

Ogilvie and RBBP (2004) recorded Egyptian Geese in 2002 in at least 43 localities in 7 counties, with a minimum of 33 breeding pairs. In Berkshire, breeding was reported from Windsor Great Park (2 pairs), the River Thames at Caversham (2 pairs), Dinton Pastures Country Park (4 pairs) and Theale Gravel-pits (2 pairs), with young hatched at each locality. In Greater Manchester, a pair fledged 2 young at Compstall Water, while in Hampshire up to 6 pairs were seen at 4 sites, but no breeding was confirmed. In Leicestershire, 5 pairs fledged 15 goslings on Rutland Water. In Norfolk, an unrecorded number of pairs bred at 17 sites, while in Suffolk up to 17 pairs nested at 14 sites. In Warwickshire single pairs were seen at two localities, but no breeding was observed.

Thus in 2003 at least 78 (and possibly 130) pairs of Egyptian Geese nested successfully in England; the totals for 2004 were 70 and 97, and for 2005 at least 82 pairs. The maximum total in 2003 exceeded the previous highest total (127 pairs) in 2000. Data from some previously recorded counties were omitted in 2005, although data for Berkshire and Nottinghamshire were included for that year only. The true position, therefore, remains unclear, and it is likely that the number of Egyptian Geese is significantly under-recorded. In general, the impression is of a stable population concentrated in Norfolk and, to a lesser extent, Suffolk, with smaller sub-populations elsewhere and some westwards expansion of range. In Norfolk the expansion of range increased from 6 per cent of tetrads in 1980–5 to 27 per cent by 2000–5.

Recent estimates of the population have varied considerably from only 575 (of which 444 were in Norfolk) to 750–800 pairs in 2002 – surely an overestimate. It is expected that the 2007–11 *Atlas* (in prep.) will provide a more accurate and definitive account of the status, distribution and population in England. Meanwhile, Tomlinson's 2006 'guestimate' of around 1,000 individuals may be not far from the mark.

[Scotland]

What may be the earliest record of the Egyptian Goose in Scotland refers to three shot in Fir Glen, Campsie, 14 km north of Glasgow in November 1832. By the mid-19th century, the species was established at Gosford House in East Lothian.

[Wales]

The map in Hagemeijer and Blair (1997) indicates the occurrence of Egyptian Geese in two localities on the Isle of Anglesey in Gwynedd (and also on the Isle of Man, England). No other *Atlas* maps show the species in either locality.

[Ireland]

Small parties of Egyptian Geese have from time to time been reported in Ireland. These have probably been escapes or releases from private and unrecorded avicultural collections, or possibly vagrants from the English population.

Characteristics

In Africa, Egyptian Geese frequent rivers, lakes, dams and marshes, and use a wide variety of nesting sites – including the ground (in reed beds, matted vegetation near water, mammal burrows and among rocks and boulders), inland cliffs, ruined buildings and trees (in natural holes or crevices, or in the abandoned stick nests of other species), where they often nest and perch up to 25 m agl. In England, they seem to breed principally in the old nests of such species as Common Buzzards and Carrion Crows, and to favour much the same habitat as Greater Canada Geese – lakes in wooded parks, broads, breckland meres, pastures and gravel-pits.

Why, then, has the Egyptian Goose been so much less successful as a colonist in England than the Greater Canada Goose, despite the fact that both were originally introduced at much the same time? It may be that, coming from warmer climes, ecological conditions in Britain are verging on marginal. Although most adults survive the winter well, breeding begins in early spring (February and March) when (especially in East Anglia) weather conditions can still be severe, and in some years the survival of goslings (which may take over 30 days to fledge) is low. Furthermore, goslings suffer from predation by Carrion Crows and other species, and adults from competition with Greater Canada Geese and Greylag Geese. Egyptian Geese probably share with another potential colonist, the Ruddy Shelduck, the problem of a shortage of sufficiently large tree cavities for nesting.

In England, Egyptian Geese moult during July and August, when family groups gather at a small number of local sites. These assemblies are fairly loose, with families and non-breeding pairs discrete. The principal moulting sites are at Holkham Hall, Blickling Hall, Sennowe Park and Lynford, all in Norfolk.

References

Atkinson-Willes 1963; Harrison 1988; Holloway 1996; Kear 1990; Lever 1977, 1987; Long 1981; Mead 2000; Ogilvie & RBBP 2004; Prater, in Gibbons et al. 1997; Rowell et al. 2000; Sharrock 1976; Sutherland & Allport 1991; Taylor, in Lack 1986; Tomlinson 2006; Tomes, in Wernham et al. 2002; Venema, in Hagemeijer & Blair 1997.

ANATIDAE (DUCKS, GEESE & SWANS)

RUDDY SHELDUCK *Tadorna ferruginea*

(BRAHMINY DUCK) (India)

Natural Distribution SE Europe, NW and NE Africa, SW and C Asia; winters in S Europe, N Africa, S and E Asia.
Naturalized Distribution England; [Scotland]; [Wales]; [Ireland].

England

Since the 1950s, there have been over 100 records of Ruddy Shelducks[1] in Britain, and Ireland, with occasional breeding of 1–2 pairs. These birds are likely to be mostly, if not entirely, releases or escapes from avicultural collections rather than natural migrants from south-eastern Europe or north Africa. In 1996, a male Ruddy Shelduck on Rutland Water, Leicestershire, successfully mated with a female Egyptian Goose. All other reports of Ruddy Shelduck breeding in England refer to Norfolk, where in 1996, 1997 and 2000 single pairs bred successfully. There were no reports of breeding in 2001 or 2002, but in 2003 up to two pairs may have bred at two sites. In 2004, a pair hatched five young, but

RUDDY SHELDUCK IN SUMMER PLUMAGE

these did not survive. In 2005 the species was observed at 22 sites, at 3 of which 3 pairs hatched a total of 21 young.

[Wales]
The species has occurred in Flintshire.

[Scotland]
From time to time Ruddy Shelducks are reported in the wild in Scotland, but no breeding is known to have occurred.

[Ireland]
Occasional breeding by introduced birds has been recorded in the wild in Ireland.

References
Blair et al. 2000; Hallman et al., in Hagemeijer & Blair 1997; Harrison 1988; Holling & RBBP 2007; Lever 2005; Mead 2000; Ogilvie & RBBP 1999.

Note
1. 'The Ruddy Shelduck has been found [in a cave in Devon] in the Ipswichian interglacial, and at Pin Hole in Derbyshire during the last glaciation' (Harrison 1988:16). The species is, however, treated here as an introduced rather than a reintroduced species.

ANATIDAE (DUCKS, GEESE & SWANS)

WOOD DUCK *Aix sponsa*

(NORTH AMERICAN WOOD DUCK: SUMMER DUCK; CAROLINA DUCK)

Natural Distribution E half of USA and S Canada, wintering in S and SE states. An entirely separate population in W North America from British Columbia to California.
Naturalized Distribution England; [Scotland]; [Wales]; [Ireland].

England

The Wood Duck was first introduced to England in the 1870s at Bicton House, near Exeter, in Devon, which then had possibly the largest collection of exotic waterfowl in the country, and from where free-flying Wood Ducks ranged as far afield as Plymouth and Slapton Ley. M.A. Mathew and W.S.M. D'Urban, in their *Birds of Devon* (1895), wrote of the 'Summer Duck' as 'breeding freely and wandering at will over the country'. The Duke of Bedford at Woburn, Viscount Grey of Fallodon in Northumberland and Alfred Ezra at Foxwarren Park in Surrey are among the landowners who kept unpinioned Wood Ducks on their waters during the early 20th century.

In the late 1960s, a pair of full-winged Wood Ducks bred at Puttenham Common in Surrey. Birds presumably from this source were reported in the Cutt Mill/Puttenham area for several years thereafter, and by 1972 there were believed to be upwards of ten pairs in the locality.

In 1969, five Wood Ducks were observed on Virginia Water on the Berkshire/Surrey border and on a number of other lakes in Windsor Great Park, and in 1973 four ducklings successfully fledged on Virginia Water.

In 1970, a pair with half-a-dozen well-grown young was seen on a sewage farm near Guildford in Surrey. Apart from a pair that bred at Cobham in the late 1940s, this was then the only breeding record for the area.

In 1964–70, small parties were occasionally seen on Vann Lake between Dorking (Surrey) and Horsham (Sussex), where an unsuccessful breeding attempt was apparently made in May 1968.

In 1972, up to 50 free-flying post-breeding birds were counted at the Tropical Bird Gardens at Rode in Somerset, where breeding in the wild is believed to have taken place.

In 1973, Wood Ducks bred on the estuary of the River Duddon near Broughton in Cumberland (Cumbria). They are believed to have originated in avicultural collections at either Drigg (19 km to the north-west) or Millom near the mouth of the estuary. In the late 1970s, Wood Ducks were unsuccessfully introduced to Grizedale Forest between Coniston Water and Windermere in north Lancashire.

In the winter of 1972–3, full-winged Wood Ducks were reported near East Dereham in Norfolk, and breeding is believed to have subsequently taken place near the Wash.

In 1972–5, adults with ducklings were seen on Bishop Offley Mill Pool near Eccleshall in Staffordshire, where the birds were thought to be nesting on a nearby private estate.

In 1974, the nest of a Wood Duck was found at Cranbury Park near Winchester in Hampshire, a few kilometres away from a bird farm at Leckford.

WOOD DUCK MALE IN BREEDING PLUMAGE

Since then, sporadic breeding in the wild has been frequently reported, the majority of records coming from Wiltshire where, in 2002, a pair fledged a single duckling at Stanton Park, which has since become the most regularly occupied site; 6–8 non-breeding birds were there in 2003–5. In 2002, a pair was reported in apparently suitable breeding habitat, and in 2003 one pair possibly bred in Lincolnshire and two pairs may have bred at one site in Cheshire and Wirral. Other breeding has been reported from Berkshire, Devon and Kent; in the last-named county, 24 Wood Ducks are said (by Mead 2000) to have been counted in 1996. Mead asserts (p.140) that in Kent 'there may be two or three hundred [Wood Ducks] currently at large'; this must surely be an over-estimate.

[Scotland; Wales; Ireland]
Non-breeding Wood Ducks have occurred in Perth and Kinross, in Monmouthshire and in Co. Dublin.

Prospects
The majority of breeding records of Wood Ducks in England seem to involve birds that have bred in the vicinity of avicultural collections to which they return, possibly for supplementary winter feeding.Why then has the species failed to become as firmly established in Britain as its near congener, the Mandarin Duck? The former has a longer fledging period (10 weeks) than the latter (8 weeks), which would make its ducklings vulnerable for a longer period to avian and terrestrial predators. Most Wood Duck breeding records have occurred in Wiltshire, Surrey and east Berkshire, where they have to compete for food and nesting holes with the well-established and far more numerous Mandarin Duck. Furthermore, the availability of old, decaying trees with suitable nesting cavities has decreased in recent years due to 'improved' forestry management. Another difference between Wood Ducks and Mandarin Ducks is that, although they are closely related (they are the only two species in the genus *Aix*), they come from different continents and different temperate latitudes. The Nearctic Wood Duck occurs at 13–51°N, while the Oriental Mandarin Duck lives at 36–55°N. The latitude of Britain and Ireland (most of which lies at approximately 51–58°N) is closer to that of the Mandarin Duck. This latitudinal variation, albeit small, is likely to affect laying date, clutch size, duckling weight, incubation and (as previously mentioned) fledging periods, and the timing of both moult and eclipse plumage.

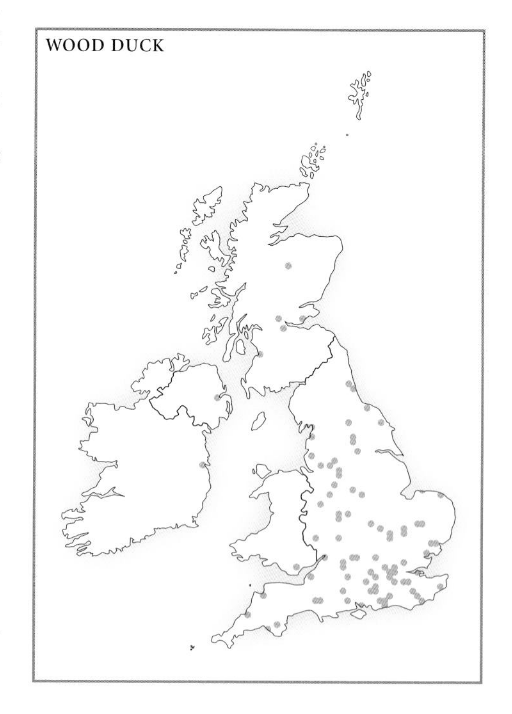

From records kept by the then Wildfowl Trust at its centres at Slimbridge in Gloucestershire (52°N) and Peakirk in Northamptonshire (53°N), the Wood Duck consistently lays its eggs earlier than the Mandarin Duck. Earlier nesting evolved to suit the Wood Duck's region of origin, but in more northerly Britain results in the birds incubating eggs and hatching ducklings when the temperature is still relatively low. Furthermore, since Wood Duck eggs are on average 5 g lighter than those of Mandarin Ducks, the former's ducklings hatch around 16 per cent lighter than the latter's, and thus lose body heat more rapidly and require more brooding by the female.

Additionally, Wood Ducks that escape or are released into the wild in Britain are the descendants of many generations of captive-bred birds, and have thus lost much of their genetic variability, whereas British Mandarin Ducks are descended from vigorous wild-caught stock. It may well be that other, at present unknown, ecological factors are also involved. On the other hand, Wood Ducks have three points in their favour. Firstly, because they nest earlier than most other hole-nesting species, they have a considerable advantage over the latter for nesting sites. (However, Mandarin Ducks sometimes 'dump' their eggs in the nests of Wood Ducks, leaving the latter to rear the young together with their own.) Secondly, since both ducks cease to lay around mid-May, the Wood Duck has a four-week longer breeding season than the Mandarin Duck. Finally, because the Wood Duck has a slightly larger clutch (on average 10–14) than the Mandarin Duck (around 9–12), it can produce more, though lighter, ducklings.

These apparent advantages, however, have yet to enable the Wood Duck to become firmly established in Britain, and it seems extremely unlikely that the fragile wild population, such as it is, would survive without regular recruitment by way of releases or escapes from avicultural collections.

References

Bellrose & Holm 1994; Fitter 1959; Holling & RBBP 2007; Kear 1990; Lever 1977, in Gibbons et al. 1993, in Hagemeijer & Blair 1997; Mead 2000; Ogilvie & RBBP 2004; Risdon 1973; Savage 1952; Sharrock 1976; Shurtleff & Savage 1996; Wheatley 1970.

ANATIDAE (DUCKS, GEESE & SWANS)

MANDARIN DUCK *Aix galericulata*

Natural Distribution SE Siberia, the Russian Federation, Korea, E China and Japan; winters S to around 40°N.
Naturalized Distribution England; Scotland; [Wales]; Ireland.

The Mandarin Duck,[1] which is known in China as *Yüan Yang* and in Japan as *Oshidori*, apparently received its vernacular name from 17th-century English merchants out of their sense of its superiority over other ducks, implied in the title 'Mandarin' – in Chinese *Kwan* or *Kwūn* – a public official, counsellor or minister of state.[2] During the period of the Ch'ing Dynasty (1644–1912), 'mandarin' was the literary, legal and official form of the Chinese language.

The earliest references to the Mandarin Duck in literature are to be found in the writings of the disciples of Confucius (551–478 BC), and the earliest representations date from the period of the Chinese Sung Dynasty (AD 960–c 1279). The birds are frequently depicted swimming beneath a peony, the emblem of spring and love. They are also well represented in Oriental prints, pottery, porcelain, textiles, embroidery and sculpture in a variety of media.

England

The earliest mention of the Mandarin Duck in Europe appears in the writings of the Italian professor of natural science at the University of Bologna, Ulisse Aldrovandi, who in his *Ornithologia* (1599–1601) describes a bird he refers to as *Querquedula*[3] *indica* that was portrayed in a painting brought to Rome by Japanese envoys.

The Mandarin Duck was first introduced to England shortly before 1745, when a drawing of *La Sarcelle de la Chine* ('The Chinese Teal') in the gardens of Sir Matthew Decker, Bt (who, as a Director of the East India Company, imported many species of alien animals and plants to England) at Richmond Green in Surrey, was made by George Edwards for his *Natural History of Birds.* Edwards also quotes from *History of Japan* (1727) by Engelbrecht Kaempfer, one of the first Europeans to visit that country:

> Of ducks there are several kinds, and as tame as the geese. One kind particularly I cannot forbear mentioning, because of the surprising beauty of its male, call'd 'Kinmodsui', which is so great, that being shewed its picture in colours, I could hardly believe my own eyes, till I saw the Bird itself being a very common one.

The *Encyclopaedia Britannica* (1777) refers to 'The *galericulata*,[4] or Chinese teal of Edwards, which has a hanging crest. ... The English in China give it the name of "mandarin" duck.' The French naturalist George-Louis Leclerc, Compte de Buffon, included an illustration of a Mandarin drake in his *Histoire Naturelle des Oiseaux* (one of the *tomes* in his monumental 44-volume *Histoire Naturelle, Générale et Particulière*, 1749–67), and in 1790 John Latham first described the female – '*Femina feminae sponsae similis*'; the female of the Wood Duck is indeed very similar in appearance to the female of its close relation, the Mandarin Duck.

In 1830, the Zoological Society of London, in whose grounds Mandarin Ducks bred for the first time in England four years later, purchased two pairs at an auction of the Earl of Derby's waterfowl collection for the considerable sum of £70. Here John Gould drew a drake from life for his *Birds of Asia* (1850–80). The first recorded reference to a Mandarin Duck in the wild was one shot at Cookham, on the Thames in Berkshire, in 1866.

In the early 20th century, the Duke of Bedford added some Mandarin Ducks to his collection of waterfowl at Woburn Abbey in Bedfordshire, where by the outbreak of the First World War they numbered over 300, and local farmers began complaining that the birds were causing damage in fields of grain. Because of the difficulty during the two world wars of supplying the birds with supplementary food in winter, the population declined by some 50 per cent to a figure of around 150, at which it remained stable.

The success of the Mandarin Duck at Woburn was due largely (as in the case of the later population in Windsor Great Park) to the ideal habitat. The numerous ancient trees in the park (Mandarin Ducks are members of the tribe Cairinini – perching ducks) provide an abundance of nesting sites, for which the birds compete with Jackdaws and introduced Grey Squirrels.

Mandarin Ducks almost invariably nest in holes in trees up to 15 m agl and up to 1.5 km from water, where they normally lay 9–12 eggs. When the ducklings are ready to leave the nest, which may be up to 2.5 m in depth, they scramble up the near-vertical sides with the aid of their needle-sharp toes and, having apparently no fear of heights, launch themselves into space. Because of their fluffy down, which acts like a 'parachute', and light weight, they bounce on landing without doing themselves any harm.

The many lakes, ponds and streams at Woburn (and Windsor Great Park), encircled by Common Rhododendrons and other marginal shrubs, reeds, sedges and other emergent vegetation, provide the birds with the shelter and refuges that are a prime requirement, while the surrounding woodland supplies a wealth of acorns, sweet chestnuts and beechmast that form the birds' staple diet in autumn and winter. (Because of their high wing-to-weight ratio and long tails, the ducks have great aerial manoeuvrability, and are adept at weaving through densely planted woodland.) Predators at Woburn have been strictly controlled, so there has been no reason for the ducks to spread outside the estate.

In the years immediately preceding the First World War, Sir Richard Graham, Bt, of Netherby in Cumberland (Cumbria) on the Border Esk, acquired some Mandarin Ducks from the Wormald and McLean Game Farm in East Anglia (to which they had been imported directly from Canton), where the birds were reared in captivity. Many failed to breed, allegedly because the Chinese caponized them to protect their trade. A few birds, however, did manage to breed successfully at Netherby and along the Esk for a number of years, but by around 1920 this population had died out.

In 1918, Viscount Grey of Fallodon added some Mandarin Ducks from Wormald and MacLean to his sanctuary in Northumberland, where although they initially flourished and spread they disappeared after Lord Grey's death in 1933.

In 1910–35, Lieutenant-Colonel E.G.B. Meade-Waldo kept a flock of free-flying Mandarin Ducks at Stonewall Park near Edenbridge in Kent where, however, they failed to become permanently established.

MANDARIN DUCK MALE IN BREEDING PLUMAGE, NORFOLK

In 1928 or 1929, the distinguished French ornithologist Jean Delacour found a large collection of Mandarin Ducks in the bird market in Paris, to which they had been shipped from Hong Kong. These birds, many of which were already dead or dying, were contained in bamboo baskets, which Delacour identified as having been made in Fujian or Guangdong, suggesting the birds' origin in China. He purchased the entire consignment of nearly 50 birds, and in 1931 or 1932 gave 4–6 pairs[5] of the healthiest survivors to a friend, Alfred Ezra, who had a collection of waterfowl at Foxwarren Park near Cobham in Surrey. These birds were wing-clipped, not pinioned, and soon bred, they and their offspring subsequently spreading to nearby Virginia Water and other lakes in Windsor Great Park on the Surrey/Berkshire border. Here they flourished, for the same reasons as they did in Woburn, and by the mid-1970s had dispersed south-west through Surrey to beyond Haslemere, northwards into Buckinghamshire and Middlesex, east and west to beyond Staines and Reading, and south to the Surrey/Sussex border.

In 1930, Ezra, J. Spedan Lewis and W.H. St Quintin unsuccessfully attempted to establish Mandarin Ducks in London, where they released 99 full-winged adults from China in the grounds of Buckingham Palace and Hampton Court, and in Regent's, Central and Greenwich Parks. At around the same time, J.O. and B. D'Eath successfully established Mandarin Ducks at Monken Hadley in Hertfordshire, from where they spread to the surrounding neighbourhood.

In 1935, Ronald and Noel Stevens added some full-winged Mandarin Ducks to their wildfowl collection at Walcot Hall in Shropshire, where the population eventually peaked at around 100.

Following the Second World War, free-flying colonies of Mandarin Ducks have been started in many places, including Tillingbourne Manor, Surrey; Leonardslee, Sussex; Leckford, Hampshire; Bassmead, Huntingdonshire; Eaton Hall, Cheshire; Apethorpe Lake and Milton Park in the Soke of Peterborough; the Zoological Society of London's grounds in Regent's Park; Leeds Castle, Kent; and Salhouse, Norfolk. From these and other centres, the species has dispersed to found many new small sub-populations, principally in southern England.

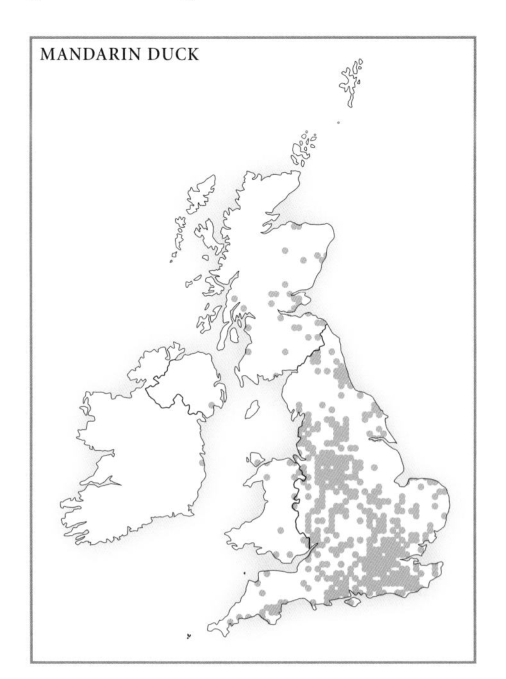

The stronghold in England today remains, as it has always been, Virginia Water and other ponds and lakes in Windsor Great Park. The success of the species here is, as at Woburn, due mainly to the ideal habitat of mature deciduous broadleaved woodland intersected by numerous streams and ponds encircled by Common Rhodedendrons, and in recent years the provision of nesting boxes, which are readily adopted by the ducks. (In these nesting boxes, as in natural sites, intraspecific and interspecific egg 'dumping' is a not infrequent occurrence.)

Recently, however, the population in Windsor Great Park seems to have declined. Whether this is due to an actual fall in numbers or to greater dispersal away from the park is hard to assess, but whereas a decade or so ago parties of 30 or more individuals could frequently be observed in winter on Virginia Water, now such numbers are seldom seen (personal observation).

Scotland

Mandarin Ducks on the River Tay in Perthshire are descended from a collection started in 1946–7 by J. Christie Laidlay at Holmwood on the outskirts of Perth. In and after 1962, half-a-dozen or so Mandarin ducklings were feather-clipped annually to keep them at Holmwood for their first breeding season. A year after Laidlay died, in 1964, when the Holmwood flock comprised 6–10 birds, his widow flushed a party of 17 from bushes on the banks of the Tay. By 1973, the population had increased to 30–40, and by 1975 it had risen to about double that figure. Most of the Holmwood Mandarin Ducks nested in some of the 50 or so nesting boxes in the trees in the garden. (Several had to be rescued from chimneys while searching for nesting sites.) Ducklings hatched in the wild usually returned to Holmwood with their parents in the winter, to take advantage of the supplementary feeding provided. By the mid-1970s, parties of 20–30 could be seen flighting up and down the Tay, or floating in on the tide at sunset. By the time of the author's visit in 1984, the population had increased to around 60–100.

In the 1970s, Mandarin Ducks became established on the Eye Water in Berwickshire, with a population in 1990 of some 30–50, and in 1996 up to 6 pairs bred at 2 places in the Loch Eck area of Argyll and 3 pairs in Strathnairn in the Highland District.

In the spring of 1974, a number of pinioned Mandarin Ducks escaped from an avicultural collection near Loch Lomond, Dunbartonshire. Since a full-winged duck was seen on the loch in the following December, it was assumed that the pinioned birds had bred in the wild.

[**Wales**]

During the Second World War, some Mandarin Ducks from Ronald and Noel Stevens's collection at Walcot Hall in Staffordshire dispersed as far as north Wales, but most of them returned to Walcot on the conclusion of hostilities. They have also occurred on a man-made lake at Dofor, near Newtown, Powys, in central Wales, and in south Wales.

Ireland

In 1987, a population of 30–40 Mandarin Ducks became established on the Shimna River in Co. Down in Northern Ireland. Elsewhere, the species occurs in Ireland only as a vagrant, presumably from England.

Summary of Status

By the early 1970s, the British naturalized population of Mandarin Ducks was estimated at around 250 pairs, a figure that had increased to 300–400 pairs by the middle of the decade. Ten years later the number had risen to perhaps 850–1,000 pairs, and by around 1990 the population may have been at least 7,000 individuals, and possibly as many as 13,000. However, because of the species' shy and secretive nature, even these figures may be underestimates.

Concomitant with this increase in the population is a corresponding increase in distribution, especially into Midland counties, northern England and Wales. However, it could be that the survival of many sub-populations in Britain away from Windsor Great Park may depend on supplementary feeding in winter by humans.

The success of the Mandarin Duck in becoming so well established in the wild in south-eastern England, and to a lesser extent elsewhere in Britain, can be attributed to a number of factors.

- It fills a vacant ecological niche for a duck that perches in trees, nests in holes in trees, and in autumn and winter subsists largely on the fruits of trees.
- The founder stock of the population in Windsor Great Park, from which it has widely dispersed, was composed of individuals chosen from the survivors of a shipment from China to Paris. They were themselves the survivors of a much larger number originally captured in winter in southern China, and were thus inherently the strongest and healthiest individuals and, coming originally from northern China and the Russian Maritime Territories, were derived from a wide genetic base that would naturally provide them with an enhanced genetic integrity.
- Mandarin Ducks in Britain have, like the also naturalized Greater Canada Goose, lost the migratory instinct that they have in the Far East, and are thus able to become established in the wild without the distraction of the urge to migrate. (The reason for this abandonment of seasonal migration is at present unknown. An obvious possible explanation is that in Britain's more temperate climate the need for migration in winter does not exist; this may well be so, but is as yet unproven.)
- The habitat into which most birds were released was ideal for their requirements.
- The latitudes of Britain and of the species' home in the Far East are not dissimilar.[6]

The future of the Mandarin Duck in the Far East has for long been in doubt. After the Second World War it appeared to be dying out, due largely to the deforestation of its two main breeding grounds – the Tung Ling and Kirin Forests. Predatory animals of the cat family are numerous in China, and birds of prey such as Black Kites also take a toll of the Mandarin Duck population. The early 21st-century estimate of the world population (including that in Britain and in continental Europe) was 25,180 pairs, of which 13,340 were in Japan, where the population has recently staged a recovery and is now stable. Nevertheless, the Mandarin Duck is listed by the IUCN as Near Threatened. The British population, which may exceed the whole of that in the Far East outside Japan, is thus of considerable conservation significance.

References

Baker et al. 2006; Blair et al. 2000; Brazil 1991; Davies 1985, 1988; Davies & Baggot 1989; Fitter 1959; Gordon 1937; Grey of Fallodon, Viscount 1929; Harrison 1988; Herber, in Lack 1986; Holloway 1996; Hutchinson 1989; Kear 1990; Lever 1977, 1987, 1990, in Gibbons et al. 1993, in Hagemeijer & Blair 1997, in Wernham et al. 2002, 2005; Lovegrove et al. 1994; Mead 2000; Ogilvie & RBBP 1999; Savage 1952; Sharrock 1976; Shurtleff & Savage 1996; Thorn 1986; Tomlinson 1976, 2006.

Notes

1. 'A[n] unexpected species … to which some of the bones from the Forest Beds [of north Norfolk in the Cromerian interglacial of some 600,000 years ago] appear assignable, is the Mandarin Duck *Aix galericulata*. In the early Pleistocene, Europe and Eurasia appear to have shared … a richer and more varied temperate [oak] forest … which would have been suitable for it. With

successive glaciations ... the ... mountain masses and the seas of southern Europe, Mediterranean and Middle East appear to have presented increasing barriers. Recolonization in warmer periods seems to have been less successful. ... The Mandarin appears to be part of what we have lost' (Harrison 1988: 14–15; 63). I am grateful to Derek Goodwin (personal communication 2007) for drawing my attention to this reference. The Mandarin Duck is here treated as a naturalized alien.
2. For much of the early history of the Mandarin Duck I am indebted to Savage (1952).
3. *querquedula* (Latin) = teal.
4. *galericulata* (Latin) = wearing a little wig or cap.
5. Savage (in Shurtleff and Savage 1996: 164) states that Delacour gave 45 birds to Ezra. In fact, this was the total number of the survivors saved by Delacour from the shipment from Hong Kong.
6. For reasons for the success of the Mandarin Duck in becoming naturalized in Britain as opposed to the relative failure of its close relation, the Wood Duck, see under the latter species, page 138.

ANATIDAE (DUCKS, GEESE & SWANS)

RED-CRESTED POCHARD *Netta rufina*

Natural Distribution C and S Europe, SW and C Asia; winters in S Europe, N and NE Africa, and S Asia. The European range is discontinuous.
Naturalized Distribution England; [Wales]; [Scotland]; [Ireland].

RED-CRESTED POCHARD MALE

England

Red-crested Pochards[1] occur in England naturally only as rare vagrants from Europe.[2] All breeding records are of individuals that have escaped or been released from avicultural collections.[3]

According to the *Handbook of British Birds*, 1939. Vol. III: 284, 'Five were seen in Lincs April, 1937, and a pair bred, but these and doubtless others among those recorded in recent years have probably wandered from Woburn [the Duke of Bedford], where they have been reared and allowed to fly for a number of years.'

In 1958, a pair nested at St Osyth in Essex, and since the following decade breeding has frequently taken place at Frampton-on-Severn, Gloucestershire (near the Wildfowl and Wetlands Trust headquarters at Slimbridge) and in other localities. In 1971, a full-winged drake paired with a pinioned duck at Apethorpe in Northamptonshire, from where their progeny subsequently dispersed. In the same year, probable breeding was recorded at Rickmansworth in Hertfordshire. In 1972, a pair bred at Kew Gardens in Surrey; these birds probably came from St James's Park or another Inner London park, where the species had been introduced in 1933, 1936 and from 1950. There have also been occasional reports of breeding on gravel-pits near Bourton-on-the Water, Gloucestershire.

In 1975, two pairs bred at the Cotswold Water Park in Gloucestershire (where the species had occurred since the 1960s). By the following decade a small but apparently stable population had become established, though initially numbers remained low, and the Cotswold Water Park has now become the species' headquarters in England: in 1989 the post-breeding population there numbered 32.

In 1997–2001, successful breeding was reported in Essex, Gloucestershire, Greater Manchester, Lincolnshire, Middlesex, Norfolk, Nottinghamshire, Oxfordshire, Surrey, Sussex and Wiltshire; the highest number of breeding pairs was six in 1998 and 2001 and seven in 1999.

In April 2002, a pair was reported at Langtoft Gravel-pits in Lincolnshire, where a dozen birds were counted in September. In Norfolk, two young were seen in May at Pensthorpe, where there was a full-winged population, and two drakes and a duck were observed in April and June at neighbouring Great Ryburgh. 'Once again', noted Ogilvie and RBBP (2004: 636) plaintively, 'our annual plea for data from the well-known population in the Cotswold Water Park has gone unheeded.'

In 2003, a minimum of 4, and perhaps 8, pairs bred successfully. At a site in Essex 2 pairs bred, and at a single site in Lincolnshire 1 pair bred and 2 other pairs may have bred. In the Cotswold Water Park in Gloucestershire, 1 pair and 3 young were observed in August, and the population in the autumn numbered 101.

In 2004, when 6–19 pairs reproduced, 2 pairs bred at a single site in Essex, and in Lincolnshire 11 pairs probably bred at 2 sites, while in Norfolk at 1 site a pair fledged 7 young. At a site in Cleveland, a drake was observed to copulate frequently with a female Common Pochard, but no young were seen. At the Cotswold Water Park 3 pairs successfully fledged 22 young, and a fourth pair probably bred.

In 2005, 5–8 pairs reproduced. At a site in Berkshire 2 pairs probably bred; in Hertfordshire 1 pair bred and 5 young were counted in May, but by late June only 1 had survived; in Norfolk 1 pair probably bred. At the Cotswold Water Park 4 pairs fledged a total of 15 young.

In the majority of autumn/winter counts, over 150 Red-crested Pochards are reported in most years, but how many of these represent the naturalized breeding population and how many are natural vagrants from continental Europe is hard to assess.

[Wales]

In early April 2002, a pair of Red-crested Pochards was seen displaying at Roath Park Lake in Glamorgan, and in late March another pair was observed at Eglwys Nunydd, Gower.

[Scotland]; [Ireland]

Red-Crested Pochards have from time to time been recorded in southern Scotland and Ireland, but the source of these individuals has not been certainly determined.

Summary of Status

The majority of reports of Red-crested Pochards in Britain and Ireland are on fresh water, especially reservoirs, lakes, ponds and flooded gravel-pits fringed by Common Reeds and emergent vegetation. The species has also been recorded from freshwater marshes and rivers, and occasionally even from tidal and offshore waters. Because of its nomadic nature and the general lack of interest among birdwatchers in introduced species, the Red-crested Pochard is, like so many other naturalized birds in Britain and Ireland, almost certainly greatly under-recorded.

References

Baatsen 1990; Berndt, in Hagemeijer & Blair 1997; Blair et al. 2000; Cox, in Lack 1986; Delany, in Gibbons et al. 1993; Fitter 1959; *Handbook of British Birds* 1939; Holling & RBBP 2007; Holloway 1996; Lever 1987, 2005; Ogilvie & RBBP 1999–2004; Pyman 1959; Sharrock 1976.

Notes

1. Harrison (1988: 14–15) records the presence of Red-crested Pochard bones in the Forest Beds of north Norfolk during the Cromerian interglacial of some 600,000 years ago. The species is treated here as an introduced alien.
2. First recorded in 1818, most subsequent records have all been in autumn and winter, and chiefly on the east coast of England (e.g. Norfolk, Suffolk, Northumberland, Yorkshire, Lincolnshire, Cambridgeshire, Essex, Thames Estuary and Sussex). The species has also occurred in Scotland (e.g. Argyll, Midlothian), Wales (e.g. Pembroke) and Ireland (e.g. Co. Cork and Co. Kerry). (*Handbook of British Birds*, 1939. Vol. III: 284.)
3. Blair et al. (2000) say the first record of an escaped Red-crested Pochard in the wild was in 1900.

ANATIDAE (DUCKS, GEESE & SWANS)

RUDDY DUCK *Oxyura jamaicensis*

Natural Distribution The nominate form (which is the one established in Britain and Ireland) breeds in Canada and the USA, and winters S to N Mexico. Other races occur from the West Indies S to W S America, as far as S Chile.
Naturalized Distribution England; Scotland; Wales; Ireland.

England

In the 1930s, a pair of Ruddy Ducks in the wildfowl collection of Ronald and Noel Stevens nested successfully in the wild at Walcot Hall in Shropshire. This collection became dispersed during the Second World War, and this wild breeding occurrence is believed to be the first record for Britain.

The Ruddy Ducks that have since become established in Britain are descended from some that escaped from the Wildfowl Trust's reserve at Slimbridge on the Severn in Gloucestershire. In 1948, the trust obtained three pairs from Pennsylvania and Salt Lake City in the United States, and these bred at Slimbridge in the following year. The trust had for some time had a policy of pinioning the young of non-native species, but as the

RUDDY DUCK MALE

stock of Ruddy Ducks at Slimbridge multiplied this policy became increasingly difficult to implement. The first two full-wing birds escaped from Slimbridge in the winter of 1952–3. From 1956, most of the Ruddy Ducks at Slimbridge were allowed to rear their own young rather than having them brooded artificially, and thus more ducklings were able to avoid pinioning. By 1957, up to 20 had departed, and by 1973 around 70 had escaped – the majority leaving before and during the severe winter of 1962–3.

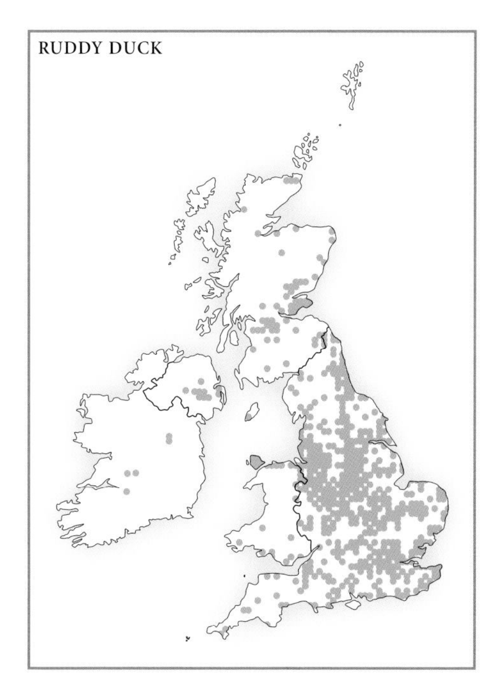

The earliest reports of Ruddy Ducks in the wild in England date from 1954, when individuals were observed at Hingham, Norfolk (possibly not from Slimbridge) and perhaps on Aqualate Mere in Staffordshire. Three years later, a single young male was seen in Villica Bay on Chew Valley Reservoir in northern Somerset (Avon). In the following winter the number of ducks on Chew increased to four, and the same number – possibly the same birds – was reported on the nearby Blagdon Reservoir. Single birds were occasionally recorded on the also neighbouring reservoir of Barrow Gurney. The birds at Chew moulted into adult plumage in the spring of 1958, and in subsequent years were joined by several other immature individuals. It was not until December 1960 that the first females, believed to be further escapees from Slimbridge, appeared on Chew Valley Reservoir, where a paid bred successfully in 1960, but where the breeding population in the 1960s remained at only 1–2 pairs. Subsequently, breeding by Ruddy Ducks was recorded in Staffordshire in 1961, and in 1963 six pairs nested successfully, including one in Gloucestershire.

Although it was at one time believed that the British population was derived solely from escapees and their progeny, it has since been known that there was at least one deliberate release, when in 1961 three or four females from Slimbridge were liberated at Chew as potential mates for the males already there.

At this time, Ruddy Ducks were also becoming established in the West Midlands, where they first definitely appeared in Staffordshire in 1959, and within two years were breeding. In the West Midlands, the population increased so dramatically that it soon outnumbered that in the whole of the lower Severn valley. Ruddy Ducks first appeared in Gloucestershire in 1957, Warwickshire and Shropshire (1962), Herefordshire (1963) and Derbyshire (1973). Away from their principal centres, they had been recorded by the mid-1970s also in Buckinghamshire, Cambridgeshire, Essex, Greater London, Hertfordshire, Huntingdonshire, Lancashire, Lincolnshire, Nottinghamshire, Wiltshire and Yorkshire.

In 1968, at least 10 pairs of Ruddy Ducks bred in the wild; by 1972, the total had risen to a minimum of 25 pairs and by the following year to around 35 pairs. In 1974, the post-breeding population numbered a minimum of 250, including 45–50 breeding pairs. By the following year, there were believed to be 50–60 breeding pairs in some 8–9 counties, with a post-breeding population of 300–350, of which around 120 were in Somerset (Avon), 190 in Staffordshire, and 10–15 in each of Shropshire, Cheshire and Leicestershire. In 1965–75, the annual rate of increase in the population averaged at least 25 per cent. By 1976 and 1977, the number had risen to 380 and 430–450 respectively, by 1978 to around 770 (an average annual rate of increase since 1975 of some 33 per cent) and by 1981 to around 1,750 (an annual rate of increase of 50 per cent since 1978). A decade later the population was approaching 2,000.

Away from the principal conglomerations, Ruddy Duck had, by the mid-1980s, been seen in England from Berkshire in the south to Yorkshire in the north, and from Cornwall eastwards to East Anglia.

It was at one time believed that a prolonged spell of hard weather would seriously deplete the population of this small duck, which is mainly aquatic and seldom visits land. Nevertheless, since its establishment in the wild it has survived relatively unscathed one of the coldest winters of the 20th century, that of 1962–3, and those of 1978–9 and 1981–2, and the losses incurred during those winters were rapidly made good.

By the late 1980s/early 1990s, Ruddy Ducks had expanded their range, mainly northwards, and had established some large breeding sub-populations away from the West Midlands, especially in Cheshire, Greater Manchester and Yorkshire. During the summer the ducks had become a common sight on reed-fringed lakes and ponds and on flooded gravel-pits throughout the Midlands, and in some northern counties in England. During the 1980s, the annual rate of increase of Ruddy Ducks slowed to approximately 10 per cent, but nevertheless the estimated 1991 population was in excess of 3,400, including around 570 breeding pairs.

Scotland

Ruddy Ducks were first recorded in Scotland, on Carsebreck Loch (Carsebreck Curling Pond), 27 km south-west of Perth, in June 1954. Twenty years later, a drake was recorded in May on Unst (the most northern of the Shetland Isles) – the most northerly British record. Breeding was first reported, in Angus, in 1979 and took place for a number of years thereafter; by the beginning of the following decade it was recorded in various other localities, mainly in the south and east. In 1984, Ruddy Ducks bred successfully on the Loch of Strathbeg in north-eastern Aberdeenshire. By the time of the 1988–91 *New Atlas of Breeding Birds in Britain and Ireland*, they had become regular breeders on Tayside in Perthshire and in Fife; they also occur on waters near Glasgow, and in parts of Roxburghshire and Berwickshire.

Wales

In Wales, Ruddy Ducks were first recorded in Flint in 1970, mid-Glamorgan (1972), south Glamorgan (1973) and Montgomeryshire (1974). By the middle of the following decade, they had become established on the Isle of Anglesey, Gwynedd, where the first record of breeding took place in 1978, and where by 1981–2 the population numbered around 100. By the late 1980s/early 1990s, Anglesey had become one of the main strongholds in Britain, with a population of some 60 pairs.

Ireland

In 1973, a single pair of Ruddy Ducks bred successfully on Lough Neagh, west of Belfast in Northern Ireland, where the birds hatched four ducklings. By the middle of the following decade, they had been recorded in Co. Cork, Co. Waterford and Co. Wexford in the south. By the late 1980s/early 1990s, 15–20 pairs were estimated to be breeding in Ireland. In the mid-1990s the post-breeding population was around 40. The present Irish pre-breeding population, in Ulster and Co. Wexford, numbers at least 130 individuals, of which 50–4 are breeding pairs, and both numbers and distribution are increasing. Many, if not most birds, leave in autumn, possibly to winter on the British mainland.

Characteristics

Ruddy Ducks belong to the tribe Oxyurini ('stiff-tails'), and the drakes have one of the most bizarre courtship displays in the sub-family Anatinae. In display, the tail is erected at a right angle to the body and is spread out like a fan. By means of air sacs beneath the feathers, the breast is puffed out like that of a 'pouter' pigeon, an inconspicuous crest is raised on the head, and the bill is lowered and drummed rapidly on the inflated breast feathers, forcing air bubbles out to produce a burbling 'popping' sound.

The legs of the Ruddy Duck are positioned well back on its body, like those of a grebe Podicipedidae, causing it to be rather ungainly on land. It is, however, a buoyant swimmer, a capable diver, able to submerge slowly without diving, and like a grebe it can swim with only its head and neck above the surface.

Even outside the breeding season, Ruddy Ducks prefer lentic to lotic waters that are surrounded by reeds and emergent vegetation. They construct a grebe-like platform (frequently floating) nest anchored to reed stems, in which, between early April and late June (sometimes even later), the female lays a clutch of 6–8 large eggs, from which emerge fat and robust ducklings that are capable of diving for food at a day old. This is one of the longest breeding seasons of all British waterfowl, and allows females the chance of rearing a second brood or renesting if the first breeding attempt fails. Eggs of clutches in these categories hatch as late as October. This, together with a high hatching success rate (around 70 per cent), a good duckling survival rate and the habit of brood parasitism by some females, which may subsequently lay a further clutch in their own nests, has contributed to the species' success in Britain and Ireland. Other factors include:

- An abundance of suitable habitats and food.
- The existence of a vacant ecological niche for a largely nocturnal, freshwater, bottom-feeding, mainly insectivorous duck.
- A temperate climate.
- Until recently, an absence of shooting pressure.

These factors have together enabled the Ruddy Duck, in the last 30–40 years, to have had one of the most explosive increases in distribution and population of any bird in Britain.

Soon after their colonization of Britain, Ruddy Ducks established a regular seasonal national migration pattern. In severe weather, these movements are almost invariably nocturnal. After the birds' late summer moult they tend to congregate in large numbers at a few meres in Cheshire and Shropshire, from where they disperse to various wintering locations, such as Blithfield and Belvide reservoirs in Staffordshire and Blagdon and Chew Valley lakes in Avon. Some remain and winter within their breeding range. Unlike other British wildfowl, Ruddy Ducks also have a full pre-breeding moult in February/March, which again leaves them unable to fly; some moult in their winter quarters while others, while still able to fly, return to their breeding grounds. Ruddy Ducks, except when dispersing or migrating, are rather reluctant flyers.

More recently, the species appears to have developed a European migratory pattern similar to the one it has in North America, with evidence of some seasonal migration to winter quarters in continental Europe. This could explain the occurrence of Ruddy Ducks in Scandinavia in the spring and early summer, and the presence of a small migratory breeding population in Iceland.

Impact

In his foreword to Lever (1977), Sir Peter Scott blames himself for the naturalization of the Ruddy Duck in Britain:

> Having been carelessly responsible myself for allowing the North American Ruddy Duck to escape and build up to what seems to be a small but viable population in England, I am in no position to pass judgement on others. To be sure the Ruddy Duck is decorative and apparently harmless, but no one can know what insidious effect it may have on the ecological web.

How prophetic Peter Scott's words have proved to be.

In Britain, Ruddy Ducks are extremely aggressive towards other animals during courtship and breeding – even to those such as Northern Lapwings, Common Blackbirds and European Rabbits that pose no conceivable threat of competition or agonism. However, only two species, the Little Grebe and Black-necked Grebe, which share the same breeding habitats as Ruddy Ducks, may be at risk: during the period of courtship and breeding, male Ruddy Ducks have been observed displaying agonistic behaviour towards both species. There may also be some competition with the Tufted Duck for chironomid larvae, which form part of the diet of both species.

The impact of British Ruddy Ducks in Europe (in particular Spain) is one of the most contentious current issues among British conservationists, polarizing opinions of laymen and scientists alike. Ruddy Ducks had not been recorded in Europe before the species became naturalized in Britain. The first continental occurrence was in Sweden in 1965, and the earliest breeding record, in the Netherlands, took place in 1973. By the early 1980s, they were being reported in Europe with increasing frequency, and by the time of the 1988–91 *Atlas*, they had been observed in a dozen countries from Iceland to Italy. By 1992, the species had occurred in 15 European countries east to the Ukraine, and by the following year there were some 600 reports of the birds from 19 countries. Although they have escaped from avicultural collections in other countries (e.g. the Netherlands and Belgium), most European and African records of Ruddy Ducks are believed to involve birds from Britain.

Soon after Ruddy Ducks first appeared in Spain in 1973, concern was expressed about the potential impact, through hybridization, on the closely related native White-headed Duck (classified by the IUCN as Vulnerable). This species, which has a world population of around 19,000 individuals, breeds in small numbers in several discrete sub-populations in Europe (Spain, Romania, Turkey and the former USSR); Africa (Tunisia, Morocco and possibly Algeria); and in Asia in China, Kazakhstan and Siberia.

Since Ruddy Ducks, almost certainly mainly if not exclusively from Britain, began to appear in any numbers in continental Europe in the 1980s, controversy has raged about the need to 'control' (i.e. eradicate) the species in Britain in order to protect the population of several hundred breeding pairs of White-headed Ducks in Spain. In the eastern part of its range, the species' total population is around 18,000, so the populations in Europe (mainly Spain) and Morocco account for less than 5.3 per cent of the global population. On the other hand, Ruddy Ducks have now appeared in Israel and Turkey, so the main central Asian population of White-headed Ducks may eventually be at risk.

In Spain, the Ruddy Duck was first reported as breeding in 1991, when hybrids with White-headed Ducks were also recorded. However, since 1980 Spanish conservationists had been expressing their concern about the infiltration of White-headed Duck breeding sites by Ruddy Ducks, but because of only circumstantial evidence that the latter came from Britain, and because of uncertainty about the degree of risk of hybridization, no action was taken by the British authorities. It has since become apparent that in captivity male Ruddy Duck x White-headed Duck hybrids can be backcrossed to female White-headed Ducks to the third generation, thus demonstrating that hybrid breakdown (when second generation hybrids become infertile) does not occur, at least in captivity.

In Spain, Ruddy Duck x White-headed Duck hybrids are reported to be dominant over all other species of waterfowl with which they have been observed, including the pure White-headed Duck, Mallard, Gadwall, Common Pochard, Common Coot and, as in Britain, Little Grebe. In Iceland, concern has been expressed regarding the interaction between Ruddy Ducks and the locally declining population of Slavonian Grebes. If Ruddy Ducks were to spread to East Africa, the local population of the Maccoa Duck (which occurs from Ethiopia south to northern Tanzania) could potentially be at risk. These two species are as closely related as the Ruddy Duck and White-headed Duck, suggesting that they too could hybridize freely.

In September 1992, at the instigation of the Wildfowl and Wetlands Trust, the UK Ruddy Duck Working Group was set up, which led, in the following March, to a conference at which it was resolved to control the number and distribution of Ruddy Ducks in the Western Palearctic in order to safeguard the future of the White-headed Duck. Walton (2001) has elegantly outlined the case for the prosecution.

Habitat destruction and over-shooting reduced the Spanish population of White-headed Ducks to only 22 individuals in the 1970s, but as a result of conservation measures it had recovered to several hundred breeding pairs by the turn of the century. Hybridization with Ruddy Ducks from Britain has been identified as the principal threat to its survival.

Research has shown that if geographical barriers that divide regionally distinct biomes are eroded, biological diversity (one of the cornerstones of conservation) will eventually be lost. Such erosion of national biogeographic barriers is precisely what happens when native species are moved, through human agency, outside their natural range.

Not every introduced species, of course, has a negative impact on alien ecosystems, and some, such as the spectacular Mandarin Duck, actually have a positive one by giving pleasure to people who see it – arguably the most important factor in gaining support for nature conservation. If, however, we wish to conserve the world's biodiversity, we must understand and admit that some naturalized species can, and do, cause ecological harm of one kind or another, and accept that sometimes unpalatable remedial measures have to be taken. After habitat destruction, introduced species pose the most serious threat to the Earth's biological diversity.

Ruddy and White-headed Ducks occur in the wild sympatrically entirely as a result of intervention by humans. Because of different mating strategies, in mixed populations the former will always dominate the latter; thus Ruddy Duck genes pass relatively quickly into the White-headed Duck population, with a consequent loss of genetic factors that make the latter a distinct species. Such introgression can eventually result in the loss of the weaker species and the survival only of the dominant one. (This, of course, is a reversal of the usual evolutionary process whereby one species splits from its 'parent' to form a new species.)

There seems no apparent reason why the above process should not occur in hybridizing Ruddy and White-headed Ducks. In addition to the threat posed by hybridization, Spanish ornithologists have voiced concern because of the agonistic behaviour of Ruddy Ducks towards White-headed Ducks (and other native species) and over inter-specific competition for food.

Turning devil's advocate, Walton (2001) summarizes the opposition's case to Ruddy Duck 'control'. Firstly, opponents of 'control' point out that it cannot be conclusively proved that Ruddy Duck x White-headed Duck hybridization (or competition) will drive the latter to extinction. However, as Walton (2001: 95) points out, 'science will never be able to *prove* [his italics] that a future environmental change will occur. The best that humans can do is to assess the available evidence and make predictions and judgements based on this: proof will only be forthcoming once the change has actually occurred.' Secondly, opponents of control question its feasibility. Successful government trials have, however, shown this fear to be groundless.

Finally, and perhaps most persuasively, opponents to control query the dilemma if the unique characteristics that make White-headed Ducks a distinct species are lost, since there would be a 'strengthening' of the gene pool and an improvement of Eurasian oxyurid ducks generally. However, as Walton points out, it is not known which White-headed Duck characteristics – in particular those not shared with Ruddy Ducks – are specifically adapted to Eurasian conditions. Furthermore, if a biological invasion of an alien species as a result of human intervention becomes acceptable, it could encourage other such deliberate introductions on the mistaken grounds that it would increase the host nation's biodiversity. In fact, however, if native species die out because man has moved exotics into their range, biodiversity will actually be lost, and if this happens 'we eradicate for ever the evolutionary potential of our planet' (Walton 2001: 97).

Gibson (2001) puts the case equally persuasively for the defence. Although Ruddy and White-headed Ducks are known to be capable of producing fertile hybrid offspring, there is no clear idea of the repercussions that this would cause in the wild. If both species share a common ancestor, and the White-headed Duck evolved millions of years ago and became especially adapted to the Eurasian environment, is it not more likely to survive as a species than either hybrids or pure Ruddy Ducks? It is possible that Ruddy Ducks would never establish a viable breeding population within the range of the White-headed Duck, and that Ruddy Ducks and their hybrid offspring would always be a minority of the 'super-species' population.

Perhaps an even more likely scenario is that after an initial period of colonization by Ruddy Ducks, with some degree of hybridization, the position would stabilize and both species would survive intact allopatrically. Alternatively, if White-headed Ducks were to die out as a result of introgression, might not evolution result in Ruddy Ducks in Europe eventually becoming genetically distinct from Ruddy Ducks in North America? Evidence already exists, says Gibson (2001), that European Ruddy Ducks have developed diagnostic DNA differences from their American ancestors, so that they may be already worthy of conservation in their own right.

The belief that, if unchecked, the Ruddy Duck would overwhelm the White-headed Duck is based on the prediction that Ruddy Duck genes will continue to flow into the White-headed Duck population. This theory is in turn based on the presumption that if uncontrolled the British Ruddy Duck population will continue to increase. Gibson (2001) claims that less than 10 Ruddy Ducks emigrate to continental Europe annually. There is also reason to believe, says Gibson, that the British Ruddy Duck population may already be near its maximum sustainability level. (With plenty of apparently suitable habitat as yet uncolonized, this must be at least debatable.)

Thus, the sole justification for controlling Ruddy Ducks in Britain is as a precautionary measure to protect the White-headed Duck. But is this a serious biodiversity problem? If the latter species were to become extinct it would be extremely unfortunate, but equally the same could be said if the Ruddy Duck were to die out in Britain at the hands of humans.

Concern has been expressed by both members of the public and some biologists about both the ethics and humaneness of destroying one species – albeit an alien – to preserve another (native) species, more especially when, as in the case under discussion, the presence of the former is not natural but engineered by humans, who are now, by eradicating it, seeking to put right their own carelessness. Might it not be more practical and feasible for the Spanish (and, if necessary, other foreign authorities) to shoot immigrant Ruddy Ducks after their arrival? Most Ruddy Ducks disperse to Europe in winter, so there would be ample time for their removal before the breeding season. By shooting Ruddy Ducks in Britain, however, we would be losing an attractive and, to many people, welcome addition to the British avifauna.

The problem can briefly be summarized as follows. On the one hand, is it ethically acceptable for man to deliberately destroy an animal that is where it is due to his actions, simply because it *may* be detrimental to a native species in another country? On the other hand, is it morally justified to deliberately put at risk a native species in a foreign country at the agency of another species that man has introduced elsewhere? That is the dilemma.

In 1998, the British government created the Ruddy Duck Task Force, which found that control was feasible and, by local trials, that shooting was the most effective method, and this policy is supported by the Royal Society for the Protection of Birds (RSPB). The policy of the RSPB Scotland on the Ruddy Duck includes the following statement: 'RSPB Scotland wishes to see the survival of *both* [their italics] the white-headed duck and the ruddy duck. To achieve this, there is an urgent need to reduce the UK population of ruddy ducks to a level at which birds are not spreading to continental Europe and threatening the white-headed duck.'

Since, however, Ruddy Ducks from Britain began to appear on the Continent from 1965, when the British population numbered only around 10 pairs, to reduce the British population 'to a level at which birds are not spreading to continental Europe' appears impossible. Only *eradication* of Ruddy Ducks in Britain will achieve this objective – something the RSPB and some other conservation organizations seem reluctant to acknowledge. In nearly all the literature on the subject, such euphemisms as 'cull' and 'control' are used rather than 'eradication'.

In 2005, the Central Science Laboratory (CSL), under contract to the Department for the Environment, Food and Rural Affairs, began an eradication project, at an estimated cost of £3.3–5.4 million, with the objective of exterminating the Ruddy Duck in Britain by 2010.

The Wildfowl and Wetlands Trust, under contract to CSL, simultaneously began conducting a five-year monitoring programme to assess the change in numbers of the British Ruddy Duck population during control operations in order to judge the success of the eradication programme, and to direct continuing control efforts.

Between 2005 (when the British population of Ruddy Ducks numbered about 4,400) and 2007, 3,691 birds were shot at 70 sites in 33 English, Scottish and Welsh counties, principally on waters supporting large wintering flocks and the best breeding grounds. Of this total, 72 per cent were adults with 27 per cent being fledged immatures. Two counts, in December 2006 and January 2007, on major wintering sites in Britain revealed 1,538 and 1,239 survivors respectively; a less-extensive census in January 2006 had totalled 3,077 birds. This eradication campaign resulted in a 56 per cent fall in the Ruddy Duck population in January–December 2006, and a 59 per cent fall in January 2006–January 2007. Although, as might be expected, reductions were greatest at sites where shooting had taken place, important declines (averaging 43 per cent) were also recorded elsewhere.

In December 2007 and January 2008, synchronised surveys of the most important sites in Britain for wintering Ruddy Ducks revealed a total of 1,007 birds at 44 of 66 sites in December and 930 at 43 of 69 sites in January. The number of waters supporting over 100 birds each was found to have decreased during the past three winters from nine in January 2006 to only one (Carsington Water, Derbyshire) in January 2008. During the same period, the number of waters with no Ruddy Ducks present increased from 9 to 27. Counts in Northern Ireland in October 2007 and January and March 2008 recorded a peak figure of only 39 individuals.

Between 1 September 2007 and 31 March 2008, a total of 1,190 birds were shot at 34 sites, 645 of which were killed since the December census.

Since the late 1990s, there has been a gradual fall in the British Ruddy Duck population, which greatly accelerated after the start of the eradication campaign in 2005. It was estimated that in 2007–8 unsurveyed sites may have held a total of 100–300 more birds, and that the 31 December 2007 total was some 1,100–1,300 individuals. Taking into account the number of birds shot in December 2007–March 2008, together with some natural mortality, the overall spring 2008 population was estimated at 400–500.

References

Anon 2005; Baker et al. 2006; Blair et al. 2000; Bremner & Park 2007; DEFRA 2003; Gibbons & Avery 2001; Gibson 2001; Gibbons et al. 2001; Green & Anstey 1992; Green & Hughes 1995; Hall et al. 2008; Harrison 1988; Holloway 1991; Holmes & Galbraith 1994; Hudson 1976a, 1976b; Hughes 1992, in Gibbons et al. 1993, in Holmes & Simons 1996, in Hagemeijer & Blair 1997, 2002, 2003; Hughes & Grussu 1994, 1995; Kear 1990; King 1959–60, 1976; Langley 2004; Lever 1977, 1987, 1994, 2005, in press; Raven et al. 2007; Sharrock 1976; Tomlinson 2006; Toms, in Wernham et al. 2002; Vinicombe, in Lack 1986; Vinicombe, in Chandler 1982; Walton 2001; Went 1975; Welch et al. 2001; Worden 2006, 2007.

PHASIANIDAE (TURKEYS, GROUSE, PHEASANTS & PARTRIDGES)

CHUKAR PARTRIDGE *Alectoris chukar*

Natural Distribution From NE Greece through Asia Minor and Arabia to NW India, W Mongolia, S Manchuria and N China.
Naturalized Distribution England; Scotland; Wales.

CHUKAR PARTRIDGE

England; Scotland; Wales

Chukar Partridges were introduced by F.R.S. Balfour to Dawyck Park, Stobo, Peeblesshire in the late 1920s or early 1930s, when they were also kept unpinioned by the Duke of Bedford at Woburn Abbey and at Whipsnade in Bedfordshire, and by Alfred Ezra at Foxwarren Park in Surrey. At none of these places, however, were they ever properly established, although a few were still breeding in the late 1940s. In the early 1970s, commercial game farms began rearing the birds for sporting purposes, and by the end of the decade they were established on the South Downs in Sussex and Kent, and in Aberdeenshire and some neighbouring counties. By the late 1990s, small populations survived only in parts of south-west and northern England, in northern Scotland and in parts of Wales; those in Scotland may have since died out.

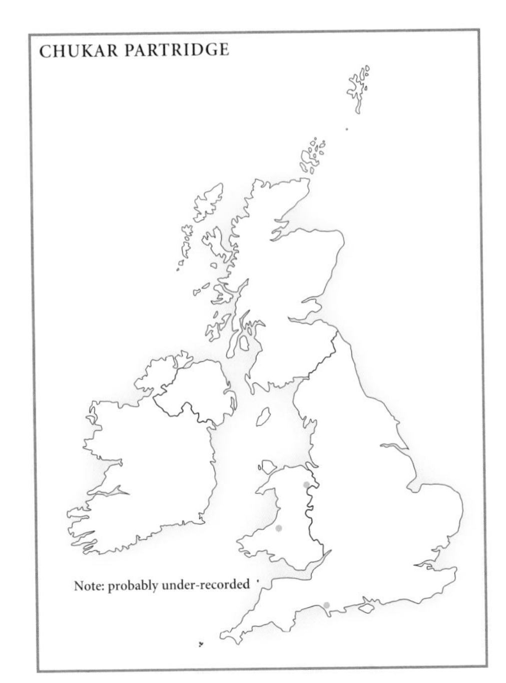

Impact

Where the distribution of the two species overlaps, Chukar Partridge x Red-legged Partridge hybrids occur.

References

Aebischer, in Hagemeijer & Blair 1997; Harrison 1988; Lever 1977, 1987, 2005; Long 1981; Raven et al. 2007; Robertson, in Holmes & Simons 1996; Sharrock 1976.

PHASIANIDAE (TURKEYS, GROUSE, PHEASANTS & PARTRIDGES)

RED-LEGGED PARTRIDGE *Alectoris rufa*

(FRENCH PARTRIDGE)

Natural Distribution From the Iberian Peninsula N to the Pyrenees and S France, E to the Balearic Islands, Corsica, Elba and N Italy.
Naturalized Distribution England; Scotland; Wales; Ireland.

England

Sir Thomas Browne (1605–82) made the earliest, albeit negative, reference to the Red-legged Partridge in England when, in about 1667 he wrote: 'Though there be here [Norfolk] very great store of partridges, yet the French red-legged partridge is not to be met with.' In their *Ornithologia*, published in 1676–8, Francis Willughby and John Ray say that 'We have been informed that the red-legged partridge (*Perdix ruffa*) is found in

RED-LEGGED PARTRIDGE PAIR, MIDLANDS

the isle of Jersey and Guernsey ... This kind is a stranger to England.' Fifty years later, Ephraim Chambers (d 1740), in his *Cyclopaedia* of 1728, confirms the position as unchanged: 'The red-legged partridge is not found in England, but is sometimes shot in the islands of Guernsey and Jersey.' These birds presumably arrived in the Channel Islands as natural immigrants from France.

The earliest reference to the original introduction of Red-legged Partridges to England in 1673 (see Leicester 1921) occurs in William Barker Daniel's *Rural Sports* (1801):

> So far back as the time of Charles II [r 1660–85] several pairs of these red-legged partridges were turned out about Windsor to obtain a stock; but these are supposed to have mostly perished, although some of them, or their descendants, were seen for a few years afterwards.

In 1682, a number of 'curious outlandish partridges' (which were probably of this species) were acquired by the Earl of Rutland at Belvoir Castle, Leicestershire, and in 1712–29 the Duke of Leeds reared some from eggs at Wimbledon in Surrey, where they 'were, after increasing for a time, all destroyed by some disobliging neighbour'.

In or about 1770, Red-legged Partridges from France were released by the Earl of Hertford and Lord Rendlesham at Sudbourne and Rendlesham Halls near Woodbridge in Suffolk; by the Duke of Northumberland at Alnwick Castle; and by the Earl of Rochford at St Osyth, Essex. The birds in Suffolk quickly became established, probably because the light, sandy soil and open heathland provided an ideal habitat for them. In 1823, Lords Alvanley and De Ros imported large quantities of Red-legged Partridge eggs from France, which were successfully hatched at Culford, Cavenham and Farnham near Bury St Edmunds in Suffolk. From there by 1839 the resulting offspring had spread into Norfolk and Lincolnshire, where by 1874 they were said to be quite common. Since about 1828, the Red-legged Partridge has outnumbered the native Grey Partridge in most parts of East Anglia.

From eastern England, Red-legged Partridges spread north to Yorkshire by 1835 and west into Nottinghamshire by 1851. A number of unsuccessful attempts were made to establish them elsewhere, e.g. in Derbyshire, Worcestershire (by the Marquess of Hastings in 1820) and Staffordshire, but by the early 1880s they were breeding in small numbers only in Cambridgeshire (first recorded in 1821), Huntingdonshire (1850), Northamptonshire and Rutland, and Berkshire (1809), Buckinghamshire (1835), Oxfordshire (1835), Hertfordshire (1815) and Middlesex (1865).

South of the Thames, an unsuccessful attempt was made in 1776 to establish Red-legged Partridges in Sussex by Sir Harry Fetherstonhaugh of Uppark. In 1841, however, the species was reared successfully at Kirdford, west of Billingshurst, whence it spread slowly throughout Sussex and east into Kent, where by the late 1860s it was said to be breeding occasionally, and west into Hampshire. It was first recorded in Cornwall in 1808, Devon (1844), the Scilly Isles (where it had died out by 1864) in the mid-1850s, and Wiltshire (1861), from where it subsequently spread into Gloucestershire, Somerset (1879) and Herefordshire (1881).

Scotland

The Earl of Orkney made the earliest recorded introduction of Red-legged Partridges to Scotland, to Orkney in 1840. Eighty years later they were apparently thriving only in Fife, and Ritchie (1920) described them as 'another introduction which has met with little success in Scotland'.

In the 1970s, they were introduced with mixed success to various localities in Scotland, including Rosehall in Sutherland, Cullen in Banffshire and Cortachy in Angus (all in 1970); and in 1973 on the Earl of Dundee's Birkhill estate near Cupar in Fife and at Killearn in west Stirlingshire. In the same year, they were recorded in Aberdeenshire, Kincardineshire, Angus (Strathcathro and Cortachy) and Galloway. In 1974 one hundred were released in Caithness.

Wales

Although Red-legged Partridges have been fairly widely introduced in the principality, the species is only currently well established in parts of Glamorganshire, Brecknockshire, Radnorshire, Montgomeryshire and Denbighshire.

Ireland

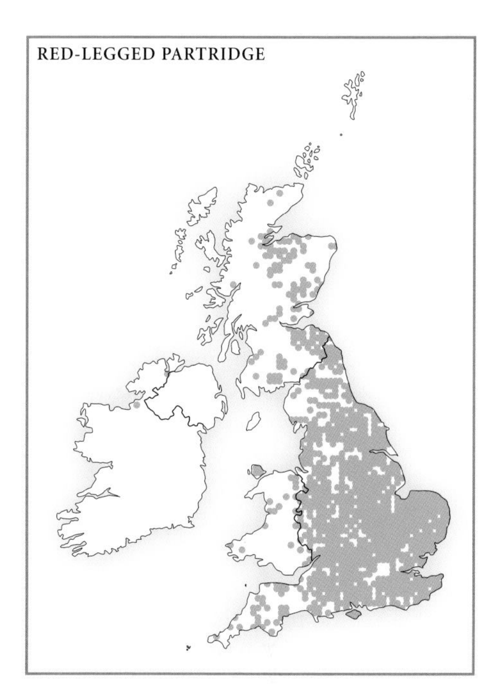

Red-legged Partridges were first introduced to Ireland at Dungannon, Co. Tyrone, by Thomas Knox in 1767, and were presumably established in the wild by at least 1810, when they appeared on an Irish game list. In 1979–80 attempts were made to introduce further birds to Ireland, including to Counties Tipperary, Carlow, Louth, Dublin, Waterford, Limerick, Wexford, Wicklow, Monaghan and Kilkenny, but post-release breeding was recorded only in Tipperary, Louth, Dublin and Wexford. The species is not, however, currently self-sustaining in Ireland.

Characteristics

Principally a bird of open ground, the Red-legged Partridge favours a warm and dry climate, with a well-drained soil and a combination of low, bushy vegetation for shelter with more open areas for foraging. Throughout much of its range it is associated with arable farming, especially where sugar beet is grown. In England it is most at home on the dry, sandy or calcareous soil with a relatively mild climate and low rainfall of parts of the south and East Anglia. A decline in the population in 1930–60, partly due to an increase in intensive farming (with the concomitant use of pesticides, mechanization, irrigation and the removal of hedgerows) and heavy shooting pressure has been reversed, mainly by large-scale annual stocking.

Impact

In East Anglia, Red-legged Partridges cause some damage to sugar-beet seedlings. Where both species occur they hybridize with Chukar Partridges.

Red-legged Partridges from time to time 'dump' eggs in the nests of Grey Partridges, which may cause the latter to abandon these nests. They are not, however, implicated in the decline of the Grey Partridge, which has been due to agricultural intensification.

Because of its tendency to run rather than fly, and when in flight to scatter rather than to pack, and because it is not a good bird for the table compared to the native Grey Partridge, the Red-legged Partridge is not highly regarded as a game-bird.

References

Aebischer & Lucia, in Hagemeijer & Blair 1997; Fitter 1959; Harting 1883; Harrison 1988; Holloway 1996; Leicester, Earl of 1921; Lever 1977, 1978, 1980, 1987, 1994, 2005; Long 1981; Mead 2000; Potts, in Gibbons et al. 1993; Robertson, in Holmes & Simons 1996; Raven et al. 2007; RSPB 2006; Sharrock 1976; Tomlinson 2006; Toms, in Wernham et al. 2002.

PHASIANIDAE (TURKEYS, GROUSE, PHEASANTS & PARTRIDGES)

COMMON PHEASANT *Phasianus colchicus*

Natural Distribution S Palaearctic and NE Oriental regions: in E Europe in parts of the Caucasus Mountains; in Asia from N Asia Minor E to Korea, China, Taiwan and Japan.
Naturalized Distribution England; Scotland; Wales; Ireland.

England

The earliest documentary evidence of Common Pheasants in Britain occurs in a manuscript (in the British Museum) of about 1177 entitled *De inventione Sanctae Crucis nostrae in Monte Acuto, et de ductione ejusdem apud Waltham*, which contains details of rations specified by the Earl of the East-Angles and West Saxons (later King Harold II) for the canons' household at the monastery of Waltham Abbey in Essex in 1058–9.

> *Erant autem tales pitantiae* [rations or commons] *unicuique cananico: a festo Sancti Michaelis usque ad caput jejunii* [Ash Wednesday] *aut xii merulae* [Common Blackbirds], *aut ii agauseae* [? Black-billed Magpies] *aut ii perdices* [Grey Partridges], *aut unus phasianus* [Common Pheasant], *reliquis temporibus aut ancae* [geese] *aut gallinae* [domestic fowl].

This document shows that one Common Pheasant was assumed to provide the same amount of food as a dozen Common Blackbirds or two Black-billed Magpies or a brace of Grey Partridges. It also proves that Common Pheasants were certainly known in Britain, if only in captivity, before the Norman Conquest, having perhaps been introduced in the early years of the reign (1042–66) of the Saxon king, Edward the Confessor.

Exactly when Common Pheasants started to become established in the wild in England is uncertain, but Sir William Dugdale (1605–86) in his *Monasticon Anglicanum*, published between 1655 and 1673, quotes from a charter of 1089 in which the Bishop of Rochester in Kent assigned to the monks of that city 16 Common Pheasants, 30 geese and 300 fowl from 4 separate manors. Two years later a licence was granted to the Abbot of Malmesbury to kill Brown Hares and Common Pheasants. In 1249, the Sheriff of Kent was ordered to produce 24 of the birds for a feast for Henry III. Fifty years later, the Bursar's Roll for the monastery at Durham lists '*In xxvj perdicibus* [Partridges] *et uno fesaund* [Pheasant]' – the latter costing four pence.

From the 12th century, the species began to appear with increasing frequency in English literature. *Sir Orfeo* (a metrical romance of about 1320) shows that by that time it had become fairly plentiful in the countryside: 'Of Game the fonde grete haunt,/Fesaunt, heron and cormerant.' Thomas Percy's *Reliquies of Ancient English Poetry*, published around 1320, first distinguishes between the sexes, when it speaks of 'partrich, fesaunt hen, and fesaunt cocke'.

The *Vision of Piers Plowman* of about 1377 makes an early, albeit negative, reference to the birds as an item of food: 'He fedde hem with no venysoun ne fesauntes ybake.' When George Neville (c 1432–76) was enthroned as Archbishop of York in 1464, no fewer than 200 'fessauntes' were served at the celebratory feast. Nicholas Upton (? 1400–57) in his *De Animalibus et de Avibus* (compiled before 1446) describes in detail how birds in captivity were reared and fattened for the table, and Alexander Barclay (? 1475–1552 describes, in his *Ecologues*, published in 1515, 'the crane, the fesant, the pecocke, and the curlewe' as birds suitable for a feast or banquet. Eighteen years later, Sir Thomas Elyot (? 1490–1546) in *The Castel of Helthe* gave it as his opinion that 'Fesaunt excedeth all fowles in swetenesse and holsomnesse.' In the domestic accounts for 1512 of Henry, Earl of Northumberland, a Common Pheasant was shown to cost 'xiid', and in the Privy Purse accounts of Henry VIII in 1532 provision was made for a salary payment to a French priest who acted as 'fesaunt breder' at the royal palace at Eltham in Kent, while in the household accounts of the Kytsons of Hengrave in Suffolk an allowance is made in 1607 for providing wheat to feed Grey Partridges, Common Quail and Common Pheasants.

Just when the species became fully naturalized in the English countryside is uncertain, but it was clearly sufficiently well established by the late 15th century to warrant legal protection from the Crown. A couplet from a ballad of this period entitled *The Battle of Otterburn*: 'The Fawkon and the Fesaunt both/Among the holtes on hee.' suggests that by this time they had spread at least as far north as Northumberland.

Scotland

The earliest mention of Common Pheasants in Scotland appears to have been made by John Lesley, Bishop of Ross, who in his *De Origine Moribus et Rebus Gestis Scotorum*

COMMON PHEASANT MALE CALLING, NORFOLK

(1578) recorded that 'fasianes with other nations are common, but are scarce with us'. An old Scots Act dated 8 June 1594 forbids the killing 'at any time hereafter [of] phesanis' and various other species in or near the royal policies.

It was, however, well over another 200 years before the species penetrated as far north as the Scottish Highlands. John Mowbray in his *Domestic Poultry*, published in 1830, said that 'in 1826 a solitary cock-pheasant made his appearance as far north as the valley of the Grampians, being the first that had been seen in that northern region'. Subsequently, the birds were introduced in the Highlands to Skibo in Sutherland in 1841, Monymusk in Aberdeenshire before 1842 and Shielding in Ross-shire around 1860.

Wales

Fitter (1959) refers to a 16th-century poem, *Asking for a Pheasant* by Gruffydd Hiraethog (in the Welsh Folk Museum), which claims that 'now they [Common Pheasants] fill everywhere in the region of Llanddwye' in Merionethshire. In the late 1580s, birds from Ireland were introduced to Haroldston West in St Bride's Bay, Pembrokeshire.

Ireland

The precise date when Common Pheasants were first introduced to Ireland is unknown, but Fynes Moryson, who travelled extensively in the country in 1600–3, wrote that there was 'such plenty of pheasants as I have known sixty served up at one feast'.

Introduced Common Pheasant Subspecies

At least five races of Common Pheasant have been introduced to Britain and Ireland, where the early introductions were of the nominate *P. c. colchicus* from Transcaucasia and Azerbaijan.

The earliest reference to the [Chinese] Ring-necked Pheasant *P. c. torquatus* (from eastern China) in Britain was by Thomas Pennant in his *British Zoology* (1766), where he recorded that: 'Mr Brooks the bird-merchant in Holborn, shewed us a variety of the common pheasant … the male of which has a white ring round its neck …'

In 1783, John Latham wrote that 'the ring-necked pheasant … is now not uncommon in our aviaries'. Four years later he continued:

> I have scarce a doubt but that these birds will hereafter become fully as plentiful in this Kingdom as the Common Pheasant. It is well known that several noblemen and gentlemen have turned out many pairs into their neighbouring woods, for the purpose of breeding.

In *A General History of Birds* (1821–8), the same author wrote of *torquatus* that:

> These were, it is said, first introduced by the late Duke of Northumberland [d 1786] by the name of Barbary Pheasants [the name given also by George Montagu in his *Ornithological Dictionary*, 1802], and many were bred and turned out at large, at His Grace's seat at Alnwick. Lord Carnarvon did the same at Highclere in Berkshire, and the late Dowager Duchess of Portland [d 1785] at Bulstrode, Buckinghamshire, besides many private gentlemen, by which means the breed is daily more common; it is true that these mix and breed with the Common Sort, and that in such produce the ring on the neck is less bright, and sometimes incomplete.

The earliest colour image of *torquatus* appears to be that in W. Haynes's *Portraits of Rare and Curious Birds … from the Menagery of Osterley Park* (1794).

In 1898, the Prince of Wales's Pheasant *P. c. principalis* (southern Turkestan and north-western Afganistan) and the Mongolian Pheasant *P. c. mongolicus* (Kirgizskaya and

Chinese Turkestan) were introduced to the Isle of Bute off the west coast of Scotland (and the former also to Norfolk and Kent) by a Colonel Sutherland and Lord Rothschild. Pallas's Pheasant *P. c. pallasi* (south-eastern Siberia and central Manchuria) was released in Norfolk before 1930, and *P. c. satscheunsis* of western Kansu in Kent in 1942. All these forms have freely interbred (many of the present stock of cocks showing some trace of the *torquatus* white neck-ring, see picture, page 165) and there are now no pure Common Pheasants in Britain or Ireland, where the populations are annually supplemented by the release of birds for sporting purposes.

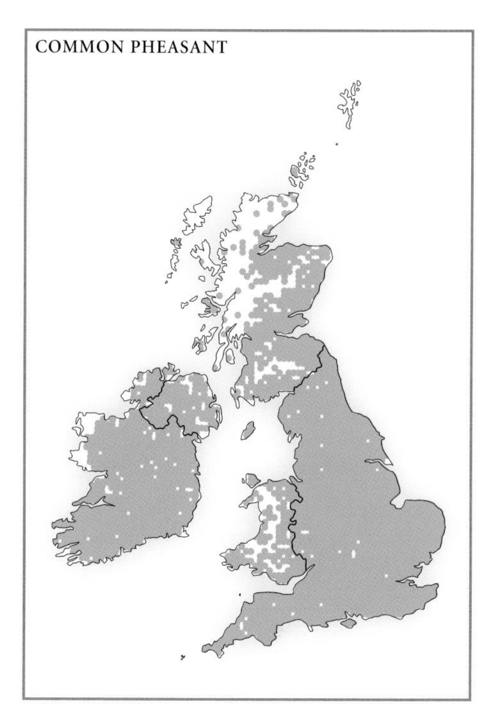

Summary of Status

Summing up the status and abundance of Common Pheasants in Britain, Bijlsma and Hill (in Hagemeijer and Blair 1977) say the country: 'has the highest population (though not necessarily the highest densities) of any European country as a consequence of the interest in game-shooting and the release of probably 15m [million] pheasants each year for shooting'.

Characteristics

Common Pheasants are extremely catholic in their choice of habitat, favouring especially partly cultivated and partly wooded country, with areas containing thick undergrowth and dense plantations well supplied with water. They thrive both on the light, sandy soil of heathland and on heavy clay soil, and in damp and marshy districts overgrown with reed and sedge beds.

Impact

The high economic and recreational value of Common Pheasants more than compensates for the small amount of agricultural damage they do, though in May–July in East Anglia they eat the seedlings of commercially valuable sugar beet.

Pheasants sometimes 'dump' their eggs in the nests of native Grey Partridges, which can cause the latter to abandon their nests. Concern has also been expressed about competition between Common Pheasants and Black Grouse on the latter's lekking[1] grounds, where cock birds have been seen disturbing the displays and attempting to mate with grey hens.[2] However, neither egg 'dumping' nor display disturbance seems prevalent enough to impact seriously on the native species.

It is possible that Common Pheasants may act as a reservoir for parasites, such as the gapeworm *Syngamus trachae* and the caecal nematode *Heterakis gallinarum*, and transmit them to other game-bird species, though this is as yet unproven.

On the credit side, shooting of the birds occurs in around 60 per cent of lowland woods, and is an important incentive for planting, management and retention of these features. Such planting of trees and shrubs, and the management of rides and woodland coppicing, are of considerable benefit to a wide variety of butterflies, songbirds and flowering plants. The control of such predators as Red Foxes, Carrion Crows and Black-billed Magpies to protect Common Pheasants is also, of course, of benefit to other species, although the decline of gamekeeping in the latter half of the 20th century has reduced the control of all predators, many of whose populations are increasing. Finally, the provision of supplementary food for the birds in winter benefits many songbirds, although it also helps to raise Grey Squirrel densities, with a consequent increase in bark stripping and egg and chick predation.

References

Fitter 1959; Baker et al. 2006; Gibbons et al. 1993; Gladstone 1923, 1924, 1926; Lever 1977, 1987, 2005; Long 1981; Lowe 1930, 1933; Matheson 1956, 1957; Mead 2000; O'Gorman 1970; Ritchie 1920; Robertson, in Holmes & Simons 1996; Sharrock 1976.

Notes

1. Patches of ground used for communal display in the breeding season by the males of such birds as Black Grouse.
2. Female Black Grouse are called grey hens; males are known as black cocks.

PHASIANIDAE (TURKEYS, GROUSE, PHEASANTS & PARTRIDGES)

GREEN PHEASANT *Phasianus versicolor*

(JAPANESE PHEASANT)

Natural Distribution Japan.
Naturalized Distribution England; [Scotland].

England

Because the Green Pheasant was for many years (and by some authorities still is) treated as a subspecies of the Common Pheasant, there are relatively few specific references to it in the ornithological literature.

The Green Pheasant is first mentioned in Britain by John Gould in his *Birds of Asia* (1857):

> About the year 1840, living examples were brought from Japan to Amsterdam, and of those a male and a female were purchased by the late Earl of Derby at a very high price: unfortunately, the female died before reaching the menagerie at Knowsley [Cheshire]. ... It is from this single male and a female of the common species that all the Green Pheasants, now becoming so numerous in the British Islands, have sprung. ... John Henry Gurney, Esq. of Norwich, and other gentlemen, became the possessors of the less pure stock [from Knowsley]. Some of Mr Gurney's birds were turned out in the woods at Easton [19 km west of Norwich].

In 1864, some Green and Common Pheasant hybrids were bred in Sussex by the Acclimatisation Society, which in the same year 'acquired by purchase three fine specimens, one male and two females, of a cross between Versicolor and Torquatus Pheasants'.

GREEN PHEASANT MALE

During the remainder of the 19th century and into the early 20th century, further consignments of Green Pheasants were imported into Britain. In England they were turned down in Bedfordshire, Cheshire, Cornwall, Cumberland, Durham, Gloucestershire, Herefordshire, Kent, Norfolk, Northamptonshire and Northumberland. In 1932, however, Seth-Smith wrote that 'for the last twenty years or more not more than some half-a-dozen pairs would appear to have reached Europe'.

The Green Pheasant is almost certainly the origin of the melanistic mutant (so-called '*tenebrosus*'), which first appeared in England in 1867, having failed to develop in the wild pheasant population during the past 800 years. These melanics began increasingly to occur in various localities during the 1920s.

Since 2000, a small and apparently expanding population of Green Pheasants, possibly of hybrid stock, has become established in Norfolk. In 2002, up to 12 cocks were present throughout the year at Worstead, and others were observed at Frettenham and North Walsham, in the north-east of the county, where the species seems to be persisting, though the localities above are up to 10 km west of the original release site. The highest count of 81 occurred in the Colby/Banningham area, some 24 km north of Norwich, in October 2005. Further undocumented populations may occur in other parts of England.

[Scotland]
In the late 19th or early 20th century, Green Pheasants were liberated on the Isle of Bute on the west coast by the Marquess of Bute, and also in Dumfriesshire.

References
Fitter 1959; Holling & RBBP 2007; Holloway 1996; Lever 1977, 1987, 1992, 2005; Mead 2000; Ogilvie & RBBP 2004; Seth-Smith 1932; Sharrock 1976.

PHASIANIDAE (TURKEYS, GROUSE, PHEASANTS & PARTRIDGES)

GOLDEN PHEASANT *Chrysolophus pictus*

Natural Distribution From N Guangxi and N Guangdon to S Gansu and S Shaanxi, China.
Naturalized Distribution England; [Scotland]; [Wales]; [Ireland].

England

The first Golden Pheasant in captivity in England was in the collection of Eleazar Albin, who describes it in his *History of Birds* (1731–8). G.C. Bompas in his *Life of Frank Buckland* (1888) suggests 1725 as the year of the Golden Pheasant's arrival in Britain.

The fourth annual report of the recently formed Acclimatisation Society recorded that 'From China were shipped, late in the year 1864, three silver and three golden pheasants, out of which only two arrived alive'.

Before 1880, Golden Pheasants were introduced as game-birds in both Norfolk and Suffolk – in the latter at Elveden Hall in Suffolk by the Maharaja Victor Duleep Singh – and in the following decade others were turned down at Tortworth in Gloucestershire where, as in East Anglia, they were soon established and breeding. In 1925, Lord Montagu of Beaulieu introduced Golden and Lady Amherst's Pheasants to Beaulieu Manor near Southampton, where they soon began to hybridize. Before 1942, Golden Pheasants were

GOLDEN PHEASANT MALE, HERTFORDSHIRE

liberated near Sevenoaks in Kent. Before 1949, the Earl of Iveagh released some at Elveden Hall, and since that date they have been introduced at Whipsnade, Bedfordshire; Exbury, Hampshire (by Edmund de Rothschild); and in Stockley Wood (New Forest), Hampshire. In 1952, they were breeding at Hinton Admiral near Bournemouth in Hampshire, having dispersed from nearby Hurn Court after a forest fire. Twenty years later, Golden Pheasants were reported in East Anglia from Santon Downham, Swaffham Heath, St Helen's Well, Shadwell, West Tofts, Rondham, West Harling and East Wretham, and at Holkham on the north Norfolk coast, and in parts of Charle and Maulden Woods south of Bedford.

In 1968–71, a pair frequented the Kingley Vale Nature Reserve on the South Downs, 4 km north-west of Chichester in East Sussex.

In 1975, the Dorrien-Smith family released some on Tresco in the Isles of Scilly. In 1996, several pairs bred on Tresco, and towards the end of the decade the species was also reported from the Brecklands of south-west Norfolk and north-west Suffolk – especially between Thetford and Brandon and in Thetford Chase. Smaller populations occurred in the Sandringham/Wolferton area of north-west Norfolk and on the South Downs of Hampshire and West Sussex, where Golden Pheasants had been introduced in the 1960s and 1965 respectively. In 1999, a small population was reported at Lytham Hall in Lancashire, where an introduction had been made in 1993.

In 2002, at least 64 Golden Pheasants occurred in 21 localities in 5 counties. In Devon, single males were observed in two localities. In Lancashire and North Merseyside, 9 cocks and 1 hen were seen in Lytham Hall Woods. In Norfolk, Golden Pheasants were regularly recorded at 3 sites – Wayland Wood (up to 11), Hockham Woods (10) and Wolferton (3); in addition, there were one or two sightings of 1–5 birds at 7 other sites. The Norfolk total of 39 birds was 10 less than that counted in 2000. In Suffolk, 6 cocks were reported in 5 separate localities. In Sussex, 3 pairs and a single bird were noted in 3 locations; however, the 3 estates that in 2000 had contained up to 45 pairs were not included in the survey for 2002.

In 2003, the English population of Golden Pheasants was estimated to number at least 42 birds, most of which occurred in only 2 counties. They were recorded at 3 regular and 9 further sites in Norfolk, with a maximum total of 32 individuals, and at 4 sites (with 4 calling cocks at one) in Suffolk. In addition, a single breeding territory was reported near Petersfield in Sussex, and in May 2 cocks were seen elsewhere. In 2004, when the total English population was a minimum of 33, the species was recorded in Norfolk at 3 regular and 8 additional sites, with a maximum total of 26 birds, and at 3 locations with at least 6 individuals in Suffolk. A minimum of 34 were counted in 2005, when they were present at 4 regular and 4 other locations in Norfolk, with a maximum total of 18, and in Suffolk a total of

16 birds was counted, with at least 10 at one site in the Brecklands. There were no reports from the Lancashire and North Merseyside site at Lytham Hall, but a single bird was reported elsewhere in May.

[**Scotland**]
J.A. Harvie-Brown and T.E. Buckley in *A Vertebrate Fauna of Argyll and the Inner Hebrides* (1892) record that Golden Pheasants were introduced to several estates in the Western Highlands, 'perhaps to none more successfully than to Gigha Island [off the coast of Kintyre] about the mansion-house [near Ardminish]'. In about 1895, the Duke of Bedford planted some Golden x Lady Amherst's hybrids at Cairnsmore near Newton Stewart, whence they spread to neighbouring estates, and in the same year the Marquess of Bute released some at Mount Stuart on the Isle of Bute, where they interbred with Lady Amherst's and Common Pheasants released at the same time. Others are said to have been freed on the island of Mull. In 1902, Sir Herbert Maxwell, Bt liberated more at Monreith in Wigtownshire. In 1974, Golden Pheasants were observed at Penninghame north-west of Newton Stewart in Galloway; on the coast at Creetown; and in Kirroughtree Forest, east of Newton Stewart, which held the largest population of some 250 birds, which by the 1990s had declined to only 10. Also in 1974, a few pairs were breeding in Cardrona Forest, Peeblesshire. All these populations have since died out.

[**Wales**[1]]
In 1963, to celebrate their wedding anniversary, the wife of Sir Richard Williams-Bulkeley, Bt gave her husband a pair of Golden Pheasants. After a heavy gale destroyed their enclosure, the birds escaped into the neighbouring woods at Baron Hill near Beaumaris on the Isle of Anglesey; the woods had a dense understorey of Common Rhododendrons that provided the birds with ideal shelter. Here they became established and bred, spreading throughout the woodland around Beaumaris and at least as far as Menai Bridge, some 10 km south-west along the Menai Strait.

A few Golden Pheasants survived in the Baron Hill woods until the mid- to late 1990s, when a combination of poachers, Weasels and Red Foxes (which in the mid-1970s had been introduced to Anglesey as cubs by workmen building the nuclear power station at Wylfa, near Holyhead) eradicated them. John Pickering is believed to have introduced Golden Pheasants (and Silver Pheasants) to the Llys Dulas estate in north-eastern Anglesey, which probably accounts for the arrival of a lone cock of the latter species at Baron Hill in 2005; this individual disappeared two years later, and both species have also died out at Llys Dulas. These populations have since died out.

[**Ireland**]
Golden Pheasants have occurred in the wild in Co. Down.

Summary of Status
There is evidence of a recent general decline in the population of Golden Pheasants in Britain, where in the 1990s the population was estimated to number 1,000–2,000 – much the same as it was in the 1980s, although due to the species' skulking habits numbers are likely to be underestimated. The birds favour coniferous woodlands – especially forestry plantations – and sometimes mixed coniferous/deciduous woods. In Galloway they were previously most numerous in 15–30-year-old stands of Sitka Spruce, Douglas Fir, Scots Pine and larches. In the Brecklands of East Anglia in England, too, they thrived after the

planting of softwoods on extensive tracts of heathland. Anecdotal evidence, however, suggests some decline in Thetford Chase, Norfolk, possibly resulting from a combination of the abandonment in the 1950s of large-scale releases, inbreeding, hybridization with Lady Amherst's Pheasant and increased predation as the number of gamekeepers declined. D. Goodwin (personal communication 2004) believes that predation by Northern Goshawks (which seldom occurred in the wild when Golden Pheasants were being introduced) and other raptors may be an important factor in the species' status in Britain where, because of its rarity in China – it is listed by the IUCN as Near Threatened – the population is of conservation significance.

Impact

Where they occur together the Golden Pheasant hybridizes freely with the closely related and genetically similar Lady Amherst's Pheasant, eventually obliterating the latter's characteristics since the offspring of hybrids eventually revert to the former. Because of their skulking habits, reluctance to fly and poor flying ability, and the pugnacity of cocks, Golden Pheasants are (like Lady Amherst's Pheasants) regarded as a poor game species.

References

Balmer et al., in Holmes & Simons 1996; Fitter 1959; Harrison 1988; Holling & RBBP 2007; Holloway 1996; Lever 1977, 1987, 1992, 2005; Maxwell 1905; Mead 2000; Raven et al. 2007; Rehfisch, in Hagemeijer & Blair 1997; Robertson & Hill, in Gibbons et al. 1993; Robertson, in Holmes & Simons 1996; RSPB 2006; Sharrock 1976; Ogilvie & RBBP 1999, 2004; Tomlinson 2006; Toms, in Wernham et al. 2002.

Note

1. For information on Golden Pheasants in Wales, I am indebted to Sir Richard Williams-Bulkeley, Bt (personal communication 2007).

PHASIANIDAE (TURKEYS, GROUSE, PHEASANTS & PARTRIDGES)

LADY AMHERST'S PHEASANT *Chrysolophus amherstiae*

Natural Distribution From N and NE Myanmar and N Yunnan to W Suchuan and W Guizhou, China.
Naturalized Distribution England; [Scotland]; [Wales].

England

Lady Amherst's Pheasant[1] was first introduced to England in 1828, although the earliest confirmed record of breeding (in captivity) was not until 1871. Subsequently, introductions in the wild were made at:

- Woburn Abbey, Bedfordshire by the Duke of Bedford in the 1890s.
- Beaulieu Manor, Hampshire, by Lord Montagu of Beaulieu in 1925 and again in 1958.
- Richmond Park, Surrey in 1928–9 and in 1931–2, when 24 were released.

- Mentmore, near Leighton Buzzard, Buckinghamshire in 1930 by the Earl of Rosebery.
- Whipsnade Park, Bedfordshire in 1931–2.
- Elveden, Suffolk in 1950 by the Earl of Iveagh.
- Later at Exbury by Edmund de Rothschild and in Stockley Wood in the New Forest, both in Hampshire.
- In 1984, more birds were released at Woburn, from where they spread some 5 km west to Brickhill in Buckinghamshire.

Especially at Beaulieu and Elveden, where the species was released with the closely related and genetically similar Golden Pheasant, hybridization between the two soon obliterated the characteristics of the former, since (as mentioned under the preceding species) the offspring of the hybrids reverted to the latter type.

At Woburn and Whipsnade, however, pure Lady Amherst's Pheasants thrived and spread along the Greensand Ridge to Heath, Reach, Battlesden, West Woburn, Aspley Heath, Woburn Park, Eversholt, Stepingley, Millbrook, Clophill, Haynes and Old Warden. In this well-wooded area the birds favoured primarily mixed plantations of Scots and Corsican Pines, Common Ash and Pedunculate Oak, with some Common Beech, Sycamore and Sweet Chestnut, with a dense understorey of Common Rhododendrons, and coniferous thicket on clay.

By the mid-1970s, Lady Amherst's Pheasants were breeding in limited numbers in the woods of south Bedfordshire, e.g. in Charle Wood, Flitwick Plantation, Eversholt, Old Warden, Maulden Wood and Breakheart Hill, and in the park at Luton Hoo, where they had been released in the 1950s by Sir Harold Werner, Bt, as well as in some neighbouring parts of Buckinghamshire (Brickhill and Mentmore) and at Ashridge in Hertfordshire, to which they may have dispersed naturally from Luton Hoo, where in 1971 at least four young birds were seen in the park. In the same year, a pair successfully reared young in

LADY AMHERST'S PHEASANT MALE

Warden Great Wood, while in 1972 a pair bred in Maulden Wood.

Since 1971, Lady Amherst's Pheasants are known to have bred successfully in Bedfordshire only at Luton Hoo (in seven years in 1971–2001), in Charle Wood (in six years in 1979–89), in Warden Great Wood (1971) and in Maulden Wood (several times in 1972–91). Around 1973, breeding was confirmed at Fakenham, Thetford, Guist and Quidenham in Norfolk. Given the difficulty, due to the species' secretive nature, in proving that breeding has occurred, it seems probable that successful nesting has taken place in other years and at other locations.

In the late 1960s/early 1970s, the British population of Lady Amherst's Pheasants (almost exclusively in Bedfordshire and, to a lesser extent, Buckinghamshire) was estimated to number some 100–200 pairs. The Bedfordshire population was said to be around 250 birds, the largest individual subpopulations occurring in Washers and Dainty Woods (25), Chicksands Wood (10) and Charle Wood (40). In the last-named locality, 30–40 young were reared in 1982, 35 in 1983 and 15–20 in 1984.

By the 1980s, the population in Bedfordshire had fallen to an estimated 100–200 individuals, and their range, especially in the north-east, had considerably contracted. By 1990, this population had apparently sunk to 60, and by the following year to only 46. Nevertheless, in this area 84 cocks and 29 hens were seen in 1995, 68 cocks and 19 hens in 1996, but only 48 cocks and 10 hens in 1997. The highest numbers were counted at Luton Hoo (up to 15 males and 3 females), Maulden Wood (5 males and 1 female), Woburn Park (10 males and 3 females) and Charle Wood (3 males). These totals are typical of the almost universal preponderance of males over females in the British population, which must have been a contributory factor to the failure of the species' long-term survival. The remainder of the Bedfordshire population was dispersed in other woods along the Greensand Ridge.

By extrapolating the results of later actual surveys, the number of birds in 1999 was estimated to have fallen to only some 75 individuals, and to only 30–40 by the turn of the century. In 2001–3, there were no records from the previous stronghold of Maulden Wood, and the last confirmed sightings at Luton Hoo and in Charle Wood were in 2002. By 2003–5, the maximum number counted at any one locality was three males. Nightingale (2005) believed that the population in Bedfordshire and Buckinghamshire amounted to no more than 20 cocks and an indeterminate number of hens.

There are no confirmed breeding records for Hertfordshire, where the very few individuals recorded have probably been either escapees or releases from avicultural collections, or stragglers from nearby Luton Hoo.

Although Tomlinson (2006) claimed that a viable population of rather less than 100 survived in the Brickhills area of Buckinghamshire, the population there had all but disappeared, and the species was apparently in terminal decline. Nightingale (2005), who comprehensively reviewed the status of Lady Amherst's Pheasants in Britain, concluded that, in spite of the fact that because of its skulking nature the species is, like the Golden Pheasant, probably under-recorded, extinction by 2010 (or even earlier) was almost inevitable. Extinction of the English population would be doubly unfortunate because in the Far East (where it is listed by the IUCN as Near Threatened) its status is uncertain. The English population is of conservation significance not only in its own right, but also because it has provided quantitative data presently lacking in Asia and has also supplied Chinese ornithologists with an opportunity to receive technical training in a native species. The further release of Lady Amherst's Pheasants in Britain is precluded by Section 14 of the Wildlife and Countryside Act 1981.

Why has the population of Lady Amherst's Pheasants in Bedfordshire (and Buckinghamshire) declined so dramatically in recent years? Nightingale (2005) suggests a combination of reasons:

- Predation by Red Foxes, Black-billed Magpies and Northern Goshawks – as pointed out by D. Goodwin (personal communication 2004), the latter were virtually absent from Britain when Lady Amherst's Pheasants were being introduced.
- Destruction of the woodland understorey (so vital for shelter and nesting) by the browsing of also introduced Reeves's Muntjac.
- Human disturbance.
- It may be that feeding in winter (when natural food stocks are low) of Common Pheasants by gamekeepers helped to maintain the population of Lady Amherst's Pheasants, and that the recent curtailment of game management on some estates has affected the latter's survival.
- As the population declined, the absence of suitable wildlife corridors would, in many localities, have inhibited intercourse between sub-populations, leading to isolation and thus inbreeding, with the probability of reduced fertility and increased susceptibility to disease.
- At winter feeding stations, Lady Amherst's cocks tend to be aggressive to Common Pheasants, which may have led gamekeepers to try to get rid of the former.
- Probably one of the main reasons for the decline of Lady Amherst's Pheasants in Britain has been (as in the case of many other species worldwide) loss of habitat, almost invariably as a result of human interference. This has been mainly through the clear-felling in the 1950s and 60s of uneconomic deciduous hardwoods and their replacement by valuable coniferous softwoods. The latter, which are now maturing, lack the understorey so necessary to Lady Amherst's Pheasants for nesting and shelter. Furthermore, some woodlands are being withdrawn from the commercial sectors and converted to amenity or recreational use, with a consequent increase in human disturbance.

[Scotland]

Lady Amherst's Pheasants (both pure-bred and hybrids with Golden Pheasants) were introduced in about 1895, without lasting success, to Mount Stuart on the Isle of Bute (Strathclyde) by the Marquess of Bute, and to Cairnsmore near Newton Stewart in

Wigtownshire (Dumfries and Galloway) by the Duke of Bedford. Golden Pheasants were released at the same time, and the latter soon obliterated the former's characteristics. In 1971, Lady Amherst's Pheasants were found to be hybridizing in Galloway with Golden Pheasants; these may have been the descendants of the late 19th-century release or possibly of more recent ones.

[Wales]

In the mid-20th century, an attempt was made to establish Lady Amherst's Pheasants at Halkyn in Clwyd, from where they spread to the neighbouring Gwynsaney estate. The population peaked at 40–150, but had apparently died out by around 1998.

Impact

Because of its skulking habits, reluctance to fly and poor flying ability, Lady Amherst's Pheasant is (like the Golden Pheasant) regarded as a poor sporting species, and is thus of little if any value as a game-bird.

References

Cannings 1999; Fitter 1959; Harrison 1988; Hill & Robertson, in Gibbons et al. 1993; Holling & RBBP 2007; Holloway 1996; Lever 1977, 1987, 1992, 2005; Lovegrove et al. 1994; McGowan & Refisch, in Hagemeijer & Blair 1997; Mead 2000; Nightingale 2005; Robertson, in Holmes & Simon 1996; Sharrock 1976; Tomlinson 2006; Toms, in Wernham et al. 2002; Trodd & Kramer 1991.

Note

1. Named after Sarah, first wife of William Pitt Amherst, 1st Earl Amherst of Arracan (1773–1857), who in 1823 was appointed Governor-General and Viceroy of India.

PHASIANIDAE (TURKEYS, GROUSE, PHEASANTS & PARTRIDGES)

REEVES'S PHEASANT *Syrmaticus reevesii*

Natural Distribution From NE Sichuan, Hubei and Anhui to EC Nei Mongol and Hebei.
Naturalized Distribution England; [Scotland]; [Wales]; [Ireland].

England

The first Reeves's Pheasant[1] to reach England was one presented to the Zoological Society of London by John Reeves in 1831. Three years later the society announced that 'A second male specimen of the Reeves's Pheasant … has also been sent to the Menagerie by John Reeves, Esq.'[2] In 1894, Newton's *Dictionary of Birds* announced that 'Other species of Pheasant have been introduced to the coverts of England [including] *P. reevesi* from China.' These introductions included those made in 1870–90 by Lord Lilford at Lilford Park, Northamptonshire; by the Duke of Bedford at Woburn Abbey, Bedfordshire; and at Tortworth in Gloucestershire and Bedgebury in Kent. Only in Woburn Park did the species become established for any length of time. In 1950, the Earl of Iveagh released a small number of Reeves's Pheasants at Elveden Hall in Suffolk, and others were liberated

REEVES'S PHEASANT, EAST ANGLIA

in Cumberland (Cumbria) in 1969. In 2002, Reeves's Pheasants were reported in six localities in three counties, but no breeding took place. In Norfolk, displaying cocks were recorded at three sites in the Stanford Training Area. In Somerset, a small number were present throughout the summer at Ash Priors, and in Wiltshire three cocks were seen in May at Druid's Lodge.

In late March and early April of 2003 and 2004, eight displaying males were observed in the Stanford Training Area. In 2005, at least three males were displaying at the same site, and single males were observed at five other sites. So far, however, breeding has not been confirmed.

[Scotland]

In 1870, Lord Tweedmouth released a pair of Reeves's Pheasants, which he had acquired direct from Peking, at Guisachan in Inverness-shire. After a further 4 cocks had been introduced, the birds bred freely – over 20 being reared in the first year – but were inclined to wander. Gray (1882) said that at Guisachan:

more than 100 had been shot in the course of a single season, and the birds were found to be as hardy (the young indeed more so than) the commoner varieties of pheasant ... it was not too much to expect that in a very few years it would become thoroughly naturalised, and be found in considerable numbers all over the country.

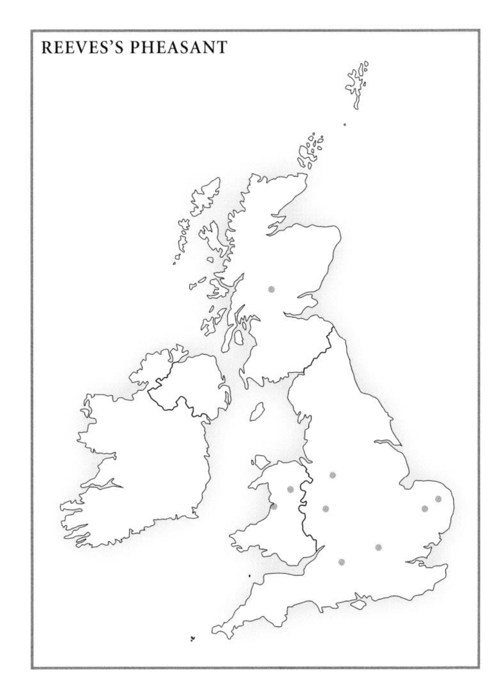

In 1892, a dozen cocks were counted, and by the middle of the decade the birds were well established at Guisachan and on the neighbouring estate of Balmacaan. Further attempts were made at this time to establish the species elsewhere, e.g. at Duff House and Pitcroy in Aberdeenshire; at Tulliallan in Fife by a daughter of Admiral Lord Keith; at Mount Stuart on the Isle of Bute by the Marquess of Bute and in Kirkcudbrightshire. In 1970, some 50 Reeves's Pheasants were liberated in Kinveachy Forest, Inverness-shire, where they were reported to be breeding until at least 1974, and around 70 were released unsuccessfully in Morayshire. More recently the species has occurred in Perth and Kinross.

[Ireland]

A single unsuccessful attempt to naturalize Reeves's Pheasants in Ireland was made at some time in 1870–90 by Colonel Edward H. Cooper at Markree Castle in Co. Sligo.

Impact

Reeves's Pheasant is a popular sporting species because it frequents hilly, densely wooded country little favoured by other game-birds, and because it is a strong and high flyer. Its failure to become properly established in Britain and Ireland has been attributed in part to the fact that the cocks, which fly long distances and lack the far-carrying call of other pheasants, became too dispersed for successful pairing.

References

Fitter 1959; Gray 1882; Harrison 1988; Holloway 1996; Holling & RBBP 2007; Lever 1977, 1987, 2005; Mead 2000; Ogilvie et al. 2004; Sharrock 1976.

Notes

1. The species is named after John Reeves (1774–1856), who in 1812–31 was in the employ of the East India Company in Canton, whence he shipped a number of species (e.g. Reeves's Muntjac) to England. In 1817 he was elected a Fellow of the Royal Society and of the Linnaean Society.
2. *Proceedings of the Zoological Society of London II* (34). 1834.

PSITTACIDAE (COCKATOOS & PARROTS)

ROSE-RINGED PARAKEET *Psittacula krameri*
(RING-NECKED PARAKEET)

Natural Distribution From S Mauretania and Senegal to Sudan and Somalia. Also NW Pakistan through India and Sri Lanka to SE China. (British individuals have characteristics of the two Indian subspecies *P. k. borealis* and *P. k. manillensis.*)
Naturalized Distribution England; Scotland; Wales; [Ireland].

England

Rose-ringed Parakeets were first recorded as breeding in the wild in Britain (at Northrepps Hall in Norfolk) in 1855, when in *Animal World* J.H. Gurney described how 'they have several times made nests and on five of those occasions the young have been brought to maturity'. They may also have occurred in the wild in south London in 1892–9. They bred at Loughton in Essex in 1930 and in the following year also in Northamptonshire.

The present naturalized population, however, dates from only the late 1960s. In 1969, a family party was observed at Southfleet, 13 km west of Rochester, in Kent, and in the same year others were reported from around Northfleet, Gravesend and Shorne in the same county. By the following decade, the species had become a familiar sight in the Gravesend area (where in 1975 a pair nested in the thatched roof of a farmhouse), from where it spread east to the Medway towns and to some of the marshland villages to the north. In 1972–3, a new colony became established based on Margate, where in 1975 up to six pairs are believed to have bred successfully in Northdown Park, and Rose-ringed Parakeets were regularly seen all over the Isle of Thanet south to Stodmarsh in the Stour Valley and as far west as Herne Bay.

In 1974, the species was observed in the Cuckmere Valley in East Sussex, and also around Chichester Harbour in West Sussex, where it may have bred in the Pagham Harbour Bird Reserve.

Also in 1969, small groups appeared in the neighbourhood of Bromley (Langley Park), Beckenham, Shirley (Monks Orchard), and Croydon (all on the Kent/Surrey border), where nesting was first reported in 1971. In the spring of 1972 and 1973, several birds were displaying and prospecting for nesting sites in Langley Park, where at least 20 were established in the latter year, and where at least one pair raised a single fledgling. In 1975, at least two pairs successfully reared three young in Langley Park.

From 1973, Rose-ringed Parakeets were reported from around Bexley, some 13 km east of Langley Park, mainly from the area near Tile Kiln Lane, Joyden's Wood and Green Street Green.

In 1970, a pair appeared between Esher and Claygate in Surrey, some 20 km west of Langley Park, where in 1971 it raised four fledglings. By the middle of the decade, most records in Surrey came from around Ashstead, Bookham, Chessington, Claygate, Esher and Surbiton, with lesser numbers from Dorking, Lightwater, Nonsuch Park, Send, Clandon Park (near Guildford), Vann Lake and Wellington.

From 1972, flocks of up to ten birds began to appear around Old Windsor in Berkshire and Wraysbury, Buckinghamshire. These are believed to have been derived from strays from a 'homing' flock in an aviary near Marlow in Buckinghamshire.

ROSE-RINGED PARAKEET FEEDING ON HORSE CHESTNUT FLOWERS, SOUTH-WEST LONDON

In or before 1971, another sub-population of up to 22 birds became established north of the Thames around Woodford Green and Highams Park in south-west Essex. Three years later, a further small group was reported at Herstmonceux in East Sussex.

By the mid-1970s, the birds had become established in Greater London, and in north-west England in the southern suburbs of Greater Manchester (first bred 1974) and Merseyside, Liverpool (first bred 1980).

By 1983, Rose-ringed Parakeets had been recorded from 50 British counties, though in few in Scotland and Wales and from none in Ireland, and breeding had been confirmed in Berkshire, Kent, Surrey, Sussex (since 1974), Greater London, Greater Manchester, Merseyside and West Yorkshire. The minimum population estimate by the early to mid-1980s was around 1,000 individuals, of which 300 occurred in Greater London and 200 in Kent.

Although by the late-1980s or early 1990s the sub-populations in north-west England had died out, perhaps due to inbreeding due to the absence of outside recruitment, elsewhere the species had increased both its number and distribution, and by the early 1990s the population had increased to several thousand. During the winter of 1994–5, a pandemonium[1] of 770 roosted at Walton-on-Thames in Surrey, and another 100 roosted on the Isle of Thanet, while several hundred more were distributed elsewhere, including new sub-populations in Berkshire, Hampshire and Hertfordshire.

In October 1996, a count at the birds' four main roosts (Esher, Lewisham, Ramsgate and Reigate) found at least 1,508 individuals, more than 1,100 of which occurred at Walton-on-Thames in Surrey. By the winter of 2000–1, the total British population, mainly west of London and north-east Kent, was 4,000–4,500. The principal west London pandemonium had moved from Wraysbury in Berkshire to Esher in Surrey, a distance of some 16 km, indicating the mobility of local sub-populations, and some birds are believed to have been commuting more than 20 km to and from roosts.

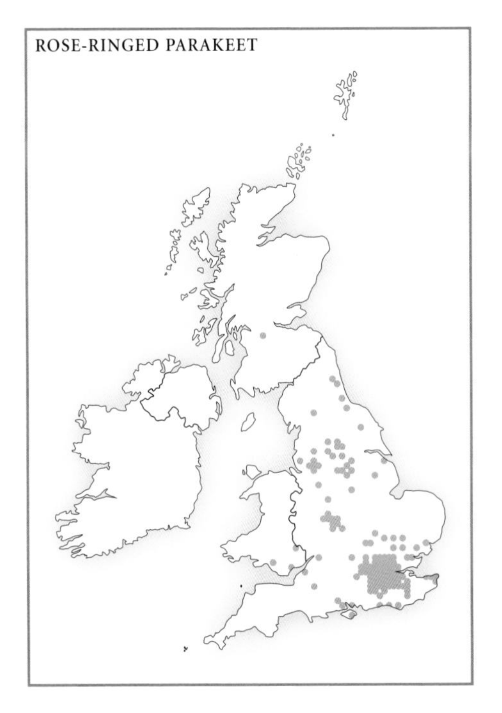

By the following winter of 2001–2, the population had increased still further to at least 5,900 birds, and the distribution had expanded westwards in Buckinghamshire and Berkshire, northwards into Middlesex, and further south within Surrey. Rose-ringed Parakeets occurred as far south as Gatwick Airport and Brighton in East Sussex, and a sub-population of a dozen had become established at Studland in Dorset. The maps in Butler (2002) indicate the south-east London sub-population to have extended from New Addington in the south to Walthamstow in the north, and from Croydon eastwards to Joyden. In south-west London, the birds ranged north from Crawley to Harrow, and from Marlow in the west eastwards to beyond Merton. In the Isle of Thanet, they occurred along the coast from between Herne Bay and Ramsgate, and inland to Brooksend and Cliffsend.

Table 10 Breeding and non-breeding records of Rose-ringed Parakeets *Psittacula krameri* in England, 2003–5

Year	County	Sites	Pairs	Breeding Results
2003	Buckinghamshire	6	46 birds	Juveniles at one site in June; nest-prospecting by 9 pairs at 2 sites in December
2003	Greater London	20	7	Breeding occurred
2003	Kent	?	?	Breeding, but no numbers available
2003	Surrey	At least 17	?	1 pair bred; 3 pairs probably bred; 26 pairs possibly bred; maximum roost count (in August) 6,818, with many coming from outside Surrey
2003–5	Sussex	?	20 birds p.a.	Believed to have bred, but no evidence
2003	Derbyshire	1	1	Probably bred (first record for county)
2004	Buckinghamshire	6	35 birds	No reported breeding
2004	Essex	1	1	2 young fledged (first breeding in county since 1980s)
2004	Greater London	4	?	?
2004	Hertfordshire	c 10	1–2 birds until autumn when 40 arrived	No reported breeding
2004	Kent	?	Maximum autumn roost (Thanet) 1,050	?
2004	Surrey	19	At least 38	8 pairs bred; at least 30 pairs possibly bred; roost counts over 6,000 pairs
2004	Derbyshire	1	1	Probably bred (but not in 2005)
2004	Lancashire	1	1	Probably bred, and at least 6 other birds also seen
2005	Berkshire	?	?	At least 1 pair bred; common along Thames in east of county
2005	Buckinghamshire	3	3	Bred at 2 sites; county total of 100–200 birds
2005	Hertfordshire	?	? 1–2	Bred
2005	Kent	?	250–300	Bred
2005	Cleveland	1	1 female	Present July to December. Prospecting nest holes

Source Adapted from Holling & RBBP 2007.

In the winter of 2001–2, the four main winter roosts had a total of 5,886 individuals, with a further 118 at a roost in Maidenhead in Berkshire. Small numbers were also reported in Cheshire, Norfolk and Suffolk.

The pandemonium in Esher, in the grounds of the local rugby club, had by 2005 reached an estimated total of 7,000. In 1994–2006, Rose-ringed Parakeet numbers in Britain are said to have increased by a staggering 302 per cent. In the latter year the birds were said to be present in every London borough. The current annual rate of increase in the population is around 30 per cent. Estimates of the present number in Britain (the vast majority in south-east England) vary from some 6,000 pairs breeding at 87 sites to 30,000 individuals. Predictions of the population by the end of the decade similarly vary from 50,000 to 100,000.

Scotland

Rose-ringed Parakeets were first recorded in Scotland in the late 1970s/early 1980s. Breeding in Scotland was recorded by Mead (2000), and sightings have been received from as far north as Glasgow.

Wales

In the late 1970s/early 1980s, Rose-ringed Parakeets were first reported in the Welsh borders, and by 1983 successful breeding had taken place on at least one occasion in Denbighshire (Clwyd).

[Ireland]

According to Marchant (2002: 715), 'Rose-ringed Parakeets have been encountered at many locations in … Ireland that are far distant from any known breeding site. Beyond the regular range, however, it is impossible to distinguish wanderers … from individuals that have recently escaped or been released from captivity.'

Sources of the Species in Britain

Various sources may have contributed to the British population:

- Birds used in the film *African Queen* at Shepperton Studios 19 km south-west of Ealing, west London, are believed to have been released after filming ceased.
- Escapes from pet shops and aviaries.
- Free-flying 'homing' birds that failed to return.
- Birds brought back by seamen returning to London, who released them when they realized the cost of enforced quarantine.
- The American guitarist Jimi Hendrix allegedly released a pair in Carnaby Street, then the centre of the trend-setting fashion trade, in central London in the 1960s.

Characteristics

Rose-ringed Parakeets in Britain nest in holes in trees (which they may enlarge) made by other birds, or occasionally in natural cavities. The nest is lined with a layer of wood detritus, on which is laid, between January and June, a clutch of 2–4 eggs, from which the young hatch synchronously after 22–4 days' incubation. Fledging takes place in 40–50 days, and the young become independent of their parents a few weeks later.

In Britain, Rose-ringed Parakeets are largely birds of the suburbs, where they feed on the bark, buds, leaves, flowers, fruits and seeds of well over 20 species of tree. When available, the fruits of the Holly, Common Hazel, Crab Apple, Black Mulberry, Common Beech, Monterey Cypress, cultivated apple, Swedish Whitebeam and Holm Oak are eaten, and various parts of Sycamores and Horse Chestnuts are taken throughout the year. Although Ring-necked Parakeets readily visit birdfeeders, because they have such a catholic diet and since their food trees are widely available, food provided by humans may be less significant for the species' survival in winter than was previously thought.

In autumn and winter, Rose-ringed Parakeets gather at dusk to roost in large pandemoniums, which may be several-thousand strong. These roosts break up at dawn as the birds disperse to feed – in autumn and early winter in parks and on farmland, but later in the season in suburban gardens.

The success of Rose-ringed Parakeets in becoming so well established in Britain can be attributed to a variety of factors:

- They have no natural predators.
- Being noisy and aggressive birds, they outcompete and even intimidate native species.
- They are tough and resilient, able to survive harsh winters apparently without the need for supplementary feeding.
- They have great longevity (up to well over 30 years).
- They reach sexual maturity at an early age.
- As one of the few species in Britain that breeds in winter, they have the choice of the best nesting sites.
- They have a high rate of fledgling survival.

Impact

The potential impact of Rose-ringed Parakeets is twofold – ecological and economic. Because they are such early breeders, they compete advantageously with native species for prime nesting sites. They mainly use the old nest cavities of Green and Great Spotted Woodpeckers, but the previous nests of other species such as Common Starlings, Common Kestrels, Little Owls, Tawny Owls and Eurasian Jackdaws have also been used, as allegedly have those of Wood Nuthatches, Great Tits and Tree Sparrows. Anecdotal evidence suggests that Wood Nuthatches may leave when Rose-ringed Parakeets enter a wood. With some of these species, and several others, the latter also compete for food. However, so far there is no evidence that any native species have been seriously affected.

Although in parts of their Asian range, Rose-ringed Parakeets are regarded as a serious agricultural pest, they have not so far become one in Britain, though the potential for them to do so in a fruit-growing county such as Kent must be considerable. In orchards they have caused damage to buds and the mature fruit of, especially, apples, pears and plums – they are wasteful feeders as they tend to spoil fruit by taking only one or two pecks before moving on. At least one English vineyard, Painshill Park, has suffered from Rose-ringed Parakeet depredations, when in one season their consumption of grapes reduced the anticipated output of 3,000 bottles to a mere 500. Two of the crops most seriously affected in India, maize and sunflowers, are being increasingly grown in Britain.

Rose-ringed Parakeets also pose a potential human health risk through psittacosis – a contagious avian disease caused by chlamydiae and transmissible (especially by parrots) to humans as a form of pneumonia.

In March 2007, the Department for Environment, Food and Rural Affairs and the RSPB commissioned the Central Science Laboratory to assess the potential threat posed by Rose-ringed Parakeets in Britain. In the meantime, individuals can apply to Natural England for a licence to control the birds: these will only be granted for purposes of conservation or the protection of crops, or for reasons of public health. (Such licences are required because, under the provisions of Section 14 (i) of the Wildlife and Countryside Act 1981, although it is illegal to release an alien animal in Britain, once such a species is deemed to have become 'wild' it is afforded protection under Part I of Schedule 9 of the same Act.)

The control of Rose-ringed Parakeets would inevitably be a bone of contention between those who value the presence of this attractive bird in their gardens, and conservationists and those whose livelihoods depend on the land. Whatever means of eradication were chosen, they would be likely to upset many local people, and the sight of dead and dying birds in a suburban setting, to say nothing of the potential danger to the public, might prove unacceptable.

References

Baker et al. 2006; Butler 2002; Chandler 2003; England 1970, 1974a, 1974b; Feare, in Holmes & Simons 1996; Gibbons et al. 2001; Harrison 1988; Hawkes 1976, in Lack 1986; Holloway 1996; Hudson 1974a, 1974b; Lever 1977, 1987, in Gibbons et al. 1993, 1994, 1994, 1996, in Hagemeijer & Blair 1997, 2005, in press; Long 1981; Macintyre 2005; Marchant, in Wernham et al. 2002; Mead 2000; Nature – Times News Service 1974; Raven et al. 2007; RSPB 2006; Sharrock 1976; Smith 2007.

Note

1. 'Pandemonium' is the collective noun for a group of parrots.

PSITTACIDAE (COCKATOOS & PARROTS)

MONK PARAKEET *Myiopsitta monachus*

Natural Distribution Bolivia, Brazil, Paraguay, Uruguay and N and W Argentina.
Naturalized Distribution England.

England

In 1987–98, a colony of up to 30 Monk Parakeets was present at Tiverton in Devon, and in 1988–93 a group of 9 individuals occurred at Barnton, Cheshire. In 1996–2001, a pair was established at Lonsdale Road Reservoir in Surrey, where it attempted, unsuccessfully, to breed in 1996 and 1997. In 1999, a small colony of up to nine individuals was observed at Castle Combe in Wiltshire, but no breeding was reported. Finally, in 2003 a pair probably bred at a site in Greater London, where it was still present in December. All these smaller colonies have since apparently died out.

Monk Parakeets were noted in the wild, in Borehamwood, Hertfordshire, in 1993, where by 1999 a colony of 17 had become established. By the following year the colony

MONK PARAKEET

had increased to 20, including 5–6 breeding pairs, and by 2001 to 32 individuals, which built a total of 7 nests. By the following year, when no breeding was reported, the colony had declined to a maximum of 32. In 2004, when the maximum count was again 32, a pair bred. In 2005, when at least 3 pairs bred, the minimum count of the Borehamwood colony was 40.

Characteristics

The Monk Parakeet is of special interest because, uniquely among some 330 species of parrot, it constructs a large communal nest of sticks, in which several pairs may nest and also roost.

References

Baker et al. 2006; Butler 2002; Gibbon et al. 1993; Lever 2005; Mead 2000; Ogilvie & RBBP 2001, 2002, 2004.

STRIGIDAE (OWLS)

LITTLE OWL *Athene noctua*

Natural Distribution Much of the Palaearctic N to around 57°N in Denmark and 50°N in Manchuria, S to about 27°N in W Africa (S Morocco) and 5°N in E Africa (S Ethiopia and Somalia).
Naturalized Distribution England; Scotland; Wales; [Ireland].

LITTLE OWL AT NEST WITH YOUNG

England

The first Little Owls[1] to reach England were some brought back from Rome by the splendidly eccentric but engaging Charles Waterton, 'Squire' (as he was affectionately known) of Walton Hall in Yorkshire. In the third series of his *Essays* (1858), Walton writes:

> Thinking that the civetta [Little Owl] would be peculiarly useful to the British horticulturalist, not, by the way in his kitchen but in his kitchen-garden, I determined to import a dozen of these birds into our own country ... I agreed with a bird-vendor in the market at the Pantheon for a dozen young civettas, and having provided a commodious cage for the journey, we left the Eternal City on the 20th July 1842, for the land that gave me birth. All went well, until we reached Aix-la-Chapelle. Here, an act of rashness on my part caused a serious diminution in the family. A long journey and wet weather had tended to soil the plumage of the little owls; and I deemed it necessary that they, as well as their master, should have the benefit of a warm bath. Five of them died of cold the same night. A sixth got its thigh broke. I don't know how; and a seventh breathed its last, without any previous symptoms of indis-

position, about a fortnight after we had arrived at Walton Hall. The remaining five ... have been well taken care of for eight months. On the 10th of May 1842 ... there being abundance of snails, slugs and beetles on the ground, I released them ... at seven o'clock in the evening, the weather being sunny and warm. I opened the door of the cage: the five owls stepped out, to try their fortunes in this wicked world.[2]

Having survived Waterton's predilection for warm baths and his altercations with an uncooperative Swiss bank official in Basle and interfering Italian customs officers in Genoa (Waterton's accounts of which in his *Essays* make entertaining reading), it is sad that his 'civettas' disappeared into the park at Walton Hall and were not seen again.

Apart from about 20 recorded examples, which were all probably vagrants from continental Europe, and one attempted introduction in the New Forest in Hampshire in the 1860s, there is no further mention of Little Owls in Britain until Lieutenant-Colonel E.G.B. Meade-Waldo released some at Stonewall Park, near Edenbridge in Kent, 'to rid belfries of sparrows and bats, and fields of mice':

We let out Little Owls first of all about 1874, and between then and 1880 about forty good birds went off. We know of one nest in 1879. In 1896 and again in 1900 I 'hacked off' about twenty-five. Since then they have been comparatively abundant all throughout the district, which is roughly between Tunbridge Wells and Sevenoaks. I know of generally some forty nests in a radius of some four or five miles [6–8 km]. With us their favourite haunts are old orchards and rocks.

From Stonewall Park, Little Owls spread north-west into Surrey, where they first bred in 1907, and south into Sussex, where nesting was first recorded in 1903. In Kent their principal early direction of expansion seems to have been to the north by way of Sevenoaks and Westerham: they arrived in Dartford by the early 1870s, Swanscombe near Gravesend by 1883 and the outer suburbs of London before 1897. In the east, they had penetrated as far as Shorne, between Gravesend and Rochester, before 1893, and had reached Cuxton, between Rochester and Maidstone, by the following year. In south and south-east Kent, they had arrived in Cranbrook by 1903, Bilsington, between Tenterden and Hythe, by 1906, and Boxley, north-east of Maidstone, a year later.

In 1876, a pair was released in the park of Knepp Castle in south-east Sussex, but by the following spring both birds had been killed. In the same year, the Earl of Kimberley released six birds in Kimberley Park near Wymondham in Norfolk, none of which survived for long. Similarly unsuccessful were Lord Rothschild, who in about 1890 tried to establish Little Owls at Tring Park in Hertfordshire, and W.H. St Quintin, who also failed with the species in the park at Scampston Hall, near Malton in East Yorkshire. Shortly afterwards, the latter wrote despairingly: 'This is not a well-wooded district and there are no rock crevices except along the coast ... I have given up any hope of naturalizing the birds here.' Subsequent introductions were more successful, including at East Grinstead in Sussex (1900–1), Essex (1905 and 1908) and Yorkshire again in 1905.

Principal credit (Coward 1920: 303. Series I unfairly describes it as 'blame') for the naturalization of the Little Owl in England belongs to that distinguished ornithologist Thomas Littleton Powys, 4th Baron Lilford, of Lilford Hall near Oundle in Northamptonshire, from whom was derived the bird's alternative early name, Lilford's Owl.

Writing to his friend, the Revd Murray Matthew, in May 1889, Lord Lilford said:

I turned down about forty little owls, about the house here and over a radius of some three or four miles [5–6 km] in the neighbourhood, early in July last. Several were too young to feed

> themselves, or, rather, to find their own food, and we recaptured more than half of those originally put out. A very few were found dead. Several were constantly seen about; during the summer and autumn of 1888 many disappeared entirely, but three or four were seen, and often heard, throughout the winter. On April 23rd, 1889, one of my keepers discovered a nest in the hollow bough of a high ash tree in the deer-park. [The first breeding record for Northamptonshire.] The old bird would not move, but on being gently pushed with a stick, two eggs were visible. On May 10th two young birds about a week old could be made out, and on the 22nd, four or five, all of different sizes. … This is encouraging, and I shall invest largely in little owls this summer, and adopt somewhat different treatment. Similar experiments have been tried out to my knowledge in Hants, Sussex, Norfolk and Yorkshire, but I do not know of a brood having been reared in a genuinely free condition in this country, till this lot of mine.

In February 1892, Lord Lilford wrote to another friend, E. Cambridge Phillips, that: 'You may be interested to hear that we have a little-owl (*Athene*) sitting on five eggs in a hollow tree not far off. I have turned out a great many of these birds during the past few years, and this is the fifth nest of which I have positive information.' Three months later, in May 1892, Lord Lilford informed Lieutenant-Colonel Meade-Waldo that: 'A pair, if not two, of little owls have taken their young off safely at no great distance.' Two days later he wrote again: 'I have not heard recently of any little-owls at a distance, and of no nests at more than two miles [3 km] from this.' In June of the following year, 1893, Lord Lilford wrote to the same correspondent: 'Last year we had a nest of little owls (*Athene noctua*), of which I have turned out a great many, in an ash-stump about two miles [3 km] off.'

In the course of an address to the Northamptonshire Field Club in February 1894, Lord Lilford said:

> For several years past I have annually set at liberty a considerable number of the little owl, properly so-called (*Athene noctua*), from Holland, [*A. n. vidalii*] and several pairs of these most amusing birds have nested and reared broods in the neighbourhood of Lilford. It is remarkable that, although this species is abundant in Holland, and by no means uncommon in certain parts of France, Belgium and Germany, it has been rarely met with in a wild state in our country. I trust, however, that I have now fully succeeded in establishing it as a Northamptonshire bird … They are excellent mouse-catchers, very bad neighbours to young sparrows in their nests, and therefore valuable friends to farmers and gardeners. The nest of this owl is generally placed either in a hollow tree at no great height from the ground, or in vacant spaces in the masonry of buildings … Never destroy or molest an owl of any sort. I consider all owls as not only harmless, but most useful.

Summarizing the then position of the Little Owl in and around Northamptonshire, Lord Lilford wrote in his *Notes on the Birds of Northamptonshire and Neighbourhood* (1895):

> I have succeeded in establishing the Little Owl as a resident Northamptonshire bird … I have for a considerable number of years annually purchased a number of these owls in the London market … We occasionally saw and frequently heard one or more of the Little Owls in the neighbourhood … In 1889 one of our gamekeepers, on April 23rd, found a Little Owl sitting in a hollow bough of an old ash-tree in the deer-park at Lilford … he found that she was sitting upon a single egg, to which she added three, and brought off four young birds in the second week in June. One, if not two, other broods were reared in our near neighbourhood in 1889. In 1890 a nest … containing six eggs was found … on April 25th: all these eggs were hatched and the young reared to maturity. On October 15th [1890] a Little Owl was … dis-

covered ... in a rabbit-burrow in the park at Deene [near Kettering] ... a bird-stuffer at Stamford [Lincolnshire] had one ... (which he called a Dutch owl) in February 1891, sent to him from Normanton, Rutland. A 'nestful' of young was discovered early in July 1891 [at] Wadenhoe Manor ... In 1892 I received ... reports of Little Owls in the neighbourhood of Lilford and on April 19th a nest containing four eggs was found at Wadenhoe ... Mr A[rchibald] Thorburn saw near our aviary ... a Little Owl ... on November 11th [1892]. In 1893 a nest containing seven young was found in the park at Lilford, on May 13th. I had authentic reports of one or two broods at Wadenhoe and another near Lyveden ... I have some reason to believe that there was a brood of these owls in Wadenhoe church-tower. On December 27th I was pleased to hear from a lady living at Stoke Doyle that several of the birds had made a settlement in some old trees in her garden. With the exception of the bird killed at Normanton and another at Elton, I have not heard of the death by human agency of any Little Owls in our district, although a few have been picked up dead from natural causes ... These Owls delight in taking the sun, and are active during the hours of daylight. They are infinitely useful in the destruction of voles, mice, sparrows, and insects of many kinds.

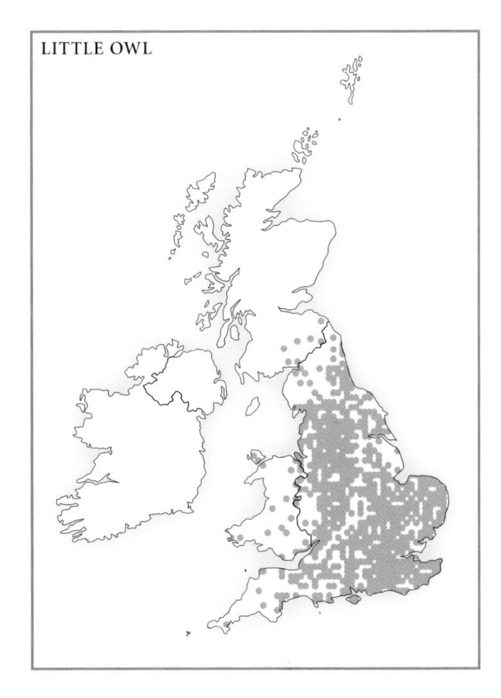

From Lilford Hall, Little Owls multiplied and spread with considerable speed, and had soon colonized the north-eastern corner of Northamptonshire. From here they travelled further east to Suffolk and Norfolk by 1901, where they first bred in 1907 and 1912 respectively; to Huntingdonshire, where they first nested near Offord Darcy in 1907; and to Cambridgeshire, where they bred first in the same year.

In Rutland, where Lord Lilford had recorded a Little Owl in 1891, the birds first nested in the county at Glaston, east of Uppingham in 1895. The first Little Owl in Nottinghamshire was caught at Newark-upon-Trent in 1896, although the earliest breeding record for the county was not until 1913. In 1902, a Little Owl was killed near Lincoln, and the birds first nested in Lincolnshire in 1907.

North-west of Lilford, Little Owls were seen on the Derbyshire/Leicestershire border in 1906, nesting in the latter county three years later. In Shropshire, a Little Owl was heard near Willey Park, Broseley, in 1899. One was killed in 1906 on the Shropshire/Staffordshire border, and in 1908 another was caught near Shrewsbury in the former county.

At Woburn Abbey in Bedfordshire, the Duchess of Bedford wrote that she:

knew the Little Owl since I first came to live here in 1892. The Keepers say that it existed here before that date. I should like to correct the impression which appears to be very prevalent that we have imported it ourselves. We have never imported them, and as they were here so early as 1892 I conclude that the birds here must have come from Lord Lilford's importation and not from the more recent one [in about 1890 by Lord Rothschild] at Tring. There are a great many of them about the Park, and they have bred freely for many years. Owing to their habit of flying about in the daytime they are frequently seen.

Elsewhere in Bedfordshire, Little Owls reached Chawston in the north-east by 1894; Turvey in the west (1897); Lidlington in the north and nearly to the Cambridgeshire border in the east (1898); and Luton in the south by the turn of the century. In 1901, nests were discovered at Great Barford and at Southill, respectively 8 km east and 11 km south-east of Bedford.

In Hertfordshire, Little Owls first nested at Ware in 1897, and by 1902 were said to be 'quite common and resident as far south as Watford'; breeding was first reported near Royston, Watton-at-Stone and St Albans in 1901.

They reached Essex in 1899, where breeding took place first at Easton Park in 1910; by 1919 the species was a common resident in much of the county.

In the wooded Chiltern Hills of Buckinghamshire and Oxfordshire, Little Owls were reported first at Turville in 1894 and Fingest in 1896 (both north of Henley-on-Thames); at Bletchley (1902); and near Goring, north-west of Reading, in 1907. Further south, in Berkshire, a pair was seen in the spring of 1907 in Windsor Forest, where breeding took place three years later.

In the last five years of the 19th century, Little Owls spread throughout southern Nottinghamshire, most of Hertfordshire, and probably into northern Oxfordshire and Huntingdonshire. From the early years of the 20th century, the range of the Little Owl in Britain grew apace, perhaps assisted by other and unrecorded introductions, and by around 1910 it had reached southern Oxfordshire, much of Lincolnshire, parts of Norfolk and Suffolk, Essex, Staffordshire, southern Derbyshire, Leicestershire, east Berkshire, Cambridgeshire and perhaps Shropshire, and thus occupied most of England westwards to the River Severn and northwards to the River Trent.

Table 11 Dates of earliest known breeding records of Little Owls *Athene noctua* in English counties in the 20th century up to the Second World War

Year	County	Year	County
1909	Derbyshire (near Nottinghamshire border)	1921	Lancashire (south)
1909	Hampshire (Southampton)	1922	Yorkshire (east)
1910	Gloucestershire (south)	1922	Cornwall (north)
1913	Leicestershire (south)	1923	Lancashire (Preston)
1913	Nottinghamshire	1923	Cornwall (west)
1914	Monmouthshire	1925	Yorkshire (south)
1915	Wiltshire	1933	Yorkshire (north)
1918	Isle of Wight	1934	Durham
1918	Dorset	1935	Northumberland (south)
1921	Cheshire	1937	Yorkshire (central)

Source Lever 1977.

Scotland

The Handbook of British Birds (1938: 326. Vol. II) lists early appearances of Little Owls in Scotland at Kincardineshire (1902), Fife (1910), Roxburghshire (1921), Peebleshire (1921), Berwickshire (1924) and Renfrewshire (1925). Breeding was first recorded, at Edrom in Berwickshire, only in 1958, and even by as late as the early 1970s there were few Scottish breeding records, none of which was north of Midlothian. It was not for another 25–30 years that Little Owls may have wintered in coastal Dumfries and Galloway (where nesting was first recorded in 1979), and even occasionally as far north as East Lothian and Dumbarton. By the mid-1980s, there were scattered reports of nesting south of a line between the River Clyde and the Firth of Forth, and breeding is still restricted to southern counties. Further northwards expansion may be limited by low winter temperatures, which restrict access to the bird's main winter food of invertebrates and small mammals.

Wales

Table 12 Dates of earliest known breeding records of Little Owls *Athene noctua* in Welsh counties up to the Second World War

Year	County	Year	County
1916	Glamorganshire	1922	Merionethshire
1918	Cardiganshire	1922	Brecknockshire
1918	Radnorshire	1923	Carmarthenshire
1919	Montgomeryshire	1930	Caernarvonshire (Great Orme)
1920	Pembrokeshire	1931	Denbighshire

Source Lever 1977.

By the early 1960s, Little Owls had bred in every Welsh county.

[Ireland]

Individual Little Owls have been recorded in Ireland, as vagrants, on only four occasions: at Kilmorony in Co. Kildare in 1903, near Larne in Co. Antrim in 1945, on Great Saltee in Co. Wexford in 1960 and at Dunlavin in Co. Wicklow in 1981. These birds presumably came from Wales or possibly north-west England.

Summary of Status

By the 1920s, Little Owls occurred in every county south of the River Humber apart from Caernarvonshire and north Denbighshire in Wales and parts of Devon and Cornwall in England, but even these had been colonized by 1930. Subsequently, the expansion of range, which in about 1900–1930 had been little short of explosive, slowed down, and it was not for another 20 years that the northwards spread to the Scottish border was complete. Even by the early 1970s, Scottish breeding records were few and far between, and reached only as far north as Midlothian.

In the 1940s, while the northwards expansion continued, there was evidence of a decrease in the population in some southern and western counties – possibly as a result of the harsh winters in the south and west at this time, in particular that of 1946–7. A similar drop in numbers was noticed after the severe winter of 1962–3. Sharp falls in the population in 1956–65, especially in south-eastern counties, may have been due to persecution by man and the contamination of prey species by the residue from chemical pesticides. In 1965–75, there was evidence of some recovery in numbers, except in the south-east. By the early 1990s, the British population may have been 6,000–12,000 breeding pairs.

Relatively few Little Owls breed above about 300 m asl, where winter survival may be a problem. This would explain the species' continued absence from, for example, the Cambrian Mountains of Central Wales, the English Lake District of Cumberland and Westmorland (Cumbria), much of northern and western Wales, the northern Pennines, the North Yorkshire moors, and upland regions of Devon and Cornwall. The Little Owl is one of the species likely to benefit from the higher temperatures that would arise from global warming. A relatively modest increase would enable it to colonize suitable habitats in many of the above regions, and also to become established in Scotland further north than its present limits.

Characteristics

The favourite habitat of the Little Owl consists of open agricultural countryside, liberally provided with nesting sites in the form of mature and pollarded trees and farm buildings. Old orchards and parks are other prime haunts, but waste ground, moorland ecotones, sand dunes, disused quarries and sea-cliffs are also frequented. Little Owls occur in relatively bare country, especially in parts of Cornwall, but also sometimes in city centres.

The Little Owl is a largely diurnal species, and is often seen perched on a fence or post, where it bobs up and down if alarmed. It has a buoyant and bounding flight not unlike that of a woodpecker. In late April or early May, the female lays a clutch of 3–5 eggs, usually in the hole of a deciduous tree (in some places a pollarded willow is a favourite of Little Owls), or less frequently in tree cavities, old buildings, stone walls, cliff faces, haystacks and cornstacks, and even European Rabbit burrows.

The principal reason for the success of the Little Owl in becoming naturalized in Britain was the existence of a vacant ecological niche for a largely diurnal and mainly insectivorous small bird of prey. Inhibiting factors to the species' establishment include:

- Lack of tree cover, shelter belts and copses in some northern and upland areas.
- Prolonged cold in winter.
- Destruction of hedgerows with their associated trees and the loss of old farm buildings, which together have reduced the availability of nesting sites.
- Competition with Tawny Owls, which may partially explain the Little Owl's absence from woodland. Little Owls are also preyed on by Tawny Owls.
- Little Owls are sometimes the victims of road-traffic accidents.

Impact

During the 1930s, when the Little Owl was rapidly expanding its range in Britain, it incurred the ire of ill-informed gamekeepers, landowners and farmers on account of its alleged predation on game-birds and the young of domestic fowl. The British Trust for Ornithology commissioned Alice Hibbert-Ware to carry out a detailed survey of the Little Owl's diet to complement that of Collinge (1921–2). She examined 2,460 food pellets[3] and the gizzards of 28 birds, and analysed the material from 76 nests, some of which were deliberately taken from game- and domestic fowl-rearing districts. Eighteen months of intensive investigation revealed only two pellets each of which contained the remains of a single game-bird poult (plus possibly one other fledgling) and the remnants of only six chicks of domestic fowl. The Little Owl's diet was found to consist of 93.5 per cent animal matter:

- Mammals (including the Wood Mouse, Field Vole, House Mouse, Common Rat, Bank Vole, Common Shrew, European Mole and small European Rabbits) comprised 19.3– 50.7 per cent.
- Birds, forming 4.96 per cent of the bird's diet and taken mainly in May–July, were largely Starlings, Common Blackbirds, Song Thrushes and Mistle Thrushes.
- Insects were found to form 49.2 per cent of the birds' prey, including Diptera (especially Tipulidae), Dermaptera (*Forficula*) and Coleoptera (*Pterostichus, Geotrupes* and *Melolontha*).
- Other invertebrates, such as woodlice, spiders, centipedes, millipedes, molluscs (snails and slugs) and earthworms (7.8 per cent), were also recorded.
- Only 6.5 per cent of the diet was herbage.

Hibbert-Ware (1937) supported Collinge (1921–2) in proving conclusively that, as Charles Waterton, Lieutenant-Colonel Meade-Waldo and Lord Lilford had predicted, the Little Owl causes far more good than harm in Britain, and that Coward's (1926: 62. Series III) condemnation that 'the intentional release of ... the Little Owl is a menace' and Ritchie's (1920: 76) stricture that 'Nature also mocked us in this country in the case of the introduction of the Little Owl ... everywhere it has betrayed its trust', were wholly without foundation. Indeed, the Little Owl is one of the few actively beneficial alien animals to have become naturalized in Britain.

References

Ainslie 1907; Baker et al. 2006; Blatwayt 1902, 1904; Bradshaw 1901; Buxton 1907; Collinge 1921–2; Ellison 1907; Fitter 1959; Genot et al., in Hagemeijer & Blair 1997; Gibbons et al. 2001; Glue, in Lack 1986, in Gibbons et al. 1993, in Wernham et al. 2002; Harrison 1988; Hibbert-Ware 1937; Hutchinson 1979, 1989; Lever 1977, 1978, 1980a, 1984, 1987, 1994, 2005, in press; Lilford, Lord 1895, 1903; Linn 1979; Long 1981; Lovegrove et al. 1994; Maples 1907; Parslow 1973; Powerscourt, Viscount 1884; Raven et al. 2007; Sharrock 1976; Steele-Elliott 1907; Thom 1986; Tomlinson 2006; Welch et al. 2001; Witherby & Ticehurst 1908.

Notes

1. Linn (1979: 9) says that 'The Little Owl ... [is] recorded from interglacial deposits in the British Isles.' Elaborating on this, Harrison (1988: 150) wrote: 'A Pleistocene record from Chudleigh in Devon, often cited, is in fact based on a [Eurasian] Sparrowhawk bone, but it [the Little Owl] appears to have been present in the Mendips in an early interglacial about 500,000 years ago.' The Little Owl is here treated as an introduced alien.
2. One of the dates given by Waterton is clearly incorrect. Witherby and Ticehurst (1908), Coward (1926) and *The Handbook of British Birds* (1938: Vol. II: 324) all give 1843 as the year in which the Little Owls were released. Fitter (1959) gives 1841 as the year when Waterton left Rome, and 1842 as the year when the birds were released. The latter dates seem the more likely.
3. All but the largest prey of owls is generally swallowed whole; the indigestible parts are subsequently disgorged as pellets.

REPTILES

EMYDIDAE (TURTLES, TERRAPINS & TORTOISES)

EUROPEAN POND TERRAPIN *Emys orbicularis*
(EUROPEAN POND TURTLE)

Natural Distribution Most of Europe except for the N, and W Asia and NW Africa.
Naturalized Distribution England; [Ireland].

EUROPEAN POND TERRAPIN

England

The earliest documented introduction to Britain of the European Pond Terrapin[1] appears to have occurred in 1890–1, when some were released by Lord Arthur Russell (a younger son of the Duke of Bedford) at Shere, near Guildford in Surrey. In 1894–5, others were freed at Bloxhall and Little Glemham south of Saxmundham in Suffolk, where in 1929 and 1932 respectively a considerable number of young terrapins and a sub-adult were reported; in 1934 another sub-adult was discovered nearby at Snape.

In about 1906, Lord Walsingham liberated some in Old Park Pond at St Lawrence on the Isle of Wight, where in the following year one was reported in the millpond at Carisbrooke, some 13 km from St Lawrence. In 1952, one was found on the beach at Chale, on the far side of St Catherine's Point from St Lawrence. In about 1905–10, a large number were released at Frensham Hall near Haslemere in Surrey. European Pond Terrapins were freed at Woburn Abbey in Bedfordshire at about the same time, and they still occurred there in 1950. Nine individuals caught in ponds in north Surrey in 1948 may possibly have been the descendants of the animals released at Frensham some 40–50 years previously.

[Ireland]
In 1906, European Pond Terrapins were liberated on Lambay Island north-east of Malahide in Co. Dublin, but apparently no more was heard of them.

Prospects
The European Pond Terrapin has for long been a favourite of the pet trade, and over the years large numbers have been imported into Britain. In addition to those mentioned above, there must have been many unrecorded releases and escapes, and it may well be that undocumented colonies of this species occur in parts of southern England, although like most chelonians it lives to a great age, so a population can exist for many years without reproducing. Although England is several hundred kilometres from the nearest breeding sites of the species in north-eastern Germany and central France, given suitably warm summers, which are likely to occur increasingly in future as a result of global warming, there must be a distinct possibility that European Pond Terrapins will become widely established in southern England.

References
Anon. 1997; Beebee & Griffiths 2000; Gent, in Poland Bowen 2003; Harrison 1988; Lever 1977, 1980b, 2003; Smith 1951a.

Note
1. In periods of continental climate during the Pleistocene, after the last Ice Age, the European Pond Turtle lived in England, where fossil remains dating from around 3000 BC have been found in East Anglia (Smith 1951a; Harrison 1988). The species is treated here as an alien introduction.

EMYDIDAE (TURTLES, TERRAPINS & TORTOISES)

RED-EARED SLIDER *Trachemys scripta*
(RED-EARED TERRAPIN; RED-EARED TURTLE)

Natural Distribution Mississippi Valley, USA, from N Illinois S to the Gulf of Mexico.
Naturalized Distribution England; Wales.

England
This species is another popular pet, and in the 1980s and early 90s, due to the 'Ninja Turtle' craze, large numbers were introduced from the United States to Britain and continental Europe. As a result, the species has appeared in ponds in many localities in southern England.

Red-eared Sliders are sometimes found on Jersey in the Channel Islands, mainly on the west coast, but also on occasion in Saint Catherine's Reservoir, but breeding has yet to be confirmed.

RED-EARED SLIDERS

Wales

A large population of Red-eared Sliders lives in a lake in a park in Cardiff, Glamorganshire, in south Wales, but breeding has yet to be confirmed.

Prospects

Like the European Pond Terrapin, the Red-eared Slider is an extremely long-lived species, and a population can survive for many years without reproducing. In its native range it occurs from around 30°N to 42°N, well south of southern England at about 51°N, where at present breeding in the wild is unlikely, although females in outdoor vivaria have been known to lay eggs. Again, as in the case of the European Pond Terrapin, if the present trend of increasingly mild summers continues as a consequence of global warming, there seems no reason why the Red-eared Slider may not eventually reproduce widely in southern England.

Impact

Although Red-eared Sliders are known to be omnivorous rather than strictly carnivorous, they may feed on native species of fish, amphibians and waterfowl, and would be likely to do so increasingly were they to start breeding.

References

Anon 1997; Beebee & Griffiths 2000; Hennig 2004; Lever 2003; Prevot-Juillard et al. 2007; Young 1987, 1988.

LACERTIDAE (LACERTID LIZARDS)

GREEN LIZARD *Lacerta viridis*

Natural Distribution Much of SC Europe from France and the Channel Islands (Jersey) and N Spain E to the Balkans and Ukraine, N to SW Russia, and the Rhine Valley.
Naturalized Distribution [England]; [Wales]; [Ireland]

GREEN LIZARD MALE

[England]

In letters XVII (18 June 1768) and XXII (2 January 1769) to Thomas Pennant, Gilbert White, in his *Natural History of Selborne* (1789), refers to sightings of 'green lizards' near Farnham in Surrey. In letter XXIII (28 February 1769) to Pennant, White states: 'It is not improbable that the Guernsey lizard and our green lizards may be specifically the same; … when some years ago many Guernsey lizards were turned loose in Pembroke college garden, in the university of Oxford, they lived a great while … but never bred.'

It seems likely that the lizards referred to by White were either the indigenous Sand Lizard or the introduced Common Wall Lizard, with both of which the Green Lizard can be confused.

In 1899, 230 Green Lizards were released at St Lawrence on the Isle of Wight, where some survived until 1936. In 1905–10, others were freed at Frensham Hall, Haslemere, Surrey, near the site of Gilbert White's sightings. A hundred were liberated at Paignton in south Devon in 1937, some of which survived until 1946.

In May 1962, five adult males and four adult females were released in an old chalk quarry near Sittingbourne in Kent. Mating was observed later in the year but no young were seen, and the last adult was recorded in June 1963. Similarly, Green Lizards freed later in the same decade in Gloucestershire did not survive for long.

[Wales]

In 1872, Green Lizards were turned loose in the Ynysneuadd Woods, Ebbw Vale, Monmouthshire, and at Portmerion, Merionethshire, where 20 more were planted in 1931, at least some of which survived until 1935.

[Ireland]

On the Burren in Co. Clare, where eight males and seven females were released in 1958, some were still there four years later.

Prospects

Even in southern England, the Green Lizard is north of its most northerly natural range in continental Europe, although in some places there have been indications of successful reproduction and sightings have persisted for over a decade. The Green Lizard is another species that could well benefit from climate change brought about by global warming.

References

Beebee & Griffiths 2000; D'Arcy & Haywood 1992; Fitter 1959; Lever 1977, 1980b, 2003; Walters 1981.

LACERTIDAE (LACERTID LIZARDS)

COMMON WALL LIZARD *Podarcis muralis*
WESTERN GREEN LIZARD *Lacerta bilineata*

Natural Distribution *P. muralis*: much of Europe between about 40°N and 50°N E to Romania. Also islands off the Atlantic coast of Spain and France (including Jersey in the Channel Islands) and in the Ligurian Sea (between Corsica and N Italy). Also NW Asia Minor. *L. bilineata*: Italy, France and N Spain.
Naturalized Distribution England.

England

Langham (2008), from whom the following account is derived, has surveyed and described in detail the history and status of Common Wall Lizard colonies in England.
Avon (Bristol) This colony, which was discovered in 2006, is based in the Royal Fort

gardens in the city centre. Its origin is uncertain, but could have been the department of zoology at the university. The form present is the brown-backed one that is normally associated with western France.

Berkshire (East Burnham Common, Beaconsfield) The site is a public recreation park. There have been only three records from this site, the last of which was in 1965. The

COMMON WALL LIZARD MALE

WESTERN GREEN LIZARD MALE

animals are believed to have been released from a pet shop in Slough. The unsuitable habitat consists of open common land with mixed woodland.

Devon (Blackawton, near Totnes) In 1954, Viscount Chaplin released a number of reptile species in his walled garden at Wadstray House. Of these only 15 lizards (probably subspecies *P. m. nigriventris* or possibly *P. m. brueggemanni*) survived, and by the 1970s the population had increased to several hundred. After Lord Chaplin's death in the 1980s, the garden fell into disrepair, and the colony is believed to have died out around 1990.

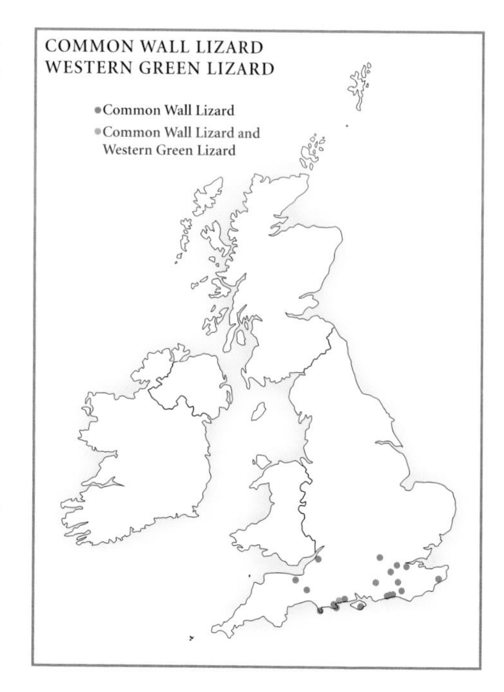

Devon (Paignton) In 1937, the founder of Paignton Zoo, Herbert Whitley, released 200 Common Wall Lizards from France in Primley Park and the adjacent Clennon Valley Park in Paignton. Although most of these animals had disappeared within a decade, a few survived until at least 1951. Occasional sightings were reported until the 1960s, but there have been none since 1970.

Dorset (Boscombe, Bournemouth) This colony is believed to have been established in 1992, when 50–60 were deliberately released. (There may have been an earlier and less successful release of brown-backed individuals from France.) Some brown-backed animals survive in the current population, though they could be natural morphs developed through natural selection. The native plant communities of extensive areas of the steep Boscombe sandstone cliffs are being replaced by alien species and scrub. The relative population estimate[1] for this colony is 1,328. In 2002, some Western Green Lizards were deliberately and illegally released on this site, where they are established and breeding (Steve Langham, personal communication 2008).

Dorset (Canford Cliffs, Bournemouth) This colony originated in 1992, when 50–60 (probably Italian) hatchlings were released at the Zig-Zag near Flaghead Chine, and has since spread along the cliffs as far as the car park at Bransome Dean Chine. The habitat is the same as at Boscombe. The relative population estimate is 3,699.

Dorset (Cheyne Weare, Fortuneswell) This colony, in Duncecroft Quarry on the Portland peninsula, was discovered in 1995. It originated from captive-bred stock, believed to be descended from wild-caught animals in Brittany in north-western France, which was deliberately released, and is entirely of the brown-backed form. The quarry consists of sparse vegetation, brambles and vines, with plenty of open places for basking – but a relative shortage of egg-laying sites on the quarry floor. The relative population estimate is 329.

Dorset (Corfe Castle) Although this colony was only discovered in 2006, it may have been established through deliberately released captive-bred stock some years previously. At present, the lizards are confined to the east of the walls of the outer gatehouse, but in view of the optimum habitat available expansion seems likely. The animals are all of the green-backed Italian form.

Dorset (Dancing Ledge, Swanage) This colony is believed to consist of the green-backed form (as at the two other Purbeck coastal locations) and to have originated in the early 1990s from deliberately released captive-bred stock. The habitat is sea-cliffs, quarry face with rocky outcrops and associated vegetation.
Dorset (Durleston, Swanage) Only three individuals, all males, were recorded in 1999–2001. The habitat is coastal cliffs and the site of the old castle.
Dorset (Longstone Ope, Fortuneswell) Almost adjacent to the colony at Cheyne Weare on the Portland peninsula, this colony of the brown-backed form originates from a deliberate release, probably in the early 1990s, of captive-bred stock probably descended from animals caught in the wild in Brittany in north-western France. Longstone Ope is a disused quarry within the coastal cliffs, in the south-east of the Portland peninsula. The colony is situated both within the quarry and on the cliffs. The habitat is the same as at Cheyne Weare. The relative population estimate is 477.
Dorset (West Weare, Fortuneswell) This colony is believed to have started in the 1990s in Tout Quarry, from where it was subsequently relocated to the sea-cliffs. Although a survey in 2007 failed to find any lizards, the colony is believed to survive and to consist of the green-backed form. The site is composed of boulder-strewn high coastal cliffs fronting abandoned quarries.
Dorset (Winspit Quarry, Swanage) The date of origin of this colony is believed to be the mid- to late-1980s, with reinforcements until the early 2000s comprising deliberately released captive-bred stock of the green-backed form. The habitat is coastal cliffs and quarry face, with rocky outcrops and associated vegetation. Reports since 2007 of Common Wall Lizards at Worth Maltravers, more than a kilometre inland from Winspit Quarry, could indicate either an expansion of range or translocation by humans.
Hampshire (Farlington, Havant) Discovered only in 2007, this individual sighting remains unconfirmed.
Hampshire (Holmsley, New Milton) This colony originates from a small number of animals that escaped from a large outdoor vivarium in the summer of 1986 into a garden, where they first became established in a pile of logs and subsequently spread to neighbouring gardens. The animals are believed to be of the Italian green-backed form. The surrounding habitat is unsuitable coniferous open forest and heath.
Ventnor (Isle of Wight) This colony is probably the largest (in terms of population and area) and longest established population of Common Wall Lizards in Britain. Although it is claimed that the colony dates from a release by a Dr Martin in 1841 this cannot be confirmed, though it does correlate with a known unsuccessful introduction of Western Green Lizards at St Lawrence some 4 km to the west. It is believed that the introduction was made in the early 1920s – according to local legend from a ship, possibly Italian, that was wrecked on the coast and from which the animals, stowaways among the cargo, escaped. By the following decade they were apparently well established on walls along Esplanade Road. They later occupied a mixture of coastal cliffs, heavily scrubbed residential gardens and walls, and landscaped terraced walls along the cliffs. Recently their range has been deliberately expanded by human translocations to the Ventnor Botanic Gardens, and there have been reports of sightings from as far east as Wheeler's Bay. It seems probable that this colony will remain restricted with the town of Ventnor and its immediate environs. The relative population estimate is 12,876.
Kent (Folkestone) This colony, believed to be descended from surplus stock of the green-backed Italian form dumped by the pet trade, originates from about 2004 or possibly earlier. The site is heavily scrubbed and grassy sea-cliffs.

Kent (Tyler Hill, Canterbury) This colony situated in a demolished residential garden has died out.

Greater London (Birdbrook, Mottingham) An escape in 1972 of half-a-dozen animals from an outdoor vivarium formed the founder stock of this colony, which was reinforced by a further dozen or more escapees some five years later. All were of the green-backed form, probably of Italian origin, and had been acquired from the local Well Hall pet shop. These animals became established on land adjacent to the Rochester Way trunk road, and at their peak are believed to have numbered over 1,000 individuals. Subsequently the land was developed into a housing estate, resulting in the deaths of large numbers of the animals and the displacement of the survivors to sub-optimal habitats. Predation by domestic cats on the new housing estate doubtless played a part in reducing the population still further. The colony managed to survive, however, and currently numbers perhaps 100 individuals. The present distribution includes an old Ministry of Defence depot at Kidbrooke (now part of the National Maritime Museum) and the small Birdbrook Nature Reserve, managed by the London Wildlife Trust. The habitat of Birdbrook comprises medium-height scrub, four ponds, Bramble thickets and a few courses of brick from a demolished house, and elsewhere partly buried and overgrown foundations and some grass-covered banks of rubble with disused storage buildings. The site backs onto a row of private gardens, which are so far uncolonized. The relative population estimate is 76.

Greater London (Hampton, Brentford) This colony became established around 1954 on two railway bridges, an associated embankment and nearby private gardens. The animals are of the Italian green-backed form, either *nigriventris* or *brueggemanni.* Although the lizards are thought to have died out since 1979 and a survey in 2007 failed to find any, it is possible that a small nucleus may have relocated locally.

Oxfordshire (Burford) This colony used to live on an old drystone wall and ha-ha in the African enclosure of the Cotswold Wildlife Park. The eastern end of the ha-ha was covered with ivy, where the lizards were most frequently to be seen basking. The colony is believed to have died out in about 1999 following pointing of the wall, which removed the interstices required by the lizards for breeding and shelter.

Shropshire (Ludlow) In about 1985, captive-bred Common Wall Lizards of the Italian green-backed form were deliberately released in a disused quarry near Overton. Even at its peak in the 1990s, the population is believed never to have reached 100 individuals, and only a small number were recorded in 1999. Possibly due to extensive clearance of the vegetation on the site by the Forestry Commission in late 2004, this colony is believed to have died out shortly thereafter – a survey in April 2007 failed to find any, despite good conditions and an ideal time of year.

Somerset (Stogumber, Williton) In about 1981, some 10 adults, believed to be of the brown-backed French form, were translocated from the colony at Wellington and released on a sandstone ridge on the top of a south-facing valley in an orchard and honey farm at Stogumber. Individuals were later seen in private gardens and on walls in the hamlet a few hundred metres down the valley.

Somerset (Wellington) In about 1981, some 10 animals (from Machecoul south-west of Nantes in western France) escaped from a vivarium near the bypass and became established on a nearby 2.4 m-high old flint boundary wall and in some adjacent greenhouses.

Surrey (Banstead, Belmont) In 2007, a colony was discovered in private gardens and walls, and in glasshouses in two neighbouring garden centres. Since neither centre nor any local private garden owners have kept or sold reptiles, it is assumed that the animals were stowaways in importations from abroad of ornamental or exotic plants.

Surrey (Farnham Castle) In May 1932 – the earliest recorded introduction of Common Wall Lizards to Britain – a dozen of the brown-backed nominate form, acquired from a London dealer who is believed to have imported them from France, were released in a walled garden at Farnham Castle; in the following year a further pair, presumably from the same source, was added. Little more was heard of these animals until Smith (1951b) announced the 'rediscovery' of the colony on some old brick and stone walls on a nearby private estate. It is believed that these animals died out shortly after 1989 – the year of the last confirmed sighting, although unconfirmed reports continued until 2000.
Surrey (Newdigate, Crawley) Fitter (1959) reported that a small colony of Common Wall Lizards was established in T.B. Rothwell's Beam Brook Aquatic Nursery at Newdigate, though whether in captivity or the wild is unrecorded. Although the population that had since become established in the wild survived until 1989, a survey in 2007 failed to find any individuals. The site was a mixture of grassy glades in woodland with numerous (artificial) ponds. There were no suitable walls for shelter or egg-deposition.
Surrey (Nutfield, Bletchingley) Between 1980 and about 1985, some Common Wall Lizards escaped from the premises of Xenopus Ltd, at Nutfield, a firm that reared reptiles and amphibians for the pet trade, educational use in schools and scientific research. The original stock was imported from Barilli & Biagi of Bologna, Italy, and was of the green-backed form. The escaped lizards became established in a nearby private garden, where they use wooden decking for basking. Xenopus Ltd ceased trading in 1989.
Sussex (Haywards Heath) In 2007, a 'small and fragile' colony was discovered at an undisclosed site at Haywards Heath. The origin may have been the population at Shoreham-by-Sea.
Sussex (Lancing, Worthing) This new colony, in a private garden backing onto a railway line, was also found in 2007. A subsequent discovery at West Worthing station lends credence to the hypothesis that the former stock also came from the Shoreham-by-Sea colony, and used the railway line as a habitat corridor.
Sussex (Shoreham-by-Sea) This colony, of both green and brown forms of Italian stock, originated in the garden of the late Professor Geoff Haslewood, who in 1975 released 25 Common Wall Lizards, followed a year later by a further 7. Since then the colony has expanded its range some 500 m along the beach as far as the Old Fort. This population could be the source of recently discovered colonies at Haywards Heath, Lancing and West Worthing. The site is a south-facing shingle beach backed by residential properties and their gardens. Although the lizards make use of most of the shingle area, the main concentrations occur on the occasional garden waste dumps and concrete remnants of coastal defences, and especially on the boundary between the gardens and the beach. The relative population estimate is 513.
Sussex (West Worthing) This colony was first reported in 2004, and has become established in the railway station, mainly in cracks in platforms and curbing, and in fenced-off areas. The source of the colony, which is of the green-backed form and probably of Italian origin, may be the population at Shoreham-by-Sea that used the railway lines as a corridor for expansion or travelled as stowaways on rolling stock; another possibility is that it is a deliberate human introduction. The relative population estimate is 17.

Prospects

The habitat and climatic requirements of the Common Wall Lizard are crucial to its survival. In England, and indeed throughout its natural range, it is in general a species of artificial disturbed habitats containing some form of vertical element, such as a wall, quarry or house with a southerly aspect for basking with, or at least near, some form of cover for protection.

The Common Wall Lizard, even in southern England at around 51°N, is north of its most northerly natural range in continental Europe, and is thus yet another species that may well take advantage of the milder climate likely to result from global warming.

Impact

Mole (2008) found that the range covered by Common Wall Lizards and Western Green Lizards on Canford Cliffs in Dorset showed that the two alien species dominated the central area, with the native Viviparous Lizard found in large numbers only on the periphery. In 2002–7, the population of Common Wall Lizards increased by some 40 per cent and that of Western Green Lizards by 36 per cent; during the same period the population of the native species declined by 75 per cent. The structure of the vegetation rather than the species was the principal factor in the lizards' distribution, and it thus seems possible for both alien species to spread unhindered along the cliffs. Where the Common Wall Lizard occurs sympatrically with the rare Sand Lizard, anecdotal evidence suggests that the alien species outcompetes the native. On the Canford Cliffs in Dorset, the population of the latter has declined dramatically since the arrival of the former, and although there is no proven causal link it must be a factor for consideration.

References

Frazer 1949, 1964; Langham 2008; Lever 1977, 1980a, 1980b, 2003; Mole 2007; Quayle & Noble 2000; Sinel 1905, 1908; Smith 1951a, 1951b; Snell 1981; Spellerberg 1975; Taylor 1948, 1963; www.surrey-arg.org.uk

Note

1. The relative population estimate is the estimated number of individuals present in the colony. Although the actual figure (which is likely to be on the low side) may not be accurate, by using the same method of calculation for each colony it should vary correctly when compared to another colony (Steve Langham, personal communication 2008).

COLUBRIDAE (NATRICINE COLUBRIDS)

AESCULAPIAN SNAKE *Zamensis longissimus*

Natural Distribution From C France, S Switzerland, S Austria, the Czech Republic, S Poland and the neighbouring former USSR, S to NE Spain, Sicily, W Sardinia and S Greece. Also from Turkey to N Iran. (This disjunct distribution is probably a remnant of an originally much wider range.)
Naturalized Distribution England; Wales.

England

Since the late 1980s, a thriving population of Aesculapian[1] Snakes has been established along both banks of the Regent's Canal in the grounds of the Zoological Society of London in Regent's Park. The founder stock of this population was said to have been released by the London Education Authority, which was at the time renting space in one of the society's buildings.

AESCULAPIAN SNAKE FEMALE WITH EGGS

In the early years, sightings of the snakes were only random and infrequent, but in the last 5–6 years they have become more regular, and increasingly juveniles have been reported, indicating breeding. At present the population, which may amount to >50 of all age classes, is established over an area of perhaps 2 ha (Richard Green, personal communication 2008).

Wales

For much of the information on Aesculapian Snakes in Wales I am indebted to Dylan Davenport (personal communication 2008).

In the mid-1960s, a population of Aesculapian Snakes became established in the grounds of the Welsh Mountain Zoo at Colwyn Bay, Conwy. Although the thriving population of around 450 individuals (85 adults and 365 juveniles) in 2006 is allegedly descended from a single gravid female escapee, VORTEX PVA (population viability analysis) software suggests that several individuals probably comprised the founder stock. Originally misidentified as either native Grass Snakes or alien

Western Whip Snakes, they were not correctly identified by the zoo director, Nick Jackson, as Aesculapian Snakes until 1978.

The Colwyn Bay population, after 40 years, remains confined within the zoo grounds and immediate environs, suggesting either a low dispersal ability or an abundance of food. Adult Aesculapian Snakes prey mainly on small rodents such as mice and voles, and since the rodent population within the zoo is high due to the amount of available food, a lack of necessity may explain the snakes' reluctance to leave the 15-ha zoo grounds.

At approximately 53°N, the Colwyn Bay colony is the species' most northerly surviving population, and the fact that individuals in the zoo tend to be both shorter and lighter than their European counterparts may be explained by the lower temperatures there rather than by inbreeding, especially since VORTEX PVA suggest that the population has the ability to survive for a further 100 years.

Peak snake activity in the zoo grounds is in June, and hibernation takes place between November and March. The main site for egg-deposition is believed to be the zoo's manure dump. In the zoo, Aesculapian Snakes are preyed on by American Alligators, Pine Martens, Arctic Foxes, Golden Eagles and other raptors, and by native Common Ravens and mustelids, and predation and road kills are believed to account for the snakes' relatively high rate of mortality.

Characteristics

Aesculapian Snakes in Europe seldom exceed 1.8 m in length, and occur in a variety of habitats such as deciduous woodland, scrub, marsh and agricultural land, and around human habitation up to 1,800–2,000 m asl. In the north of their range they are confined to sheltered, south-facing slopes on light soils. They are skilled climbers, able to ascend vertical tree-trunks. Powerful constrictors, Aesculapian Snakes also prey on small birds and other reptiles, but apparently not on amphibians.

The species has a wide distribution in Europe but is dwindling in numbers, so in 1997 it was recognized by the Council of Europe as a Species of Special European Concern; the two British populations are therefore of some conservation significance. It is not beyond the bounds of possibility that one or other could, at some future time, provide the founder stock for reintroduction to continental Europe.

References

Anon 1997; Beebee & Griffiths 2000; Gent, in Poland Bowen 2003; Lever 2003.

Note

1 Aesculapian (Asclepian) = of or relating to medicine or physicians. Aesculapius was the Greek and Roman god of medicine and the earth. He is portrayed bearing a caduceus (wand) entwined by a snake.

OTHER REPTILES OCCURRING IN BRITAIN

The precise status of reptiles (and amphibians) can sometimes be difficult to determine. In addition to the above species the following exotic reptiles may be encountered in the wild, though breeding has not yet been recorded: Italian Wall Lizard *Podarcis sicula*, Garter Snake *Thamnophis sirtalis*, Tesselated or Diced Snake *Natrix tessellata*, king or milk snakes *Lampropeltis* spp., Stripe-necked Terrapin *Mauremys caspica*, Snapping Turtle *Chelydra serpentina* and Painted Turtle *Chrysemys picta*.

AMPHIBIANS

Serious Threats to Amphibians

Cutaneous chytridiomycosis is an emerging pathogenic, highly transmissible and virulent fungal disease of amphibians caused by the chytrid *Batrachochytrium dendrobatidis* (*Bd*), and is one of the principal reasons for the decline and local extinction (perhaps even international extinction) of some amphibian populations around the world. Although the main reasons for the emergence of *Bd* are unclear, anecdotal evidence suggests a close link with the expanding global trade in amphibian species. *Bd* can cause up to 100 per cent mortality in the post-metamorphic life stages of some species, and it has been estimated that if unchecked it could eradicate up to one-third of the world's amphibians by the end of the century. Although the precise mechanism whereby *Bd* affects individuals is as yet undetermined, it attacks keratin, a schleroprotein in an animal's skin, resulting in an increase in epidermal sloughing and ulceration, and hyperaemia of the ventral and digital skin; this may affect both respiration and osmoregulation. In 2007, an Australian herpetologist, Lee Berger, claimed that 'the impact of chytridiomycosis on frogs is the most spectacular loss of vertebrate biodiversity due to disease in recorded history'.

Cunningham et al. (2005) have described the emergence of *Bd* in Britain. In the summer of 2004, *Bd* was discovered in 2 out of 14 tested juveniles in a population of introduced American Bullfrogs near Edenbridge at Cowden on the Sussex/Kent border. Although *Bd* had previously occurred in exotic amphibians in captivity in Britain, this was the first known case of the disease in the wild. The index site also contained a population of introduced African Clawed Toads, and native Common Frogs, Common Toads, Smooth Newts and Great Crested Newts. It is presumed that *Bd* was introduced with either American Bullfrogs or African Clawed Toads.

In view of the widespread mortality and catastrophic decline of amphibian species elsewhere (especially Australia, Central America and Spain), the arrival of *Bd* in Britain could pose a grave threat to native amphibians. It seems unlikely that the infection will remain localized in south-eastern England, and it is possible that it may eventually spread into native species nationwide.

Amphibians (and reptiles) are even more susceptible to a variety of ecological threats than are homoiotherms, including:

- Loss of habitat.
- Ultraviolet radiation as a consequence of erosion of the ozone layer due to global warming.
- Pollution (especially atmospheric acid deposition – so-called 'acid rain').
- Commercial exploitation for meat, skins and traditional medicine in Asia.
- Disease (such as the mass mortality of Common Frogs that occurred in south-eastern England in the 1990s, and was caused by a ranovirus similar to pathogens that have resulted in widespread deaths elsewhere).
- Predation and competition from introduced species.

The *Bulletin of the American Museum of Natural History* No. 297 proposed new scientific names for a number of amphibian species. Since, however, these new names have yet to be widely accepted, the previous names have been retained here. (Per Jim Foster of Natural England, 2008.)

DISCOGLOSSIDAE (MIDWIFE TOADS, FIRE-BELLIED TOADS & PAINTED FROGS)

MIDWIFE TOAD *Alytes obstetricans*

Natural Distribution W Europe S to the Alps and the Iberian Peninsula, E to Germany and Switzerland.
Naturalized Distribution England, Wales.

England

In 1919, in the course of an address to the South London Entomological and Natural History Society entitled 'British Batrachians', G.A. Boulenger said:

> The Midwife Toad has established itself, no one knows how, in a former nursery garden in Bedford; it has been there for many years ... [It] furnishes an interesting example of parental solicitude, the male taking charge of the eggs, which are large and few [20–100] and strung together like a rosary, immediately after oviposition on land, not in the water as in most other Batrachians. ... the male fastens the string of eggs round its hind limbs and carries them for a period of about six weeks, when he betakes himself to the water for the purpose of releasing his progeny.

There is some doubt as to whether the Midwife Toads that formed this colony arrived in 1878 or 1898; as they were first seen and heard calling in 1903 the latter date seems the more likely. The nursery belonged to the firm of Horton & Smart, by whom the Midwife

MIDWIFE TOAD MALE CARRYING EGGS

Toads are believed to have been introduced accidentally (probably as eggs) in a shipment of ferns and water plants from southern France. In 1922, Bedfordshire County Council acquired part of the nursery, and the breeding pond was filled in. W.S. Brocklehurst removed about a dozen to his private 0.4-ha walled garden nearby. In 1950, another small colony was discovered in a neighbouring garden, and a remnant population was found to have survived in the original nursery garden site. A survey in the 1980s located several small garden ponds in and around Bedford that were inhabited by Midwife Toads, but there seemed to be no evidence that the species was further extending its range; one colony in a greenhouse was doing exceptionally well, almost certainly due to the warmth of the microhabitat.

In 1947, Robert Brocklehurst brought five adults and a dozen tadpoles from his father's garden in Bedford to his own home, Woodsetts Grange, near Worksop in Nottinghamshire. In 1965, his widow moved south to Northamptonshire, bringing with her a number of toads that became established in a garden pond near a disused gravel-pit, which were 'rediscovered' in 1985.

In 1933 or 1953 (accounts differ), a small colony of Midwife Toads became established in a large private walled garden near York, which was said to be 'still going strong' at the turn of the century.

In about 1954, a colony (which survived until the 1970s) was started by Viscount Chaplin at Blackawton near Totnes in south Devon with two egg-bearing males from London Zoo.

Today, Midwife Toads are believed still to occur around Bedford and in York, and possibly also near Worksop and in Northamptonshire.

Wales

In about 2004, male Midwife Toads were heard calling in a garden at Howey, 3 km south of Llandrindod Wells in Powys, where the species still occurs, though in depleted numbers. A small colony also became established some 4 km west of Howey at Newbridge-on-Wye, where an egg-carrying male was caught in 2007 (Ray Woods, personal communication 2008).

References

Anon 1997; Beebee & Griffiths 2000; Blackwell 1985; Fitter 1959; Frazer 1964; Gent, in Poland Bowen 2003; Lever 1977, 1980b, 2003; Muir-Howe 2007; Smith 1949–50, 1951b; Taylor 1948, 1963.

DISCOGLOSSIDAE (MIDWIFE TOADS, FIRE-BELLIED TOADS & PAINTED FROGS)

YELLOW-BELLIED TOAD *Bombina variegata*

Natural Distribution Most of C and S Europe apart from the Iberian Peninsula, S Greece, much of Sicily and most other Mediterranean islands.
Naturalized Distribution England.

England

In about 1954, Viscount Chaplin released a number of Yellow-bellied Toads in a pond in his walled garden at Blackawton near Totnes in south Devon, where they became established and survived for at least a decade. In 1964, W.L. Coleridge imported five Yellow-bellied Toads from Switzerland to his garden at Bishopsteignton near Teignmouth, also in south Devon, where they bred annually and by the mid-1970s had increased to around 30 individuals.

In several localities in south London, Yellow-bellied Toads were apparently colonizing garden ponds up to at least the late 1980s, though it seemed likely that increasing urban development would probably restrict any future expansion.

Future Prospects

Being a species of undisturbed habitats, the Yellow-bellied Toad probably finds much of suburban southern England a less than optimum environment, and this, together with the fact that southern England is north of the species' natural range in continental Europe, is likely to hinder its long-term establishment here. This is yet another species that could well benefit from the likely increase in global warming, and it is possible that hitherto unrecorded colonies already exist here.

References

See under Fire-bellied Toad, opposite.

YELLOW-BELLIED TOAD (UNDERSIDE)

DISCOGLOSSIDAE (MIDWIFE TOADS, FIRE-BELLIED TOADS & PAINTED FROGS)

FIRE-BELLIED TOAD *Bombina bombina*

Natural Distribution E Europe W to Denmark and S to N Bulgaria.
Naturalized Distribution England.

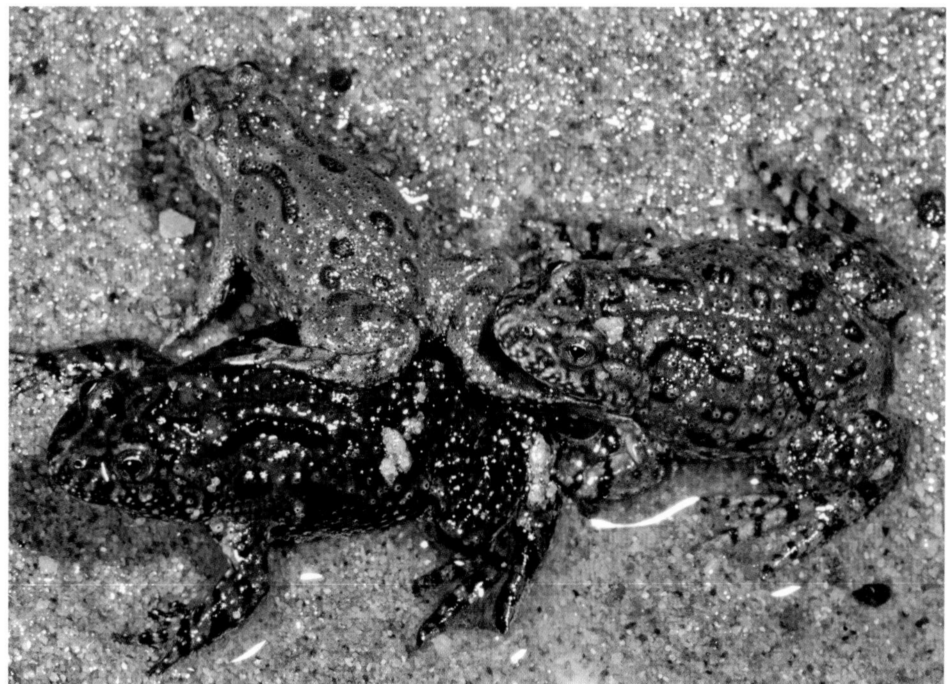

FIRE-BELLIED TOADS

England

The Fire-bellied Toad was at one time established and breeding in the Beam Brook Aquatic Nursery at Newdigate in Surrey, and in the 20th century numerous but unsuccessful attempts were made by the Duke of Bedford to naturalize the species at Woburn Abbey in southern Bedfordshire.

Prospects

Since southern England falls within the latitude occupied by the species in continental Europe, there seems to be no reason why it should not be able to survive here in suitable habitats, e.g. edges of rivers and streams, shallow ponds, and marshy pools and drainage ditches. Indeed, as in the case of the Yellow-bellied Toad, there may be undocumented colonies of the Fire-bellied Toad in parts of southern England.

References

Anon 1997; Beebee & Griffiths 2000; Coleridge 1974; Fitter 1959; Lever 1977, 2003.

PIPIDAE (PIPID FROGS & 'CLAWED' TOADS)

AFRICAN CLAWED TOAD *Xenopus laevis*

Natural Distribution Most of sub-Saharan Africa, from Cameroon E to E Africa and S to South Africa.
Naturalized Distribution England; Wales.

AFRICAN CLAWED TOAD

From the 1930s until relatively recently, *Xenopus* spp. were widely used for human pregnancy testing, and for over half a century for biological laboratory research, including physiology, biochemistry, genetics and developmental biology. Their ease of maintenance in captivity also made them a popular pet, and in the 1950s/1960s, they were widely introduced to aquaria worldwide. These toads were all of the (largest) nominate form from southern Africa, and were imported from the Cape, South Africa. Many escaped or were deliberately released, and the 'durability' that made them a suitable laboratory and pet species enabled them to adapt to alien conditions in the wild, usually disturbed or artificial habitats such as man-made ponds, flooded excavations, sewage farms and fish farms, and canalized watercourses and, in the case of natural water bodies, sites that are subject to considerable seasonal variability.

England

In 1967, Frank Boyce liberated a number of African Clawed[1] Toads in some ponds at Brook, about 5.6 km from Freshwater on the south-west coast of the Isle of Wight. Tadpoles were found in the ponds in 1970, and four years later others were discovered in a pond nearby, suggesting either emigration or a second introduction. By 1976, the population of this apparently thriving colony was estimated to number 40–50 individuals.

These ponds are located on maritime cliffs of unstable clay that is subject to continual erosion by wave and rainfall, and the resulting subsidence causes some ponds to be ephemeral while new ones are formed in settled areas. Droughts in summer may cause the ponds to evaporate, but the animals presumably retire into refugia in the clay substrate until the ponds are replenished. These ponds and nearby streams have been regularly surveyed in recent years, and it seems likely that this colony has either declined or died out.

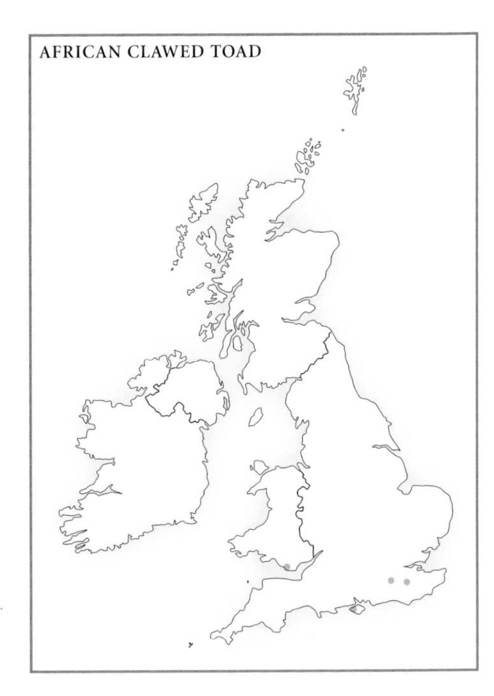

African Clawed Toads and their tadpoles were seen in some ornamental ponds in Kent in 1987, and in 1990 a large breeding population was found in a pond in south-east London, but none of the toads has been seen in either locality in recent years, and they may have been eradicated by fish predation. From time to time individual adults have been caught in watercourses in south-west England, but no breeding has yet been confirmed. In about 1997, the species temporarily colonized the Scunthorpe Police Pond at Flixborough, Humberside, in northern Lincolnshire.

In the 1980s, some escaped from a firm called Xenopus Ltd at Nutfield in Surrey, where some of their descendants are believed to remain (Steve Langham, personal communication 2008).

In 2004, a population was found living sympatrically with introduced American Bullfrogs at Cowden on the Kent/Sussex border. It is believed that one or other of these species was probably the agency whereby the deadly fungal disease chytridiomycosis (see page 209) was introduced to Britain.

Wales

In 1979, two large, extensive and well-established populations of African Clawed Toads, probably descended from escapes or releases from local medical laboratories (see opposite) or schools, were found in the catchment of Afon Alun near Castle-upon-Alun, Bridgend, in the Vale of Glamorgan. One occurred in a 3–4 m-wide stream that is fast flowing in winter but liable to dry out in summer, from which the animals apparently commute to and from a number of artificial ponds. The second occupied the grounds of a derelict castle and house – here the creatures occurred in subterranean water cisterns beneath the castle, and also in nearby streams and in a series of ornamental weirs built to form alternating pools and cascades. This site, like the one on the Isle of Wight, is near the coast and is subject to spray from the sea. The two drainages are separated across a less than 1-km watershed, so both colonies may have resulted from a single introduction.

In 1982–96, by which time the population had considerably declined, probably due to a loss of breeding sites through the evaporation of ephemeral ponds, and predation and/or cannibalism of eggs and larvae, a total of over 350 adults were caught, dye-marked, released and recaptured during a study of the creatures' ecology by Westfield

College, University of London. This research revealed, inter alia, that the African Clawed Toad is a relatively mobile and long-lived species, some individuals being repeatedly caught and recaught for up to nine years.

Prospects

The growth of the English and Welsh populations has been limited by seasonal drying out of their ponds and by temperature, although in Africa they are fairly tolerant of both heat and cold. In Britain, conditions favouring reproduction – warm and wet summers – have, until recently, been of irregular occurrence. If recent trends in climate change persist as a consequence of global warming, the African Clawed Toad is yet another species that could benefit from such climatic alterations.

Characteristics

The success of African Clawed Toads in disturbed environments, where prey resources for adults (principally macroinvertebrates) are frequently scarce or even absent, may be explained by the advantages of cannibalism. Their high fecundity gives rise to large numbers of phytoplankton-feeding tadpoles that exploit seasonal algal blooms, and adults may rely on their own offspring as a food source, thus enabling older individuals to survive periods of food shortage: the nutrient resource represented by algal populations is exploited by adults through their young. This strategy ensures rapid population growth under optimum conditions for reproduction and dispersal. On the other hand, where reproduction is irregular and populations are localized, cannibalism may be partially the cause of low annual recruitment, as has been the case in England and Wales.

Impact

Because the African Clawed Toad is almost exclusively aquatic, seldom venturing onto land except after heavy rainfall, and because, unlike other anurans, it is not a noisy species, large populations can both go unnoticed and be confined within a small area with a high density of as many as 8.86 per square metre. A detailed examination of the population demographics, diet and growth of the species in Wales, before it showed signs of a decline, raised concerns that the size and potential impact of the animals was rapidly increasing. They consumed a wide variety and size range of prey, of which zooplankton[2] and zoobenthos,[3] both by weight and numerically, formed the bulk of their diet. Terrestrial vertebrates comprise only a small percentage of the diet in spring and summer. Nektonic[4] prey is taken throughout the year. Native species most at risk are other amphibians, fish and a large range of aquatic vertebrates. When eggs and larvae are present, cannibalism is widely practised.

References

Anon 1997; Ashton 2005; Beebee & Griffiths 2000; Calado & Chapman 2006; Coote 1998; Frazer 1964; Halliday 2001; Kiesecker 2003; Lever 1977, 1980b, 2003; Measey 1998a, 1998b; Measey & Tinsley 1998; Simmonds 1982–3; Tinsley & McCoid, in Tinsley & Kobel 1996.

Notes

1. So-called because of the horny black claws on the outer three toes of the hind feet.
2. Zooplankton = minute aquatic organisms, such as protozoans, small crustaceans and larvae of larger organisms, which drift with water movement.
3. Zoobenthos = organisms attached to or resting on the bottoms of water bodies.
4. Nektonic = aquatic animals able to swim independently, rather than just drift.

HYLIDAE (TREE FROGS)

EUROPEAN GREEN TREE FROG *Hyla arborea*

Natural Distribution Most of Europe apart from the north, the Balearic Islands and parts of S France and Iberia.
Naturalized Distribution England.

England

The earliest recorded introduction of the European Green Tree Frog to England took place in the 1840s, when a number were turned out on the Undercliff at St Lawrence on the south coast of the Isle of Wight. In 1906, Lord Walsingham released some more at the same site, where they bred successfully for a number of years. At about the same time some escaped from a greenhouse at Freshwater Bay on the other side of the island, where they bred in a pond until it was filled in during the First World War. On Lundy Island in the Bristol Channel, some escaped in 1933 when a bullock overturned their container, and at least one survived until 1939. They have also been introduced with varying degrees of success to the Scilly Isles; at Paignton in south Devon, where 50 turned out in 1937 soon disappeared; and in a garden of a college at Cambridge University. T.B. Rothwell had a colony in his Beam Brook Aquatic Nursery at Newdigate in Surrey in the early part of the 20th century, but no breeding was recorded.

In 1952, Oliver H. Frazer released about 50 near Mottistone Mill at Brightstone on the Isle of Wight, where although two clumps of spawn were found in the following year the frogs subsequently disappeared. Another attempted introduction at Freshwater Bay was similarly unsuccessful. In 1952, Viscount Chaplin freed some near Totnes in south Devon, and three years later two dozen were unsuccessfully turned out at Boxley, near Maidstone, in Kent.

EUROPEAN GREEN TREE FROGS

In 1962, a colony was discovered in a small ephemeral pond on the edge of the Beaulieu Abbey Estate in the New Forest in Hampshire, where a Mr Jones had apparently established it in the early years of the century. (It is said that the original introduction was made to a pond some distance away from the final site, to which the frogs eventually dispersed.) By the 1970s, the population only numbered around a dozen or so, but was nevertheless viable and self-sustaining. The pond, which measures some 25 m in diameter, is surrounded on three sides by low trees and bushes on the top of a hill; the water, which dries out in late summer, is 60–120 cm in depth, and in summer is appreciably warm to the touch. This may have been a significant factor in the success of this colony, which apparently died out in the late 1980s. Possible reasons for its demise include removal by collectors, the development of dense beds of the alien weed New Zealand Stonecrop and the establishment of a population of Great Crested Newts. However, it is possible that another colony may have become established in the same general area (Tony Gent, personal communication 2008).

In Essex, a colony of European Green Tree Frogs is established in a pond at Kingswood near Basildon (John Cranfield, personal communication 2008). Another colony at Kidbrooke near Lewisham in south-east London disappeared when its habitat was destroyed by construction work in 1989–90.

Many years ago, tree frogs from Cannes in southern France, where the Stripeless Green Tree Frog replaces the European Green Tree Frog, were released in a garden in Suffolk and were probably of this latter species. The introduction of this southern species, which occurs naturally from North Africa to southern France, and which even in southern England would thus be well north of its natural range, may explain why many other attempts to naturalize tree frogs in England have been unsuccessful. Another reason may be that frequently few, if any, females are released. Tree frogs are usually collected at their breeding ponds, where although the males remain throughout most of the breeding season the females only appear to deposit their eggs and then depart. Thus for most of the year the population of a pond is almost exclusively male.

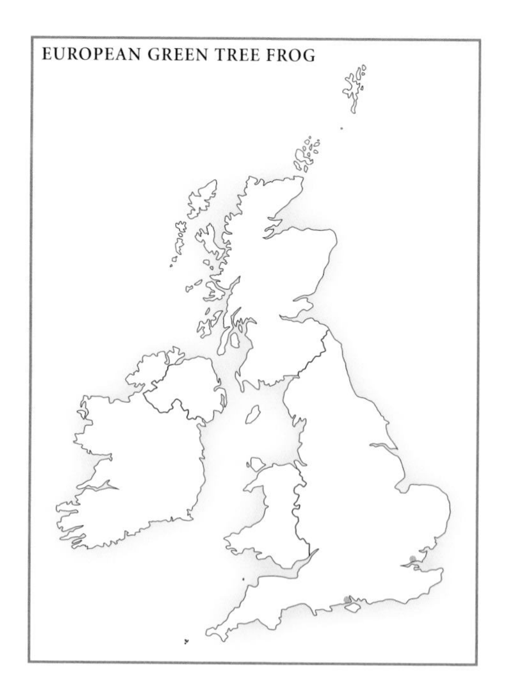

Characteristics

The two species of tree frog are the only European frogs that have disc-like adhesive pads on their toes. These enable them to clamber about – usually at night – in trees, bushes, reeds and other vegetation, where they feed mainly, if not exclusively, on insects and spiders. They share with chameleons (Chamaeleontidae) the ability to change their colour quite extensively.

References

Anon 1997; Beebee & Griffiths 2000; Dalton 1950; Fitter 1959; Frazer 1948, 1964; Lever 1977, 1980b, 2003; Snell 1991; Taylor 1948, 1963; Yalden 1965.

RANIDAE (TYPICAL FROGS)

AMERICAN BULLFROG *Rana catesbeiana*

Natural Distribution E North America, from Nova Scotia in SE Canada S to C Florida and the Gulf of Mexico, W across the Great Plains to around the 100th meridian E of the Rocky Mountains, and S into N Mexico. [The American Bullfrog has been widely translocated within the USA W of the 100th meridian and the Rocky Mountains, especially to California.]
Naturalized Distribution England.

AMERICAN BULLFROG

England

The earliest record of American Bullfrogs in Britain seems to be by Fitter (1959), who referred to their presence in a collection of reptiles and amphibians at Frensham Hall, Shottermill, in Surrey, in about 1905–10; whether these were in captivity or in the wild is unclear, but it was probably the former.

Banks et al. (2000) and Jim Foster (personal communication 2008), from whom much of the following account is derived, have documented the more recent establishment of the species in southern England.

In the late 1970s, the number of American Bullfrogs imported into Britain annually increased, as sources of imported Goldfish switched from Italy to North America and Israel. Goldfish farms in the United States are infested with flourishing populations of American Bullfrogs, whose adults eat mature fish and whose tadpoles compete with the fish for food. Initially, small quantities of tadpoles were introduced accidentally or as free 'test samples' with shipments of Goldfish. Later, as a result of the popularity of the 'test samples', aquatic centres and coldwater fish retailers began stocking American Bullfrogs for sale to the public, resulting in the importation to Britain of batches of tadpoles of 'up to low thousands'. Tadpoles continued to be readily available until June 1997, when live importations of the species were controlled by EU Regulation 338/97, designed to combat the potential threat of alien species to European wildlife.

From the 1980s, the species began to be widely recorded in garden ponds and wetlands in southern England, and although no breeding was reported the potential for successful breeding was predicted by Cunningham and Langton (1997).

In 1996 and 1997, two adults were shot at two fish ponds and a wet flush area near Edenbridge on the East Sussex/Kent border, where in the summer of 1999 hundreds of large tadpoles, which metamorphosed successfully, were observed. How the animals arrived in the ponds is unknown. A decision was made to attempt to eradicate this population while it was still fairly localized. This was in accordance with the guidelines on the elimination of non-native terrestrial vertebrates under the provisions of Article 11 of the Convention on the Conservation of European Wildlife and Natural Habitats (the Bern Convention); the work was undertaken by Herpetofauna Consultants (Froglife) and the then English Nature, with help from the Environment Agency. The two ponds were surrounded by a 1 m-high plastic net fence, and were visited at night to capture the animals, which tended to congregate around the inner perimeter where they were easy to catch with a torch and a long-handled net – 477 froglets were captured during a single night. A number of immature individuals were also captured outside the fence, suggesting that some animals had dispersed before the fence was erected or perhaps that they were foraging; in either case, given the proximity of a fast-flowing stream, a matter of some concern. Another successful means of control in daytime was shooting with either an air-rifle or a shotgun. Less successful methods were the use of pit-fall and crayfish traps, catapults and electrofishing.

In December 1999, both ponds were drained in an attempt to remove any surviving adults and tadpoles. It was found that the remaining animals had not, as expected, congregated in the central hollows, but had stayed in burrow refugia in the mud throughout the ponds, among emergent vegetation, under carpets of green algae, and in water plants growing on silt and litter; they were most numerous in deeper mudslide areas of silt below rocky banks. Finally, the ponds' silt was excavated, buried and covered with compacted soil. A few American Bullfrogs were discovered hibernating on land surrounding the margin of the enclosure fence. By the end of 1999, a total of 4,744 tadpoles and 2,269 froglets had been captured.

Surveys carried out in the spring of 2000 of a further 21 ponds within a kilometre of the 2 originally infested revealed that immature American Bullfrogs had dispersed to a further 5 waters up to 500 m from the Edenbridge ponds, from which during 2000 a total of 108 subadults were removed. Since 2000, 53 ponds have been surveyed, at 7 of which American Bullfrogs were found. Although it was believed that most of the subadults in the Edenbridge area were captured or killed in 2000, it was thought probable that a small number (<20) survived. In the late summer of 2001, spawning was confirmed at two ponds, in one of which tadpoles were observed in September. Numbers of adults were by then thought to be low – up to perhaps only 1–2 at each of four ponds. No breeding activity was reported at three ponds with small numbers of adults and immatures.

In 1999–2004, Natural England removed 11,830 American Bullfrogs (and an adult and several hundred tadpoles of the African Clawed Toad) from the Edenbridge site and nearby ponds. In 2004–5, a further 167 were removed, and since May of the latter year there have been no further sightings, suggesting that the population may have been eradicated. The total cost to date has been in the region of £100,000.

In the summer of 2004, chytridiomycosis (see page 209) was discovered in the Edenbridge American Bullfrog population. This outbreak was the first detected case in Britain outside captivity. It is believed that either American Bullfrogs or African Clawed Toads were the agency whereby the fungus causing the disease first appeared in the wild in Britain.

In 2000, as a result of media publicity about the Edenbridge animals, Froglife received reports of the species in 30 English and Scottish counties, from Devon in the south-west north to Aberdeen. The highest number of sightings came from Kent (43), Surrey (11), London (7) and East and West Sussex (5 each). No breeding at any site was reported.

In 2006, a population believed to have originated in 2003 was reported in a pond on a golf course near Romford in south Essex. At least 200 mainly adult American Bullfrogs have been removed from this population, and individuals have been discovered in at least eight other ponds within a 3-km radius. Surveillance and control of this population by Natural England is continuing.

The only other place where breeding by American Bullfrogs in Britain has been confirmed is in a fishing pond at Sway in the New Forest in Hampshire in the mid-1990s, where the population was successfully eradicated.

Impact

The potential ecological threat posed by this large (up to 20 cm long) and efficient predator, were it ever to become widely naturalized in Britain, is considerable. In North America it eats, among other species, small birds, voles (Microtinae), young American Mink, ducklings and fish. Where it has become naturalized in continental Europe (e.g. Italy, Spain, France, the Netherlands), it has been implicated in the decline of several native species. Post-mortem examination of American Bullfrog stomachs in England has revealed the presence of small native mammals, other amphibians and dragonflies (Odonata). Potential prey in Britain includes voles, Freshwater Crayfish, Common Frogs, Common Toads, young birds and other insects. American Bullfrog tadpoles would be likely to compete for food and habitat with native amphibian larvae, and the adults might be vectors of new diseases to which native species have no inbuilt resistance. Bullfrog predators include native Grey Herons and possibly adult introduced American Mink. The damaging potential of an animal with such a high colonizing ability away from its point of introduction must be a point of particular concern.

References

Anon 1997, 2000c, 2000d, 2001; Banks et al. 2000; Beebee & Griffiths 2000; Chapman 2000; Cunningham & Langton 1997; Lever 2003; Morgan 2000.

RANIDAE (TYPICAL FROGS)

MARSH FROG *Rana ridibunda*

(LAUGHING FROG; LAKE FROG)

Natural Distribution E and SE Europe from France, the Netherlands and Germany into Russia and S to the Balkans. A disjunct population occurs in SW France and the Iberian Peninsula.
Naturalized Distribution England.

England

The most important and best-documented introduction of Marsh Frogs to England was made by E.P. Smith (the playwright Edward Percy), who describes it as follows:

> In the winter of 1934–35 I introduced twelve specimens of the European edible frog[1] (the Hungarian variety) into a pond beside a running stream in my garden at Stone-in-Oxney, East Kent. The site ... abuts upon a tract of land where the flats comprising the Walland, Romney and Denge Marshes [hence the usual English name] usurp the place of the salt water. It is a maze of dykes, canals, meres and streams intersecting rich pastures.

In June 1935, two of the frogs dispersed overland to a mere about 800 m away, where in the following October they were joined by the remaining ten. This mere had dried out two years previously, and was thus free of large aquatic predators. Here the frogs became so excessively noisy that local villagers were roused to complain to their local Member of Parliament and to the Minister of Health!

By the autumn of 1936, the mere had become colonized by the Marsh Frogs, which later in the year appeared 5 km away at Appledore, which is connected to the mere by a circuitous series of dykes and sewers. 'In May 1937,' wrote Smith (1939), 'began what I might call the Great Year of Rana esculenta.' Large quantities of spawn appeared, followed in due course by numerous tadpoles and froglets. A few managed to reach Great Lake at Bedgebury, 22 km from the mere and unconnected to it by water, having presumably been translocated overland by human agency. At about this time Marsh Frogs appeared at Lydd, the same distance from the mere but in the opposite direction, and connected to it by water. Following further complaints about the creatures' vocal activity, some were caught up and transferred to Essex, Hertfordshire, Sussex and even Scotland.

By 1938, Marsh Frogs had succeeded in colonizing an area of Romney Marsh covering some 50 sq km. By the outbreak of the Second World War in 1939, they had occupied a lengthy stretch of the Royal Military Canal, and were beginning to disperse widely into many other parts of Kent and eastern Sussex. In 1943, they crossed the River Rother at Thornsdale, Iden, and by 1946 had spread as far upstream as Maytham and downriver to near the coast at Rye.

Brackish water was no barrier to the species' advance – the first individuals being reported in Rye harbour in 1951 and in the following year some appeared in the Pett Level and Winchelsea Marsh. By 1956, Marsh Frogs had travelled 7 km along the New Sewer from Appledore to Breznett, and two years later they appeared at Baynham on White Kemp Sewer.

Until the 1950s, the main spread of the Marsh Frog was along the waterways that intersect the area of its introduction. Thereafter it extended its range apace, and had soon

MARSH FROG

colonized most of Romney, Walland and Denge Marshes, and much of the Pett Level, occurring from west of Hythe to south-west of the Pett Level near Hastings, and as far up the Rother and Brede valleys as Newenden and Brede Bridges. The lower valley of the Tillingham and the land between the Royal Military Canal and the coast (apart from stretches of Denge beach) were also occupied.

By the mid-1970s, Marsh Frogs were widely but patchily distributed over an area of more than 250 sq km of Romney Marsh and the Rother Levels, extending as far west as the eastern end of Pevensey Levels near Bexhill. They also occurred around Sittingbourne in the southern half of the Isle of Sheppey and on the Iwade marshes in north Kent, having probably arrived there by human agency. In 1974, some Marsh Frogs were released by an agricultural labourer at Lewes Brooks in East Sussex, and by the end of the decade they had spread out across the marshes between Lewes and Newhaven on both banks of the River Ouse, and thereafter advanced as far as Barcombe, north of Lewes. The Marsh Frogs' colonization of some farm ponds and gravel-pits around Ashford in Kent appears to have been achieved by natural dispersion rather than human agency, and from here they spread, also naturally, along the Stour valley as far as Canterbury. The River Stour may also have been the route by which Marsh Frogs reached the Sandwich Bay area of north-west Kent.

Away from Sussex and Kent, largely unsuccessful attempts have been made to naturalize Marsh Frogs in various localities in, mostly southern, England. Fitter (1959) traced only one pre-20th-century introduction, when around 1884 St George Mivart released some at Chilworth in Surrey.

In 1939, 70 Hungarian Marsh Frogs were turned out by a son of Professor (later Sir) Archibald Vivian Hill in Reach Lode, which runs into the River Cam at Upware near Wicken Fen in Cambridgeshire. In 1948, Alfred Leutscher liberated some in Wanstead Park in Essex, and a year later, 16 were freed at Thorne, between Doncaster and Goole in

Yorkshire. In 1958, others were released near Minehead in Somerset and in Devon, and four years later a pair was placed in a sewage farm at Beddington, near Croydon in south London. All these attempts were apparently unsuccessful.

The only other well-documented and successful release of Marsh Frogs in Britain was made by W.L. Coleridge, who around 1954 transplanted about a dozen from Romney Marsh to his water garden at Bishopsteignton near Teignmouth in south Devon, where they bred almost annually and where their descendants may still survive.

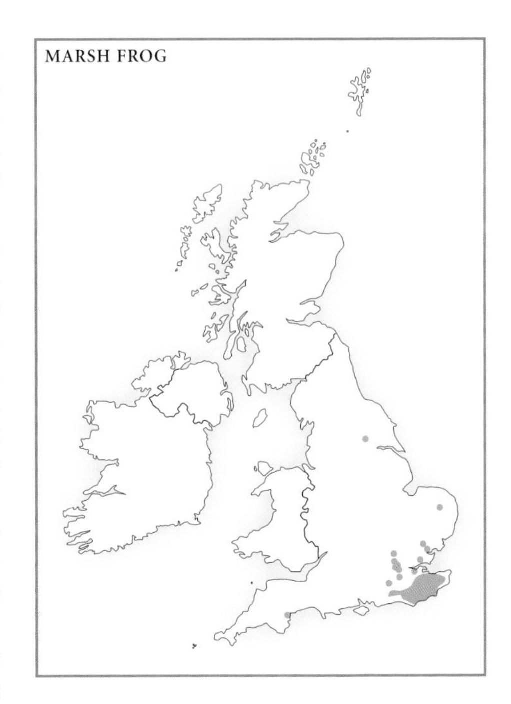

Wycherley (2003 and personal communication 2008) and Wycherley et al. (2003) have summarized the distribution of the Marsh Frog in England.

Bedfordshire Marsh Frogs that are believed to have spread upstream in the River Lea from a site near Harpenden, Hertfordshire (see below), have become established near East Hyde in Bedfordshire near the Hertfordshire border.

Essex In 1990, a few tadpoles from Romney Marsh in Kent were translocated to the Borrow Dyke by Two Tree Island around Hadleigh Castle, in south-east Essex. By the middle of the decade the population had increased explosively, and it has now dispersed over 3 km to the north. Other colonies are known at Cudmore Country Park, West Mersea, Langenhoe Marsh, at Ardleigh near Colchester, and in the catchment of the River Mardyke and Raynham Marshes. Marsh Frogs survive well in mildly saline waters, so they may well spread to other estuaries in Essex.

Greater London What are almost certainly Marsh Frogs established in ponds and other watercourses around Heathrow Airport are believed to have escaped from a consignment from Turkey confiscated by Customs and Excise. Large numbers occur in the Staines reservoirs, on Kempton Park racecourse and in the Wildfowl and Wetlands Trust's London Wetland Centre Reserve at Barnes. There has also been a colony since 1999 in the Lee (or Lea) Valley Navigation.

Hertfordshire In 1999, a small colony, believed to be still in existence, was discovered in a pond constructed three years earlier on the golf course at Aldwickbury, near Harpenden. Another population occurs at East Hyde in a backwater of the River Lea, which connects these two sites to the one in Bedfordshire (see above).

Kent Sites currently occupied by Marsh Frogs include Sandwich Bay, Dungeness, Lydd Airport, the Royal Military Canal, the Gravesend-Stroud and Thames-Medway Canals, Higham and Chetney Marshes, Minster dykes, Rye clay pits, Appledore, Ashford, Lenham Heath, Iwade and the Isle of Sheppey. Frogs in north-east Kent appear to have a distinctive call and colouration compared to those in the south, and may thus be of different origins although the same species.

Norfolk Buckley (1986) identified as Marsh Frogs a water frog colony near the Steam Museum at Forncett St Mary, south of Norwich.

Surrey Females have been recorded at the Beam Brook Aquatic Nursery near Newdigate, and the species has also been noted at Ewhurst, Pyrford golf course, the Royal Horticultural Society's gardens at Wisley, Old Woking and in the River Wey.
Sussex Marsh Frogs spread rapidly following their release in 1974 in Lewes Brooks, and are now established along tributaries of the River Ouse in the marshes between Lewes and Newhaven, and north to Ardingly Reservoir. They are also found in places along the River Adur as far north as Storrington and Partridge Green, and at Warnham Mill, Horsham, dykes in the River Brede Levels near Icklesham, Pel Levels and dykes, and the Chennells and Boulding Brooks.
Yorkshire Populations of water frogs may be a mixture of Marsh and Edible Frogs.

Prospects

As a colonist, the Marsh Frog has undoubtedly been the most successful alien amphibian in Britain. A number of factors have helped it to become established, the most important of which (as indeed for all introduced species) are probably the suitability of the habitat in Kent and Sussex and the availability of a suitable supply of food, together with an absence of major predators. Romney Marsh, an area devoted almost entirely to sheep farming, consists principally of small fields divided by dykes and ditches bordered by 'headlands' of ungrazed grass, with no need for hedges. This provides an ideal environment and source of food. The aggressive nature of the male Marsh Frog and the species' colonizing instinct help it considerably in becoming established in new localities. The animals' survival in the Romney Marsh area, however, is almost entirely dependent on the continuation of sheep farming. Were this ever to be abandoned in favour of arable farming, which would result in the infilling of waterways and their replacement by hedges, the population of Marsh Frogs would probably soon die out.

Characteristics

Zeisset & Beebee (2003) investigated the genetic outcome of the establishment of the Marsh Frog in Sussex and Kent, which provided an opportunity to test for bottleneck effects during colonization by an introduced species. The sub-populations in Romney and Lewes were found to possess similar genetic diversity to those in Hungary (the source of the founder stock) at 14 random amplified polymorphic DNA marker (RAPD) and five microsatellite loci. Introduced sub-populations were, however, differentiated genetically from each other and from the reference population in Hungary. Fitness assessments (larval growth and survival) revealed no differences between the Lewes and Romney groups. Despite starting with a low founder stock, significant bottleneck effects on the Marsh Frog in England were thus undetectable, presumably because the sub-populations expanded rapidly after introduction and translocation.

Impact

Early fears that the Marsh Frog might displace the Common Frog in Romney Marsh were not supported by studies in the Rivers Ouse and Adur valleys, but concern about the species' possible impact remains. Marshland, formed of pastureland and watercourses, is sometimes held to be a less than optimum habitat for Common Frogs and Common Toads, and the Marsh Frog may thus be merely occupying a near-vacant ecological niche. On the other hand, the Common Frog is known to breed in some freshwater dykes where its tadpoles are eaten by Marsh Frogs.

References

Arnold 1973; Beebee 1981; Beebee & Griffiths 2000; Buckley 1986; Bunting 1957; Burton 1973; Coleridge 1974; Fitter 1959; Frazer 1964; Harvey et al. 2001; Kelly 2004; Knight 1948; Langton & Burton 1997; Lever 1977, 1980b, 2003; Menzies 1962; Morrison 1994; Smith, M. 1951a, 1953; Smith, E.P. 1939; Taylor 1948, 1963; Wycherley 2003; Wycherley et al. 2003; Yalden 1965; Zeisset & Beebee 2003.

Note

1. The Marsh Frog (here described) used to be regarded by some as a subspecies of the Edible Frog. (For the relationship between the Marsh Frog, Pool Frog and Edible Frog see under the latter two species, below.) Smith's specimens had been imported from Debrecen in eastern Hungary for research at University College, London, from where he acquired them.

RANIDAE (TYPICAL FROGS)

EDIBLE FROG *Rana* kl. *esculenta*
POOL FROG *Rana lessonae*
SOUTHERN MARSH FROG or IBERIAN WATER FROG *Rana perezi*
ITALIAN POOL FROG *Rana bergeri*

Natural Distribution Edible and Pool Frogs: from France E through C Europe to W Russia, N into S Sweden and S into Italy and the N Balkans. Southern Marsh Frog: Iberian Peninsula and SW France. Italian Pool Frog: peninsular Italy south of Genoa/Rimini, and Sicily and Corsica.
Naturalized Distribution England.

Beebee & Griffiths (2000), Harvey Pough et al. (2001), Arnold & Ovenden (2002) and Barry Clarke (personal communication 2008) have outlined the complex relationship between *R.* kl.[1] *esculenta, R. lessonae* and *R. ridibunda* (Marsh Frog, see also page 222).

R. lessonae and *R. ridibunda* sometimes hybridize, the offspring being kl. *esculenta.* Unlike most other hybrids, kl. *esculenta* is able to reproduce successfully with either *lessonae* or *ridibunda.* Instead of such matings being either non-productive or resulting in a range of intermediate forms, as might be expected, the offspring are all kl. *esculenta.* That a hybrid can thus reproduce itself by breeding with either of its parental species is due to an unusual chromosomal mechanism. In common with most other animals, first generation kl. *esculenta* have a set of chromosomes inherited from each parent. In nearly all other species the two sets would become mixed (by the chromosomes 'crossing over') when the sperm and eggs were formed, so that each would contain a mixture of genes from the two parental species. In kl. *esculenta,* however, one set of chromosomes is destroyed before this can occur, so their sperm and eggs contain the chromosomes and genes of only one of the two parental species. For example, when kl. *esculenta* hybridizes with *ridibunda,* the sperm and eggs usually have only *lessonae* chromosomes, those of *ridibunda* having been destroyed. Thus the offspring of such matings will have one *lessonae* and one *ridibunda* set of chromosomes, and are consequently kl. *esculenta.* Conversely, when kl. *esculenta* mates with *lessonae,* the latter's set of chromosomes disappear. This bizarre situation persists from generation to generation, so that kl. *esculenta* persists indefinitely.

EDIBLE FROGS

In fact, the situation is even more extraordinary and complex. Occasionally, kl. *esculenta* seems able to maintain a pure population without the presence of either parent species. In addition to the usual diploid kl. *esculenta* with two sets of chromosomes, there are triploid individuals with three sets, the third being that of one or other of the parent species. The Edible Frog is thus a hybridogenetic form of the water frog group.

England

The precise status of the Pool Frog in Britain, whether as native or alien, was for many years in doubt. This problem was resolved by the research of Wycherley (2003) and Beebee et al. (2005).

The Pool Frog has been the subject of documented introductions from southern and central Europe since the early 19th century, and it has been generally accepted that the present British populations are all descended from these introductions. Early references in the literature, however, and the recent discovery of isolated native populations in Norway and Sweden, have suggested that the species may have originally been native to Britain. Research was commenced on four main lines of inquiry – genetic, bioacoustic, archaeological and archival. A high degree of convergence in the genetic and acoustic investigations showed that the possibly native British Pool Frogs were closely related to those in Scandinavia, thus ruling out as a potential origin the possibility of introductions from further south. Sub-fossil evidence of Pool Frogs was found from c 1,000 years ago, showing that the species occurred here long before the earliest introductions. Archival research revealed no support for introductions from northern Europe. Evidence for the native status of the Pool Frog in eastern England is thus compelling: furthermore, the British population seems to be part of a distinct northern clade.[2]

In view of the close relationship between Edible and Pool Frogs, and because both species have frequently been imported together, the early history of both species is here described together under the name of '(Palaearctic) Water frogs'.

In addition to the documented introductions in Table 13, between 1853 and about 1966, colonies of water frogs were found in Cambridgeshire, East Sussex, Essex, Kent,

POOL FROG

SOUTHERN MARSH FROG

Middlesex (now Greater London), Norfolk, north London, Oxfordshire, Suffolk, Surrey and Warwickshire.

Although the identity of some introductions of water frogs is known (e.g. the Marsh Frog to Kent in 1934–5), most were of uncertain origin and species. All water frogs were formerly lumped together under the collective name of 'Edible Frog', and only relatively recently has it been shown that in fact there are at least five further species. As a result of bioacoustic analysis, it has been possible to identify from their individual calls the three species involved in the other early introductions, and also to identify two further

Table 13 Introductions to Britain of Palaearctic Water Frogs, 1837–1961

Year	Introduced By	Locality	Comments
1837	George Berney	Morton Hall, Norfolk	More than 1,500 adults and spawn from Paris, Brussels, St Omer; persisted for >70 years
1840	Colonel Durant	Tong Castle, Shifnal, Shropshire	Disappeared by end of century
1841	George Berney	Morton Hall, Norfolk	-
1842	George Berney	Morton Hall, Norfolk	-
c 1884	St George Mivart	Chilworth, Surrey	From Brussels
c 1884	G.E. Mason	Brandon, Norfolk/ Suffolk border	Never seen again
1882/1892	?	Blaxhall, east Suffolk	From Normandy; survived only a few years
1880s/1890s	Ld. Arthur Russell	Shere, Surrey	From Berlin; survived at least 20 years
c 1895	Mr Cotton	R. Itchen, Hampshire	Disappeared by 1914
Before 1897	?	Oxfordshire	From Italy; soon died out
1905-10	Hon. Charles Ellis	Frensham Hall, Surrey	Survived until 1939–40
c 1900-1950	Duke of Bedford	Woburn Abbey, Bedfordshire	Large numbers imported, mostly from Berlin; bred, but not self-sustaining
1907	?	Queensferry, Midlothian	Survived c 40 years
c 1930	E.J. Rope	Snape, Suffolk	Bred for a few years
1938–9	Mrs M.G. De Udy	Bratton, Wiltshire	From Cannes, France, 1938 or 1939. Still breeding in late 1950s, but then died out
1948	?	Golders Hill Park, London	5 from Ham, Surrey
1949	?	Thorne, Yorkshire	12 from Ham, Surrey (and 16 Marsh Frogs)
1959/1961	Frank Boyce	Freshwater, Isle of Wight	36 introduced; bred for c 4 years, but then died out

Sources Lever 1977, 2003.

naturalized British species, the Southern Marsh Frog or Iberian Water Frog, and the Italian Pool Frog.

Since the 1960s, populations of water frogs have proliferated in south-eastern England, especially since the 1980s in Kent, Surrey and Sussex.

Wycherley et al. (2003) have identified introduced water frogs in Britain by the characteristics of male advertisement calls, and they, and Wycherley (2003) have summarized their current distribution although, as they concede, the number of colonies and their distribution are probably greater than indicated (for data on the Marsh Frog see under that species).

Essex A colony of Pool Frogs is established at Ardleigh near Colchester and in a nature reserve near Basildon. As yet unidentified water frogs have been reported in ponds at a golf course at Witham. There are some 11 sub-populations in the county.

Greater London Unspecified water frogs are present in Richmond and Bushey Parks, and along the Rivers Longford and Crane. Further east along the River Thames, various water frog species occur in Birdbrook Nature Reserve near Woolwich. There are presently nine sub-populations in Greater London.

Hampshire Pool and Edible Frogs have been established since at least the late 1990s in a flooded gravel-pit on Bramshill Common, from which they have colonized several neighbouring ponds.

Herefordshire An isolated colony of water frogs is said to occur on Bedenham Moor.

Kent Water frogs on the Isle of Sheppey are distinct from the Marsh Frogs established in the county elsewhere; they have been identified as Marsh Frogs, Southern Marsh Frogs, Edible Frogs and perhaps Pool Frogs.

Lancashire There is a single record of water frogs in the Brickcroft Nature Reserve.

Norfolk At Wolterton Hall near Itteringham, there is a mixed colony of Edible and Pool Frogs. The native Pool Frog population on Thompson Common became extinct in the 1990s, following drainage of its fenland habitat. In 2005, Pool Frogs were reintroduced by English Nature, which imported 50 from Uppsala in south-central Sweden for release near Thetford. Sweden was selected because genetic research had revealed that the native Norfolk Pool Frogs were, along with those in Sweden and Norway, part of a distinct north European clade.

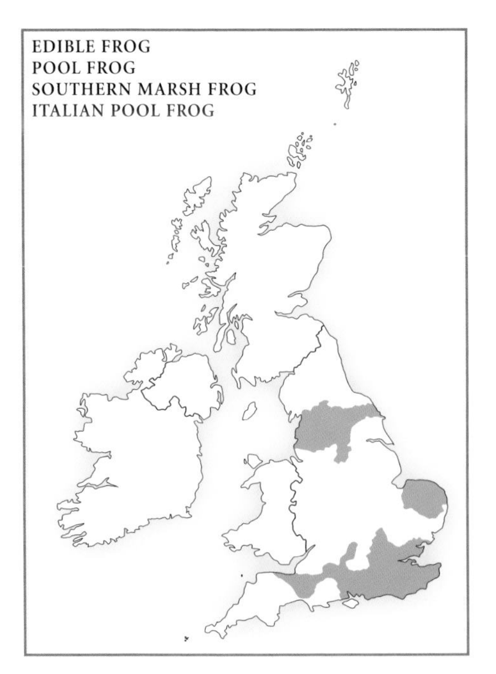

Oxfordshire As in Lancashire, there appears to be only a single instance of water frogs (believed to be mainly Pool Frogs) in a garden pond in Towerley.

Somerset 'Edible Frogs' (and perhaps other water frog species) are well established in a pond and neighbouring dyke in peat workings on Shapwick Heath in the Somerset Levels, and also at West Sedgemoor.

Surrey Since its formation by T.B. Rothwell in 1903, and subsequently under the ownership of Gilbert Taylor, the Beam Brook Aquatic Nursery at Newdigate has imported large numbers of water frogs

from Belgium, France, Germany, Italy and perhaps elsewhere, for commercial purposes. The present population comprises mainly Pool, Edible, Italian Pool and Southern Marsh Frogs, and possibly some other species. This population had remained relatively contained until an exodus took place in the early 1980s along the upper tributaries of the River Mole. The current range extends over 100 sq km, as far as beyond Capel on the Surrey/Sussex border. Elsewhere in Surrey, Pool and Edible Frogs have been identified at a number of sites. An introduction in the Horley/Gatwick area in the early 1990s resulted in the spread of water frogs via the Gatwick Stream, Salford Brook and the River Mole.
Worcestershire Water frogs in gravel-pits and fishing lakes in wetland areas near Holt and Ombersley on either bank of the River Severn seem to be slowly extending their range. There are currently some six sub-populations in the county.
Yorkshire Water frogs were first heard calling in Yorkshire in the 1980s, but the increase in range since 1997 suggests a recent rapid dispersal and/or recent introductions. Reported from angling waters near Hedon in 1999, water frogs are now established over a much wider area of the valley of the River Hull that, via numerous connecting drains and canals, eventually debouches into the tidal River Humber. The largest concentrations of water frogs, at Tophill Low Nature Reserve, are adjacent to the tidal limit of the Humber. Additional populations occur between Driffield and Hedon. The Yorkshire water frog populations may be a mixture of Edible and Marsh Frogs. The most northerly population occurs on Swinemoor Common. There are some 19 sub-populations in the county.

Since the extinction of the native Pool Frog in Norfolk in the 1990s, it is now probable that all current British water frog populations are a result of introductions.

References

See under Marsh Frog, page 222.

Notes

1. 'The kl. stands for klepton, a word derived from the ancient Greek for thief. This refers ... to the way hybridogenetic frogs "steal" half their chromosomes in each generation from members of other species' (Arnold & Ovenden 2002: 88–9).
2. The Pool Frog is here treated as an introduced alien species.

SALAMANDRIDAE (NEWTS & EURASIAN SALAMANDERS)

ALPINE NEWT *Triturus alpestris*

Natural Distribution W, C and E Europe, from W Russia to N and E France, and from S Denmark S to N Italy and C Greece. A disjunct population occurs in Asturias and Cantabria in N Spain.
Naturalized Distribution England; Scotland.

England

The oldest known naturalized population of Alpine Newts in England, numbering several hundred individuals, was established in 1903, when T.B. Rothwell released some in a series of ponds at the Beam Brook Aquatic Nursery at Newdigate in Surrey. (For information on Alpine and Italian Crested Newts at nearby Nutfield see under the latter species,

ALPINE NEWT ADULT MALE IN BREEDING CONDITION

page 234.) Other, more recent, populations have since occurred in various localities, including south-east London (Kent), Sussex, Birmingham, Warwickshire, Shropshire, Sunderland and Co. Durham.

The Shropshire population is derived from four adult males and three adult females released in ponds in a village garden near Market Drayton in April 1970. This garden is situated in open countryside on the edge of the village, where native amphibians such as the Smooth Newt and Great Crested or Warty Newt frequented the ponds. The introduced Alpine Newts were supplied by a retailer named de Rover of Ermelo in the Netherlands, though whether they were of Dutch origin is unrecorded.

In 1976, adults and larvae from this Shropshire site were transferred to another garden pond 3.2 km to the south-west, to which native amphibians (and predators) also had access. Bell (1978) reported that the Alpine Newts from the Market Drayton pond had spread barely 20 m during the first seven years.

Between April and September 1993, 23 and 17 years respectively after the animals' release at the 2 sites, a total of 52 ponds nearby was surveyed, when the presence of Alpine Newts was revealed in only 10 (19 per cent). At both release sites the Alpine Newt remained common; from the 1970 release site the newts had dispersed to ponds up to 70 m away, and from the 1976 release site to a pond 50 m distant. The species had thus spread very little from its original ponds, whereas in its natural range it is capable of quite wide dispersal. This suggests the existence in Shropshire of ecological barriers, such as perhaps buildings, and/or the relatively small founder stock compared to the existing stock of native species, which might out-compete the invasive. However, at a site in Kent much larger-scale movements of Alpine Newts have taken place. In 1994, a single adult female was caught in a pond containing Smooth and Great Crested Newts – she had almost certainly come from a garden pond containing a long-standing colony 1 km to the south. Three years later a male joined this female, presumably from the same source. The possibility of translocation by humans, however, cannot be ruled out, since the Alpine Newt is a highly aquatic species.

In about 1980 and again in 1984, Alpine Newts from Brighton in Sussex were introduced to two ponds in Sunderland, Tyne & Wear (one in the grounds of the Sunderland Polytechnic at Droxford Park). These populations died out when the ponds were infilled in 1988 and 1987 respectively.

Scotland

As long ago as 1957, Alpine Newts of Swiss and/or French origin that came from a University of Edinburgh experimental laboratory were introduced in at least two local ponds, Goldenacre quarry and Aberlady, near Edinburgh, at the first of which they were said to be 'plentiful' by 1968. In 1994, a further colony was discovered in another pond near Edinburgh, on the Mortonhall golf course, the origin of which is unknown. No emigration from these ponds has yet been recorded.

Impact

As both a congener of and a competitor with native newt species, the Alpine Newt has the potential to cause ecological perturbation in England and Scotland, despite its successful coexistence with other *Triturus* spp. in continental Europe. Since aquatic predators play a major role in the formation, composition and structure of larval amphibian communities, the presence in the newts' ponds in Britain of such species as Three-spined Sticklebacks, Grass Snakes and carnivorous water beetles *Dytiscus marginalis*, and the visitation of avian and mammalian predators, are together likely to have an impact on the long-term survival and spread or extinction of Alpine Newt populations.

References

Banks 1989; Banks & Laverick 1986; Beebee 1995; Beebee & Griffiths 2000; Bell 1978; Bell & Bell 1995; Fitter 1959; Frazer 1964, 1983; Gillet 1988, 1991; Langton & Burton 1997; Lever 1977, 1980b, 2003; Welch et al. 2001.

SALAMANDRIDAE (NEWTS & EURASIAN SALAMANDERS)

ITALIAN CRESTED NEWT *Triturus carnifex*

Natural Distribution From S Italy N to the Alps, and along the W side of the Baltic peninsula S to N Greece.
Naturalized Distribution England; ? Wales.

England

A small population of Italian Crested Newts (formerly regarded as a subspecies, *T. c. cristatus*, of the threatened native Great Crested or Warty Newt) became established after 1903 in ponds at the Beam Brook Aquatic Nursery at Newdigate in Surrey, which they share with introduced Alpine Newts and native Great Crested Newts. Hybridization between the Italian Crested Newt and the Great Crested Newt has been recorded at the Beam Brook site, but there is so far no morphological evidence for the spread of either hybrids or Italian Crested Newts to neighbouring ponds.

In the 1980s, Italian Crested Newts, imported from Barilli & Biagi in Bologna, escaped from the premises of Xenopus Ltd at Nutfield in Surrey, where they became established in an artificial pond with Alpine Newts obtained from the Beam Brook Aquatic Nursery at Newdigate. Both species thrived until a tree that fell in the gales of 1987 damaged the clay lining of the pond, which subsequently dried out.

Another colony of Italian Crested Newts near Birmingham, however, has extended its range across two roads to colonize a number of garden ponds up to 600 m from the point of introduction.

? Wales

Reports of populations of Italian Crested Newts in Wales (and elsewhere in Britain) are hard to confirm without genetic testing, as both Italian Crested and Great Crested Newts vary considerably both morphologically and in colouration.

References

Beebee & Griffiths 2000; Brede et al. 2000; Gillett 1991; Langton & Burton 1997; Lever 1977, 2003.

FISH

ACIPENSERIDAE (STURGEONS)
SIBERIAN STURGEON *Acipenser baerii*
RUSSIAN STURGEON *Acipenser gueldenstaedtii*
STERLET *Acipenser ruthenus*

Natural Distribution Siberian Sturgeon: rivers in Siberia. Russian Sturgeon: Black, Azov and Caspian Seas, and their tributaries. Sterlet: basins of the Black and Caspian Seas.
Naturalized Distribution England.

England
The Common Sturgeon is the only species of sturgeon that is native to British waters. However, in recent years the presence of at least three other species (*A. baerii*, *A. gueldenstaedtii* and *A. ruthenus* x *gueldenstaedtii* hybrids) has been recorded in at least 30 small (<2-ha) commercial still-water fisheries in England, ranging from Kent north to Lancashire and East Yorkshire, and from Norfolk westwards to Warwickshire and Somerset. About half of these waters are landlocked, the other half either lying in a floodplain or connected to river systems such as the Great Ouse, Severn, Trent and Stour. Although the introductions, using stock obtained from ornamental fish dealers and each averaging 5–25 fish weighing 1–4 kg, were designed to improve the quality and variety of the fishing, all were illegal under present legislation.

So far, there is no evidence to suggest either reproduction or natural dispersal from the sites of introduction. However, since many of the introductions were made either in floodplains or in still waters connecting with river systems, and because sturgeon are potentially long-lived (up to 60 years), it is very possible that some individuals will disperse into rivers and breed.

SIBERIAN STURGEON

RUSSIAN STURGEON

STERLET

Characteristics

Sturgeon are bottom-living fish that in spring migrate into rivers to spawn, where the young remain for a period of up to three years, feeding on insect larvae and small crustaceans. Although in England the largest rod-caught individual, from a lake in East Yorkshire, weighed only around 13 kg, in its native range the Russian Sturgeon can weigh up to 115 kg and grow to a length of 2.35 m. In their natural range, sturgeon are prized for the delicacy of their flesh, and the females are the source of caviar.

Impact

Should sturgeon become established away from their introduction sites, the potential impact on the native fish fauna is likely to include an increase in competition for resources such as food (zooplankton and benthic invertebrates); predation on eggs, larvae, juveniles and adults; and the transmission of new parasites such as the non-native nematode *Cystoopsis acipenseris*. Since sturgeon breed only in flowing water, it is possible that their access to suitable spawning sites in rivers could result in the establishment of self-maintaining populations, as has occurred elsewhere outside their natural range.

References

Anglers' Mail 2005; Britton & Davies 2006; Colclough 2006; Copping 2007; Davies et al. 2004; Hickley & Chare 2004; Maitland 2004; Wheeler 2001; Wheeler et al. 2004.

CYPRINIDAE (CARPS)

GOLDFISH *Carassius auratus*

Natural Distribution From the Lena River system of E Europe E to S Manchuria, the Amur River Basin, the Tym and Poronoi Rivers of Sakhalin Island, and China.
Naturalized Distribution England; Scotland; Wales.

England

Samuel Pepys may have been the first person to write of the Goldfish in Britain, when in his *Diary* on 28 May 1665 he recorded that at the house of Lady Pen he and his wife 'were shown a fine rarity: of fishes kept in a glass of water, that will live so for ever; and finely marked they are being foreign'. (These are commonly believed to have been Goldfish, although Hervey and Hems (1968) are sceptical of this attribution.) The German ichthyologist M.E. Bloch, in his *Oeconomische Naturgeschichte der Fische Deutschlands* (1784), claimed that he could trace the arrival of Goldfish in England back to the reign (1603–25) of James I.

There is circumstantial evidence to show that Goldfish may have been on board a ship belonging to the East India Company that sailed from Macao in September 1691 and berthed in London early in the following year. In his *Natural History of Birds* (1745), George Edwards wrote: 'The first account of these fishes being brought to England may be seen in Petiver's works, published about *anno* 1691.' This statement is repeated by Thomas Pennant in his *British Zoology* (1768–70) and by William Bingley who, in his *Animal Biography* (1813), wrote that 'Gold Fish ... were first introduced into England about the year 1691.' James Petiver's book, *Gazophylacium Naturae et Artis*, in which the author mentions two live Goldfish imported direct from China, was, however, published not in 1691 but in 1711. Edwards also mentions that the Duke of Richmond (1701–50) possessed a large Chinese earthenware vessel full of Goldfish that had been imported from the Far East. Two Goldfish are figured in Petiver, and there is some evidence to suggest that the drawings for the illustration were made in 1705 or 1706. The species is not mentioned by Izaak Walton in *The Compleat Angler* (1653), by Francis Willughby in *De Historia Piscium* (1686) or by John Ray in his *Synopsis Methodica Piscium* (1710).

During the 1730s, increasingly large shipments of Goldfish were imported into England. Both George Edwards and the Dutchman Job Baster in his *Opuscula Subseciva* (1765) say that in 1728 some specimens, which had been given to Captain Philip Worth, Master of the East India Company's vessel *Houghton*, by Sir Matthew Decker, Bt, a director of the company, were landed in England from China. In the same work, Baster mentions that Goldfish 'were placed in fish-ponds in England, and were increased, so that, sent into other parts of Europe, they became well known'.

In a letter dated 19 July 1746 to Henry Fox, then MP for Windsor, Horace Walpole refers to Goldfish in the 'purling[1] basons' at Vauxhall. Thomas Gray wrote of the species in his *Ode on the Death of a Favourite Cat, Drowned in a Tub of Gold Fishes* (1742) – the animal referred to being Walpole's 'Selima'. The pond (Po-yang) in Walpole's garden in Strawberry Hill was celebrated for its Goldfish. 'They breed with me excessively and grow to the size of small perch', wrote Walpole to the naturalist George Montagu on 6 June 1752. On 16 August of the following year he wrote to Montagu again: 'You may get your pond ready as soon as you please, the goldfish swarm. Mr Bentley carried a dozen to town t'other day in a decanter.'

Thus by the 1750s, Goldfish were being widely kept in captivity in England as pets, and Walpole used to give them away as presents to his friends.

Precisely when Goldfish first escaped or were deliberately released into the wild is unknown, but Bolton (in Davies et al., 2004: 65) records that they are today:

> widely naturalized, being found in ornamental ponds in parks and on large estates and, increasingly, in still waters used for commercial sport fisheries. Goldfish are commonly released into the wild as unwanted pets ... and more recently also through mass stocking of sport fisheries.

Most populations of naturalized Goldfish occur in still waters in central and southern England, where their distribution is increasing, and also in the Channel Islands (and perhaps in parts of southern Wales).

Scotland

Although Adams and Maitland (2001: 37) refer to the Goldfish in Scotland as 'very local – 1 known population', Bolton (in Davies et al. 2004: 65) says that 'a population ... in the Forth-Clyde Canal disappeared when an outfall of warm water from a factory ceased'.

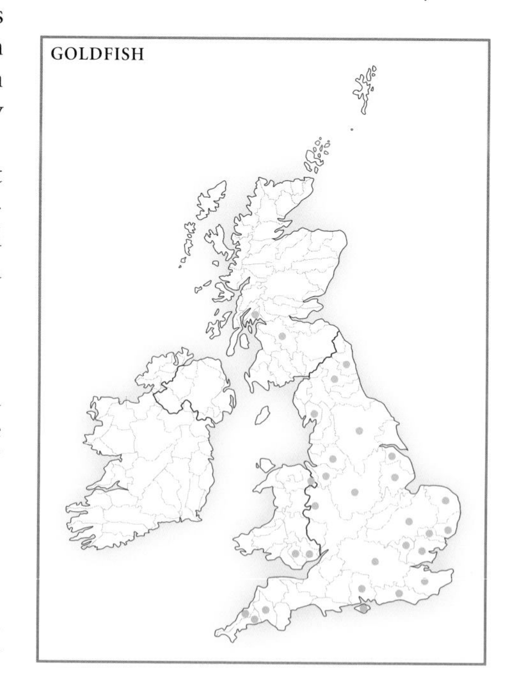

The Goldfish is yet another species that would be likely to benefit were the temperature of water bodies in southern Scotland (and northern England) to increase as a result of global warming.

Wales

The maps in Davies et al. (2004: 65) and Maitland (2004) indicate the occurrence of Goldfish in localities in Glamorgan/Monmouth in south Wales.

Characteristics

In the wild the colouration of the Goldfish differs little from that of the closely related Crucian Carp (see opposite), which is now usually accepted as indigenous in south-eastern England, from where it has been widely translocated anthropogenically to other parts of England and to parts of southern Scotland and to Wales. Both species favour small, still water bodies, with an abundance of macrophytes, although they also occur in slow-flowing lowland rivers. They are hardy species, well able to tolerate low levels of dissolved oxygen, and can survive temperatures of near 0°C to 38°C. In the absence of males, Goldfish are able to reproduce by gynogenisis[2] – the sperm of other carp species (e.g. the Crucian Carp) can stimulate egg division to form young fish genetically identical to the maternal parent. This may result in all-female populations. Hybrids between Goldfish and other species (e.g. Crucian Carp and Common Carp) are also likely to be capable of gynogenisis, which will lead to further competition with pure-bred parents.

The taxonomic status of the Goldfish is a matter of some contention among ichthyologists, and the correct identification of both species is a worldwide problem due to morphological similarities. Some authorities regard it as a 'domesticated' form of the wild Gibel or Prussian Carp, allegedly indigenous to Eastern Europe and Asia, whereas others consider the Gibel Carp as a distinct subspecies, *C. a. gibelio*. On the other hand, Gibel Carp may simply be 'wild' Goldfish or perhaps Goldfish x Crucian Carp hybrids.

Impact

Wild Goldfish pose a significant threat to British aquatic ecosystems through competition, hybridization and the spread of disease. The species most at risk is the Crucian Carp, with which Goldfish readily interbreed, and indeed populations of pure Crucian Carp are now rare in waters in which Goldfish (and Common Carp) now occur.

References

Adams & Maitland 2001; Bolton, in Davies et al. 2004; Fitter 1959; Hervey & Hems 1968; Hickley & Chare 2004; Hill et al. 2005; Lever 1977, 1996; Maitland 2004; Smartt 2007; Smith 2006; Welch et al. 2001.

Notes

1. Purling = rippling.
2. Gynogenisis: process by which an egg develops parthenogenetically (without fertilization) after being activated (but not fertilized) by sperm.

CYPRINIDAE (CARPS)

CRUCIAN CARP *Carassius carassius*

Natural Distribution E and C Europe (and perhaps Belgium, France, Spain and SE England) and much of C Asia.
Naturalized Distribution England; Scotland; Wales.

Although a long-standing member of the eastern and central European ichthyofauna, the Crucian Carp may have spread naturally into Western Europe (including southern England when Britain was still joined to the Continent), or have been introduced as a source of fast-day food with Common Carp in the Middle Ages.

CRUCIAN CARP

England; Scotland; Wales

The status of the Crucian Carp as alien or native in England has long been in doubt. Although most authorities now generally agree that it was in all probability native in south-eastern England, this is not certain. Bolton (in Davies et al. 2004: 67) writes: 'Crucian carp are considered by some authors to be native to south-east England. Others regard the species as having been introduced to this area, possibly in medieval times.' Writing of the Crucian Carp, Hill et al. (2005: 14) say:

> This species was previously considered to be non-native to the British Isles (e.g. Maitland 1972), and indeed continues to be classed as such in rare cases (e.g. Maitland 2004). However, following assessment of archaeological evidence and pre-1960's [*sic*] distribution of freshwater fishes (Wheeler 1977) [Crucian Carp is now] recognized as native to southeastern England.

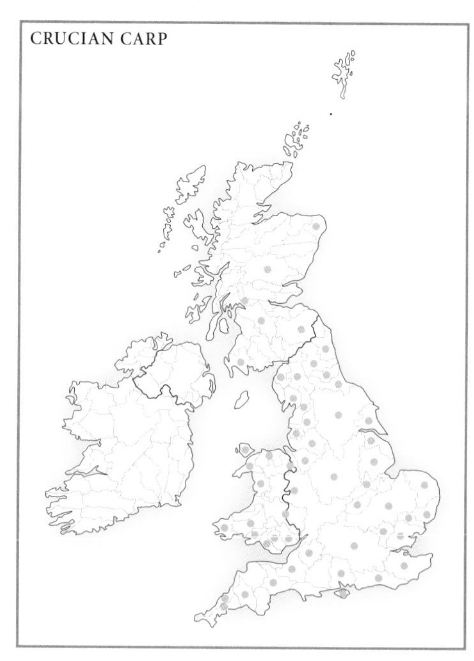

Crucian Carp now occur widely in lowland waters of Britain and Ireland, especially in southern, eastern and north-western England and northern Wales (including the Isle of Anglesey), and at a few sites in southern and central Scotland, where they first appeared (in Loch Lomond) in 1991 and where they are now 'local but spreading, mainly through translocations' (Adams and Maitland 2001: 37).

The distribution of Crucian Carp is complicated by cases of confusion with Goldfish and Crucian Carp x Goldfish hybrids. The distribution has probably been largely, if not entirely, due to translocations for commercial sport angling, sometimes accidentally accompanied by juvenile Goldfish into waters holding only pure Crucian Carp, resulting in both hybridization and interspecific competition.

Characteristics

Crucian Carp tend to favour much the same habitats as Goldfish and Common Carp (with which they also hybridize freely) – small, rich still-water bodies in lowland areas with an abundance of macrophytes. Like the two other species they are extremely hardy, and can survive in temperatures of just above freezing to 38°C and in acidic waters with a pH as low as 4.

Impact

For the impact on Crucian Carp by Goldfish and Common Carp see under the two latter species (below and page 237).

References
Adams & Maitland 2001; Adams & Mitchell 1992; Bolton, in Davies et al. 2004; Hill et al. 2005; Lever 1996; Maitland 1972, 2000, 2004, 2006; Newdick 1999; Smartt 2007; Welch et al. 2001; Wheeler 1977, 2000.

CYPRINIDAE (CARPS)

COMMON CARP *Cyprinus carpio*[1]

Natural Distribution Originally restricted to C Asia E of the Caspian Sea, from where it spread naturally E during the later glaciations to Manchuria and W to the rivers of the Danube basin and the Black and Aral Seas.
Naturalized Distribution England; Scotland; Wales; Ireland.

England

In his delightful treatise on fishing, *The Compleat Angler or the Contemplative Man's Recreation*, first published in 1653, Izaak Walton wrote:

> The Carp is the Queen of Rivers: a stately, a good, and a very subtle fish, that was not at first bred, nor hath been long in *England*, but is now naturalized. It is said, they were brought hither by one Mr *Mascal*, a gentleman that then lived at *Plumsted* in *Sussex*, a County that abounds more with this fish than any in this nation.
>
> ... doubtless, there was a time about a hundred, or a few more years ago [i.e. before c 1550], when there were no Carps in *England*, as may seem to be affirmed by Sir *Richard Baker* [1586–1645], in whose Chronicle [of the Kings of England, 1643] you may find these verses.
>
> Hops and Turkies, Carps and Beer,
> Came into England all in a year.

Hops (and therefore beer) were introduced to southern England from Flanders in about 1520–4, and Wild Turkeys from North America around 1530.

COMMON CARP

In his *History of British Fishes* (1859) William Yarrell wrote:

> Leonard Mascall [kitchen-clerk to Matthew Parker, Archbishop of Canterbury] takes credit to himself [in *A Booke of Fishing with Hooke & Line* ... (1590)] for having introduced the Carp ... but notices of the existence of the Carp in England occur prior to Maskall's time, or 1600. In the celebrated *Boke of St Albans*, by Dame Juliana Barnes, or Berners, the Prioress of Sopwell Nunnery ... Carp is mentioned as 'a deyntous fisshe, but there ben but fewe in Englande, and therefore I wryte the lesse of hym'. ... in the Privy Purse expenses of King Henry the Eighth in 1532, various entries are made of rewards to persons for bringing 'Carpes to the King'.

The *Boke of St Albans*, which was only allegedly written by Dame Juliana Barnes, was originally published in 1486. The edition that refers to the Common Carp, which was printed ten years later, contains a *Treatyse of fysshynge wyth an Angle* (said to date from about 1450), which was not included in the earlier edition.

Thus the arrival of the Common Carp in England would seem to date from before 1496 to around 1530. The species is today fairly widely distributed (except in upland areas), especially in slow-flowing and still waters with a muddy bottom and surrounded by thick vegetation to provide food, shelter and spawning substrates, throughout much of central and southern England (including the Channel Islands) and in parts of northern England, and it appears to be spreading slowly northwards. Even in southern England, summer water temperatures in some places are barely high enough for successful spawning (a minimum of 18°C is required), and in many places populations are only maintained by regular stocking for angling purposes. Nevertheless, the species is both hardy and adaptable, occurring in most freshwater habitats, from cold, upland waters and deep gravel-pits to warm, nutrient-rich ponds. If the water is sufficiently warm it will breed in many different conditions, and can tolerate water temperatures of nearly 0°C to 40°C, brackish water, and water with a low oxygen content and of relatively poor quality.

Common Carp were originally introduced to Britain to stock medieval monastic fish stews as a source of fast-day food. Although highly prized in continental Europe (and elsewhere) for the quality of their flesh, in Britain and Ireland they are today valued only as catch-and-return sport fish or, in their *Koi* form (see Note 1) as an aesthetic asset. This is yet another species in Britain that should benefit from the increase in water temperature through global warming.

Scotland

The date of the arrival of Common Carp in Scotland appears to be unrecorded, and it is presently largely confined to a few waters around Glasgow and Edinburgh. Adams and Maitland (2001: 37) describe the fish in Scotland as 'Local but spreading, mainly through translocations'. Welch et al. (2001: 38) say that they 'breed in Scotland only in warmer than average summers but breeding may increase if water temperatures were to increase'.

Wales

The date of introduction of Common Carp to Wales is similarly unknown, but they are now widely distributed in suitable lowland waters throughout the country, including the Isle of Anglesey.

Ireland

The history of the introduction of the Common Carp to Ireland has been described by Went (1950).

The Journal Book Entry of the Royal Society for 29 April 1663 records 'Mr Boyle was desired to communicate his papers, concerning the manner how my Lord of Corke, his ffather, had Carpes transported into Ireland, where they were not before.' In the course of his address to the Royal Society, Robert Boyle read the following extracts from his father's diary:

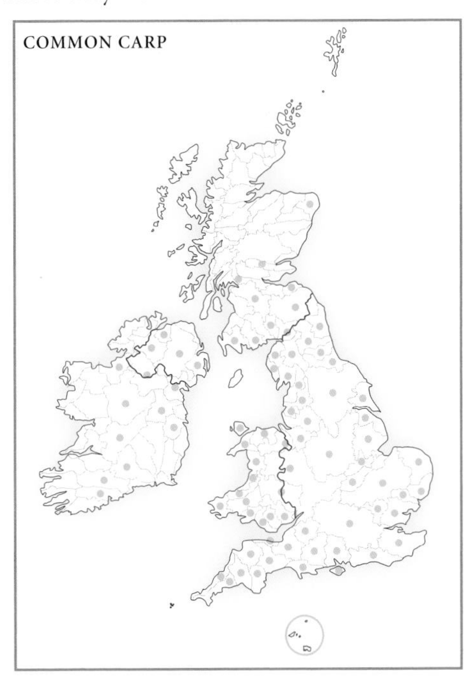

> 3 Sept. 1634: I wrott to the Lord President of Mounster, and as he desired, gave order to Sir John Leek to deliver to his Lap. 20 yonge carpes ...[2]
>
> 26 Sept. 1640: I wrott by Badnedg to Sir John Leek to furnish Sir Philip Percival with 40 yonge carpes, Mr Henry warren second Remembrancer of Exchequer with 20 Carpes and my daughter the countess of Kildare, with 20 carpes to stoar the new pond withall.[3]
>
> 7 Oct. 1640. Letters to Sir John Leek to deliver to Sir Philip percival 40 yonge carpes ... towards the storing of his ffishe pond.[4]

Since it is known that Lord Corke lived in Ireland after 1588, the 'carpes' referred to were almost certainly to be translocated from his Irish estate rather than introduced from England; it is thus necessary to seek further back in time for their arrival in Ireland. Charles Smith in his *History of Cork* (1750) claims that Common Carp (and Tench) were to be found in the River Awbeg in Co. Cork during the reign (1603–25) of James I, and that both were commonly kept in captivity in ornamental waters, although Tench were less plentiful. It has thus come to be accepted that Common Carp (and Tench) were introduced to Ireland around 1600. In his *Natural History of the County of Dublin* (1772), John Rutty reports the suggestion that Common Carp were introduced to Ireland at some time during the reign of James I.

In his *Statistical Observations Relative to the County of Kilkenny* (1802), William Tighe found Tench 'in the Barrow [River] together with the *Cyprinus carpio* or carp; it is said they first came there upon the breaking down of some ponds at Low-Grange by a flood'. W.B. Daniel in his *Rural Sports* (1807) mentions that the 'Lakes of Killarney' held large numbers of fish, including Common Carp – a statement that was repeated by James Windele in his *Historical and Descriptive Notices of the City of Cork and its Vicinity* (1849).

An entry in the register of a guest-house at Clifden, Corofin, Loch Inchquin, appears to be the earliest reference to the species in Co. Clare:

> 23rd. June 1815. About 25 years ago his [William Burton's] father put into the lake 24 Carp. It is surprising that after such a lapse of time in so fine a piece of water, so prolific a fish as *Cyprinus carpo* [*sic*] has not multiplied to a sufficient degree to afford proof of their existence therein, not one of them has been taken.

John Templeton of Cranmore, under the heading of 'Irish vertebrate animals', in the *Magazine of Natural History* (1837), describes the Common Carp as a 'naturalized species' in Ireland.

Summing up, in his *Natural History of Ireland* (1849–56), William Thompson wrote: '[The Common Carp] which was introduced into the British Isles has long been in Ireland. Localities noted: Montalto and Killyleagh, Co. Down; and Markethill, Co. Armagh; Co. Dublin; Counties Galway and Sligo.'[5]

Common Carp occur today in a number of widely scattered localities in Ireland.

Impact

Because of their roiling[6] activities when feeding (usually on the bottom) and spawning (in shallow margins), Common Carp can considerably alter natural aquatic habitats, especially as the population increases, since disturbed silt inhibits the growth of submerged macrophytes. They also readily hybridize with (perhaps native) Crucian Carp and with naturalized Goldfish, the offspring of which are fertile. Indeed, hybridization and competition with Common Carp and Goldfish threaten the survival of Crucian Carp in waters containing one or both of the alien species.

In 1977 the blood fluke *Sanguinicola inermis* and in 1986 the carp tapeworm *Khawia sinensis* were first identified in Common Carp in England. Since then their range has greatly expanded, despite legal prohibition of the movement of infected fish.

References

Adams & Maitland 2001; Adams & Mitchell 1992; Balon 1969; Bolton, in Davies et al. 2004; Fraser & Adams 1997; Hill et al. 2005; Lever 1977, 1996; Maitland 1964, 2004; Went 1950; Wheeler 2001; Yeomans et al. 1996, 1997.

Notes

1. Includes six different forms, i.e. Common, Mirror, Leather, Ghost, King and Nishikigoi or *Koi* Carp.
2. Lismore Papers, 1st series, 4: 44.
3. Ibid 5: 161.
4. Ibid 5: 162.
5. 'Introduced by the Great Earl of Cork in the South of England – *vide* Robert Boyle in a paper to the Royal Society: R. Ball.'
6. Make water turbid by disturbing the sediment.

CYPRINIDAE (CARPS)

CHINESE GRASS CARP *Ctenopharyngodon idella* (WHITE AMUR)

Natural Distribution Lowland rivers of China and the Amur River Basin in E Siberia.
Naturalized Distribution England; Wales.

CHINESE GRASS CARP

The Chinese Grass Carp has been widely introduced around the world, principally for the control of aquatic vegetation. Although its strict breeding requirements – especially the necessity for a water temperature of 27–29°C for a period of some nine weeks and a current flow of 60–150 cm per second – preclude it from breeding in British and Irish waters, where acclimatized (but not naturalized) populations are maintained by restocking, its widespread occurrence justifies the inclusion of a brief account of it here.

England; Wales
Originally introduced to England from farmed stock in Hungary in the 1960s to control excessive growth of macrophytes in eutrophic waters, mainly in the Fens of East Anglia, the Chinese Grass Carp's size in Britain (up to 1.25 m in length and 18 kg in weight; in Europe up to 35 kg) has recently made the species a popular sporting quarry. It is now widely distributed in rich lakes and slow-flowing rivers in England, mainly in the south-east and north-west, and in parts of southern and north-eastern Wales, including the Isle of Anglesey.

Impact
Because adult Chinese Grass Carp feed exclusively on aquatic vegetation (the young eat small invertebrates), they are capable of significantly altering ecosystems.

References
Chare & Musk, in Davies et al. 2004; Gibson 1994; Lever 1996; Wheeler 2001.

CYPRINIDAE (CARPS)
SUNBLEAK *Leucaspius delineatus*
(SUNDACE; BELICA)

Natural Distribution Widely distributed in Europe from the Caspian Sea W to the rivers of Belgium and the Netherlands. May have entered parts of N France by natural dispersal through the canal system of S Belgium.
Naturalized Distribution England.

England
Much of the information on the Sunbleak in England is derived from Farr-Cox et al. (1996) and Pinder and Gozlan (2003).

Sunbleak[1] are believed to have been first imported to England from Germany in 1986–7 by the Crampton Fishery near Romsey in Hampshire, which was then owned by an ornamental-fish dealer. The nearby Two Lakes Fishery was at the time under the same ownership, and fish were transferred between the two fisheries. Both sites drain into streams that, downstream of Romsey, join and enter the River Test, which flows into Broadlands Lake in the Test floodplain a further 2 km downriver. Sunbleak were first observed in Broadlands Lake in about 1987, where they were originally misidentified as native Bleak. Until 1990, fry of Roach and Bream from Broadlands Lake were sold for stocking other waters (e.g. in Somerset), and inevitably Sunbleak fry were inadvertently included in some of these consignments.

The origin and route of dispersal of Sunbleak in Somerset can only be conjectured. The Somerset Levels are low-lying with interconnecting artificial drains and semi-natural waterways. The King's Sedgmoor Drain is fed by water from the lower reaches of the River Parrett, and Sunbleak could have moved between these two systems in times of flood or high water flow. The Bridgwater and Taunton Canal is fed from the River Tone, which runs into the River Parrett where both are tidal, but these tidal reaches have a large

Table 14 Occurrence of the Sunbleak *Leucaspius delineatus* in England to 1995

Year	County	Locality
1990	Somerset	King's Sedgmoor Drain
1994	Somerset	Bridgwater and Taunton Canal; River Parrett
1995	Somerset	Wych Lodge Lake (Blackdown Hills, east of Pitminster); Combe Lake (a recently excavated angling water near Langport, filled by floodwater from the River Parrett in the winter of 1994–5); River Brue; Whites River (a tributary of the Brue)
c 1987 and 1994–5	Hampshire	Broadlands Lake (near Romsey)
1995	Hampshire	Two Lakes Fishery (near Romsey); Stoneham Lake (near Eastleigh)

Source Lever 1996, adapted from Farr-Cox et al. 1996.

component of fresh water in winter and during other times of high flow, and Sunbleak may have moved between the canal and the Parrett by this route. Near the Parrett flows the River Brue, although the two are not connected; the estuaries of both rivers join at Burnham-on-Sea, where the water is permanently saline. However, research by Scott et al. (2007: 70) found that 'Despite the fact that exposure to saline water is stressful ... short-term use of brackish waters is a feasible dispersal route for sunbleak ... in the wild.' Humans may also have transferred Sunbleak between waters accidentally or deliberately.

In both Somerset and Hampshire, the Sunbleak expanded its range, both naturally and anthropogenically, fairly rapidly and widely, and was said by Farr-Cox et al. (1996: 196) to be 'present in enormous numbers'. It now occurs in several other still waters in Somerset and probably also in the River Wye, as well as in fishing lakes near Sherborne in Dorset, and possibly in waters in north-western England.

Further dispersal is likely to be anthropogenic rather than natural, either through deliberate release into the wild or accidentally through transfers of other similar species. If, as Pinder and Gozlan (2003) point out, Sunbleak were to be introduced to the Bristol Avon, which borders the Somerset Levels, this would give them access to the English canal network, and therefore to such rivers as the Thames and Great Ouse and the fenland drains of Lincolnshire and Norfolk, which would provide them with an abundance of apparently ideal habitat.

Characteristics

The Sunbleak is a small (up to 8 cm) cyprinid whose biology has undoubtedly helped it to become such a successful colonist. Most cyprinids take 2–4 years to reach sexual maturity and only spawn annually, the females scattering thousands of eggs over a wide area, where they are left untended and thus vulnerable to predators. Sunbleak, however, are sexually mature at one year old, and lay their smaller number of eggs in several batches between April and July. The male zealously guards the eggs until they hatch some four days later. These characteristics enable Sunbleak to have a high hatching success rate, and because the fry emerge over a prolonged period they are less vulnerable to changing climatic and environmental conditions. Sunbleak prefer slow-flowing lowland rivers and associated backwaters, ponds, lakes, drains and canals, where they form pelagic shoals that in winter disperse to deeper water. They are hardy fish, tolerating high water temperatures and low levels of dissolved oxygen.

Their small size enables Sunbleak to escape easily from both captivity and natural water bodies, and their catholic choice of spawning substrata gives them a wide selection of ova-deposition sites.

Impact

Anecdotal evidence suggests that Sunbleak may have a significant impact, through sheer weight of numbers, on native cyprinids, largely through dietary overlap. Because of the Sunbleak's small size, it is not only young-of-the-year that compete for food with native juveniles, and Gozlan et al. (2003a) showed how the slow growth of Sunbleak can result in four year-classes competing for the same food as young-of-the-year Roach, Bream, Bleak and Rudd at a time when the native species are most vulnerable.

Although Sunbleak are beneficial in providing an additional source of food for predators such as Perch and Zander, they have the potential to themselves prey on the eggs and larvae of these and other species, with which they also compete for habitat.

In Britain, aquaculture poses the greatest threat to freshwater biodiversity, and the deliberate release of unwanted pet fish is one of the main means of dispersal for some exotic species. If British Sunbleak are found to prey opportunistically on the eggs and larvae of native species, their appearance in such natural waters as Windermere, in the Lake District of Cumbria, could have a devastating impact on the already threatened populations of indigenous Whitefish.

References

Atherton 1994; Farr-Cox et al. 1996; Gozlan et al. 2003a, 2003b; Henderson 2005; Lever 1996, in press; Pinder & Gozlan 2003; Shelley & Pinder, in Davies et al. 2004; Wheeler 2001.

Note

1. Known by anglers as the 'Motherless Minnow', the name is derived from its common German name, 'Moderliesсhen', from the Low German *moder-loseken* (motherless). It refers to the species' tendency to appear suddenly and unexpectedly in large numbers in waters from which it was previously absent.

CYPRINIDAE (CARPS)

ORFE *Leuciscus idus*

(IDE)

Natural Distribution E and C Europe E of the River Rhine, from E Scandinavia E to the Arctic Ocean, White Sea and the River Volga, and rivers of the N coast of the Black Sea.
Naturalized Distribution England; Scotland; Wales.

ORFE

England

The earliest known introduction of Orfe to England appears to be one reported to Frank Buckland (1881: 271) by Lord Arthur Russell (son of the Duke of Bedford) in 1874:

> Ever since I first saw these splendid fish *Cyprinus orfus* in the ponds of the Imperial Palace, Laxenburg, Vienna, I determined to introduce them if possible to England. ... My first attempt a year ago was unsuccessful, two gold[1] orfes only survived of the batch my brother [Lord Odo Russell, British Ambassador at Berlin] had obtained in Berlin. They are still living in a pond at Woburn Abbey [Bedfordshire]. My second and successful attempt has been accomplished with the assistance of Mr. Kirsch, Director of the Association for Pisciculture at Wiesbaden; he sent me one hundred and fourteen golden orfes of last year's breed, about two inches [5 cm] long each, and two large specimens. They travelled from Wiesbaden in two tin cans in charge of one of the clerks of the association, and favoured by the cold weather, they were all deposited, without a single loss, in one of the Duke of Bedford's ponds at Woburn Abbey.

Many other attempts have been made over the years to naturalize Orfe in England, but few have succeeded in establishing a breeding stock.

Table 15 Some English waters in which Orfe *Leuciscus idus* are known to have become established

County	Locality
Avon	Winford Brook, Bristol
Bedfordshire	Woburn Abbey
Berkshire/Buckinghamshire	River Thames, Henley
Cambridgeshire	Oakington
Cheshire	Lymon Vale
Cornwall	Hayle Kimbro and Stithians
Devon	Lundy Island, Bristol Channel
Dorset	Moigne Combe Pond, Moreton; Owermoigne, Dorchester
Essex	Wivenhoe, Colchester; Loughton
Gloucestershire	Torthworth Lake, Falfield
Hampshire	Cemetery Lake, Southampton Common; River Kennet, Cookham; River Test
Huntingdonshire	Hartford
Lincolnshire	River Ouse, Ely; Bullwants Pond, Mablethorpe
Middlesex	River Ember, Hampton Court
Surrey	Shottermill, Haslemere; Westcott, Dorking; River Thames, Chertsey
Worcestershire	Spetchley Lake

Sources Fitter 1959; Maitland 1972; Lever 1977, 1996.

According to the Duke of Bedford, the original stock of Orfe at Woburn died out as a result of competition from other coarse fish in the lakes, and this may be one of the causes of the species' failure to become established in other apparently suitable English waters. Nevertheless, according to Wellby and Vickers in Davies et al. (2004: 78), 'Feral [*sic*] populations now occur at scattered sites, mainly in [north-western] England, and the species is becoming increasingly recorded in many areas ... as it is introduced for both sporting and ornamental purposes.'

Scotland

Maitland (1972, 2004) and Lever (1977, 1996) record established populations of Orfe only in Perthshire (Killiecrankie) and Dumfriess-shire. The map in Wellby and Vickers, who say that there are only a small number of records for Scotland, indicates the species' presence also in Midlothian and Argyll. Adams and Maitland (2001: 37) record Orfe in Scotland as 'very local – in ornamental ponds but 1 known wild population'.

Wales

Maitland (1972) and Lever (1977, 1996) record the presence of Orfe only in Bodnant Gardens in Denbighshire. The map in Wellby and Vickers, who say that Orfe are being increasingly recorded in Wales, indicates their presence also in Glamorganshire and Carnarvonshire. The map in Maitland (2004) also shows the species' occurrence on the Isle of Anglesey.

Characteristics

Orfe favour clean, slow-flowing middle and lower reaches of rivers, and lakes, with a temperature range of 4–20°C, although they seem able to become accustomed to higher temperatures and to a wide range of habitats.

Impact

Orfe are known to have hybridized with other cyprinid species to produce infertile offspring. Although no hybrids have yet been recorded in Britain, Orfe pose a potential threat to native (and introduced) British species.

References

Adams & Maitland 2001; Buckland 1881; Fitter 1969; Hill et al. 2005; Lever 1977, 1996; Maitland 1972, 2004; Perring 1967; Welch et al. 2001; Wellby & Vickers, in Davies et al. 2004; Wheeler 1971; Wheeler & Maitland 1973.

Note

1. The golden (and blue) forms of the Orfe are widely bred for the ornamental pet trade. They can be trained to become habituated to humans and to be hand fed.

CYPRINIDAE (CARPS)

TOPMOUTH GUDGEON *Pseudorasbora parva*

(FALSE HARLEQUIN)

Natural Distribution China, Japan, Korea and the Amur River basin.
Naturalized Distribution England; Wales.

England

Topmouth Gudgeon[1] were accidentally introduced from the Far East to Romania in 1960 in a shipment of other cyprinid species destined for a fish farm in the lower River Danube. From Romania they rapidly dispersed throughout the Danube system, by 1990 occurred in much of Europe and Russia, and by the turn of the century had even reached North Africa.

Table 16 Status of Topmouth Gudgeon *Pseudorasbora parva* in England and Wales, 2008

Location	National Grid Reference	Water	1st Known Record	Report Date	Inflow/ Outflow	Number	Comments
Whickham, Gateshead	NZ186004	-	-	-	-	-	-
Kendal, Cumbria	SD484958	Club pond	-	-	? R. Kent	Many	Eradicated March 2005
Yorkshire	SE5654066300	Commercial	June 2001	-	No	Many	-
Thirsk, Yorkshire	SE537797	Commercial	-	-	R. Swale	-	Eradicated 2006
Doncaster, Yorkshire	SE541767	-	-	-	-	-	-
Wolverhampton, Salop	SJ8506701646	Estate ponds	-	-	No	-	-
Newcastle-under-Lyme, Staffordshire	SJ853451*	Club pond	-	May 2004	Lyme Brook; R. Trent	-	-
Ormskirk, Lancashire	SJ243871	Commercial	-	-	No	Many in upper lake	-
Leigh, Lancashire	SJ630803	Club pond	2000	2004	Yes		-
Kidderminster, Worcs	SO82805890*	Club pond	-	-	R. Severn	Many	-
Buxton, Derby	SK07700710	Commercial	-	July 2006	Occasionally	Many in top pool	-
Worcester	SO765578	Club pond	-	March 2005	R. Severn	2	-
Rothley Brook, Loughborough, Leics	SK5426607104	River	-	-	Yes	1	-
Crampmoor Fishery, Romsey, Hampshire	SU386224	Aquaculture	1985–6	2002	Tadburn L.; R. Test	Many	Original site
High Wycombe, Bucks	SU740940	Estate ponds	1996	-	No	Many	None when sampled November 2005
R. Test	SU355157	River	Probably mid-1980s	2004	Many	-	Caught in Nursling smelt trap
Tadburn L.	SU352207	Lake	-	-	-	Many	-
Chandler's Ford, R. Itchen	SU462200*	River	2003	2004	-	Many	-

Location	National Grid Reference	Water	1st Known Record	Report Date	Inflow/ Outflow	Number	Comments
Monks Brook, Winchester, Hampshire	SU423218*	River	2003	2004	-	Many	-
Broadlands Lake, Romsey, Hampshire	SU350161	Estate	-	-	-	-	-
Shamley Green, Guildford, Surrey	SU9987346992	Council owned	-	July 2006	R. Wey	Many	-
Eversley, Hampshire	SU778610	Private	April 2007	April 2007	R. Blackwater, Loddon, Thames	Many	Eradicated March 2008
Shalford, Guildford, Surrey	SU99854700	Council owned	- -	- -	Occasionally	Many	Eradicated March 2007
Huntingdon	TL17810342	Club pond	-	-	R. Colne	1 dead	-
R. Lee, Ponders End, Greater London	TQ362955	River	-	-	-	1	-
R. Otter and nearby lake	SY0705486143 727285850	Syndicate	- -	- -	Yes	Many	Eradicated March 2007 (both)
Goldings Hill Pond, Epping Forest, Greater London	TQ429981	Pond	1998	2003		No	Individuals caught 1988, 2001, 2003; eradicated 2006 (Copp et al. 2007)
Royal Tunbridge Wells, Kent	TQ702378	Aquaculture	April 2004	May 2004	No	2	-
Frittenden, Kent	TQ816004060	Aquaculture	-	-	No	-	-
Marden, Kent	TQ777469	Aquaculture	-	-	-	-	-
Turkey Brook, Brighton, East Sussex	TQ3407398881	River	-	-	Yes	Few	-

Source Environment Agency National Fish Population Database (per Phil Hickley, 2008).
* = approximate NGR based on best information available.

Topmouth Gudgeon first arrived in England in the mid- to late 1980s, when they were accidentally imported, probably with a shipment of Golden Orfe from Germany, to the same fish farm, the Crampmoor Fishery near Romsey in Hampshire, as the Sunbleak. Their very small size (up to c 8–10 cm) enables them easily to escape, like the Sunbleak, from enclosed waters and rapidly to colonize connecting water bodies. From Crampmoor, they spread to Tadburn Lake, which drains the original introduction site and joins the River Test some 4 km downstream. A survey of this catchment area in 2003 revealed that their distribution was limited and patchy, though the presence of juveniles suggested successful establishment.

The presence of Topmouth Gudgeon elsewhere has almost certainly been achieved anthropogenically, since all occupied waters appear to be landlocked. If, however, undetected populations become established via the canal/river network, Pinder and Gozlan (2003) predicted that Topmouth Gudgeon will become as widely distributed in Britain as they have throughout continental Europe. As is the case with Sunbleak, Scott et al. (2007: 70) found that 'Despite the fact that exposure to saline waters is stressful ... rapid recovery suggests that short-term use of brackish water is a feasible dispersal route for topmouth gudgeon in the wild.'

Although the Topmouth Gudgeon has become widely naturalized and abundant in Europe, not all of its invasions have been successful. Copp et al. (2007) describe their appearance and subsequent disappearance (following drainage and repeated electrofishing) from the small (c 0.38-ha) Goldings Hill Pond in Epping Forest. They have also been eradicated from Ratherheath Tarn near Kendal in Cumbria. By about 2005, they were believed to be present in some 25–30 waters in England and Wales (but see Table 16).

Wales

The Environment Agency National Fish Population Database (per Phil Hickley) refers to the presence of two colonies of Topmouth Gudgeon in commercial waters near Llanelli, Carmarthenshire (National Grid References SN464008 and SN4936004950). The first, which was sampled in November 2005 when it held c 100 fish, is on the coast and has regular intrusions of salt water from the sea; the second, at Five Roads, some 4.5 km north of Llanelli, was first recorded in 2001 and reported in April 2006. It contains a large number of Topmouth Gudgeon, and has both an inflow and outflow. It is suspected that the fish at Five Roads were deliberately translocated from the coastal site at Llanelli.

Characteristics

The successful establishment of Topmouth Gudgeon is a consequence of a number of factors. Being, like the Sunbleak, extremely small, they can easily escape from captivity undetected. Whereas most cyprinids take 2–4 years to reach sexual maturity and spawn only annually, the females scattering large numbers of eggs that are left untended over a wide area, and thus vulnerable to predators, Topmouth Gudgeon (and Sunbleak) become sexually mature at one year, and the females deposit fewer eggs in batches several times between April and July, which are zealously guarded by the males until they hatch after about seven days. Thus a smaller number of eggs tends to produce a higher number of surviving fry, and since the breeding season is longer than that of other cyprinids the larvae are less vulnerable to changing weather and environmental conditions, thus improving their chances of survival.

Topmouth Gudgeon live principally in the bentho-pelagic zone, where they eat algae, benthic invertebrates, zooplankton, small molluscs, and the eggs and larvae of other fish.

Spawning substrata, which had previously been known to include the undersides of rocks and broadleaved floating macrophytes such as the White Water Lily, Yellow Water Lily and other Nymphaeaceae, were found by Pinder and Gozlan (2003) also to include almost any flat-surfaced object. After hatching, the fry disperse from the nest site to the benthos, where they hide among stones until their fins are fully formed. Topmouth Gudgeon are less specific than Sunbleak in their habitat requirements, and although they prefer shallow, still or slow-flowing waters with an abundance of macrophytes, they can also flourish in faster-flowing waters and in deeper still waters.

Impact

Widely regarded as a pest species, the Topmouth Gudgeon in Europe is reported to eat the eggs and larvae of other fish species, especially those of the Zander. Although there have as yet been no reports of Topmouth Gudgeon in England acting as vectors of new pathogens, the recent discovery by Gozlan et al. (2005) that the species 'is a healthy host to an emergent infectious disease (Rosette-like Agent, closely related to *Sphaerothecum destruens*), which threatens European fish diversity, is the first real evidence that *P. parva* is capable of causing real ecological damage to freshwater ecosystems outside its native range' (Pinder et al. 2005: 411). In Europe, Topmouth Gudgeon are known to spread such parasites as the swim-bladder nematode *Anguillicola crassus* and the fluke *Clinostomum complantum*, which have severely affected populations of Sunbleak, and to be carriers of Northern Pike fry rhabdovirus (PFRW).

For other impacts refer to the Sunbleak (page 246).

References

Allen et al. 2006; Beyer 2004; Britton & Brazier 2006; Britton et al. 2006; Copp et al. 2007a, 2007b; Domaniewski & Wheeler 1996; Gozlan et al. 2002, 2005; Hickley & Chare 2004; Lever 1996, in press; MacRae 2005; Pinder 2005, Pinder, in Davies et al. 2004; Pinder & Gozlan 2003; Pinder et al. 2005; Salkeld 2006; Scott et al. 2007; Weir 2008; Winfield & Durie 2004.

Note

1. The name Topmouth Gudgeon is derived from the species' upturned lower jaw. It is also known to aquarists as the Stone Moroko, False Harlequin, Japanese Minnow, Sharpnose Gudgeon and Clicker Barb. This last name is derived from its habit of feeding on terrestrial invertebrates from the water surface, which is accompanied by an audible clicking sound.

CYPRINIDAE (CARPS)

BITTERLING *Rhodeus sericeus*

Natural Distribution E from NE France through Europe to the basins of the Black and Caspian Seas. Also C and NE Asia.
Naturalized Distribution England; Wales; [Scotland].

BITTERLING

England

The earliest evidence for the suspected presence of European Bitterling[1] in the wild in England appears to date from the early 1900s, when Hardy (1954: 13) was told that, under the name 'Prussian Carp',[2] 'strange types of fish' were being caught in a pond at Moss Lane, St Helens, Lancashire. Wheeler and Maitland (1973) dispute this identification, and it may well be that Bitterling were not established in the wild in England until the 1920s.

Before the Second World War, Bitterling (known locally as 'Pomeranian Bream') were found being used as live bait in the Carr Mill Dam near St Helens. Their source was the disused arm of a canal at Blackbrook, and others are known to have been present in Leg of Mutton Dam, also near St Helens. After the war the canal was found still to contain large numbers of Bitterling, and they were also known to occur in 10–12 other waters in south Lancashire; the population in Leg of Mutton Dam had, however, declined.

It was generally believed that the species' original appearance in the wild was a result of the disposal by aquarists of unwanted stock, and that its spread was due to the release by anglers of surplus live bait.

In 1948, H. Norman Edwards removed a dozen Bitterling from Carr Mill Dam to use as live bait in Black Harry's Pit at Moreton near Wallasey in the Wirral, Cheshire, where a number were released.[3] Six years later, Edwards introduced some more to the Station Pond at Meols in Lancashire, where they became established; after Black Harry's Pit was filled in, the Bitterling there were transferred to the Station Pond.

Hardy (1954) recorded Bitterling at the following places in Lancashire: Collins Green Flashes; Duckery Flashes; Derbyshire Hill; part of the Southport Sluice; the rock hole at Bold; ponds in the Haydock and Knowsley areas (where they had been introduced by Edwards in 1950) and near the Black Horse Inn at Rainhill – all within 6 km of St Helens. In 1953, large numbers of Bitterling were transported as live bait by anglers fishing in Esthwaite Water in the Lake District of Lancashire, and to Rydal Water and Grasmere in Westmorland (Cumbria), where they apparently failed to become established.

In the mid- to late 1970s, Bitterling are believed to have been introduced (again probably as unwanted live bait) into the Great Ouse River system at Ely in Cambridgeshire, where in places they are now the second most abundant fish after Roach. Bitterling are continuing to expand their range in the Great Ouse and its associated waters. They have also become established, sometimes only temporarily, elsewhere in England, e.g. at Hadley Green in Hertfordshire in 1956, and in the Praes Branch of the Shropshire Union Canal on several occasions since 1969. Orkin (1957: 125) recorded that they were 'said now to be established in southern Yorkshire', a claim repeated by Torbett (1961: 256), who said the species was to be 'found wild in one or two small ponds in Yorkshire ...' These references to Yorkshire, however, probably resulted from confusion with waters in Lancashire, and today only in Lancashire/Cheshire and Cambridgeshire do Bitterling occur in large numbers.

Wales

A large and self-maintaining population of Bitterling in Llandrindod Wells Lake, Powys, is believed to be the only one of its kind in Wales (T. Hatton-Ellis, personal communication 2008).

[Scotland]

In 1925, John Berry released some Bitterling in ponds at Newport-on-Tay in Fife, but by 1934, following a flood, they had all disappeared.

Characteristics

The Bitterling has a bizarre commensal method of reproduction that is unique in the animal kingdom. In early spring, males defend territories near freshwater mussels such as the Swan Mussel, to which they lead responsive females, who deposit their eggs through an extended pendulous ovipositor (which may be almost as long as the fish herself), which is inserted into the mussel's exhalant siphon. The large eggs are laid either singly or in very small batches into the mussel's gill cavity, after which the male Bitterling ejaculates his milt into the mussel's inhalant siphon, so that fertilization occurs within the host's gills. Bitterling larvae develop safely within the mussel for 40 or more days, emerging only when the yolk sac has been fully absorbed. Because the means whereby the eggs are deposited ensures a high rate of survival, only about 40–100 are normally laid.

This commensalism is counterbalanced by parasitism by the mussel when it breeds itself. Its tiny and mobile larvae (glochidia) drift about in the water until they come into contact with a fish – frequently a Bitterling – to whose gills they adhere by means of a sticky filament. The fish's tissues become irritated by the glochidium and form a cyst, in which the glochidium remains to feed on its host for up to 16 weeks. Then, having grown a new shell, it falls from the cyst, possibly in an entirely new area to which it has been borne by its unwitting host.

Because the life cycle of the Bitterling depends entirely on the presence of freshwater mussels, it can only live in waters occupied by the mussels; these include still waters, slow-flowing lowland rivers, fens, drains and canals.

Bitterling share with Goldfish and Crucian Carp the ability to survive in waters that suffer periodic falls in oxygen that would prove fatal to most other species – their metabolism enables them to breathe in conditions of very low dissolved oxygen without contaminating themselves. This unusual trait also enables Bitterling in their first year to overwinter buried in mud.

Bitterling are tiny fish, seldom exceeding 5 cm in length in English waters. They are omnivorous, and their exceptionally long gut enables them to digest even filamentous algae, such as blanket weeds *Cladophora* spp., a commonly used food.

In the early years of the 20th century, Bitterling were occasionally used in human pregnancy testing; the female's ovipositor would elongate in human urine containing oestrogenic hormones.

References

Aldridge 1999, in Davies et al. 2004; Ellison 1959; Ellison & Chubb 1963; Fitter 1959; Hardy 1954; Lansbury 1956; Lever 1977, 1996; Orkin 1957; Perring 1967; Torbett 1961; Wheeler 1974; Wiepkema 1961.

Notes

1. So called because of the reputedly bitter taste of its flesh.
2. A name more commonly used for the Gibel Carp, which may be a hybrid between the Goldfish and Crucian Carp.
3. A.C. Wheeler subsequently identified these as apparently belonging to the European subspecies *R. s. amarus*.

ICTALURIDAE (AMERICAN CATFISH)

BLACK BULLHEAD *Ameiurus melas*

(BLACK CATFISH)

Natural Distribution C and E N America, from S Ontario, the Great Lakes and the St Lawrence River S through the basin of the Mississippi and Ohio Rivers to the Gulf of Mexico, and from Montana E to the Appalachian Mountains.

Naturalized Distribution England.

England

Wheeler (1978) unravelled the problem of the identity of North American bullheads introduced to continental Europe and England.

For many years it was generally accepted that the introduced species was the Brown Bullhead. Study of the literature, however, revealed the presence in Europe of both Brown and Black Bullheads, and leaves no doubt that the latter is the one that is more widely distributed, and that the majority of fish imported to England come from Italy, where the Black Bullhead predominates.

The National Fish Culture Association in South Kensington, London, where 48 of a total shipment of 50 arrived alive, made the earliest introduction of what were probably Black Bullheads to England in 1885. Wheeler and Maitland (1973) recorded the species' occurrence discontinuously in England, probably as a result of isolated plantings, e.g. near North Weald in Essex; in Kingsdown Road Flash, near Abram, Wigan, Lancashire; in a pond in Sussex; and in the Grand Union Canal near Alperton, Middlesex. Welcomme (1988) claimed that both Black and Brown Bullheads were breeding in the wild in Britain. Since, however, he says that the stock of '*nebulosus*' (Brown Bulheads) came from Italy, it was probably in fact the Black Bullhead.

Today, the only known population of Black Bullheads in England occurs in a lake in Warwickshire. Since, however, the species is becoming increasingly popular for keeping in garden ponds, the possibility of accidental escapes or (illegal) releases into the wild elsewhere cannot be ignored.

BLACK BULLHEAD

Characteristics

Black Bullheads favour rich, still or sluggish lowland rivers and lakes. The omnivorous and voracious adults feed on benthic organisms; large individuals may be piscivorous.

Impact

Where they occur in large numbers in continental Europe, Black Bullheads are regarded as a pest species because they compete with native benthic fish.

References

Chare & Musk, in Davies et al. 2004; Lever 1992, 1996; Maitland 2004; Welcomme 1988; Wheeler 1978, 1998; Wheeler & Maitland 1973.

ICTALURIDAE (AMERICAN CATFISH)

CHANNEL CATFISH *Ictalurus punctatus*

Natural Distribution Central drainage of North America from the prairie provinces of Canada and the Hudson Bay region to the Great Lakes and the St Lawrence River basin, S to Florida and N Mexico.
Naturalized Distribution England.

England

In 1968, 900 10-cm juvenile Channel Catfish, imported from the USA supposedly to stock aquaria, were released for sporting purposes in the following waters in Surrey: the conjoined North and South Lakes (200 and 250 fish respectively) at Wraysbury near Staines; the Fleet Lake (250) at Thorpe near Chertsey; the Sailing Club Lake (100) at Paper Court Farm near Ripley; and the Car Park Lake (100) at Yateley near Sandhurst. As the stocking density was too low (the recommended figure is 3,750–5,000 per hectare), and since the Channel Catfish needs warm, flowing water with a summer temperature of 27–32°C for successful spawning, breeding was unlikely. The species is included here because it is said to have been stocked more recently in a number of other commercial fisheries, including the River Trent in Nottinghamshire. No successful breeding has so far been reported, but it might well benefit if global warming were to continue.

Characteristics

In North America the Channel Catfish lives in deep, large rivers and still waters with a cool, clear and clean sandy or gravelly bottom with no thick vegetation, from which it ascends small streams to spawn between April and June. The adults are mainly piscivorous, while the young feed on small invertebrates.

References

Lever 1977, 1996; Wheeler 1974, 2001; Wheeler & Maitland 1973.

SILURIDAE (EUROPEAN CATFISH)

WELS *Silurus glanis*

(EUROPEAN CATFISH; DANUBE CATFISH; SHEAT FISH)

Natural Distribution C and E Europe, mainly E of the River Rhine, to the S former USSR.
Naturalized Distribution England; [Scotland]; Wales; [Ireland].

England

Although the naturalist John Fleming includes *Silurus glanis* in his *British Animals* (1828), apparently on the slender evidence of discovering a reference to '*Silurus Sive Glanis*' in the Scottish physician Sir Robert Sibbald's *Scotia Illustrata* (1684), Francis Willughby does not mention its presence in Britain in his *De Historia Piscium* (1686). Sibbald may have been referring to the superficially similar, though much smaller Burbot, although a number of old authorities used the name *Silurus* when referring to the Common Sturgeon.

The ichthyologist Francis Day in *The Fishes of Great Britain and Ireland* (1880) says that: 'One or more, *fortunately unsuccessful*, attempts have been made during the last few years to introduce this hideous monster into British rivers. *Silurus glanis* has a voracious appetite, is a foul feeder, inferior as food, and almost rank when of large size, its presence would be of exceedingly questionable advantage.'

The earliest of the attempts to introduce the Wels to Britain referred to by Day appears to have been one enigmatically mentioned by Llewellyn Lloyd in his *Scandinavian Adventures* (1854), in which he wrote that: 'Through the indefatigable exertions of

WELS

Mr George D. Berney of Morton, Norfolk, the silurus was last year [i.e. 1853] introduced into England'. Since no more was heard of this introduction, it must be assumed to have failed.

The Acclimatisation Society of the United Kingdom, founded under the auspices of Frank Buckland in 1860, had for some time been actively pursuing, through Dr A. Günther of the British Museum, attempts to introduce the Wels to Britain. In 1863, a pair, packed only in wet moss by Professor Jager in Vienna, not surprisingly failed to survive the journey. 'It is highly to be regretted,' said the society's Annual Report for 1863–4 rather sternly, 'that any attempt was made to transport these fish by a mode so certain of failure as packing them in wet moss.'

In 1863 or 1864, the society announced that Samuel Gurney, MP, had offered the use of a lake he owned at Carshalton in Surrey for the introduction of Wels which, the society claimed, was 'extremely well-fitted to harbour the fish', and that 'the arrival of the fish may be expected'. However, nothing seems to have come of this proposal.

Under the heading 'The arrival of the Silurus Glanis in England', a second attempt, irresponsibly sponsored by *The Field* magazine, was described, with typically Victorian verbosity, in that journal on 17 September 1864[1] by James Lowe, joint honorary secretary with Frank Buckland of the Acclimatisation Society:

> That much desired fish, the Silurus, has at last been brought alive to this country, after various failures. The success is entirely due to the intelligent enterprise and perseverance of Sir Stephen B. Lakeman, who himself accompanied the fish all the way from Bucharest, a distance of 1800 miles; and on Thursday night I had the pleasure of assisting Mr. Francis in placing fourteen lively little baby-siluri in a pond not far from the fish-hatching apparatus belonging to the Acclimatisation Society on Mr. Francis's grounds at Twickenham. …
>
> The fourteen little siluri (or siluruses) which have arrived are what remain of thirty-six of the same species, which started from Kopacheni, where Sir S. Lakeman's estate is situated. This place is on the banks of the Argisch, a tributary of the Danube, and is about ten miles [16 km] from Bucharest. …
>
> On arriving at Folkestone, there were fourteen survivors of the thirty-six which started from Kopacheni, and I am happy to say that every one of these reached Mr. Francis in the most lively and promising state.
>
> This is (so far as I am aware) the first time that this valuable fish has been brought to our shores …

In January 1865, Buckland 'took down ten of them to my friend, Higford Burr, Esq.; Aldermaston Park, Reading, and turned them out into a large pond in front of the house.

Some three years afterwards this pond was let dry – the silurus had entirely disappeared,' doubtless 'from their habit of buring themselves in the mud'.

In *The Field* of 8 September 1894, A. Gunther wrote:

> A week or two ago, the gamekeepers [of Mr Nocton, of Langham Hall, Colchester) ... caught a fish unknown to them, 4 ft. 3 in. [129 cm] long, and weighing over 30 lbs. [13 kg] in the River Stour at Flatford Mill, Suffolk. ... Mr. Nocton suspected at once the real nature of the fish. ... How did the fish get into the River Stour? I am informed that Sir Joshua Rowley put, about twenty-nine years ago [i.e. in about 1865], young Siluri into a lake communicating with that river, and distant some six or seven miles [10–11 km] from the place of capture. There is, therefore, no doubt that this fish was a survivor of Sir Joshua's experiment.

Fitter (1959) records that in 1872, the Marquess of Bath introduced a few small Wels to his lake at Frome in Somerset, where they ate so many trout that in 1875 the lake was drained to remove the three survivors, the heaviest of which weighed 28 lb (13 kg). In 1881, Frank Buckland wrote:

> There are several casts of Silurus in my museum in South Kensington.[2] The first I received from Mr. [T.R.] Sachs, in January, 1875. Herr Max von dem Borne was kind enough to catch this fish for us at Witsterwitz, near Berlin. He weighed 18 lbs. [8 kg] and measured 3 ft. 10 in [1.16 m] in length ...
>
> In February, 1876, Lord Odo Russell, Ambassador at Berlin, kindly sent me, through Lord Arthur Russell [his brother], a magnificent specimen of the silurus. It measured 5 ft. 6 in. [1.67 m] long, and weighed about 100 lbs. [45 kg].
>
> On the 27th October, 1880, Lord Odo Russell brought to Woburn Abbey, from Berlin, seventy of the *Silurus glanis* in their second year. Herr von dem Borne bred these Siluri at Berneuchen, near Custrin.

This is by far the largest and most successful importation of Wels to Britain, although Fitter (1959) suggested there may have been some later introductions to Woburn from the Danube. The Wels at Woburn flourished, and in 1951 a few small ones were transferred to the lake at Claydon Park in Buckinghamshire. Wheeler and Maitland (1973) chronicled the spread of Wels from Woburn to other waters on the Bedfordshire/Buckinghamshire border – notably at Little Brickhill and Tiddenfoot Pit – although the same authors add that 'one of the Woburn Abbey lakes – Lower Drakeloe Lake – has been netted in order to remove the wels and other fish before the release of sea-lions in the water as part of the Game Reserve!'

Glegg (1951: 131) quotes in A.C. Williams's *Angling Diversions* (1946) that in: 'about 1906 Mr. Walter Rothschild (afterwards Lord Rothschild) placed two or three large catfish – up to 30 lbs. [14 kg] in weight – in Marsworth Reservoir, at Tring. Nothing more was seen of them, but in 1928 a 3 lb. [1.4 kg] specimen was caught from this water ... There can be little doubt that it was a descendant of the fish placed in the reservoir some twenty years before.'

Since the 1970s, the popularity of Wels among sports anglers has resulted in their introduction to many still waters throughout England (apart from the far north and west), from which their spread into rivers may have been a result of flooding. Wheeler (2001) recorded their presence in some 166 English waters.

[Scotland]

None of the Wels introduced by John Berry into his ponds at Newport-on-Tay in Fife in 1925 survived the severe flood of 1934.

Wales

Since the 1970s, the popularity of Wels among sports anglers has led to their introduction to a few still waters in Wales. They are believed to exist in at least five, including one on the Isle of Anglesey, where spawning has been reported.

[Ireland]

William Yarrell, in his *History of British Fishes* (1836), whose account is repeated by William Thompson in his *Natural History of Ireland* (1849–56) and subsequently by Francis Day in his *The Fishes of Great Britain and Ireland* (1880), mentions the capture in Ireland of an unusual species of fish. According to Day:

> an unique example of a fish which some have considered may be the *Silurus glanis* is stated to have been captured about 1827 or 1828 from a tributary of the Shannon, near its source, about three miles [5 km] above Lough Allen. A fisherman asserted that a fish at least 2½ ft. [76 cm] long, and 8 or 9 lbs. [3–4 kg] weight, was seen struggling in a pool in the river as a flood subsided; that it had worm-like feelers to its mouth, while its appearance was so hideous that those who first saw it were afraid of touching it. The mouth of the figure of *Silurus glanis* in Yarrell's *British Fishes* was not considered large enough for that of the Irish specimen. ... The captured fish was not eaten but adorned a bush for two or three years until the skeleton fell to pieces, and with it all evidence to connect *Silurus glanis* with Ireland.

Characteristics

The Wels is a solitary species, living mainly in lakes and the still waters of marshes and lagoons, and sometimes in the lower reaches and backwaters of large, sluggish rivers with a muddy bottom. In some parts of the Baltic and Black Seas, it occurs in brackish waters. Wels spend the day skulking in mud or under stones, or in hollows in the lake banks or riverbanks, only emerging to forage after dark. Adults are voracious predators, mainly of fish, but also of benthic molluscs and crayfish, and will readily take water birds, small mammals and amphibians from the surface. Juveniles feed on fish fry and invertebrates. Wels spend the winter in a torpid condition, emerging from semi-hibernation to spawn between May and July.

The demand among anglers for large and unusual fish has resulted in the illegal stocking of Wels in both England and Wales. Although in continental Europe they grow to a weight of over 45 kg, in Britain they seldom exceed 1.5 m and around 20 kg. However, in 2007 a Wels was caught in Cambridgeshire that weighed 29.9 kg [66 lb] – a new record for a British rod-caught freshwater fish, and beating by 0.871 kg (1.9 lb) the celebrated Atlantic Salmon captured on the River Tay in Perthshire by Miss G.W. Ballantine in 1922.

Impact

In addition to preying on a variety of British mammals (including the threatened Water Vole), birds, fish and amphibians, the Wels is also known to be a carrier of the European Union notifiable carp disease, Spring Viremia.

References

Britton et al. 2006; Chare & Musk, in Davies et al. 2004; Dandy 1947; Fitter 1959; Gibson 1994; Glegg 1951; Gunther 1894; Hickley & Chare 2004; Lever 1977, 1992, 1996; Maitland 2004; Lowe 1864; Perring 1967; Wheeler 1971, 2001; Wheeler & Maitland 1973.

Notes

1. Although the contemporary account by Lowe clearly states that the year of arrival of Wels in England was 1864, Buckland (1880) says it was 1862 or 1863, and Fitter (1959) says 1865.
2. The Fisheries Museum, founded by Buckland in 1865 in the Science Museum, expanded in 1883 into the International Fisheries Exhibition.

SALMONIDAE (SALMON)

RAINBOW TROUT *Oncorhynchus mykiss*

Natural Distribution Coastal drainage of N America mainly W of the Continental Divide, from about 25°S in Mexico to a maximum of 64°N.
Naturalized Distribution England; Scotland; Wales; Ireland.

England

The earliest shipment of Rainbow Trout eggs to Britain, amounting to around 3,000, arrived at the National Fish Culture Association at Delaford Park, near Iver, in Buckinghamshire, on 14 February 1884. This was followed a year later by further consignments of 10,000 and 5,000 ova. These early introductions came from the Baird Trout Hatchery on the McCloud River in northern California. Some of the resulting fish were despatched to several fish farms in England (and Scotland).

RAINBOW TROUT

In almost every winter in 1888–1905, Rainbow Trout eggs from the United States arrived in Britain and were forwarded to hatcheries at Bridgnorth in Shropshire and Malvern Wells in Worcestershire (and to Scotland and Ireland). In 1890, the first ova from Europe arrived from S. Jaffé's hatchery at Osnabrück west of Hannover in Germany. Subsequent shipments from Europe may have been migratory sea-going 'steelheads' (corresponding with the migratory native Sea Trout), which have been widely introduced on the Continent.

No eggs were imported from the United States in 1905–31, but in the winter of 1931–2, D.F. Leney of the Surrey Trout Farm and United Fisheries at Haslemere in Surrey imported 50,000 ova through the United States Bureau of Fisheries and from the government hatchery at White Sulphur Springs in West Virginia, followed by a similar number in 1938–9.

Table 17 Waters in England where Rainbow Trout *Oncorhynchus mykiss* bred, 1900–1936

County	Locality
Berkshire	Monastic stew-pond near Newbury fed by River Lambourn
Buckinghamshire	Two lakes at head of River Gade; River Misbourne; River Mimran in Chiltern Hills
Cambridgeshire	Rivers Cam, Granta, Rhee
Derbyshire	River Wye
Herefordshire	River Garron
Hertfordshire	River Chess; Bury Lake, near Chesham
Kent	River Malling, Maidstone; flooded gravel-pit
Norfolk	?
Nottinghamshire	Lake at Newstead Abbey
Rutland	River Guash, Empingham
Somerset	Blagdon Reservoir
Surrey	Five small lakes near Haslemere; River Wey
Sussex	Stream near Chichester

Sources Worthington 1941; Lever 1977.

In all, Worthington recorded the presence of Rainbow Trout in 50–55 waters in Britain and Ireland.

Fitter (1959) reported successful breeding in the River Beane, Hertfordshire; River Wye, Buckinghamshire; and River Anton in Hampshire.

It is interesting to note that all the above waters are spring-fed and alkaline.

Waters listed by Worthington where Rainbow Trout failed to breed in England included the River Coquet, Northumberland; the River Lambourn, Berkshire; the River Test, Romsey, Hampshire; and the River Tamar near the Devon/Cornwall border.

Frost (1974) carried out a follow-up to the survey by Worthington. She found that in 1971 there was a total of at least 491 waters in Britain and Ireland that were known to contain Rainbow Trout. Few introductions were made in 1940–50 (presumably because of the Second World War and its aftermath). A few were made in the following decade.

Since then an increasing number of waters have been, and are continuing to be, stocked with Rainbow Trout to satisfy the ever-increasing demands of sport anglers.

The 1971 Frost survey revealed that of the English waters mentioned by Worthington as containing breeding stocks, only those in Blagdon Reservoir in Somerset, the River Wye in Derbyshire, the River Misbourne in Buckinghamshire and the River Wey in Surrey definitely still contained a breeding stock. Only the population in the Derbyshire Wye was unassisted by artificial stocking in both 1941 and 1971. The Frost survey further found that of the 40 or so English waters in which Rainbow Trout were reproducing, only in the Derbyshire Wye and a tributary, the Lathkill, and a feeder of the Leigh Brook on the Herefordshire/Worcestershire border, were the populations self-sustaining. In total, spawning has been recorded in over 40 waters in England.

Scotland

Some of the fish resulting from the shipment of eggs from California in 1884–5 were sent to Sir James Maitland's hatchery at Howietoun in Stirlingshire, where by 1887 breeding had been recorded. Some of the eggs imported in 1888–1905 were despatched to hatcheries at Loch Uisg and several small lochans near Lochbuie on the Isle of Mull, and in parts of Inverness-shire.

In the 1960s, some Rainbow Trout that were believed to have come from a small lochan in the valley of the Blane Water – the main tributary of the River Endrick, which was regularly stocked with Rainbow Trout – were caught in Loch Lomond between Stirlingshire and Dunbartonshire.

In 1967–9, Rainbow Trout were reared experimentally in the rehabilitated hatchery at the Lake of Menteith, at the edge of the Trossachs, west of Stirling, in Perthshire. In April of the former year, a total of 11,217 hatchery-bred yearlings were released in the lake, which soon spread out over its 263 ha. In autumn of the same year, a run of (all-male) Rainbow Trout was reported in inflowing burns. In March 1968, a second run occurred; 50 males were caught in the Malling Burn, and 15 males and a single ripe female in the Portend Burn, where in April a further 11 males and 1 ripe female were captured. On the same day, seven redds with ova were found in the Malling Burn and a week later redds were discovered in the Portend Burn. The first eggs hatched on 30 May, and a number of fry were subsequently seen. In October 1968, Rainbow Trout (and native Brown Trout) were observed making their way up the Malling and Portend Burns. In the former, the complete pattern of spawning behaviour by Rainbow Trout (territory guarding, courting, redd-cutting and ovaposition) was observed. There has been no successful spawning in the Lake of Menteith for many years, although large numbers of females continue to be stocked.

In May 1975, A.F. Walker of the Freshwater Fish Laboratory at Faskally found a self-maintaining colony of Rainbow Trout in a 1-ha slightly acid hill lochan 425 m asl near Loch Lyne in Inverness-shire.

Adams and Maitland (2001: 37) describe the Rainbow Trout in Scotland as 'widespread but not established'. This is confirmed by Welch et al. (2001: 207) who say it is 'widespread and abundant except in Northern Isles, Jura and Islay', and that at only a single Scottish site, in Inverness-shire, is there a self-maintaining population.

A questionnaire survey in 2001–2 (Walker 2003) revealed the presence of over 300 fisheries in Scotland that are stocked with Rainbow Trout. Most are in lowland areas close to conurbations, and there are relatively few in the Grampians and Western Highlands and Islands. Rainbow Trout that had escaped from fish farms or dispersed from stocked water were found in 48 Scottish rivers. Walker also reported spawning in 51 localities, in both lentic and lotic waters, but with very limited success. With the increasing use of all-female triploid stock, spawning may be less widespread than formerly. Nevertheless, successful spawning was reported in 16 waters.

Table 18 Waters in Scotland where Rainbow Trout *Oncorhynchus mykiss* may spawning, 2001–2

Locality	Comments
Airthrey Loch	Probably stream spawning; young seen once
Ardochy	Present for several decades; present status uncertain
Auchintaple	Small fish seen
Ballo Reservoir	Running water; limited success
Beecraigs	Small fish seen
Brother Loch	Sites unknown; little success
Butterstone Loch	Small fish seen
Coldingham	Small fish seen
Corby Loch	Small fish seen
Crakaig Pond	Fry in burn and loch
Crannoch	Small fish seen
Craufurdland Fishery	Still water, first 3 years (February–April) no success noted
Dargarvel Loch	Spawning site(s) unknown
Dowally	Spawning seen; still water, no success
Drummond Farm	Running water December–January; no success noted
Duntanlich Lochan	Spawned in flowing water, and young fish common; now died out
Fincastle Loch	Small fish seen
Fisheries Laboratory, Faskally (raceway)	Spawned in running water; some success
Glensherup Reservoir	Fine, clean gravel beds; small fish seen
Heatheryford Fishery	Still water; small fish seen annually
Howwood Fishery	No data
Kingennie Fishery	Inflowing running water, November–December; no success noted
Kirkchrist Fishery	Running water, February, one year before release of triploids

Logan Burn During the 1990s, Rainbow Trout were observed spawning in this burn, which joins the Glencorse and Loganlea Reservoirs in the Pentland Hills, south-east of Edinburgh. Juveniles have been caught in recent years in both reservoirs and in the burn.
Heatheryford In most years, Rainbow Trout that are believed to have been spawned naturally are caught in this commercial fishery (c 4 ha) near Kinross.
Criggie Fishery In 1989, gravel was added to a small burn that flowed into the mill of Criggie Fishery near St Cyrus in Kincardineshire to form experimental redds, which were seeded with eyed ova. These eggs hatched successfully, and naturally grown-on fish spawned in the burn in 1992–8, when the redds were intentionally destroyed. The spawning fish were all fin-perfect, suggesting they were wild individuals.
Moffat Rainbow Trout are said to spawn annually near springs that feed two artificial ponds of c 0.4 and 2 ha near Moffat in Dumfries-shire, albeit in smaller numbers than when the fishery was first formed. The spawning fish include survivors from natural spawning, although some mature fish are included among the triploids that are regularly stocked. American Brook Trout also spawn successfully in these two ponds.

Table 18 *Continued*

Locality	**Comments**
Lake of Menteith	High likelihood of successful spawning
Linlithgow Loch	Spawning noted in mouth of burn flowing into Hatchery Bay
Loch Bhac	Some successful still-water spawning
Loch Awe	Small fish seen
Loch Fad	Small fish seen
Loch Fithie	Still-water sites; young fish common
Loch Leven	Small fish seen
Loch Lochy	Small fish seen
Loch Rusky	Small fish seen
Loganlea Reservoir	Spawning in Logan Burn
Mill of Criggie Fishery	Spring, running water, 1989–98
Moffat Fishery	Spring percolation; successful spawning annually
Monzievaird	Small fish seen
North Third Reservoir	In 1999 small fish near burn at Bluffs on south side
Orchill Fishery	Small fish seen
Portmore Reservoir	Small fish seen
River Earn	Small fish seen
River Tay (Stanley)	Small fish seen
Rothiemurchus Fishery	Small fish seen
Rotmell	Spawning at edge regularly, but no young fish seen
Stoneyfield Lochs	Small fish seen
Straloch	Spawning in still water, but no young fish seen
Watch Reservoir	Small fish seen
Unnamed loch	Still present after 1993 stocking

Source Walker 2003.

Rossdhu Pond In about 1993 this small, 1-ha fishery was constructed on a golf course near Luss, Loch Lomond. Every spring since the first stocked fish matured, Rainbow Trout have spawned successfully around the gravel shores, and naturally spawned juveniles are common. The majority of mature cock fish are believed to result from natural spawning, since stockings in recent years have been with triploids, although a few fish have matured.
Stoneyfield Lochs Rainbow Trout are said to spawn annually in the three Stoneyfield Lochs (totalling c 5.5 ha) near Invergordon in Easter Ross, and juveniles are common. American Brook Trout also spawn successfully in these lochs.

In addition to the spawning sites revealed by Walker's questionnaire, the following Scottish waters where Rainbow Trout had spawned successfully were already known to the Fisheries Research Service.
Ardochy Hill Loch This small, 1.2-ha loch lies at an altitude of 440 m asl in a partially afforested moorland setting above Loch Loyne in Inverness-shire. It is believed to have been stocked with Rainbow Trout in the mid-1920s, having previously been devoid of fish. Initially they prospered, the heaviest reaching over 2.5 kg, later declining in size when the population outgrew the food supply. In 1975, the water was found to have a pH of 6.8 and alkalinity of 1.1, though the pH would be likely to fall after heavy rain. It is believed that Rainbow Trout continue to survive in the loch.
Corby Loch Lying immediately north-east of Aberdeen, this loch is fairly shallow (normally 1–2 m in depth) and extends to over 12 ha. It was constructed in the 1930s to provide adequate water power during the summer months for Mundurno Mill. The water in the loch is extremely peaty, the bottom being largely deposited peat with stony margins and some gravel. During the winter months, over 3,000 wild geese often roost on the loch, providing some natural fertilization. The loch appears to support a variety of invertebrate species. Until about the turn of the century, non-triploid Rainbow Trout were regularly stocked and thrived in the loch, where individuals of over 4.5 kg have been caught, some of which have been full of large eggs, and many fin-perfect smaller fish have been seen in recent years.
Crakaig Pond This less than 0.5-ha pond near Loth in Sutherland was stocked with Rainbow Trout in the late 1980s, and a small number of fry was observed in 1993.
Loch Bhac Covering about 12 ha and surrounded by coniferous forestry plantations, Loch Bhac lies to the north of Loch Tummel in Perthshire. Since it is situated on intrusions of metamorphic limestone, the water is mildly calcareous. Since 1975, the loch has been stocked with Rainbow and Brook Trout, and also supports a natural population of small Brown Trout. In the early years Rainbow Trout of both sexes were used for stocking purposes, and some survived to maturity. From time to time fin-perfect juveniles have been caught, though none in recent years when all the stocked fish have been female. Rainbows are said also to have spawned successfully at the time in a number of neighbouring lochs.
Duntanlichan Lochan This very small (<1-ha) and shallow (<1.5 m-deep) artificial lochan, which is fed by two narrow burns, lies in the hills to the south of Loch Tummel. In 1975 it was treated with the piscicide Rotenone to remove an excessive population of Brown Trout, and subsequently stocked with 22 Rainbow Trout and 35 Brook Trout, all of which were one or more years old. The Brook Trout became established and self-maintaining through natural spawning, and although some of the Rainbow Trout also reproduced successfully they did not become self-supporting, despite a further light stocking

in 1984. Recently stocked Brown Trout seem to have displaced Rainbow Trout and may even be ousting the Brook Trout. In 2003, juvenile Brown Trout were the most common species in the out-flowing burn, and although the lochan contained a mixed Brown/Brook Trout population, the latter were mainly older individuals, suggesting an absence of juvenile recruitment.

Loch Fad Successful spawning by Rainbow Trout has been recorded in this loch on the Isle of Bute, where a commercial cage farm has been situated since 1976, and in its main tributary the Woodend Burn. The burn, which is fast flowing and has extensive areas of gravel, has a mean pH of 6.9.

Loch Fithy In the early 1970s, Rainbow Trout spawned successfully on shelving gravel banks in this large (c 18-ha) gravel-pit near Forfar in Angus. Since there were numerous Rainbow Trout in the loch that were smaller than those stocked, it was clear that many fry had survived.

Monzievaird Loch This approximately 20-ha loch is situated in light deciduous woodland near Crieff in Perthshire. In about 1890, Brook Trout from North America were introduced to it, and still survive there. In 1965, Monzievaird Brook Trout were removed from the loch, which was then treated with Rotenone to eradicate the coarse fish population, the Brook Trout subsequently being replaced. Thereafter, Rainbow Trout were stocked for only a few years, and are believed to have spawned; their current status in Monzievaird is apparently uncertain.

Fisheries Research Services Freshwater Laboratory (FRSFL) In the early 1970s, some mixed-sex Rainbow Trout were kept in concrete 'ponds' at Faskally, near Pitlochry, in conjunction with the stocking experiment at Loch Straloch (see below). In January 1975, mixed gravel was placed in the concrete raceway that fed the ponds, and with the arrival of milder weather in early February extensive spawning was observed in the raceway, and fry began to emerge and to start feeding in early May.

Straloch In May 1973, 250 two-year-old Rainbow Trout were obtained by the FRSL from a hatchery at Almondbank, west of Perth, and released in Straloch, a shallow (<2-m deep), 9-ha artificial and biologically productive water near Pitlochry, Perthshire, which holds a population of native Brown Trout. With the arrival of milder weather in early February 1974, ripe male Rainbow Trout began to be caught regularly in box-traps placed in the loch. The first ripe female was captured in late February, and by mid-April Rainbow Trout were observed spawning in gravel areas that had been previously used by Brown Trout. During the following two seasons, only an occasional naturally spawned Rainbow Trout was reported by anglers, and since no subsequent stockings were made the species disappeared.

Wales

The maps in Maitland (2004: 196) and Davies et al. (2004: 107) show Rainbow Trout to occur throughout much of Wales, particularly in the south (Pembrokeshire, Carmarthenshire, Swansea, Bridgend, Newport and the Vale of Glamorgan) and west (Cardiganshire). In the north, Rainbow Trout are found principally in Gwynedd, Conwy and on the Isle of Anglesey. All these waters contain only stocked populations, none of which is believed to be self-maintaining.

Ireland

Some of the eggs shipped from the United States in 1888–1905 were sent to hatcheries at Inishannon, where Morton Trewen introduced them to the River Bandon in Co. Cork, and at Ballymena in Co. Antrim. In about 1905, C. Maude of Glen House on the island of Aran off the coast of Co. Donegal, acquired some Rainbow Trout, probably through his brother who was manager of Hanlon's Hatchery at Dungloe. He released them in Lough Shure (*Loc a Cabanaig* – 'Loch of the Shepherd'), a peaty moorland lough measuring about 800 x 400 m and about 180 m asl on the western side of Aran, where they bred successfully. This was the only recorded non-alkaline water into which the fish had then been introduced.

In her survey of 1971, Frost recorded 21 waters in Northern Ireland and 8 in the Irish Republic that were known to hold Rainbow Trout. Although no introductions were made to the Irish Republic in 1940–50, there were a few in 1955–64, after which stocking considerably increased, as it did in England, Scotland and Wales, to satisfy the demands of sport anglers. The Frost enquiry revealed that of those waters containing Rainbow Trout in 1941, only Lough Shure still held a self-perpetuating breeding stock in 1971. There have been no further introductions to Lough Shure since that of 1905 where Frost (1974) was told that fishing for them in Lough Shure was 'still good'.

The only other Irish water in which self-maintaining breeding Rainbow Trout were recorded by Frost was in Lough na Leibe, near Ballymote in Co. Sligo, which had been stocked by Michael Kennedy of the Ireland Fisheries Trust in Dublin in 1955. This spring-fed lough, which has a limestone bottom, extends to some 4–5 ha: in the outflowing stream, redds and post-alevins were found, but because of the angling pressure every few years takeable-size Rainbow Trout were introduced, although a nucleus of self-perpetuating fish survived. Frost (1974: 236) concluded that 'the rainbows when first introduced to Lough na Leibe established a self-perpetuating population which still exists contemporaneously with the stocked rainbows, which are added infrequently and in small numbers'.

Somewhat similar conditions to those in Lough na Leibe occurred in White Lake, Castlepollard, in Co. Westmeath, which had been stocked from the same source and in the same year. The Rainbow Trout again reproduced naturally, but because of the limited number of suitable spawning sites the quantity of fish was insufficient to satisfy the requirements of anglers, and both Rainbow and Brown Trout were extensively stocked. Frost wrote (1974: 238):

> It would seem that if there is any spawning going on it may well be that the brown trout dominate the much limited spawning grounds ... this would suggest that on White Lake, the small natural breeding population [of Rainbow Trout] at first established, has not been able to survive the predation of the large stock fish (Rainbow and Brown Trout) on any small native [*sic*] rainbows and the heavy angling pressures on any native [*sic*] specimens which may have survived to become larger size.

Since the Frost report, Rainbow Trout have become established in several other Irish waters. In 1964–8 some, imported from Denmark to the Movanagher Fish Farm in Northern Ireland, were released in a total of 17 lakes, where they bred in Shaws Lake, Co. Armagh and Lough Eye, Co. Fermanagh. In the Irish Republic, they are believed to have spawned in Woodenbridge River, Co. Wicklow, and in the Golden Grove Brook and Little Brosna River, Co. Tipperary.

Prospects

The reasons for the failure of Rainbow Trout to establish more viable populations in Britain and Ireland, despite widespread and repeated introductions and breeding, remain somewhat obscure. One factor may be interspecific competition between Rainbow and native Brown Trout fry: the latter hatch earlier in the year than the former, and are thus able to dominate and harass them in the search for food, territory and refugia at a critical time in their development.

Although Rainbow Trout grow well in the wild, long-term rearing in hatcheries may reduce the vigour and fitness of captive-bred stock for life in the wild, and thus further limit successful long-term reproduction. Many individuals that do manage to survive to become yearlings probably fall prey to larger cannibal Rainbow and Brown Trout. Thus heavy artificial stocking of large fish to satisfy the demands of sports anglers, frequently accompanied by over-fishing, may be another contributory factor preventing the establishment of more self-maintaining Rainbow Trout populations. Other factors may include water flow and temperature, stream gradient and channel geomorphology, and parasites and diseases.

The most likely factors preventing the widespread naturalization of Rainbow Trout in British waters are biotic resistance from native salmonids, or parasites or diseases such as 'whirling disease', and over-fishing. This situation could well change if environmental modifications trigger a new interaction that allows establishment. These could include:

- Decline in native species due to habitat degradation or overexploitation, thereby favouring the survival of Rainbow Trout fry.
- Introduction of a new strain of Rainbow Trout that is resistant to disease.
- Climatic change or other human-caused alterations that alter water flow and temperature, thus causing changes in parasites, diseases and native species that currently prevent the reproduction of Rainbow Trout.

Once successful recruitment occurs, evolution that increases the probability of establishment and spread can occur within only a few generations.

'Throughout Britain,' wrote Walker in Davies et al. (2004: 107), 'self-maintaining populations [of Rainbow Trout] were never very common, but appear now to be restricted to a very few localities in England [but see under Scotland], including the Derbyshire Wye.' This is confirmed by Maitland (2004: 160), who wrote: 'In the British Isles there are very few [self-maintaining] populations, though the species is stocked widely each year.'

Impact

Fausch (2007) has summarized the potential impact were Rainbow Trout to become widely naturalized in Britain.

Because the species is regarded as an important and valuable non-native asset that is believed to pose a low risk to native aquatic biota due to its inability to establish widespread naturalized populations, over two million are legally stocked annually to satisfy the demands of anglers. Because of concerns about the potential ecological effect of triploids, the vast majority of deliberately stocked fish have been all-female diploids. However, although a high proportion of the numerous fish that accidentally escape from hatcheries are females, some males also abscond, and if the females are ripe this could increase the potential for Rainbow Trout to become naturalized, since a small number of males can fertilize the eggs of many females. If only a few self-sustaining populations become established, there would be a temptation for anglers to translocate illegally any large anadromous forms ('steelheads') to neighbouring waters, where eradication might be impractical.

If Rainbow Trout were to become widely naturalized they could change the behaviour, abundance and distribution of native species, which in turn might reduce the predation pressure on zooplankton or bethnic invertebrates that graze on algae and change carbon and nutrient dynamics.

Empirical evidence suggests that naturalized Rainbow Trout in Britain could potentially have a negative impact on native salmonids (Brown Trout and Atlantic Salmon) and other indigenous fish species – mainly through predation and competition.

Rainbow Trout can also have a strong indirect impact on food webs in both lentic and lotic waters, which may alter ecosystem processes. For example, their direct negative effect on benthic invertebrates and frog tadpoles (both of which graze on algae) can result in an increase in algal biomass and in a trophic cascade. Furthermore, introduced Rainbow Trout in still waters can recycle phosphorus, which also indirectly increases algal biomass. Thus, introduced Rainbow Trout can have a strong direct and indirect impact not only on other fish species, but also on entire aquatic communities, and even on such riparian predators as birds, bats, lizards and spiders.

Finally, invading salmonids can fragment indigenous fish populations, and this increases the possibility of locally reducing or eradicating such native species.

On the positive side, Rainbow Trout provide an additional source of prey for mammalian and avian predators like Eurasian Otters and Ospreys, both of which are staging a welcome resurgence in Britain (nesting Ospreys are an important tourist attraction).

In the early 1950s, a viral disease of young salmonids – subsequently named infectious pancreatic necrosis (IPN) – made its appearance in the United States. It arrived in continental Europe (allegedly in a shipment of Rainbow Trout ova from North America) in the early 1960s. It first appeared in Britain (in Scotland and Ireland) in 1971, and in England (Hampshire) in 1972. IPN is believed to have been endemic in fish stocks around the world long before it was diagnosed, and its 'discovery' occurred because of the enormous expansion of intensive trout farming. The chronological order of the discovery of the virus in different countries closely mirrors the establishment in those countries of laboratories with diagnostic capabilities.

The IPN virus, which is easily transferable by the movement of either infected fish (as carriers or infected fry) or contaminated eyed ova, kills the young fry of both Rainbow and Brown Trout, whose resistance increases with age until, at 3–5 months, they are believed to become immune. IPN has appeared in several strains with differing degrees of virulence – some causing few if any mortalities, others having a death rate of up to 90 per cent in affected fish.

References

Adams & Maitland 2001; Barnard 1997; Cambray 2003; Clark 1885; Ellison & Chubb 1963; Frost 1940, 1974; Fausch, in Kováă et al. 2007; Hill et al. 2005; Hunt 1972; Kottelot & Freyhof 2007; Lever 1977, 1996, in press; MacCrimmon 1971; Maitland 1966, 1969, 2004; Richmond 1919, 1932; Ritchie 1920; Stuart 1967, 1969–71; Tew 1930; Walker & Petterson 1898; Walker 2003, 2004, in Davies et al. 2004; Wheeler 1971; Wheeler & Maitland 1973; Worthington 1941.

SALMONIDAE (SALMON)

AMERICAN BROOK TROUT *Salvelinus fontinalis*
(BROOK CHARR)

Natural Distribution NE North America, from N Quebec S to about 60°N and Labrador S of around 57°N, S through Newfoundland and Nova Scotia to about 40°N in Pennsylvania (with an extension S to N Georgia) and W to Wisconsin, Michigan and Manitoba. (Sea-going Brook Trout, known as 'coasters', occur mainly off the Canadian coast, and also in some rivers off Maine and Cape Cod.)
Naturalized Distribution England; Scotland; Wales.

England

The first shipment of American Brook Trout ova to reach England was despatched to Frank Buckland by Livingstone Stone from his Cold Spring Trout Ponds at Charlestown, New Hampshire in the late winter of 1868, arriving at its destination early in the following spring where, Stone recorded, the trout were 'favourably noticed in the London Times'. This was subsequently confirmed by Buckland (1880: 221), who wrote:

> American brook trout brought over by Mr Parnaby of Troutdale Fishery, Keswick ... the first specimens ever seen in this country were sent to me beautifully packed with moss ... by some friends in America. The parent fish were obtained from Lake Huron, in Canada. Since that time the import of eggs of *fontinalis* has become a regular business.

The migratory instinct, which was strongly implanted in the American Brook Trout imported to Britain in the 19th century, was presumably derived from introduced migratory stock. Armistead (1895: 319) records that some American Brook Trout in Britain had been caught 'in salt water in some of our bays and estuaries'. (These fish would have been the equivalent of migratory American 'steelhead' Rainbow Trout, or native British Sea Trout.)

This initial introduction was followed in November 1871 by a second shipment, of which Stone wrote that: '10,000 trout eggs were packed in sphagnum moss ... at

AMERICAN BROOK TROUT

Charlestown, N.H. ... They made a long passage of eighteen days to Liverpool and a considerable journey by rail from Liverpool to Keswick. … two-thirds were found to be in good condition, although some had hatched on the way and died.'

These are generally accepted as the earliest introductions of the species to Britain. In a somewhat petulant letter to *The Times* (29 October 1885), however, Parker Gillmore claimed that in 1866–7 he had been the first to suggest the introduction of the fish, and that late in the latter year he had collected some in North America which were shipped to England early in 1868, where he claimed they were distributed before his return and without his knowledge. In his letter, Gillmore complained that:

> The Prince of Wales had a pond stocked with *fontinalis*. When I visited it the person in charge was ignorant of my being the introducer, as was also His Royal Highness, until I informed him by letter. ... at the Zoological Gardens, I also saw some *fontinalis* in a glass tank with another person's name given on the label as presenter and introducer. ... I have not only been left a pecuniary loser, but also have never been credited with my work.

Writing in the *Zoologist* (1876: 5111), John T. Carrington said he had seen, at James Forbes's fishery at Chertsey Bridge, on the Thames, 'a large number ... of several ages, of the American brook trout (*Salvelinus fontinalis*) ... I cannot help thinking that owing to Mr Forbes's efforts, this fish will soon obtain permanent hold in the Thames.'

Summing up the position of the species in England in the late 19th century, Day (1887: 341) wrote:

> It has, during the last twenty years, been acclimatized in this country, and thrives in some of the places where it has been turned out, in south, east or west. ... in Mr Andrews' ponds [at Guildford] in Surrey, and [in the River Wey] it is said to have done well, while in Bagshot Park [Surrey] it is likewise stated to have thriven. ... at Tehidy, near Cambourne, ... Mr Basset stocked his ponds [with *fontinalis*] some nine years since.

In the early years of the 20th century, American Brook Trout were introduced to four recently constructed artificial tarns (Scale, Wise Een, Wraymires and Moss Eccles) on the Claife Heights above the western shore of Windermere in Lancashire. The fish, which came from the Wraymires Hatchery, thrived in the tarns, though by the mid-1970s only a few survived.

In 1935, the Troutdale Fishery at Keswick was acquired by D.F. Leney of the Surrey Trout Farm at Haslemere, where a thriving population of the species became established. In 1936, Leney introduced 200–300 American Brook Trout to a chain of spring-fed pools at Deepdene, near Haslemere, the property of Sir Hildebrand Harmsworth, Bt, where he attributed their successful establishment to the absence of lime in the water.

In the late 1950s/early 1960s, Leney sold a number of American Brook Trout (and Brown Trout hybrids) to the Bristol Waterworks Company for stocking their reservoir at Chew, where although the conditions appeared suitable they failed to establish. At about this time, the lake at Newstead Abbey near Nottingham held a thriving and apparently self-maintaining population; unfortunately, this colony died out as a result of pollution during the Second World War and later subsidence due to coal-mining.

In the 1970s, fry from Llyn Taws in Powys were stocked in two waters in the Forest of Dean in Gloucestershire.

Elsewhere in England, in 1870–1931 fish were released unsuccessfully in the River Nene, Northamptonshire (1870–90); at Fallodon in Northumberland (1890–2); at Glastonbury, Somerset (1891); in the River Eden at Appleby (1889) and at Yew Tree Tarn,

Westmorland (Cumbria) (1925); and in a man-made water on the Dorset/Wiltshire border (1931).

Scotland

The earliest reference to American Brook Trout in Scotland appears to be that made by Armistead (1895: 319), who wrote that 'at Howietoun [hatchery], to which Sir James Maitland imported 10,000 ova in 1878, it has done fairly well, but does not often seem to live over its fifth year. ... The Maclaine of Lochbuie has acclimatized this fish in a moor loch about a thousand feet [300 m] above the sea, near Loch Uisk, in Mull.'

Harvie-Brown and Buckley (1892: 420) recorded that the fish 'have been introduced by Mr M'Fadyen into the lochs of the Cuilfail district [above Kilmelford near Oban] over many years. They have also been successfully placed in the Lochbuie lochs on Mull. The Maclaine writes "they have done better here than (so far as I can learn) any other part of the United Kingdom". '

In the 1880s, the Duke of Sutherland released some American Brook Trout in Loch Brora and the Kintradwell Burn in Sutherland, and Ritchie (1920) believed that over 1,000 fish, reputed to be Rainbow Trout, which were introduced in 1898 to the River Buchart, a tributary of the Aberdeenshire Don, were actually American Brook Trout.

Scott and Brown (1901: 177) record that the American Brook Trout had been:

> introduced in many small lakes throughout the [Clyde drainage] district, also in Loch Lomond where it still maintains its identity, but has not thriven. Has been distributed throughout Renfrewshire and Ayrshire, and is thriving in the Rivers Ayr and Irvine, and in the Waters of Borland, Kilmarnock, Cossnock, Carmel and Alnwick [Northumberland] ... it has been widely distributed throughout Scotland, even in islands such as Mull.

In the late 19th century, and again shortly after the First World War, American Brook Trout were stocked in a 1-ha hill loch (*Lochan an Eireannaich* – 'The Loch of the Irishman') at the head of Kirkton Glen, near Balquhidder in Perthshire, where they survived until at least the mid-1970s.

The population in Monzievaird Loch on the Ochtertyre estate, near Crieff in Perthshire, was introduced in the early 20th century by Sir Patrick Keith Murray, Bt., who in about 1890 had acquired some from John Berry of Tayfield at Newport-on-Tay in Fife. At Ochtertyre the fish became established in St Serf's Water ('The Serpentine'), a small and apparently unsuitable lochan above Monzievaird. The fish in St Serf's Water died out in the mid-1970s, but survived for longer in Monzievaird.

The original population at Tayfield became established before 1890, but died out after the pond became polluted in 1934; attempts to restock it with fish from Monzievaird in 1974–5 were unsuccessful.

In 1960, fish from Monzievaird were used to stock a number of neighbouring waters, including the spawning burn of Fincastle Loch near Faskally and in 1970 a small limestone lochan above Fincastle, but they had all disappeared by the mid-1970s. Others from the same source were planted in Loch Dunmore near Pitlochry, Lindores Loch near Newburgh in Fife, and several small lochans owned by the Milton Park Hotel at Dalry west of Dumfries in Kirkcudbrightshire.

In 1968, American Brook Trout were successfully introduced to a small burn near Pitlochry, from which some rapidly dispersed downstream to the River Tummel, and in 1972 and 1975 some adult fish were planted in Loch Bhac, near Fincastle, where they soon became established.

Four small hill lochans in the Torridon Mountains of Wester Ross, where they were said to have been originally introduced in the 1890s, held established populations in the 1970s, in one of which they co-existed with native Arctic Charr. In the 1970s, 6 lochs in Sutherland, 1 in Angus and 11 in Perthshire were stocked with American Brook Trout by the Freshwater Fisheries Laboratory at Faskally, which also supplied fry to fish farms in Morayshire, Fife, Wester Ross, Stirlingshire and East Lothian, and brood stock fish to the Highlands and Islands Development Board.

In 1887–1936, fish were unsuccessfully released in Scotland in the River North Esk, Midlothian (1887–9); Loch Lomond, Stirlingshire (before 1895); New Abbey, Dumfriesshire (before 1895); Loch Coulin, Ross and Cromarty (1924); and Linlochewe, Ross and Cromarty (1936).

Wales

In 1889, American Brook Trout were unsuccessfully stocked in waters in Cardiganshire in south-western Wales.

One of the most successful British populations, the origin of which is unknown, became established in Llyn Tarn (Llyn-y-Tarur) on the Plas Dinam estate of Lord Davies at Llandinam, near Newton, in Montgomeryshire (Powys). Spawning has occurred in the lake itself, as there are no feeder streams. In the 1970s, fish from Llyn Tarw were stocked in a small hill water, Lyn Bugail, and in a minor tributary of the upper River Severn. Whether any survive in Llyn Tarw is uncertain, as none is known to have been caught recently (Lord Davies, personal communication 2008).

The map in Maitland (2004: 198) shows American Brook Trout as occurring (or as having occurred) on the Isle of Anglesey, Conwy, Denbighshire, Gwynedd and Powys. The map in Walker, in Davies et al. (2004: 117) omits Anglesey.

Prospects

In spite of repeated stockings in apparently suitable waters in many localities, American Brook Trout have succeeded in becoming naturalized (rather than only acclimatized) in only a few widely scattered places in England, the Scottish Highlands and Wales.

According to Welch et al. (2001: 38), 'There are approximately nine populations of brook charr (*Salvelinus fontinalis*) in Scotland', where Adams and Maitland (2001: 37) describe the species as 'very local – 8 known established populations'. Summing up, Walker, in Davies et al. (2004: 116) says:

> Self-sustaining populations occur in only a few locations, nearly all of which were established from introductions made before 1900. In Scotland, long-established and highly colourful brook charr still occur in the wild including in some small lochs near Lochbuie in Mull ... in

lochs and associated streams above Torridon, in Wester Ross, and in a small hill loch in the Trossachs [Stirlingshire]. A population in Monzievaird Loch, near Crieff, was established in about 1890... .The population in a small pond and stream on the south side of Loch Tummell, Perthshire, originated in the mid-1970s, but brown/sea trout have been reintroduced there recently and appear to be supplanting the brook charr. In upland areas of Britain, successful spawning occurs sporadically where brook charr are stocked, but the species is seldom able to become self-sustaining in the presence of brown/sea trout.

Characteristics

American Brook Trout favour cool, clear, acid water or only mildly alkaline water in medium-sized and fast-flowing rivers and well-oxygenated lakes, with a temperature range of 11–16°C. Although they readily acclimatize in British waters, they seldom become naturalized, especially in waters containing Brown Trout. This may be due to interspecific competition for spawning sites – certainly, in the absence of Brown Trout, they spawn well in Britain in both flowing and still waters.

References

Adams & Maitland 2001; Anon 1925; Armistead 1895; Buckland 1880; Carrington 1876; Day 1887; Harvie-Brown & Buckley 1892; Lever 1977, 1996; MacCrimmon & Campbell 1969; Maitland 2004; Scott & Brown 1901; Walker 1976, in Davies et al. 2004; Welch et al. 2001;Wheeler & Maitland 1973.

CENTRARCHIDAE (SUNFISH)

PUMPKINSEED *Lepomis gibbosus*

(COMMON SUNFISH; YELLOW SUNFISH; SUN BASS; SUN PERCH)

Natural Distribution S Canada and the USA from the Great Lakes and North Dakota E to the Atlantic coast and S to Texas and Florida.
Naturalized Distribution England; [Scotland].

England

There appear to have been comparatively few deliberate attempts to naturalize the Pumpkinseed in England, although it has become established and bred in several waters, e.g. during the First World War at Groombridge in East Sussex; after 1938 near Crawley in West Sussex; in 1953 near Bridgwater in Somerset; and in 1974 in the Hollow Pond at Whipp's Cross, Leytonstone, Essex. The population at Crawley was still extant in 1974, but was thought to be threatened by encroaching development, and by the following decade both the Sussex populations are believed to have died out.

In 1986–7, Pumpkinseeds that are believed to have been imported from Germany with Sunbleak and Topmouth Gudgeon by the Crampmoor Fishery near Romsey in Hampshire, escaped into the River Test and thence into the Broadlands estate some 2 km downstream, where they became established in at least two lakes.

In recent years, Pumpkinseeds have also become established in several other waters in southern England, mainly in east and west Sussex and Surrey.

Table 19 Waters in England where Pumpkinseeds *Lepomis gibbosus* have been taken, 1989–2007

Date	County	Site Location	National Grid Reference
May 1989	Surrey	River Mole (upstream of Horley sewage treatment works)	TQ2660043700
September 1992	Surrey	River Mole (downstream of Gatwick)	TQ2625441023
August 2001	Sussex	River Ouse (Sloop)	TQ3851625417
August 2002	Sussex	River Ouse (Sharpsbridge)	TQ 4400920779
August 2002	Sussex	River Ouse and tributaries (Batts Bridge Stream, Batts Bridge)	TQ4530023400
July 2003	Sussex	River Ouse and tributaries (Batts Bridge Stream, Batts Bridge)	TQ4530723321
August 2003	Sussex	River Ouse and tributaries (Shortbridge Stream, Maresfield)	TQ4530024700
September 2003	Sussex	River Ouse and tributaries (Cackle Street)	TQ4560026400
August 2004	Sussex	River Ouse (Sharpsbridge)	TQ4400920779
August 2007	Sussex	River Ouse (Sharpsbridge)	TQ4404820824
August 2007	Sussex	River Ouse (Sloop)	TQ3844624473
September 2007	Sussex	River Ouse (Sheffield Bridge)	TQ4058723615
September 2007	Isle of Wight	Nettlecombe Farm	SZ5274178143

Source Environment Agency National Fish Population Database (per Phil Hickley).

PUMPKINSEED

[Scotland]
In the 1920s, John Berry succeeded in establishing a small breeding population of Pumpkinseeds at Tayfield near Newport-on-Tay in Fife, which lasted until a severe flood in 1934.

Characteristics
The Pumpkinseed's deep, laterally compressed body with large scales is totally unlike that of any other northern European freshwater fish, and its attractive colouration makes it a popular aquarium species. It is found mainly in small, cool, clear, shallow, sheltered and weedy still waters and in the lower reaches of slow-flowing rivers. The males are extremely territorial, and zealously guard the eggs deposited by the female until they hatch.

The Pumpkinseed is a voracious and omnivorous predator, feeding on small invertebrates, fish eggs and larvae, and on young fish, including those of its own species. In England and continental Europe, it seldom exceeds around 15 cm in length and a weight of 100 g. Although the juvenile growth rate in English, continental and North American waters is much the same, the mean adult body size (up to 23 cm) and growth rate are both significantly greater in North America, where the species is harvested for food. In England and Europe, the species' growth rate appears to be compromised by limited, though adequate for survival, food resources, probably as a consequence of intra-specific interactions, which seem to be especially acute in adults.

Impact
In continental Europe, the Pumpkinseed is regarded as an undesirable 'trash' fish that competes with and preys on the eggs and larvae of such local species as the Common Perch and Bleak. In England, fears have been expressed about its possible impact also on such native species as Common Bream, Roach, Rudd and Brown Trout.

References
Chare & Musk, in Davies et al. 2004; Copp et al. 2004; Farr-Cox et al. 1996; Henderson 2005; Hill et al. 2005; Lever 1977, 1996; Maitland 2004; Perring 1967; Tortonese 1967; Welcomme 1988; Wheeler & Maitland 1973.

PERCIDAE (PERCHES)

ZANDER *Sander (Stizostedion) lucioperca* (PIKEPERCH[1])

Natural Distribution C and E Europe, from Sweden and Finland S to the former Yugoslavia and the Black and Caspian Seas, and E to the Ural Mountains.
Naturalized Distribution England; [Wales].

England

On several occasions in 1860–1, the newly formed Society for the Acclimatisation of Animals, Birds, Fishes, Insects and Vegetables within the United Kingdom (the Acclimatisation Society) (see Lever, 1992) considered and rejected, partly on the grounds of expense but also because of the potential threat it would pose, the introduction of the '*Lucio perca*' to British waters. Nevertheless, in the society's Annual Report for 1865–6, the council declared its intention of implementing 'their former arrangements to secure the introduction of ... this valuable fish'. It was not, however, until 1878 that the society's 'arrangements' came to fruition, as described by Frank Buckland (1881: 274), one of the founders of the society:

> Many attempts have been made to transport this fish alive to England. My friend, Mr. T.R. Sachs has taken immense pains in this matter, and the following is his report to me of successful experiments to bring this fish over alive for His Grace the Duke of Bedford.
>
> 'Mr. Dallmer, chief fishing master of Schleswig-Holstein, relates in the "*German Fishery Circular*" of April, 1878, that he had the honour of being requested by the President of the German Fishery Association to supply His Grace the [9th] Duke of Bedford with about one hundred small Zander, he having many opportunities of observing the peculiarities of these fish in Schleswig-Holstein ...
>
> 'Mr. Dallmer ... would select fish of a large sort: first, because small fish were weak; second, they would more easily fall prey to larger fish; third, larger fish would soon after their arrival in England produce a family of English Zander.
>
> 'Mr. Dallmer selected 24 two-pounders [907 g] – twelve male, twelve female [from the lake at Bothkamper], so as to procure as many marriages as possible... The next ship to London was the steamer *Capella*, and [due] to depart on Thursday night January 31, from Hamburg... arriving in London by Sunday mid-day. There they were met by servants of His Grace the Duke, who took charge of the carriers, and conveyed them to the railway station, about 3 English miles [5 km] distant ... On arrival [at Bedford] carriages were waiting to convey the fish to the estate, 4 English miles [6 km] more ... four fish, two male, two female, were placed in one sheet of water, the others in a lake of about twenty-three acres [9 ha], which was full of small fish but no pike, the gravelly bottom being eminently suited for Zander.'

Although Sachs's account states clearly that this 1878 introduction was successful, Fitter (1959: 297) claimed that it failed, and that 'The successful introduction was made in 1910.' In fact, the 1910 introduction by the Duke of Bedford was merely to reinforce the existing stock at Woburn Abbey.

In March 1909, Frank Batterson sent to G.A. Boulenger at the British Museum a fish for identification that had been taken in 'the eel trap at Kings Mill, Skefford, Beds. Seven

ZANDER

others of the same species having been caught in the trap during the last few months.' Boulenger's reply is unfortunately not extant, but Batterson subsequently wrote in acknowledgement: 'I beg to thank you most sincerely for your kind and valuable answer to my enquiry re Pikeperch.' This specimen and the other seven mentioned by Batterson were presumably Zander.

In 1947, the Woburn lakes were netted when it was found that the Zander had not prevented the establishment of a large stock of native fish. About 50 Zander, ranging in size from fingerlings to individuals of nearly 2 kg, were given to the Leighton Buzzard Angling Club, which released the majority into Firbank's Pit at Leighton Buzzard, and a smaller number into the River Ouzel, which flows into the Grand Union Canal near Tring, in Hertfordshire. Three years later, in 1950, the Leighton Buzzard anglers received a further 30 young Zander, which they released in Claydon Lake near Steeple Claydon south of Buckingham. This water subsequently became the source of Zander for at least one other neighbouring water. In 1952, Zander from Woburn were used to stock a pond near Sevenoaks in Kent.

Wheeler and Maitland (1973) documented the extension of the Zander's distribution in England from the 1960s. In the winter of 1959–60, the lakes at Woburn were again netted and 97 Zander were removed and released into the Great Ouse Relief Channel at

Stowbridge in Norfolk, 5 km downstream of Downham Market. The release was made on the instructions of the fisheries offices of the Great Ouse River Authority, but contrary to the advice by the then Ministry of Agriculture, Fisheries and Food, to provide an additional quarry for anglers. This introduction was again successful, and in 1969 the Great Ouse River Authority was able to report that 'zander are now prevalent in the Relief Channel between Denver [south of Downham Market] and King's Lynn [a distance of some 15 km] and also in the cut-off channel upstream of Denver sluice'.

In 1965, 100 or 500 Zander from Sweden (accounts differ about the exact number) were liberated in the Relief Channel at Stowbridge. In 1969, Zander may have been introduced to a private fishery in the fens, possibly Landbeach Lakes or Mepal Pit near Cambridge, where in 1965 (or 1969) it was reported that 'a few pike-perch were put in as fingerlings three seasons ago'. By the late 1960s, Zander were well established and numerous in the Great Ouse Relief Channel. In flood conditions the Relief Channel connects with the Great Ouse River system, and with the Rivers Little Ouse and Wissey and connecting waterways.

Table 20 Waters other than the Relief Channel in which Zander *Sander lucioperca* were first caught, 1967–70

Year	Water
1967	Hundred Foot Drain (near Ely)
1967	River Delph (near Downham Market)
1967	Old Bedford River (near Downham Market)
1969	River Cam (Dimmock's Cote)
1969	River Wissey (Stoke Ferry)
1969	Old Bedford River (Welney)
1969	River Nene (Milton Ferry; Peterborough)
1970	Old Bedford River (Sutton Gault)
1970	River Great Ouse (near St Ives, Huntingdonshire)
1970	River Great Ouse (Cresswells Reach, near Ely)
1970	Middle Level Drain
1970	River Cam (500 m upstream of the confluence with Swaffham Lode)
1970	River Great Ouse (Littleport)

Sources: Wheeler & Maitland 1973; Lever 1977.

In 1974, it was reported in the angling press that Zander were 'showing up in good numbers throughout the whole length of the River Delph ... the Old Bedford River ... holds fair numbers ... the Great Ouse itself holds zander ... the advance is getting dangerously close to the Nene system too'. A year later, 'Zander are now being taken in rapidly increasing numbers in the Forty Foot, Sixteen Foot and Pophams Eau. They have also appeared for the first time in the Twenty Foot and Old River Nene as they continue their spread in a westwards direction.'

In the winter of 1973–4, Zander were illegally released in a flooded gravel-pit at Moxey near Peterborough, from which less than two years later the Anglian Water Authority removed 6 adults, 55 yearlings and 730 fry.

Away from East Anglia, reports of Zander were received in the late 1960s from the River Mease near Tamworth in Staffordshire and from the lake at Stoke Poges Golf Club near Slough in Buckinghamshire. In 1973, a specimen was taken in the River Severn below Tewkesbury Weir, and two others were released in the River Lee in Hertfordshire. Wheeler and Maitland (1973: 64–5) wrote that:

> There seems little doubt that the zander will spread through at least the rivers of East Anglia. Its spread has been considerably assisted by the stocking of fishing waters, and the presence of zander in numerous accessible waters will probably lead to misguided attempts to introduce the species elsewhere for angling purposes. It is noticeable that not until it was released into a river system (the Ouse Relief Channel) did the expansion of its range become uncontrolled and apparently uncontrollable.

This prediction proved prophetical. Although by the mid-1970s the Zander had still only partially colonized the extensive system of interconnecting waters in East Anglia (it did not, for example, occur in the Broads), it was steadily increasing its range, and had been further illegally introduced by anglers to other localities in England, including those in the Rivers Severn and Trent water systems and in the lower River Avon, and at isolated sites in Berkshire, Essex and Surrey. In 1976, Zander were discovered in the West Midlands, in Coombe Abbey Lake near Coventry, and in the Oxford Canal near Rugby, both in Warwickshire. By 1978, they had spread to the Ashby Canal, and in the following year to gravel-pits at Wanlip, Leicestershire.

By the early-1980s, Zander occurred in the Coventry, Grand Union, North Oxford and Stratford Canals, as well as in the connecting Rivers Avon (1981), Severn (1980), Soar and Teme (1981). By the following decade, they were also established in the Birmingham Fazeley and Trent and Mersey Canals and in the Rivers Thames, Trent and Lee (or Lea) and probably in many other lotic and lentic waters elsewhere.

By the mid-1980s, Zander had travelled further upstream in the River Cam; invaded the Reach and Burwell Lodes; fully colonized the main channels of the Middle Level drainage system, including numerous small interconnecting drains; spawned successfully in the River Stour in Suffolk (where they are believed to have arrived in the mid-1970s), and been introduced, illegally, into Stanborough Lake near Welwyn Garden City, Hertfordshire. In 1980, large numbers of fry were removed from Abberton Reservoir, which is fed from the lower Stour.

While the spread within East Anglia has, since the last stocking of the Great Ouse Relief Channel in 1963, been mainly by natural dispersal, their appearance elsewhere has resulted from illegal releases by anglers, each of which has provided a fresh locus for the further natural expansion of the species' range, and this process is a continuing problem. As long ago as 1979, Rickards and Fickling listed no fewer than 15 counties in England where Zander were considered to be established.

Table 21 Main populations of Zander *Sander lucioperca* in England in the late 1990s

Canals	Still Waters	Rivers (Midlands)	Rivers (East Anglia)
North Oxford	Woburn Abbey	Trent	Great Ouse Relief Channel
Coventry	Coombe Abbey	Severn	Cut-Off Channel
Ashby	Wanlip Gravel-pits	Avon	Ely Ouse
Birmingham and Fazeley	Old Bury Hill	Teme	Cam
Gloucester and Sharpness		Soar	Old West
		Stour	Lark
			Little Ouse
			Wissey
			Bedford Level
			Middle Level
			Great Ouse

Source Smith et al. 1998.

More recently, Zander have been found in south-east England in the River Thames at Teddington, Molesey and perhaps Sunbury, west of London, and in three still waters.

[Wales]
Copp et al. (2003) received unconfirmed reports of small (15–20 cm) Zander in October of that year in the Welsh Dee at Farndon, on the Cheshire/Denbighshire border, and at Crook of Dee. These may have been as a result of illegal introductions or of natural dispersal from England. Neither Maitland (2004) nor Hickley in Davies et al. (2004) show them as currently present in Wales.

Prospects
Zander are clearly continuing to expand their present range both within and without their East Anglian stronghold, and they now have access to most English (and many Welsh) rivers. Further natural expansion of range now seems inevitable, and virtually unpreventable – control, though theoretically feasible, would in practice be impractical on grounds of expense. It has been suggested that the species' spread may be restricted by the canal locks system, but the increasing use of back-pumping around locks will probably help further spread. Although Zander seem able in most cases to become rapidly established in waters into which they have been introduced, their natural spread into new localities can frequently be slow.

Characteristics

Zander live in both deep and shallow still and slow-flowing waters. Although they can survive in mildly brackish water (e.g. in the Baltic Sea) they are sensitive to low levels of oxygen. They thrive best in eutrophic and turbid waters with few macrophytes. The young feed on zooplankton, while the adults prey on other fish species, which they actively pursue rather than ambush; they are also cannibals. Their relatively small mouth restricts them to taking only smaller fish. Their specially adapted eyes (in which cells beneath the retina reflect light back to increase sensitivity) enable them to hunt effectively in poor light conditions and murky waters. Spawning takes place between April and June, the eggs being zealously guarded by both parents.

Impact

Although there is evidence that as predators the diets of the alien Zander and native Northern Pike do to an extent overlap, the two species do not appear to be serious competitors. Northern Pike occupy a different habitat niche, such as clear weedy and mesotrophic waters, can consume larger prey species and, as ambushers, have a different hunting strategy.

In continental Europe, Zander have a much wider range of fish prey than they do in England, which helps to explain why, in their presence, the survival of British native species can be locally at risk. The impact on such indigenous fish communities depends on the rate of spread of the Zander population (which, as mentioned above, can be slow), and the ability of prey populations to cope with a predator with whose method of hunting they may be unfamiliar. Because even adult Zander can cope with only fairly small prey, the species can all but annihilate the juvenile cohort of some prey populations.

As long ago as the early 1970s, it was revealed that in some localities there was a relatively poor recruitment to cyprinid stocks, resulting in a lower than expected biomass. In addition, stocks of cyprinids in waters occupied by Zander had generally fallen to a lower level than that prevalent in Zander-free waters, and the degree of decline appeared to correlate with the period during which Zander had been present. The earliest open waters to be infested, the Relief Channel and the Ely Ouse, had fallen to exceptionally low biomass levels, and comparison of the composition of diminished stocks in those waters occupied by Zander with those as yet unoccupied clearly suggested a reason for degeneration. A very strong 1975 year-class for most cyprinids in East Anglia, in particular Roach, was noticeably subdued in waters holding Zander. A comparison of the combined Pike/Zander biomass in some waters with that of cyprinids clearly showed an imbalance in the predator/prey relationship, and the degree of imbalance seemed loosely related to both the duration of the presence of Zander and the amount by which the cyprinid biomass had declined. The recent revival of Smelt (which in some waters in Europe can comprise nearly 100 per cent of the Zander's diet) in such Zander-occupied rivers as the Thames, Trent, Great Ouse and Nene (a result of improved water quality) could relieve the pressure on cyprinid species and also on the Bullhead or Miller's Thumb, a UK Biodiversity Action Plan Species of Conservation Concern. However, the evidence for impacts by Zander on prey species is equivocal, and in some places where Zander occur there has been no diminution in the abundance of prey species. Where they are found in small numbers in clear waters, or where large fish such as Common Carp are regularly stocked, Zander may even benefit the larger species by reducing the number of small fish.

In canals, the impact of Zander on prey species is closely correlated with the amount of boat traffic. In canals with a large amount of boat traffic, which causes turbidity and inhibits the growth of macrophytes, forming a habitat favourable for Zander, the species develops large populations, alters the structure of fish communities and reduces the abundance of small Roach and Gudgeon. Interestingly, as Smith and Briggs (1993) point out, a reduction in the populations of Roach and Gudgeon results in an increase in the numbers of Perch, because the two former species apparently outcompete Perch in turbid waters and because Perch, which have a spiky dorsal fin, are avoided by Zander.

Populations of Zander in England appear to be dominated by a single age class for a number of years, with larger individuals preventing recruitment by cannibalism. Only when the dominant age class falls to a low level, when cannibalism is no longer effective at obstructing recruitment, does such recruitment become successful and a different age class becomes dominant.

'In parts of East Anglia,' wrote Hickley (2004: 134), summing up the relationship between Zander and their prey in Britain, 'zander populations have settled into an uneasy balance with native stocks whereas, in the canal systems of the Midlands, this species remains a significant threat to roach populations.'

The control of Zander by culling, usually by electro-fishing or chlorination and dechlorination of water, has been found to be effective in the long term only when it is of high intensity and of long duration, and when over 75 per cent of adults are removed annually. The culling, or attempted eradication, of Zander is very expensive, and has been found to be neither guaranteed of success nor cost-effective.

Until the early 1980s, the spread of Zander in England was helped by anglers' catch-and-release policy, thus reducing mortality through angling to a minimum. Thereafter, anglers were encouraged to kill all Zander caught, but because of the relatively small number of fish involved this policy had little impact. Unfortunately, whereas in continental Europe, Zander are highly regarded both as a sporting quarry and for the table, in England they are held in rather less esteem both by anglers and as a source of food.

References

Anon. 2002; Barr 1976; Cawkwell & McAngus 1976; Copp et al. 2003; Deedler & Willemsen 1964; Fickling 1985; Fickling & Lee 1985; Gibson 1994; Hickley 1986; Hickley, in Davies et al. 2004; Hickley & Chare 2004; Hickley & North 1983; Kell 1985; Klee 1981; Lever 1977, 1996, in press; Linfield 1984; Linfield & Rickards 1979; Maitland 1969, 2004; Maitland & Campbell 1992; Mansfield 1958; Perring 1967; Rickards & Fickling 1979; Sachs 1878; Smith & Briggs 1999; Smith & Eaton 1998; Smith et al. 1996, 1998; Taylor 1994; Wheeler 2001; Wheeler & Maitland 1973.

Note

1. So-called because of its resemblance to a cross between a Pike and a Perch.

FERAL DOMESTIC SPECIES

WHAT EXACTLY IS A DOMESTIC ANIMAL? Mason (1984) makes no distinction between 'domestic' and 'domesticated', defining it as one that is breeding under human control, provides a product or service useful to humans, is tame and as been selected away from the wild type.

The ultimate distinguishing characteristic of a domestic animal is, Mason continues, the existence of a range of genotypes resulting from natural selection. The early genetic changes may be by natural selection, such as adaptation to captivity or to poor fodder, and increasing tameness and ease of breeding in confinement. Artificial selection demands control of breeding, which is why this criterion is of prime importance. This human selection has evolved more productive and specialized breeds on the one hand, and more 'ornamental' or 'fancy' breeds (of e.g. pigeons, much studied by, among others, Charles Darwin) on the other. Nevertheless, some animals with a lengthy history of domestication continue to resemble the wild type (e.g. Reindeer, and Bali cattle and their ancestor the Banteng).

As C.A. Reed (in Mason 1984) points out, although taming is a necessary pathway towards domestication, a tamed animal is not necessarily domesticated. Taming may eventually merge into domestication. If the originally wild-caught individuals were selectively bred in captivity for such attributes as docility, conformation, colour and quality of pelage, meat production and size of horn, they can be regarded as domestic animals. Such selective changes, brought about by artificial selection, have a genetic base. Thus, for example, early in the history of domestication the legs of sheep became shorter and goats' horns became twisted.

A domestic animal cannot, by definition, revert to being a truly wild animal. Domesticated animals that return to the wild – either by escaping or being deliberately released from captivity – and establish breeding populations in the wild are properly termed 'feral'. It is thus incorrect to refer, as so many people do, to species such as American Mink and Sika as 'feral'.

All the animals in this section fulfil the above criteria, including the Reindeer that in their native range are 'managed' by the local people who depend on the animals for their survival. Similarly in Scotland, the Reindeer are 'managed' and are thus, strictly speaking, 'acclimatized' rather than 'naturalized'.

MAMMALS

FELIDAE (CATS)
FERAL CAT *Felis catus*

Natural Distribution The most likely ancestor of *Felis catus* is the African and Asian form of the Wildcat *F. silvestris lybica.*
Naturalized Distribution England; Scotland; Wales; Ireland.

England; Scotland; Wales; Ireland
The cat is known to have been domesticated in continental Europe since at least the time of the Bronze Age. In Britain it was first recorded, in England (Harcourt 1979) and Scotland (Smith 1994), from Iron Age sites, but was still scarce in the Roman period and did not become common until the Middle Ages.

What appears to be the earliest known written record of the domestic cat in Britain occurs in a code of laws attributed to the Welsh king Howel (Hywel) Dda, or the Good (d 950), quoted by Wade-Evans (1909):

> Whoever shall kill a cat which guards a barn of a King or shall take it stealthily, its head is to be held downwards on a clean level floor, and its tail is to be held upwards; and after that wheat is to be poured about it until the tip of the tail is hidden [and that is its worth]. Another cat is four legal pence in value [the same amount as for a sheep].

Over the years, many domestic cats have been abandoned by their owners or have deliberately chosen to lead a feral existence. In *Hampshire Days* (1903), the naturalist W.H. Hudson stated that of the estimated total of 400,000 domestic cats in London, no fewer than 80,000–100,000 (c 22.5 per cent) were living in a feral state, and Matheson (1944) suggested that there were around 6,600 feral cats in the city of Cardiff out of a total population of some 23,500 – i.e. about 28 per cent.

There have been a number of instances of deliberate releases of cats on islands in various parts of Britain and Ireland. About a dozen were freed on St Kilda in the Outer Hebrides in 1930, and although they all soon disappeared it is commonly believed that domestic and feral cats contributed to the decline of the endemic St Kilda House Mouse. In the 1890s, cats were liberated on Noss in the Shetland Islands to control the population of Common Rats, and they were also freed on a number of occasions on South Havra and on Holm of Melby, also in the Shetlands. In *Highland Days* (1963), Seton Gordon said that cats were set free on the uninhabited island of Monach (where they still survive today), off the west coast of South Uist in the Outer Hebrides, to control the population of introduced European Rabbits. In Ireland, some three-dozen domestic cats were released in 1950 on Great Saltee Island off Ballyteige Bay, Co. Wexford, again to control introduced European Rabbits.

There are few urban areas today without a stock of feral domestic cats, which are most closely associated with such developments as hospitals, hotels, clubs, warehouses, dockyards, city centres and similar locations. Little is known about the degree of their dependency on provisioning by humans, but it is likely to be considerable. Feral cats also occur

FERAL CAT

in rural areas, especially where there are crofters as in the Western Isles of Scotland. There may be over 9 million domestic cats in Britain and Ireland today, of which perhaps 20 per cent are classed as feral.

Characteristics

Feral cats, living in varying degrees of independence from humans, occur in a variety of habitats ranging from urban environments to woodlands and farmland. Those found in urban localities depend, to a greater or lesser extent, on scavenging for food provided by humans. Maritime feral cats feed primarily on colonial seabirds, while those living away from the coast eat mainly rodents and European Rabbits. Feral cats established around farms are believed to be able to prevent the recovery of Common Rat populations after poisoning, but to be unable to reduce untreated infestations. Although domestic cats, especially in residential localities, can be significant predators of garden songbirds, feral cats, apart from those that prey on colonial seabirds, do not consume much avian prey.

Impact

The principal impact of feral cats is caused by their hybridization with the critically threatened native Wildcat, where the two animals occur sympatrically in the Scottish Highlands. The offspring, whose body usually closely resembles that of the Wildcat but whose tail is normally more similar to that of a feral cat, are fertile, and the genes of the native Wildcat are thus being seriously diluted. It is believed that this hybridization in Scotland has been occurring for more than 150 years.

Since 1988, the Wildcat has been included on Schedule 5 of the Wildlife and Countryside Act 1981, and it is also on Annex IV of European Directive 92/43/EEC (the Habitats and Species Directive), both of which list species in need of special protection. Such protection is at present ineffective, due to weak implementation and the difficulty of differentiating between Wildcats and hybrids (Balharry and Daniels 1998). Recent genetic research, however, has identified a DNA marker which, combined with an analysis of coat markings, will enable researchers to distinguish between pure-bred Wildcats and hybrids. Scottish Natural Heritage will use the results of this research to establish the status of the surviving population of Wildcats in Scotland, with a view to a possible restocking programme in sparsely populated or unoccupied habitats.

References

Balharry & Daniels 1998; Fitter 1959; Harcourt 1979; Hope & Fox 1899, 1905; Kitchener 2005; Lever 1977, 1985; Long 2003; Macdonald, in Corbet & Harris 1991; Macdonald & Kitchener, in Harris & Yalden 2008; Matheson 1944; Mivart 1881; Robinson, in Mason 1984; Smith 1994; Stelfox 1965; Tegner 1976; Universities Federation for Animal Welfare 1981; Wade-Evans 1909; Zeuner 1963.

MUSTELIDAE (WEASELS & ALLIES)

FERAL FERRET *Mustela furo*

Natural Distribution The ferret is a domesticated form of the Polecat (e.g. Davison 1998), which is widely distributed in W Europe, though more patchily in the S. The two species may be conspecific.
Naturalized Distribution England; Scotland; Wales; Ireland.

England; Scotland; Wales; Ireland

According to Thomson (1951), the domesticated ferret was first mentioned in 350 BC by Aristotle, and as early as the 1st century AD Greek and Roman writers refer to the use of ferrets for bolting rabbits from their burrows in the Balearic Islands. Ferrets were introduced to England, possibly from Spain, by the Normans in the 11th or early 12th century. According to Owen (in Mason 1984), the first reference to the ferret in Britain dates from 1223. In 1272, the capture of rabbits with ferrets at a village called Waleton is mentioned by Thorold Rogers in his *History of Agriculture and Prices in England* (1866), and a decade later Richard le Forester was paid 3s. 6d. for catching rabbits and keeping ferrets for the king at Rhuddlan Castle in Flintshire.

An illustration of a ferret being inserted into a warren is shown in the *Luttrel Psalter* of around 1343 (Backhouse 1989). On Mull, where Polecats are not indigenous, both they and ferrets were kept in captivity in the mid-1930s. Both species escaped and interbred in the wild, soon becoming pests throughout the island where, as well as preying on European Rabbits, they also took a toll of domestic poultry and ground-nesting birds. In 1944, a trapper at Oskamull near Ulva Ferry killed 22 feral ferret x Polecat hybrids with long and shaggy coats, and others were caught elsewhere on the island.

FERAL FERRET

Table 22 Viable populations of feral ferrets *Mustela furo* on British offshore islands in the 1990s

Island	References
SCOTLAND	
Arran	Gibson 1970; Gibson, J.A. personal communication*
Benbecula, North and South Uist	Harman, M. personal communication*
Bute	Gibson 1970; Gibson, J.A. personal communication*
Islay (10–15 introduced in 1979 to control European Rabbits)	Ogilvie, M. personal communication*
Lewis (spread naturally from Harris)	Fergusson, J. personal communication*
Mull	Pocock 1932; Tetley 1945
Shetland (introduced (although already present) in 1986 to control European Rabbits)	Kitchener, A.C. and Birks, J.D.S. personal data
ENGLAND	
Isle of Man	Pooley, E. personal communication*
Jersey	Magris, L. personal communication*

Source Kitchener & Birks (in Harris & Yalden 2008). *To Kitchener & Birks.

On the British mainland, apparently viable populations of feral ferrets are believed to occur in North Yorkshire, Renfrewshire, Argyll and possibly Caithness, and on Speyside (Kitchener & Birks, in Harris & Yalden 2008). Many records since the 1970s, especially in central England, are likely to be misidentifications of reintroduced or naturally recolonizing Polecats. In Ireland, a single population of feral ferrets is reported in Co. Monaghan. The species has also occurred in the wild on the Isle of Anglesey in North Wales.

Many domestic ferrets escape from captivity or are abandoned, though their prospects for survival in the wild – especially in the presence of Polecats – is poor. Particularly large numbers were abandoned in the 1950s after myxomatosis had reduced the European Rabbit population by some 99 per cent, making the keeping of ferrets no longer economical, although some are used today for line-carrying in electrical and plumbing conduits.

Harris et al. (1995) estimated the British population of feral ferrets to be around 2,500.

Characteristics

Feral ferrets in Britain show a wide range of pelage colouration – those in Shetland, for example, varying from almost melanistic to true albino – but are always distinguishable from Polecats. There is little, if any, evidence that even long-established populations develop greater morphological uniformity, as has occurred in New Zealand (where they were introduced from 1867), and no evidence for a reversion to the wild type.

In Britain, feral ferrets occupy a wide range of habitats, generally in lowland localities. On islands they occur on moorland, heathland, shores and other more open areas. They feed largely on birds, European Rabbits and small rodents.

Impact

In Shetland, feral ferrets prey largely on ground-nesting seabirds. On Harris, in the Outer Hebrides, Seton Gordon in *Highland Days* (1963) stated that ferrets, released to control European Rabbits, were responsible for exterminating the Ptarmigan on the Hill of Clisham – its last haunt on the archipelago. In New Zealand, feral ferrets are suspected of being an important vector of bovine TB.

References

Backhouse 1989; Birks & Kitchener 1999; Blandford & Walton, in Corbet & Harris 1991; Davison et al. 1998; Fitter 1959; Gibson 1970; Kitchener & Birks, in Harris & Yalden 2008; Lever 1977, 1985; Long 2003; Matthews 1952; Miller 1933; Owen, in Mason 1984; Pocock 1932; Porter & Brown 1997; Tetley 1939, 1945; Thomson 1951; Walton, in Corbet & Southern 1977; Yalden 1999; Yeaman 1932.

EQUIDAE (HORSES)

FERAL HORSE *Equus caballus*

Natural Distribution All domesticated horses are ultimately derived from the Wild Horse or Tarpan, of which Przewalski's Horse of central Asia and the Polish Konik (see under Impact) are the nearest survivors.
Naturalized Distribution England; Scotland; Wales; Ireland.

The Wild Horse or Tarpan, the ancestor of the domestic horse, which was closely related to, or may have been conspecific with, Przewalski's Horse, was well established in Britain between the Devensian Glaciation and the Late Glacial period, and was recorded from at least 21 Late Glacial sites, but not in Ireland. After the ice had retreated, the Wild Horse continued to survive for 1,000 years until Britain became divided from the Continent by the formation of the English Channel. Since the Wild Horse was associated with open grasslands, it was unable to survive in the forests of the Mesolithic, which provided little if any suitable grazing. The most recent records of the Wild Horse in Britain are from the Boreal phase of the early post-glacial in the Darent Gravels in Kent (9770 BP) and Seamer Carr in Yorkshire (9330 BP). After this time, the Wild Horse dwindled in numbers until it eventually died out. Suggestions that the Wild Horse may have survived through the

DARTMOOR PONY

Mesolithic to become an ancestor of, for example, the Exmoor pony are not supported by convincing evidence.

The earliest proof of the presence of the domestic horse in Britain and Ireland comes from the Beaker Settlement at Newgrange in Co. Meath, Ireland, with a likely date of around 3975 BP. It is probable that the Newgrange remains refer to introduced animals. The remains of horses of around the same date (c 3740 BP) have also been discovered in the flint mines at Grime's Graves in Norfolk, but other such remains do not become frequent in the archaeological record until the Iron Age.

England; Scotland; Wales; Ireland

Domestic horses have thus occurred in Britain since the Neolithic period, some 4000 BP. In Britain and Ireland, there are today a number of free-ranging semi-feral populations of relatively unimproved stock, all of which are to some extent 'managed' by humans. Today, such populations are found as follows:

England New Forest, Hampshire; Dartmoor and Exmoor, Devon; around Ullswater and Haweswater, Lake District, Cumbria; Lundy Island, Bristol Channel; Northumberland.

Scotland Shetland and such Hebridean islands as Rum.

Wales Gower Peninsula, Swansea and Carneddau Hills, Gwynedd.

Ireland Around Connemara in Co. Galway in western Ireland, where they are traditionally held to have been introduced by the Celts in the Late Bronze Age.

Those on Dartmoor and Exmoor and in the New Forest probably have the most ancient lineages, with those on Exmoor (together with the Polish Konik and Przewalski's Horse) and perhaps those in Wales and the Hebrides being considered to be relatively direct descendants of the Tarpan, occurring in the archaeological record from c 100,000 BP. According to Fitter (1959), there were 'wild horses' on Dartmoor in the early 11th century, where they may have been present since the Dark Ages (c AD 500–1100), and the population in the New Forest is believed to date from the afforestation of the region by William the Conqueror in 1079.

Differences in origin and widespread subsequent attempts at 'improvement' have resulted in extensive variations both within and between populations:

- Fell ponies of the Lake District and Northumberland tend to have a dark pelage, and to measure up to 125–130 cm at the withers.
- Dartmoor ponies are bay, black or brown, and measure up to 125 cm.
- Exmoor ponies are bay or dark brown with black points, and a noticeably mealy (pale) muzzle (c 125 cm).

- Welsh ponies tend to be dark in colour and of a variable height, depending on their ancestry.
- New Forest ponies are traditionally bay with dark manes and tails, but vary a lot in both colour and height as a consequence of extensive 'mongrelization'.
- Semi-feral ponies in the Western Isles of Scotland are dun in colour with a dorsal stripe, and usually black points and a silver-grey mane and tail. They tend to be smaller than the mainland 'garron' type, with a height at the withers of up to 134 cm (Putman, in Harris and Yalden 2008).

Characteristics

All British populations of free-ranging ponies occupy essentially marginal habitats such as open moorland and rough grassland. They are preferential grazers (in the New Forest feeding especially on Purple Moor-grass), but as the availability of grass declines in hot and dry summers and in winter, their diet changes to Common Gorse, tree leaves, moss and Common Heather, sometimes supplemented by humans with hay, straw and roots.

Impact

Feral horse populations may exhibit competitive or facilitative relationships with other large ungulates. Thus on the island of Rum in the Inner Hebrides there is considerable competition for food in summer, when resources are comparatively abundant, with domestic cattle and Red Deer hinds, but less so in winter. In the New Forest, feral horses show a similar high degree of overlap in diet (and habitat) with free-ranging domestic cattle throughout the year, and some overlap in diet with both Roe and Fallow Deer – the populations of both deer may be adversely affected by competition for grazing with feral horses. Also influenced may be the structure and composition of vegetation communities, with a consequent negative impact on species diversity and the abundance of small rodents, and thus on the density and breeding performance of dependent predators such as Tawny Owls, Common Kestrels and Common Buzzards.

On the other hand, feral horses can, like feral cattle, be a useful conservation tool. In 2006, two herds of Koniks[1] – the nearest living relation of the extinct Tarpan – were imported from the Netherlands to help manage the 120-ha Stodmarsh and Ham Fen Nature Reserves close to the River Stour near Canterbury in Kent. The objective was to restore riverine water meadows by grazing, and by deliberately allowing the horses to poach the land to increase the extent of rushes, reed beds and other wetland plants, making the habitat more attractive to threatened animals such as the Black-tailed Godwit, Great Bittern, Water Vole and Eurasian Otter. Koniks have also been used as a conservation agent in parts of East Anglia and elsewhere in southern England.

References

Baker 1990; Bökönyi, in Mason 1984; Clutton-Brock, in Corbet & Harris 1991; Clutton-Brock & Burleigh 1983; Fitter 1959; Gates 1979, 1981; Gill 1994; Long 2003; Putman 1986, 1987, 1996, in Corbet & Harris 1991, in Harris & Yalden 2008; Putman & Sharma 1987; Russell 1976; Smith 2006d; Speed & Etherington 1952, 1953; Vilà 2001; Wijngaarden-Bakker 1974; Yalden 1999; Yalden & Kitchener, in Harris & Yalden 2008.

Note

1. Konik = little horse (Polish). It is also used to refer to a certain breed of feral horse originating in Poland (Krystyna Mayer, personal communication 2009).

CERVIDAE (DEER)

REINDEER *Rangifer tarandus*
(CARIBOU in Nearctic range)

Natural Distribution The tundra zone of the Palaearctic S in the taiga zone to the Altai Mountains and N Mongolia. In the Nearctic from Alaska W through N Canada and the Arctic Islands to W and E Greenland.
Naturalized Distribution Scotland.

REINDEER BULL, CAIRNGORMS, GRAMPIAN HIGHLANDS, SCOTLAND

In the Late Glacial the Reindeer was common and widespread in Britain and Ireland – in Britain it has been recorded from at least 30 archaeological sites and from 22 in Ireland. The latest record in Ireland, at Roddan's Port, Co. Down, dates from 10,250 BP, but the species survived in Britain into the Mesolithic, the latest confirmed dates including Darent Gravels, Kent (9760 BP), Anston Stones (Dead Man's) Cave, Yorkshire (9940–750 BP), Green Craig, Pentland Hills, Midlothian (9710 BP) and Creag nan Uamh, Inchnadamph, Sutherland (8300 BP). These early dates suggest, as Yalden and Kitchener (2008) point out, that climatic and vegetational changes, rather than hunting by humans, were responsible for the species' extinction in Britain and Ireland: the well-wooded Mesolithic environment would have been inimical to an animal of largely open tundra habitats. There is no evidence for a popular belief, based largely on a misinterpreted reference in a 12th-century Norse text, the *Orkneyinga Saga*, that the Reindeer survived in Scotland until medieval times.

A number of unsuccessful attempts have been made over the years to re-establish Reindeer in Britain and Ireland. In 1738, William Hamilton brought back from Sweden to Beltrim Castle in Co. Tyrone a pair of Reindeer, which he released in the Sperrin Mountains. These animals clearly bred before dying out because in 1770 three were shot 'by persons unknown'.

In 1790, the Duke of Atholl introduced 14 Reindeer to Blair Atholl in Perthshire, where they 'failed to find a permanent home in the forest' (McConnochie 1923). *The Sporting Magazine* for August 1799 records that 'Mr Brooks … the celebrated collector of foreign animals brought to this town [London] from Lapland … twelve reindeer on their road to the Duke of Norfolk's estate near Penrith [Cumbria]': what became of them is unknown. Early in the following century, Sir Thomas Liddell (later Lord Ravensworth) failed to establish the species at Ravensworth Castle and Eslington Park in Northumberland. In 1816, three Reindeer from Archangel in the former Soviet Union were released near Kirkwall in the Orkneys, and some four years later the Earl of Fife freed a number in the Mar deer forest near Braemar in Aberdeenshire; as in the case of their predecessors, both these attempts were unsuccessful.

Scotland[1]

In April 1952, eight domesticated Reindeer from Jokkmokk in Arctic Sweden, under the care of Mikel Utsi, were placed in a 120-ha fenced enclosure near Loch Morlich on Glen More (Queen's Forest), Aviemore, Inverness-shire, by the recently formed Reindeer Council of the United Kingdom. The object of the introduction was to try to establish the species as an economic source (of meat and hides), rather than as an aesthetic amenity. In 1953, a much larger area on the slopes of Airgiod Meall (Silver Mount) was enclosed, and a year later 17 animals (the balance of the 25 authorized by the Secretary of State for Scotland in 1951) were placed in the new enclosure. Subsequently further Reindeer (of both the mountain form *R. r. tarandus* and forest race *R. r. fennicus*) were imported from Sweden, Russia and southern Norway.

Until 1990, the Reindeer were restricted to the original 2,400-ha sites, but since then a second reserve herd has become established on a 1,200-ha site near Tomintoul. Although now free-ranging in these two localities, the deer are owned and managed by the Reindeer Company Limited, which aims to maintain the population of the two herds at 140–50; the late-2007 total was 130.

Characteristics

Reindeer in Scotland occur on upland heather moorland (wet heath, dry heath and blanket bog) well above the tree-line. They graze principally on heathers, other dwarf shrubs such as Bilberry and Cowberry, sedges (mainly *Scirpus* spp.), grasses and lichens (largely *Cladonia* spp.). There are strong seasonal variations in diet, with *Boletus* fungi and lichens featuring especially in autumn and winter. Throughout the year the diet is supplemented by artificial feed. The Reindeer are culled annually, other causes of mortality including road-traffic accidents and dogs.

Impact

The Scottish Reindeer have become a major tourist attraction.

References

Baker 1990; Campbell 1977; Clutton-Brock, in Corbet & Harris 1991; Clutton-Brock & MacGregor 1988; Dansie et al., in Harris & Yalden 2008; Fitter 1959; Lever 1985; Long 2003; MacConnochie 1923; Mitchell 1941; Ridley 1981; Skjenneberg, in Mason 1984; Yalden 1999; Yalden & Kitchener, in Harris & Yalden 2008.

Note

1. Strictly speaking, the Reindeer in Scotland represent a reintroduction rather than an introduction, and because they are managed by humans are acclimatized rather than naturalized. Since they have not occurred naturally in Scotland for thousands of years, they are here included as an introduction.

BOVIDAE (CATTLE, ANTELOPES, SHEEP & GOATS)

FERAL CATTLE *Bos taurus*

Natural Distribution Domestic cattle are derived from the Aurochs that lived in Europe and W Asia until dying out in the wild in the 17th century, due to loss of habitat and hunting.
Naturalized Distribution England; Scotland; Wales.

The now-extinct ancestor of domestic cattle, the Aurochs, was widespread and common in post-glacial Britain (but not Ireland), with several hundred records of uncertain date. Cattle were first domesticated in the Middle East, and were introduced to Britain by Neolithic farmers around 5500 BP. Troy et al. (2001) offered convincing evidence that native Aurochsen were not involved in the development of White Park Cattle or any of the other current domestic breeds. Aurochsen remains in Britain date from the early Mesolithic, with a few accurately dated records in the Bronze Age, most recently from Galloway (3315 BP) and Charterhouse Warren Farm, Somerset (3245 BP).

England

The best-known and most important herd of feral cattle in Britain today is that at Chillingham Castle in Northumberland. In May 2007, the author paid a visit to this herd as a guest of Sir Humphrey Wakefield, Bt, the owner of Chillingham Castle, and the Chillingham Wild Cattle Association Ltd. The following account is based on notes made during that visit and subsequent correspondence.

Although Chillingham was imparked as long ago as the 13th century, almost certainly with the white breed of cattle that then roamed the forests of southern Scotland and northern England, the earliest written records of Chillingham Cattle date only from the late 17th century. Chillingham Cattle are derived from other White Park Cattle, but by

CHILLINGHAM BULL, CHILLINGHAM PARK, NORTHUMBERLAND

the mid-20th century had diverged from the latter sufficiently to give them the status of a separate breed. Until the late 17th century, new breeding stock was probably introduced from time to time to the Chillingham herd, and these cattle were considered to be a part of the White Park breed well into the last century.

In the second half of the 20th century, the cattle at Chillingham were a 'managed' herd, being fed hay, minerals and supplements in winter, and the land was dressed with lime. Although originally large animals, nowadays an adult bull weighs only around 300–425 kg and a cow 280 kg, and the latter stands only about 110 cm at the withers. Bulls live to a maximum of nine years and cows to almost twice that age, though the average is much less. Chillingham Cattle breed throughout the year. The herd is ruled by a so-called 'king bull' who alone has breeding rights during his reign of perhaps three years. Since the 'king bull' is not succeeded by one of his sons, heterozygosity is maintained.

The Chillingham herd breeds true for colour (white with rufous ears, some rufous markings on the face, with a small and variable amount of black spotting on the shoulders), and although much smaller in size the animals bear a remarkable resemblance to the ancient wild Aurochs – especially in the head. All individuals are horned.

Chillingham Cattle are thus derived from other breeds of White Park Cattle, but by the middle of the 20th century had diverged sufficiently to effectively give them the status of a separate breed. They are unique in their remarkable uniformity as a result of close inbreeding. There is no negative inbreeding problem because the cattle, being essentially wild by nature rather than 'domestic', have a natural selectivity against it.

Today, the Chillingham Cattle are not managed at all, other than by the provision (since at least 1721) of hay in winter and magnesian limestone. Hay was necessary because until recently the herd coexisted with a flock of around 300 domestic sheep, which not only competed with the cattle for winter grazing, but also put them at risk from liver fluke. In 2007–8, the Chillingham Wild Cattle Association acquired the sheep-grazing lease, which had been in existence for some 30 years. The cattle now have the park to themselves, apart from a few Fallow Deer (which, although preferential grazers, also consume a considerable amount of browse, whereas the cattle are near-obligate grazers), and have quickly responded positively to these improved conditions by increasing from a total of 66 in May 2007 to 80 in June 2008 – the highest number since 1838 (Peter Steel, personal communication 2008).

The Chillingham Cattle provide a potentially important gene pool of a genetically 'strong' breed which could, should the need ever arise, be invaluable in the prevention of the possible degeneration of domestic stock.

A well-established and free-ranging but 'managed' population of semi-feral cattle is established in the New Forest in Hampshire. Another semi-feral herd, of primitive-type cattle, occurs in a medieval hunting-park at Chartley in Derbyshire.

Scotland

In 1970, two yearling heifers and a yearling bull from Chillingham Castle were transferred, as an insurance against the Northumberland herd having to be destroyed because of foot-and-mouth and other diseases (to which they may be highly vulnerable), to form a reserve herd on a small and partly wooded permanent pasture on Crown land near Elgin on the Moray Firth. Subsequently a further 10 females and 3 males were transferred from Chillingham to Elgin, where by 2005 a total of 50 calves had been born. In June 2008, this reserve herd numbered a total of 27 animals. The cattle in the reserve herd are much 'wilder' than those at Chillingham, which are more accustomed to visitors (Peter Steel, personal communication 2008).

Free-ranging feral cattle have also been established since 1978 on the 113-ha Swona Island, one of the most southerly of the Orkney Islands. These animals, which are completely unmanaged, are derived from Shorthorn/Aberdeen Angus commercial beef cattle crosses. Swona is very exposed and devoid of trees and shrubs; shelter is provided by the lee of numerous stone buildings (secured against the entry of livestock) and walls. On the seashore, the cattle feed mainly on Serrated (Toothed) Wrack and brown seaweeds *Laminaria* spp. Most animals are black, the remainder being brown. These feral cattle on Swona are of particular interest because, as Hall and Moore (1986) point out, they are one of the very few truly feral herds of cattle anywhere in the world. Semi-feral cattle also occur on the island of Rum.

A semi-feral herd of primitive-type cattle occurs in a medieval hunting park at Cadzow in Lanarkshire.

Wales

Nineteenth-century herds of semi-feral cattle are found at Vaynol (Glan Faenol) in north Wales and at Dynevor (Dinefwr) in south Wales. Because the Vaynol herd, which was established in 1872, had become increasingly remote from the mainstream of White Park Cattle breeding, it was in 1989 accorded the status of a distinct breed.

Characteristics

Feral cattle in England and Scotland occupy a wide range of differing habitats. In Northumberland, Chillingham Park is a medieval wooded pasture of 135 ha, with the remains of a designed landscape of the early 19th century featuring Common Rhododendrons and other ornamental shrubs. In the New Forest in Hampshire, the semi-feral cattle are confined to open unenclosed forested areas, with a mosaic of ancient deciduous woodland, heath and mires and acid grasslands, alluvial riverine grasslands and artificially improved pasture (Pratt 1986; Putman 1987). Swona consists of maritime heathland with overgrown pasture and arable land. On Rum, feral cattle range over a mosaic of wet heath, *Calluna* heath, herb-rich heath, *Juncus* marshes and *Agrostis/Festuca* grassland (Gordon 1989).

In the New Forest, feral cattle primarily graze on *Agrostis curtisii*, *A. capillaris*, *A. canina* and *Festuca rubra*, with lesser amounts of *Calluna* and *Erica* heathers. In winter they are fed supplements of hay and straw. On Rum, feral cattle are also preferential grazers on *Agrostis/Festuca* in summer, switching in winter to coarser plants such as *Juncus*, *Molinia*, *Scirpus* and heather (Gordon 1989).

Present-day White Park Cattle are a mixture of the semi-feral stocks at Chartley, Cadzow, Vaynol, Dynevor and elsewhere, with lesser contributions from other domestic breeds, but are themselves only semi-feral, as is the British White breed – superficially similar to the White Park but polled (genetically hornless) – which is descended from an 18th-century herd at Middleton Park in Lancashire.

Impact

In common with all ungulates, feral cattle have had a significant impact on the natural environment. As preferential grazers and bulk feeders, their principal effect has been on grasslands, and they are consequently frequently used as a conservation tool.

On Rum, feral cattle improve the availability of forage for Red Deer (and perhaps feral sheep) in spring because those species preferentially graze on swards grazed by cattle during the previous winter (Gordon 1988). On the other hand, changes in patterns of resource usage and the division of resources among Red Deer, feral cattle, feral horses and feral goats implies potential competition when the food supply is limited (Gordon and Illius 1989). In summer, when forage is abundant, a high degree of overlap in resource use occurs between all four species; in winter, when fodder is in short supply, greater partitioning of diets occurs.

In the New Forest, Putman (1986) observed a high degree of overlap throughout the year in the diets of cattle and feral horses and of cattle and Fallow Deer. Grazing by feral cattle and horses appears to have been at least partly responsible for the decline in the populations of both Fallow and Roe Deer and, as in upland regions, to impact on species diversity and population densities of mice, voles and their predators.

References

Alderson 1997; Clutton-Brock 1981; Epstein & Mason, in Mason 1984; Fitter 1959; Gordon 1988, 1989; Gordon & Illius 1989; Hall 1985, 1989, in Corbet & Harris 1991; Hall & Clutton-Brock 1988; Hall & Moore 1986; Hall & Putman, in Harris & Yalden 2008; Hall et al. 2005; Harris & Kitchener, in Harris & Yalden 2008; Ingham 2003; Lever 1985; Long 2003; Porter 1991; Pratt 1986; Putman 1987; Storer 1877; Troy 2001.

BOVIDAE (CATTLE, ANTELOPES, SHEEP & GOATS)

FERAL GOAT *Capra hircus*

(WILD GOAT)

Natural Distribution *Capra hircus* is descended, solely or principally, from the Wild Goat or Bezoar *C. aegagrus*, which is widely distributed from the E Mediterranean to C Asia.

Naturalized Distribution England; Scotland; Wales; Ireland.

FERAL GOAT BILLY ON SHORE OF LOCH LINNHE, WESTERN SCOTLAND

The Wild Goat was first domesticated some 11,000 years ago in south-west Asia, possibly in western Iran or south-eastern Turkey (Davis 1987). Domestic goats were first introduced to Britain in the Neolithic period – the earliest fossil record being one at Windmill Hill in Wiltshire, which is carbon-dated at 4530 BP (Jope and Grigson 1965).

The majority of feral goat populations in Britain and Ireland became established well over a century ago, and exhibit phenotypic characteristics (small size, both sexes horned, prick ears, dark forelegs' stripe, very variable pelage pattern and colour, with a few all white, and absence of gular tassels or toggles) that accord with those of the so-called 'old native' domesticated breeds. These disappeared in the early 20th century as 'improved' breeds were gradually introduced from abroad and the interest in sheep farming increased. Feral populations thus tend more closely to resemble their domestic forebears of the medieval period than their wild ancestors.

Whitehead (1972) cites a number of examples of goat husbandry in Britain within historic times. In 1229, when he was at Stamford, Henry III received a number of petitioners who claimed that Hugh de Neville, Keeper of the Forest of Rockingham, had refused them permission to graze their goats in Clifford Forest. In 1323–4, no fewer than 56 people were tried by the justices of Epping in Essex for illegally grazing goats within the forest. The inhabitants of Broughton in Amounderness, Lancashire, claimed for themselves at the Forest Eyre[1] of 1334–6 the ancient right of 'common pasturage' for all their livestock except goats in Fulwood Forest. In about 1460, Syr Dafydd Trefor of the Isle of Anglesey composed a poem requesting that a number of goats should be sent to him from Snowdonia, while one of his neighbours, Gruffydd ab Tudor of Howell, in a second poem, objected to the proposed introduction on the grounds that goats would cause damage to Anglesey's agriculture and forestry. At a swainmote[2] held in 1479 in the Forest of Wyresdale, eight goatherds were arraigned for allowing their charges – some of which were owned by the Prioress of Seton, who was fined 4d. for the offence – to enter the forest. In a survey of 1615, it was reported that 'sheepe and goates, most pernitious cattle, intolerable in a forest, make a far greater show than his Majestie's Game' in Kingswood Forest in Gloucestershire.

As evidence that by the late 16th century feral goats were already established in many localities, John Manwood, a Justice of the Peace in the New Forest, wrote in *A Brefe Collection of the Lawes of the Forest* (1592) that 'there be some wilde beastes ... that so long as they are remaining within the bounds of the Forest, the hunting of them is punishable by the lawes of the Forest, such as are wilde Goats, Hares and Connies'.[3]

As one of the earliest of domesticated 'utility' animals, the uses made of goats since very early times have been numerous. Their skin has been used in the manufacture of 'morocco' leather and parchment. In his *Tours in Wales* (1772), Thomas Pennant wrote that goatskin is:

> well adapted for the glove manufactory, especially that of the kid; abroad it is dressed and made into stockings, bed-ticks, bolsters, bed-hangings, sheets and even shirts. In the army it covers the horseman's arms and carries the foot-soldiers provisions. It takes a dye better than any other skin, it was formerly much used for the hangings in the houses of people of fortune being susceptible of the richest colours; and when flowered and ornamented with gold and silver, became an elegant and superb furniture.

Pennant also records that on the hills of Caernarvonshire (Gwynedd), goats were hunted in autumn to provide tallow for candles, and that by grazing on cliff faces goats removed succulent herbage that might cause sheep or cattle to become crag-fast or to fall.

In Ireland, according to John Rutty in *An Essay Towards a Natural History of the County of Dublin* (1772), kidskin was popular for the making of fans and firescreens.

In the 18th century, goat hair was much in demand for making wigs, and it has also been extensively used in rope making. Domesticated Angora and Cashmere Goats have been kept in Britain for their wool – known respectively as mohair and cashmere. Goat horn has for long been used for numerous purposes.

On the consumption of goat meat in Wales, Pennant (ibid.) recorded that:

> the haunches of the goat are frequently salted and dried, and supply all the uses of bacon: this by the natives is called *Coch y wden* or hung venison … the meat of a splayed goat of six or seven years old (which is called *Hyfr*) is reckoned the best; being generally very sweet and fat. This makes an excellent pasty; goes under the name of rock venison, and is little inferior to that of the deer.

In Ireland, Rutty (ibid.) recorded that:

> the flesh of the male goat, castrated and fed, makes a good venison, and lately kids are reared about the mountainous part of the south of Dublin for the delicacy of the flesh preferable to that of the lamb: for this purpose they are taken into the house presently after they are dropt, *viz*: before they have tasted their mother's milk, and fed with cow's milk spouted into their mouths.

The milk of goats has for long been popular, both medicinally and as a source of nourishment; Cheddar and Somerset cheeses, among others, were originally made from the milk of sheep and goats.

With the decline of lactation and as they grazed further afield when the herbage coarsened in late summer, domestic goats became lost and were allowed to become feral. In the Scottish Highlands, many goats were turned loose when in the first half of the 19th century the crofters were evicted from their holdings. In *A Tour in Scotland & Voyage to the Hebrides* (1771), Pennant recorded that the goat was 'the most local of our domestic animals, confining itself to mountainous parts of these islands'.

England

Whitehead (1972) traced the history of England's most celebrated herd of semi-feral goats (analogous to the Chillingham Park Cattle, see pages 300–3) at Bagot's Park, Blithfield, Hall, Uttoxeter, Staffordshire, ancestral home for over 600 years of the Bagot family, which for the same period has had as its crest a goat's head with two goats as supporters. The colouring of the Bagot Goat is unique among British goats, the head, neck and shoulders being black and the rest of the body white; both sexes are long-haired and horned. The original home of Bagot Goats was in the Rhône Valley of Valais in Switzerland. How they first came to Bagot is uncertain; theories include their introduction from Normandy in the early 12th century; as a present to John Bagot by Edward II in the early 14th century, or Sir John Bagot may have imported them, possibly as a gift from Richard II, around 1387. In 1957, when the Forestry Commission assumed responsibility for the woodland at Bagot, the goats were confined to a 44-ha enclosure. After Lord Bagot's death in 1961, all but a dozen were acquired by Robin Bagot of Levens Park, Kendal, Cumbria.

Feral goats first occurred on Lundy Island in the Bristol Channel in the 18th century, but by the outbreak of the First World War in 1914 they had died out. In 1926, more were reintroduced to Lundy by Martin Harman, where they still survive.

According to Mills (2005), the presence of 75 (presumably feral) goats is referred to on the Manor of Lyntonia in north Devon in the *Domesday Book* of 1086. In the late 19th century, when they are mentioned in R.D. Blackmore's novel *Lorna Doone* (Bullock 1996), they were removed and replaced by white goats, reputedly from the royal herd at

Sandringham in Norfolk. These goats died out during the hard winters of the early 1960s, and in 1976 were in turn replaced with three feral goats from the Cheviot Hills in Northumberland. Within about 30 years, the population had increased to around 150, of which 100 were culled in 1997. The present population of some 120 lives in a 120-ha site of Special Scientific Interest (SSSI) in the Valley of the Rocks at Lynton. During the autumn rut, damage to gardens, trees and fences has been caused by wandering billies seeking nannies in oestrus. A cull, proposed by the local town council and by some local residents, was opposed by others and by Natural England on the grounds that the goats are an important component of the valley's ecosystem, and that by controlling pervasive Bracken they contribute to the survival of rare plants on a designated SSSI.

There have for long been populations of feral goats on the Cheviot Hills in Northumberland. Those on the Isle of Man may be descended from some imported as a source of food and as pets by lighthouse keepers in 1818–75. Bullock (1995) mentions as among the prime localities to see feral goats in southern England the south side of Brean Down in Somerset, Lundy Island and the Valley of the Rocks, and in northern England Yeavering Bell and Newton Tors in the College Valley of Northumberland.

In 1992, to combat an invasion of Holm Oak on the butterfly-rich Ventnor Downs on the south-east of the Isle of Wight, nine feral goats were introduced to 40 ha of Bonchurch Down. The unpalatability of Holm Oak to other species, and its continual invasion of established fruiting trees, makes it likely that the goats will remain on the Down indefinitely (Tutton 1994).

Scotland

The feral goats at Inversnaid on the eastern bank of Loch Lomond reputedly date from the time of Robert the Bruce (1274–1329), who decreed that they should never be molested and provided a sanctuary for them at Pollochthraw. According to legend, in 1306 the king was hiding in a cave when a tribe of feral goats lay down outside the entrance; his enemies, thinking that the animals would not go so near a man, omitted to search the cave. Another fable claims that the goats provided the king with sustenance during his time as an outlaw. Tradition also records that feral goats were once established on Inchlonaig in Loch Lomond, where they destroyed all the trees.

In the forest of Blair Atholl, feral goats featured at the tinchel[4] organized for Mary Queen of Scots during her visit in 1564, and in 1590 'Mad' Colin Campbell and his men encountered a herd of feral goats on Stuchd an Lochain in Glenlyon.

A tribe of large white goats in the forest of Mamlorn in Perthshire is mentioned in a report on the rebellion of Hew Murray in 1661–79, when it was said that 'he and his men would never be subdued while they could get a goat on Creag Mhor'.

By the 18th century, feral goats were becoming widely distributed on the Scottish mainland. A report of around 1745 tells of goats on Eilean nan Gobhar (Isle of Goats) on Loch Ailort in Inverness-shire, and others were said to be established at Benderloch, Easdale, Eorsa, Aonach Mor (or Innean Mor) in Morven, Argyllshire, and Aird-Ghobhar (Ardgour), Stob Ghabbhar (Hill of Goats) and Blackmount. A petition of 1763 by crofters of Dalclathick asked for a reduction in their rent because they were no longer allowed to graze domestic goats in the forest.

In the 19th century, reports of feral goats in Scotland grew apace. In 1819, Robert Southey, the Poet Laureate, saw feral goats around Loch Achray and on the slopes of Ben Venue, Callander, Stirlingshire and subsequently wrote of them that 'the extirpation of wild beasts from this island is one of the best proofs of our advancing civilization: but in losing these wild animals, from which no danger could arise, the country loses one of its great charms'. A story of about 1829 relates that 'the only novelty from the hill this season is the murder of half-a-dozen goats by an English party in Atholl, who had mistaken their prey for deer'.

Later in the 19th century, populations of feral goats were reported in the Morar hills of Inverness-shire; at Wester Ross, Strathpeffer and Rothiemurchus; on the Perthshire hills from Glen Lyon to Rohallion; at Altyre and Douchfour and in Glen Moriston and Glen Urquhart in Inverness-shire (1835); at Gruinard and Fisherfield and on the Black Isle cliffs in Ross and Cromarty (1837); near the Bridge of Dulsie on the River Findhorn, where in *Short Sketches of Wild Sports and Natural History of the Highlands* (1846) the naturalist Charles St John described them as being 'long-horned, half-wild, and with shaggy hair and long, venerable beards'; in the Cairngorms on the slopes of Lochnagar; in Glen Callater, Braemar, Aberdeenshire; at Abriachan, Balmacaan, Schiehallion and Bennachie in Perthshire; and at Loch Lairig Eala, Glenogle, Glen Falloch and Ledard near Loch Ard.

In the 20th century, the already flourishing population of feral goats in Scotland was augmented by a number of releases, including some liberated on the northern side of Strath Beag near Dundonnell in Wester Ross in 1911, and others freed on An Tallach, also near Dundonnell, in 1927. Some of the latter subsequently spread to Little Loch Broom, Kildonan and Rhirevoch on Beinn Ghobhlach.

In the Lowlands, a tribe of feral goats near St Mary's Loch in Selkirkshire was mentioned by Sir Walter Scott (1771–1832), and other populations are known to have existed at Saddleyoke, Brandlaw, Megget Water, Moffat Water, Ewes near Langholm, Dumfries and Galloway, Kells, New Galloway, Glenopp, Cairnharrow and on the Cheviot Hills south of Jedburgh.

Feral goats are also well known in the Hebrides, where in *A Description of the Western Islands of Scotland* (1703) Martin Martin says that there were some before 1697 on the Isle of Lewis. Harvie-Brown and Buckley (1892) record feral goats on Canna off Rum, where they had been introduced before 1793; on Ardinaddy (before 1812); on Tiree and Coll; near Uamh nan Gabhar (Cave of Goats) on Colonsay; and 'on another small island near Oronsay'.

Thomas Pennant (ibid. 1772) suggested that in Skye 'goats might turn to good advantage if introduced to the few wooded parts of the island', and 'Skie' is mentioned by Martin as 'producing' goats among the 'cattell' at the biannual fair. Other Hebridean islands on which feral goats are known to have occurred include Scalpay; Sleat; Priest; Horse; Rum (on Sgurr nan Goibhrean – 'Peak of Goats'); Eigg; Little Colonsay; Staffa (1802); Mull (at Lochbuie and Pennyghael; Ulva, west of Mull; Jura (where Pennant saw c 80 in the 1770s); Islay (Kinnabus near Mull-of-Oa, 'from time immemorial'); Smaull near Kilchoman (translocated from Kinnabus in 1787); and Gortantaoid on Loch Gruinart (introduced from Mull and Jura in the early 20th century).

In the Clyde area, feral goats became established on the Mull of Kintyre, Davaar, Arran (where they were seen near Kilpatrick by Pennant in the early 1770s), Holy Island (east of Arran) and Bute (Garroch Head). On Ailsa Craig, there were some magnificent white goats (reputedly the descendants of survivors from a wreck of the Spanish Armada in 1588) until 1925. The *Old Statistical Account of Scotland* (1791) states that Ailsa was uninhabited 'save for the wild goats', and in about 1836 the tenant of the island 'went there for a few days' shooting of the wild goats which abound'.

Today, feral goats occur in many small and discrete populations in mountainous areas of, mainly, western Scotland from the Highland region southwards to Dumfries and Galloway, including the southern half of the Mull of Kintyre and such islands as Bute, Cara, Colonsay, Holy Island, Islay, Jura Mull and Rum. Bullock (1995) mentions as among the best places in Scotland to see feral goats the Grey Mare's Tail in the Moffat Hills of Dumfries and Galloway; the south of the Mull of Kintyre; between Glen Harris and Glen Guirdil on the south-western side of Rum; between the Rowardennan Hotel and Inversnaid Lodge on the east side of Loch Lomond; the Highlands on the north side of Loch Morar in the valley of the Findhorn; and Craig Dubh near Newtonmore. Bullock (in Harris and Yalden 2008) estimated the sub-populations in southern Scotland and the Scottish/English border to number at least 2,265; in the Clyde area, 355; on Rum, 200; on Islay, Jura, Mull and on the Scottish west coast mainland, >400; and in the central and northern Highlands, >500 – a total of >3,720. This compares with a total of at least 3,150 given by Bullock (1995). Heavy culling is now reducing the size of the Scottish population, although the sub-population in the southern Highlands is from time to time augmented by released or escaped domestic stock. During the autumn rut, feral males occasionally appear in Scottish villages and farms, having probably been attracted by a domestic female in oestrus.

Wales

Feral goats are first mentioned in Wales in the late 1760s/early 1770s, when Pennant (ibid: 1772) refers to their presence in considerable numbers on the Rhinogs, north of Dolgellau; 30 years later, Bachwy in Radnorshire (Powys) was mentioned as one of their few strongholds in south Wales. George Barrow in *Wild Wales* (1854) described a herd that he saw between Festiniog and Beddelgert in Snowdonia: 'they were … white and black, with long silky hair and long upright horns; they were of large size and very different from the common race'.

Some of the feral goats in Wales today may be descendants of those that escaped from Irish herds which in the 19th century were driven through that country en route to sales in England. One of the last of such drives took place in 1891, when three men, three boys and five dogs drove a herd of 300 goats from Cardigan to Kent.

Merionethshire and Snowdonia (Gwynedd) are today the main strongholds of feral goats in Wales, where they occur mainly on Tryfan, Glyder Fawr, Moelwyn and Rhinog Fawr. Here, Bullock (1995) mentions as among the best places to see goats the vicinity of the Roman steps at Cwm Bychan and the Llanberis Pass. Goats on the Great Orme off the Carnarvonshire (Conwy) coast have never been truly feral, but are the descendants of Cashmeres imported from Windsor Park in about 1900 to provide regimental mascots for the Royal Welch Fusiliers.

Drystone wall enclosures for kids (*cwt myn*) in Snowdonia, designed to attract and trap the nursing nannies so that half their milk could be used for human consumption, indicate a lengthy association between feral (rather than domestic) goats and local people in Wales (Caffell 1995).

Ireland

Exactly when feral goats became established in Ireland does not seem to have been recorded, but Whitehead (1972) listed them in Co. Clare (the Burren Hills); Co. Donegal (Glenveagh); Co. Dublin (Howth); Co. Galway (various localities); Co. Kerry (various localities); Co. Mayo (Achill Island); Co. Sligo (in the hills); Co. Tipperary (Slieve-na-Mon); Co. Waterford (Comeragh Hills); Co. Wicklow (Wicklow Mountains); and in Northern Ireland in Co. Antrim (various localities); Co. Armagh (various localities); and Co. Fermanagh (islets in Lough Erne and around Lough Navar).

Today, feral goats in Ireland occur in small and discrete sub-populations mainly on the west coast from Co. Mayo south to the Burren in Co. Clare and in Co. Kerry, and on the east coast in Co. Antrim, Co. Wicklow and Co. Waterford, as well as on such offshore islands as Achill, Co. Mayo, Rathlin, Co. Antrim, and Great Blasket and the Skelligs, Co. Kerry. Bullock (1995) cites as good places to view feral goats the boulder slopes and woodland beneath the cliffs in Co. Antrim, Bray Head, Co. Wicklow, and the Burren in Co. Clare.

Estimates for the sub-populations in Ireland are unavailable, but apart from the Burren, where there are over 2,000 that are from time to time augmented by new released or escaped stock, are probably <300 (Bullock, in Harris and Yalden 2008).

Current Status

In Britain and Ireland, the current total population of feral goats is estimated at 5,000–10,000; this compares to the estimate of c 5,275 by Bullock (1995) and >3,565 for Britain alone given by Harris et al. (1995).

Characteristics

The colour of feral goats in Britain and Ireland varies from white to black, the majority of individuals being grey, fawn or greyish-brown with paler patches. Black animals frequently have a fringe of long, reddish hair running along the back and on the hindquarters. Many Welsh individuals have dark forequarters and whitish hindquarters. Goats in south-western Scotland are generally paler and greyer than those from further east. Both sexes bear horns, those of the males growing backwards and then outwards in a smooth curve. Wide-spreading 'dorcas' horns are prevalent in Wales, east of the Scottish Highlands and in some Highland and Irish populations. In south-western Scotland, closer-set, backwards-sweeping 'scimitar' horns are more common (Bullock 1995).

Feral goats in Britain and Ireland are mainly confined to dry and well-drained mountainous, hilly or maritime areas with cliffs. Steep ground, frequently above the 300-m contour, is used for refuge, shelter (in caves, crags and gullies) and for foraging in associated dwarf shrub communities. In severe weather, feral goats move to lower ground. Many populations (e.g. in the Wicklow Mountains, and western Scotland and north Wales) occur in native and exotic coniferous woodlands, and (e.g. in the Burren) in scrub. Where they live sympatrically with domestic mountain sheep, such as Welsh Mountain or Scottish Blackface, feral goats are often found on steeper and rockier terrain. Adult mortality is due mainly to starvation or culling by man. Kids can die from hypothermia associated with starvation and fall prey to Golden Eagles, Red Foxes, Wildcats and probably Common Ravens.

Impact and Foraging

Bullock (1995) and in Harris and Yalden (2008) has summarized the foraging behaviour of feral goats.

As selective but versatile feeders that browse or graze depending on availability, feral goats browse and bark strip (thus inhibiting tree regeneration) more than other livestock, which brings them into conflict both with managers of commercial coniferous plantations and foresters in semi-natural ancient woodlands such as the Sessile Oak woods of Wales, the Lake District and western Scotland. Bark stripping usually occurs to a maximum height of 2 m, but goats' tree-climbing ability enables them to inflict damage much higher up.

In upland acidic grasslands, heath and woodlands, goats exhibit a similar seasonal diet pattern to that of domestic sheep, but with more dwarf shrubs such as Bog Myrtle, Common Gorse, and especially Common Heather, bark and other woody items in winter, more rushes in spring and a surprisingly large amount of Common Bracken in late summer and early autumn. In Sessile Oak woodlands, small-girth Common Ash, Rowan, oaks, European Holly and willows are preferentially stripped, followed by Common Hazel and Hawthorn. Birches and Common Alder are not favoured. The bark of conifers, especially pines and spruces, is stripped by goats. On Rum and Arran, and doubtless on other islands also, seaweed is extensively eaten in winter, and as goats are poor swimmers, cases of drowning occur. Where they live together, goats are accused by farmers of competing for forage and shelter with sheep and of damaging crops and retaining walls.

On the credit side, the propensity of goats for bark stripping and browsing can make them a useful tool in the control of scrub on grasslands of conservation importance. In calcareous grasslands invaded by trees and scrub, goats will browse in summer on Common Ash, Common Hazel, Blackthorn, Traveller's Joy, Ivy and Yew. In the Cheddar Gorge, shrubs and woody climbers comprise nearly 80 per cent of goats' diet. On the Ventnor Downs on the Isle of Wight, feral goats browse in winter and spring on Holm Oak (for at least 40 per cent of their diet), Common Gorse and Bramble, and bark strip Blackthorn, Common Ash and Sycamore. When at moderate to high densities and in the absence of scrub, goats graze extensively in calcareous grasslands in spring on Tor Grass and in summer on False Brome. On hill pastures, they can control such invasive weeds as Soft Rush and Mat Grass.

As one of the few large mammals in the British and Irish uplands and on maritime cliffs, feral goats are in some places a tourist attraction. They produce high-quality cashmere wool, the economic potential of which is being explored. Goat meat (of both adults and kids) finds a ready market among Afro-Caribbean and Asian communities, and roast kid is eaten at Easter in the Burren. In some localities, the stalking of males for their trophy heads is a minor source of income. Feral goat carcasses are an important source of carrion for Common Ravens and the few surviving Golden Eagles in parts of southern Scotland. They readily readapt to captivity, becoming docile and tractable, providing they have shelter and an adequate variety of forage, although even then the billies can be difficult to contain (personal observation).

Although some feral goat populations in Britain are now being properly and successfully 'managed', others unfortunately are not and are being controlled by random culling. Feral goats in Britain and Ireland (and the 'wild' goats on Crete in the

Mediterranean that are descended from feral goats) share ancestry from early Neolithic animals, and may well be of considerable genetic and historical importance. Feral populations in Britain and Ireland share the phenotypic characters of old breeds now extinct in domestication, and appear to be unique in Europe. A coordinated management plan in Britain and Ireland should be a priority.

References

Buchanan Smith 1927, 1932; Bullock, in Corbet & Harris 1991, 1995, in Harris & Yalden 2008; Bullock & Kinnear 1988; Caffell 1995; Caseby 1936; Crook 1969; Davis 1987; Greig 1969, in Corbet & Southern 1977; Harris 1962; Jope & Grigson 1965; Lever 1985, 1994; Long 2003; Mason, in Mason 1984; Matheson 1933; Matthews 1952; McDougall 1975; Mills 2005; Milner & Goodier 1968; Niall 1952; Richmond 1955; Schwarz 1935; Simmons & Tooley 1981; Tegner 1952, 1965; Tutton 1994; Watkins-Pitchford 1963; Watt & Darling 1937; Whitehead 1945, 1952, 1957, 1972; Yalden 1999.

Notes

1. Itinerant court.
2. Forest assembly, held triennially, in accordance with a Forest Charter of 1217, originally to enable verderers to supervise the clearance from the forest of pigs, cattle, sheep and goats.
3. European Rabbits.
4. Game-drive.

BOVIDAE (CATTLE, ANTELOPES, SHEEP & GOATS)

FERAL SHEEP *Ovis aries*

(SOAY SHEEP; BORERAY SHEEP; NORTH RONALDSHAY SHEEP)

Natural Distribution Domestic sheep *O. aries* are descended from the Urial or Asiatic Mouflon of SW to C Asia.
Naturalized Distribution Scotland; England; Wales.

Sheep were first domesticated in south-eastern Turkey, northern Syria and Iraq in the late Mesolithic (Middle Stone Age) around 8500 BC. Domestic sheep were first introduced to Britain by Neolithic people before 4000 BC. The exact date is uncertain because of the difficulty in differentiating between sheep and goat bones in archaeological sites – the earliest claimed date for sheep being in a long barrow of about 5635 BP on the Lambourn Downs in Berkshire (Yalden 1999).

Scotland

Two forms of feral sheep, the Soay and the Boreray, occur in Britain (plus one semi-feral form), each of which is named after islands in the St Kilda group in the Outer Hebrides.[1]

Soay Sheep bones are very similar to those discovered in late-Neolithic, Bronze Age and Iron Age sites, and it seems likely that the sheep had arrived on St Kilda in the Bronze Age by 2000 BC, when there is evidence for the presence of pastoral people.

In historical times, the historian Hector Boece was one of the first people to see and refer to the presence of Soay Sheep on St Kilda when he visited the islands around 1520; in *Scotorum Historiae* (1527) he records that:

SOAY SHEEP RAM, HIRTA, ST KILDA, OUTER HEBRIDES

> Beyond this island [Hirta] is yet another isle, but it is not inhabited by any people. In it are certain beasts, not far different from sheep, so wild that they cannot be taken except with a snare; their hair is long ... neither like the wool of sheep or goat... This last island is named Soay,[2] which in Gaelic is called a sheep; for in this island are great numbers of sheep, each one bigger than any male goat with horns longer and thicker than those of an ox, and with long tails hanging down to the earth.

Half a century later, these sheep sorely perplexed John Leslie (or Lesley), Bishop of Ross, who in *De Origine Moribus et Rebus Gestis Scotorum* (1578) records seeing them during a pastoral visit to the islands: 'Near here [Hirta] lies another island [Soay], uninhabited, where no cattle are found except some very wild, which whether to call them sheep or goats, or neither sheep nor goat, we know not. ... They have neither wool like a sheep nor hair like a goat.'

In medieval times a four-horned breed of sheep, known as the St Kilda or Hebridean Sheep, was introduced to the island of Hirta (but not to Soay) from northern Europe to improve the island's domestic stock. It was this breed that Martin Martin described in *A Late Voyage to St Kilda* (1698), and in *The History of St Kilda* (1764) Kenneth Macaulay[3] described the same breed as present when he was the minister on the islands in 1758, adding that 'every one of these sheep has two horns and many of them have four'.

The sheep on Soay were first distinguished as a distinct form from others in the archipelago in the 1830s, when the population varied between c 200 and 500. Even then there was little management of the population, though rams of 'improved' breeds may have occasionally been introduced. At around the same time, Scottish Blackface Sheep (a cross between the now-extinct Old Scottish Shortwool and the English Blackface of the Pennines) were introduced to Hirta and Boreray (another island in the St Kilda group), where they remained the main breed on the islands until the time of human evacuation a century later.

When in 1930 the inhabitants of St Kilda were removed, at their own request, to the Scottish mainland, they took with them 500–600 Blackface Sheep from Hirta, leaving behind the flocks of Blackface on Boreray and Soays on Soay. In 1931, the Marquess of Bute bought St Kilda (acquired in 1957 by the National Trust for Scotland) from the Macleods of Macleod. In the following year a number of islanders returned to St Kilda, and on Lord Bute's instructions transferred 107 Soay Sheep (20 rams, 44 ewes, 21 ewe lambs and 22 wether lambs) from the 99-ha island of Soay to the 638-ha island of Hirta. Thereafter the St Kildans returned annually to tend the sheep on 77-ha Boreray and Hirta until the outbreak of the Second World War in 1939, when the Soay Sheep on Hirta were said to number around 500. Since then the Soay Sheep on Soay and Hirta and the Blackface Sheep (now recognized as a smaller distinct form, the Boreray) on Boreray have been unmanaged.

Elsewhere in Scotland, Soay Sheep have been introduced to Holy Island (east of Arran in the mouth of the Firth of Clyde) and to Sheep Island (between Sanda and Kintyre). A small population on Ailsa Craig in the Firth of Clyde, to which Soays were introduced from St Kilda in the 1930s, was evacuated in the late 1950s/early 1960s. Stevenson and Bullock (in Harris and Yalden 2008) estimated the population on Sheep Island in 1998 at c 35.

In the Orkney Islands, a population of around 2,000 semi-feral sheep lives on North Ronaldsay, where the sheep may have existed since the Neolithic. Although they do have owners they receive very little management, and since they are restricted to the 458 ha of seashore by a wall that encircles the island, they survive almost entirely on a diet of seaweed – mostly *Laminaria* spp. The ewes are brought onto more nutritious pastures to lamb, but are returned with their offspring to the coast in August. In 1974, about 400 sheep were transferred to Linga Holm west of Stronsay, and 1 ram, 12 ewes and 11 lambs were also moved to the Island of Lihou north-west of Guernsey in the English Channel Islands (Wilberley 1979).

England; Wales

In 1910, a breeding nucleus of Soay Sheep was transferred from St Kilda to Woburn Abbey in Bedfordshire by the Duke of Bedford. In 1934, the Duke sent 2 young rams and 4 ewes – all of the 'Dark' colour-type (see opposite) – to R.M. Lockley on the 96-ha island of Skokholm off the coast of Pembrokeshire, where a decade later they had increased to only 40. In 1944, 1 ram and 3 ewes from Woburn, and the same number from Ronald Stevens's collection at Walcot Hall, Shropshire, were sent to the 16-ha Cardigan Island off the coast of Cardiganshire (Ceredigion). In the following year, 1 ram from Cardigan and 1 ewe from Skokholm were transferred to the 8.4-ha island of Middleholm, where by 1959 they had increased to around 70. In 1952, 1 ram and 3 ewes were moved to the 8-ha St Margaret's Island near Tenby, Pembrokeshire, where by 1959 they had multiplied to 20–30. On Christmas night of that year, the entire population was killed by lightning. In 1958, a number of Soay Sheep were released on the island of Skomer off the Pembrokeshire coast. By the turn of the decade there were about 40 Soays on Cardigan,

Skokholm and Skomer, but by 1975 they survived only on Cardigan, where the population, with a high ratio of rams, numbered around 80.

In 1927, Martin Harman introduced Soay Sheep to Lundy Island in the Bristol Channel, where within some 30 years they had built up a population of over 80. In 1973, a dozen rams from Lundy were despatched to a number of flocks on the English mainland. Stevenson and Bullock (in Harris and Yalden 2008) gave a figure of c 188 Soays on Lundy in October 1998.

In the 1990s, a population of Soay Sheep became established in the Cheddar Gorge in Somerset, where in 1998 Stevenson and Bullock (in Harris and Yalden 2008) gave the number as 138.

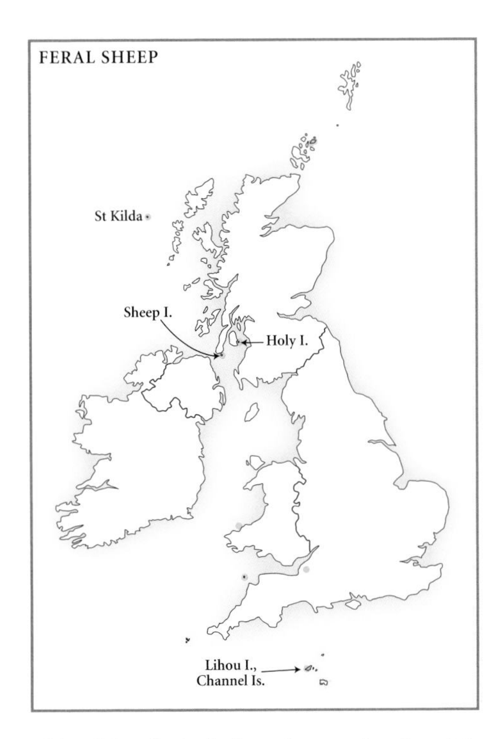

Characteristics

The Soay Sheep is the most primitive and, standing barely 50 cm at the shoulder, the smallest form of domesticated sheep in the world; with relatively long legs and a shortish tail, it bears a fairly close resemblance to 'wild' sheep. Soay fleece (which moults naturally in late spring or early summer but for commercial use in captivity is plucked rather than shorn) is divided into two colour forms – 'Dark' or 'Light' with 'Wild' or 'Self' markings.

- 'Dark' animals vary in colour from deep brown (known to St Kildans as *lachdann*) to near black, and 'Light' individuals from oatmeal to buff.
- 'Wild'-type markings are similar to those of the Mouflon, with whitish underparts.
- In 'Self'-marked animals the white markings are replaced by the colour of the body.

In the wild, 4–5 per cent have white markings – in captive flocks this percentage can be much higher and can cover 50 per cent or more of the fleece (personal observation). Most Soay rams and about 30 per cent of ewes are horned. Soay Sheep scatter rather than flock, and thus cannot be gathered by sheepdogs. 'Managed' mainland captive populations usually have Dark-Wild-type woolly fleeces, and most ewes are horned.

The fleece of the Borerays varies in colour from creamish-white to blackish, and is shorter and finer than that of modern Scottish Blackface Sheep. Both sexes grow a single pair of horns.

Soay Sheep on Soay and Hirta, and Borerays on Boreray, live on *Agrostis/Festuca, Holcus/Agrostis, Festuca* and *Poa* grasslands and *Plantago* sward in summer. In autumn and early winter, Soays on Hirta favour *Calluna* wet heathland, which does not exist on Boreray. In St Kilda, *cleits* or *cleitans*[4] provide shelter from both severe weather in winter and the sun in summer.

Although domestic sheep are mainly or entirely grazers, Soay Sheep in captivity in 'managed' flocks will readily browse on Bramble and other species (personal observation), and the population in the Cheddar Gorge browses on Ivy and Yew. On Hirta, Soays exhibit well-defined seasonal diet variation similar to that of domestic hill sheep on the Scottish mainland, with grass predominating in summer and an increasing proportion of Common Heather being consumed in autumn and winter, when on occasion energy intake may be below survival levels. This diet is likely to be similar to that of other Soay island populations. On Boreray, where Common Heather does not occur, Borerays graze perforce throughout the year.

On Hirta, most deaths occur between early February and early April, when after a hard winter ram lamb mortality can be as high as 99 per cent. Since males generally suffer a much higher mortality rate than females, there may be spanandry in the population with a skewed sex ratio of as high as 1:14.

In 1952–2000, the populations of Soays on Soay and Hirta and of Borerays on Boreray fluctuated dramatically, as shown below.

Table 23 Maximum and minimum numbers of feral sheep *Ovis aries* on the St Kilda Islands, 1952–2000

Island	**Hirta**	**Soay**	**Boreray**
Area (ha)	638	99	77
Breed	Soay	Soay	Boreray
Maximum number (year)	2,022 (2000)	c 360	699 (1980)
Minimum number (year)	610 (1985)	c 100	215 (1990)

Source Stevenson and Bullock (in Harris and Yalden 2008). **Note** Because of the inaccessibility of the island, numbers for Soay are subject to a high degree of error.

Irregular population 'crashes' on Hirta (and probably on Soay), when up to 70 per cent of animals may die, occur at 3–15 yearly intervals, and since they coincide with those of Boreray Sheep on Boreray, only 6 km from Hirta, may be weather related. The extremely high peak density of Boreray Sheep may be a result of high-quality grazing due to deposition of seabird guano.

Harris et al. (1995) estimated the total number of feral sheep in Britain at 2,100.

The feral sheep of the St Kilda Islands – and the islands themselves – are unique in a number of ways.

- The sheep live in a habitat isolated from outside influences, and in conditions that afford excellent opportunities for scientific observation.
- There have been no additions or subtractions for around 180 years.
- There are no ground predators and the few avian ones take a toll of only a small number of lambs and sick or injured adults, and are useful scavengers on carcasses.
- There is an abundance of rich summer pasturage, for which there is no competition from other herbivores.
- Apart from deer and feral goats, the feral sheep is Britain's sole surviving large wild mammal.[5]

A long-term study of the ecology and behaviour of Soay Sheep on Hirta, largely funded by Scottish Natural Heritage and the National Trust for Scotland, has been conducted since 1959 and continues to provide fresh insights into population dynamics, genetic diversity, parasite/host relationships and plant/herbivore interactions. The current policy for both breeds on St Kilda is for a minimum of interference that is compatible with the integrity of the ecosystem. Both breeds may be of genetic and historical significance, and the Soay especially is an important link with the domesticated sheep that were first introduced to Britain in Neolithic times.

An interesting phenomenon has recently emerged regarding the Soay Sheep of St Kilda. Research by scientists at Imperial College London (Pelletier et al. 2007) has revealed that the higher temperatures associated with global warming enable a larger number of smaller animals to survive the winter, which has resulted in the average size of individuals in the population falling. This establishes a direct link between genetic changes in an animal population and climate change. It has been revealed that in harsh winters more of the larger individuals survive whereas in milder winters more of the smaller animals survive, thus reducing the average size.

In the 1980s, big sheep were genetically favoured on Hirta (and presumably on Soay) because large individuals had more resilience in withstanding severe winters. However, as climate change ameliorates the St Kilda winters there will be a reduction in the natural selection for larger individuals, which could have a significant impact on the population dynamics of the Soay Sheep population. Pelletier et al. (2007) found that the distribution of body sizes within the population can have a marked influence on population dynamics, accounting for up to 20 per cent of population growth. They conclude that 'there is substantial opportunity for evolutionary dynamics to leave an ecological signature and vice versa'. Thus humans are seen to be having not only an ecological impact on the natural environment, but an evolutionary one as well.

References

Boyd 1953, 1964a, b, 1981; Boyd & Jewell 1974; Bullock 1983; Campbell 1974; Clutton-Brock 1991; Elwes 1912; Ewart 1913; Fisher 1948; Grubb & Jewell 1966; Hall 1975; Jewell 1974, 1980, 1984, in Corbet & Southern 1977, in Corbet & Harris 1991; Lever 1977. 1980a, 1985, 1994; Lockley 1960; Long 2003; Pelletier et al. 2007; Poore & Robertson 1949; Ryder 1968, 1975, in Mason 1984, 1995; Simmons & Tooley 1981; Smith 2007a; Stevenson & Bullock, in Harris & Yalden 2008; Taylor 1967–8; Wilberley 1979; Williamson & Boyd 1960; Yalden 1999.

Notes

1. Long (2003: 527) includes St Kilda under the heading 'Orkney Islands'.
2. It is believed that visiting Norsemen may have named the 'Island of Sheep' (Old Norse Sauoa-ey) after the sheep they found there.
3. According to the lexicographer Samuel Johnson (1709–84), Macaulay may have done no more than pass on information to the real author, John Macpherson of Skye (1710–65).
4. Drystone cells measuring 4–6 m long and about 2 m high, built by the St Kildans as drying chambers for a variety of products.
5. Soon, perhaps, to be joined by Wild Boar (see page 326).

BIRDS

ANATIDAE (DUCKS, GEESE & SWANS)

MUSCOVY DUCK *Cairina moschata*

Natural Distribution S Mexico S through Colombia, Ecuador and Peru to N Argentina.
Naturalized Distribution England.

MUSCOVY DUCK MALE

The Muscovy Duck was domesticated long before Europeans arrived in the New World in the late 15th century, and in common with many other long-domesticated birds the plumage of *forma domestica* is predominantly white. It seems probable that the *forma domestica* was, with the Wild Turkey, the first bird to be introduced from the Americas to Europe in the early 16th century. It is possible that it is the most widely distributed of the world's exotic wildfowl and, because it is regarded by many ornithologists as mere 'farm-yard poultry', the least studied.

England

In 1968–72, there were only five successful breeding records by feral Muscovy Ducks in four 10-km squares (TL15S; TL57B; TM59B; TQ40B). From at least the early 1980s, a population of up to 130 (November 1991) has been established on the River Ouse near Ely in Cambridgeshire, where breeding was first confirmed in 1987. In 1996–2001, successful breeding in the wild was recorded in Bedfordshire, Cambridgeshire, Cheshire, Derbyshire, Devon, Dorset, Greater Manchester, Norfolk, Northumberland, Nottinghamshire, Suffolk and Surrey. In 1996, flocks of up to 75 individuals were recorded, and in the following year 99 Muscovy Ducks were counted on Lothing Lake and Oulton Broad in Norfolk; 24 birds were present at Ely in 1999, and up to 30 in 2001. In 1999, feral Muscovy Ducks occurred in the wild on shallow lakes, drainage channels, village ponds and reservoirs over much of north-eastern, east-central and south-eastern England, although only around 15 pairs bred annually.

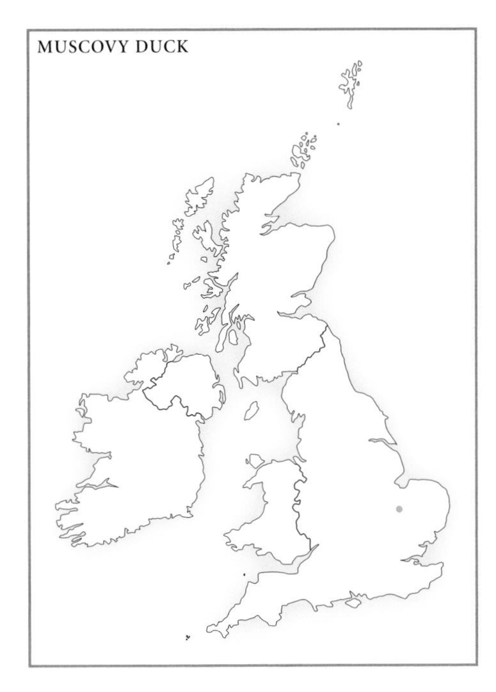

Recently the population away from Ely has declined; in 2003 breeding was only recorded in Devon, where a single pair reared 13 ducklings. At Ely there was a minimum of 5 broods with at least 41 young, and in 2004 the same number of broods reared a minimum of 59 ducklings. The Ely population is said to depend largely on the public for food, and the local council is attempting to control the numbers by oiling eggs.

Characteristics

In Central and South America, the Muscovy Duck is a bird of tropical lakes, lagoons, marshes and slow-flowing rivers in lowland forested areas. In the dry season it sometimes occurs on brackish coastal marshes and lagoons. It normally nests in tree holes or hollows, and occasionally on the ground in dense waterside vegetation. A dabbling duck, it also sometimes grazes on waterside meadows.

Impact

Hybridization in the wild between Muscovy Ducks, wild Mallard and feral *forma domestica* Mallard is not uncommon – the offspring being known in France as *mulard*. The Muscovy Duck is a very efficient food converter and produces a carcass with an exceptionally high lean-meat content.

References

Blair et al. 2000; Clayton, in Mason 1984; Gibbons et al. 1993; Lever 1985, 2005; Ogilvie et al. 1999–2003; Sharrock 1976.

REINTRODUCED SPECIES

EUROPEAN ECONOMIC COMMUNITY COUNCIL DIRECTIVE 92/43/EEC of 21 May 1992 (usually known as the Conservation of Natural Habitats and of Wild Fauna and Flora Directive) requires, under Article 22, member states to: 'study the desirability of reintroducing species ... that are native to their territory where this might contribute to their conservation ... provided that such reintroduction contributes effectively to re-establishing these species at a favourable conservation status and that it takes place only after proper consultation of the public concerned'.

Member states thus have a legal as well as a moral responsibility to restore their country's biodiversity as much as possible by the reintroduction of formerly native species that have died out. Apart from general stakeholder acquiescence, as required under Article 22 quoted above, two other prime requirements must be met before the reintroduction of a species can be contemplated:

1. That the reason or reasons for that species' disappearance is known and no longer occurs.

2. That sufficient suitable habitat exists to make such reintroduction at least theoretically practicable.

In all the deliberate reintroductions made since 1992 and described in the following section (those of the Wild Boar and Eurasian Eagle Owl could be described as 'self-introduced'), the necessary criteria have been met. The Red Squirrel in Ireland and the Western Capercaille in Scotland were reintroduced in the 19th century.

This is not to say that all reintroductions have received universal acclamation. In particular, the restoration of birds of prey such as the Red Kite, Northern Goshawk and in the Republic of Ireland the Golden Eagle have met with some local opposition. Even that distinguished ornithologist and proponent of exotic species, the late Derek Goodwin, declared his opposition to the restoration of the White-tailed Eagle. It is true that some White-tailed Eagles do occasionally take (mostly sick or injured) lambs, but perhaps not unnaturally, some west coast Scottish crofters tend to exaggerate their losses. They should, nevertheless, receive adequate compensation for any losses that do occur – a small price to pay, surely, for the sight once more of these magnificent birds in British and Irish skies. Similarly, the reintroduction (currently only to a single location in western Scotland) of the Eurasian Beaver has been the subject of considerable contention among biologists and landowners alike.

MAMMALS

SCIURIDAE (SQUIRRELS, MARMOTS & ALLIES)

RED SQUIRREL *Sciurus vulgaris*

Natural Distribution Widely distributed in the Palaearctic from W Europe E to Korea and Kamchatka.
Reintroduced Distribution Ireland.

RED SQUIRREL

Ireland

Although certainly native in Britain, the status of the Red Squirrel in Ireland is more ambiguous, and its presence may originally have been due to prehistoric introductions for fur. The species had apparently died out by the late 17th century, since Fairley (1983) refers to its notable absence from a list of furs exported between 1697 and 1819. The current population, which is widely distributed except in parts of the far west and northeast, results from reintroductions listed below.

Table 24 Reintroductions of Red Squirrels *Sciurus vulgaris* to Ireland, 1815–76

Date	Locality	Reintroduced By	County
1815–25	Ashford; Castle Howard	Mr Synge	Wicklow
1833	Garbally	Lord Clancarty	Galway
Before 1836	Castleforbes	?	Longford
1850	Oakpark	Colonel Bruen	Carlow
1851	Ravensdale Park	Lord Clermont	Louth
1864	Birr Castle, Parsonstown	Lord Rosse	Tipperary
1870–4	Moneyglass	Mr Jones	Antrim
Before 1875	Ramelton	Lord George Hill	Donegal
1876	Anna Liffey, Lucan	Joseph Shackleton	Dublin
?	Donore, Multifarnham	Miss Nugent	Westmeath

Source: Barrington 1880.

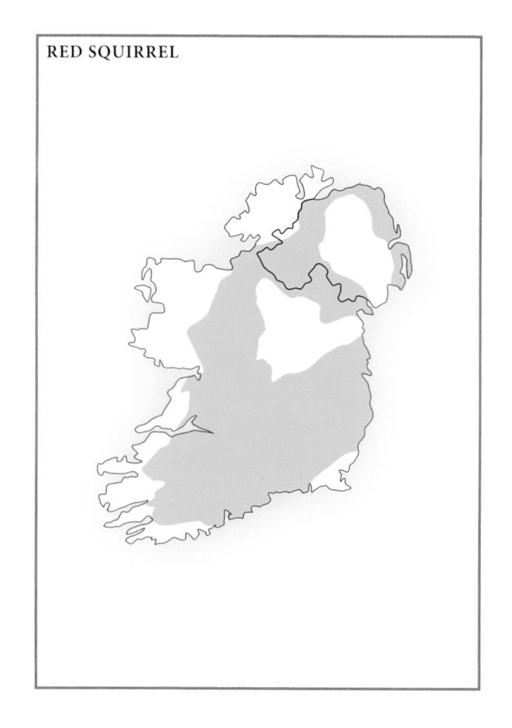

References

Barrington 1880; Fairley 1983; Gurnell et al., in Harris & Yalden 2008; Shorten 1954; Yalden 1999.

CASTORIDAE (BEAVERS)

EURASIAN BEAVER *Castor fiber*

(EUROPEAN BEAVER)

Natural Distribution Formerly widespread in Britain and Europe, but by 1900 restricted to only five European populations (including Scandinavia) and three in Asia. Now re-established in 18 European countries from France eastwards to the Ukraine, largely as a result of deliberate reintroductions.
Reintroduced Distribution Scotland; ? England; ? Wales.

There is documentary evidence for the survival of Eurasian Beavers in Wales until 1188 and in Scotland until the 16th century. They have never occurred in Ireland. Coles (2006) tentatively suggested that they may have survived until as late as the 18th century. Beavers were exterminated in Britain as a result of over-hunting for their flesh and fur and for their castoreum, a glandular secretion with medicinal properties, and widespread deforestation in Scotland.

Much has been written (e.g. Sitwell 1979; Lever 1978, 1980a, b, 1981, 1994) about the advantages and disadvantages of reintroducing the species to Britain.

Kitchener (2001) sums up the case for reintroduction:

- Beaver dams sustain water flow and help to prevent flooding. In times of drought there is a gradual release of water, thus reducing seasonal variations. Most of the energy dissipated by the stream is at the dams, below which plunge pools develop.

EURASIAN BEAVER

- The diversity of aquatic habitats in the riparian zone increases in complexity with varying water-table levels, soil-water regimes and stream flow paths.
- A complex network of channels may develop, especially in low-lying districts. Stability of riverine channels is improved.
- More sediments are accumulated in streams above dams, thus reducing sediment loss by the river catchment. The stream sediments become sorted by size, with finer sediments above dams and coarser ones below them.
- Organic matter entering the river is stored and processed, and decomposes near to its point of origin rather than being swept downstream.
- There is a several hundred-fold increase in the surface area of the river, leading to a pH increase in forest soil.
- Water depth is increased and the water table rises, possibly improving pastures, crops and wetland vegetation.

Somerville (2001), although in favour of reintroduction, also sums up the case against it. The principal problem is shortage of suitable habitat. Since Eurasian Beavers became extinct land use, especially in Scotland, has altered dramatically. Changes in climate, the amount of woodland, the species of tree used in forestry, demography and the drainage of wetlands have reduced the amount of suitable habitat. Foresters have expressed concern about damage to trees – both commercial and amenity – and conservationists are worried about the suite of rare insects associated with old Aspens, which are one of the Eurasian Beaver's favourite foods, which only occur locally in Scotland. Anglers are concerned that dams could adversely affect river geomorphology and thus have a negative effect on Brown Trout and Atlantic Salmon fishing – the source of millions of pounds annually to the Scottish rural economy – by silting up spawning redds or by hindering fish returning to their natal rivers. Large Eurasian Beaver ponds could result in high evaporation and a reduction in water flow. The construction of dams and the excavation of burrows could lead to localized flooding and the undermining of riverbanks, field margins and even roads, while animals on farmland could eat valuable crops.

Scotland

Most of the previous attempts to establish beavers in Scotland (and England) have been made with the North American or Canadian Beaver. However, in 1874 the Marquess of Bute released four Eurasian Beavers in a large, walled enclosure at Kilchattan on the Isle of Bute. These animals did not survive for long, so a further five from Scandinavia were released in the following year. At first the animals did well, and by 1877 had increased to 12 and by the following year the population had more than doubled. During the winter they fed mainly on the bark of trees, especially that of the Scots Pine and English Elm, but also of larches. However, by about 1887 the last animal had died.

More recently, enclosed Eurasian Beaver colonies have been established at the Aigas Field Centre near Beauly in Inverness-shire, as well as at Bamff near Alyth in Kinross and Perthshire.

In May 2008, the Scottish Wildlife Trust and the Royal Zoological Society of Scotland received permission for an experimental reintroduction of Eurasian Beavers into the wild in Scotland. In late 2008, four families, totalling about 15–20 individuals, were acquired from the Telemark region of Norway. After six months quarantine, they will be radio-tagged and released into the wild in the spring of 2009 in five lochs in Knapdale Forest (owned by Forestry Commission Scotland) west of Lochgilphead in Argyllshire, where for a period of five years they will be monitored by Scottish Natural Heritage.

Knapdale Forest is bounded by water on two sides – the Crinan Canal to the north and the Sound of Jura and Loch Sween in the west. The Eurasian Beavers' release site is an unenclosed area of about 15–20 sq km of low hills and valleys interspersed with small- to medium-sized lochs connected by burns. The dominant riverside woodland is composed of birches, willows and oaks, with some Common Alder, Rowan and Common Hazel, and a well-developed ground vegetation. Because the area is surrounded by large water bodies and by unsuitable Sitka Spruce plantations, it is hoped that the animals will not disperse (Kitchener 2001). Depending on the outcome of the experiment, subsequent reintroductions may be made elsewhere.

?England

Gurnell et al. 2009 confirmed the feasibility of reintroducing Eurasian Beavers to England. Experimental sites suggested include rivers in the Weald of Kent; the New Forest, Hampshire; Bodmin Moor, Cornwall; the Peak District, Derbyshire/Staffordshire; the Lake District, Cumbria and the Forest of Bowland, Yorkshire/Lancashire. Enclosed colonies already occur in Devon, Gloucestershire, Kent and Lancashire.

?Wales

The Wildlife Trusts of Wales are currently assessing the feasibility of reintroducing the Eurasian Beaver to Wales (The Welsh Beaver Assessment Initiative). On completion of this feasibility study, the Countryside Council for Wales will review the findings to determine whether the reintroduction is appropriate. Any reintroduction project would require a licence from the Welsh Assembly Government (CCW 2008; Liz Howe, personal communication 2008).

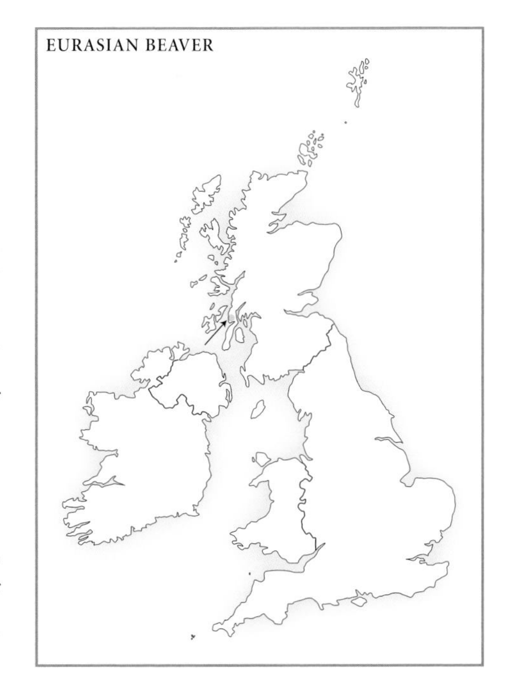

Characteristics

Beavers require year-round access to fresh water. Their optimal habitat is wooded, slow-flowing rivers in valley bottoms, especially floodplains. They modify sub-optimum habitats by dam building, which extends the water protection around lodges (although Eurasian Beavers prefer to breed in natural holes or burrows in riverbanks) and foraging areas. They feed mainly on herbaceous vegetation in summer and on bark (principally that of Aspen, willows, poplars and Common Alder) in winter.

Impact

The Natural England Report (see under England, above) confirmed the findings of Kitchener (2001), and additionally found that dams positively influenced invertebrate populations by increasing the extent and diversity of aquatic habitats and providing additional food resources for vertebrate species. Although dams can form an impenetrable barrier to anadromous fish such as Atlantic Salmon and Sea Trout (the anadromous form of the Brown Trout), for non-migratory species they create new foraging habitats and refugia, and in pools formed by dams some fish species grow more rapidly and achieve a larger size. Dams additionally improve water quality, which benefits plants, animals and humans.

References

Cole et al., in Harris & Yalden 2008; Coles 2006; Conroy et al. 1998; Countryside Council for Wales (CCW) 2008; English 1998; Evans 1996; Fitter 1959; Harting 1980; Kitchener 2001; Kitchener & Conroy 1996; Lever 1978d, 1980a, b, 1981, 1994; Macdonald et al. 1995; May 1998; Mouland 2005; Nevard 1991; Nuttall 1994; Pierce 2005; Rose 2008; Sage 1981; Scott Porter Research & Marketing 1998; Scottish Government Licence Decision 2008; Scottish Government News Release 2008; Scottish Natural Heritage 2005; Sitwell 1977; Sommerville 2001; Sykes 1979; Webster 1983; www.snh.org.uk/speciesactionframework

SUIDAE (PIGS)

WILD BOAR *Sus scrofa*

Natural Distribution A Palaearctic species extending from W Europe E to Japan and from S Scandinavia S to N Africa; in Asia extends further S to Sri Lanka and Malaysia.
Reintroduced Distribution England; Wales.

The Irish population of native Wild Boars is believed to have died out as long ago as the Neolithic. There is documentary evidence, however, that the species survived in England until around 1300, and probably until about the same date also in Scotland. In the 16th, 17th and 18th centuries there are many references to the presence of Wild Boar in parks in England and, to a lesser extent, Scotland, all of which certainly refer to animals reintroduced from Europe as an amenity or as a source of food. The animals' extinction was probably due to a combination of habitat loss due to deforestation, over-hunting for sport and food, and assimilation into the sounders of domestic pigs pannaged in the surviving woods.

England; Wales

Martin Goulding (2003 and personal communication 2008) and Charles Wilson (2005 and personal communication 2008) have summarized the origin of populations of Wild Boar in England today.

The present population is descended from escapes or releases from commercial farms, the first of which was started in Cambridgeshire in 1981 with surplus stock imported from France by London Zoo. Within a little over a decade, the industry had mushroomed to some 40 farms holding around 400 breeding sows. Later importations were of German and Swedish stock.

Escapes from some of these farms inevitably began to occur, and in 1983–94 over 60 animals absconded from 6 farms in 6 counties. The big storm of 1987 destroyed a section of a perimeter fence of a farm at Tenterden in south-west Kent, and several Wild Boar escaped into the wild, although it was not until 1994, when two were shot at Beckley in neighbouring East Sussex, that their presence in the wild was acknowledged. This was the first occurrence of the species outside captivity for some 300 years. Two years after this, several Wild Boar were shot at Aldington in Kent, having escaped from an abattoir at Ashford a few years earlier.

Goulding (2003) and Wilson (2005) received reports of escaped or released Wild Boar in England in Cheshire; Gateshead, Tyne-and-Wear (2001); Dereham, Norfolk; near York (2003–4) and the Bodmin area of Cornwall (2002), and in Scotland in East Lothian. In 1993, up to 27 animals were released or escaped at Catterick in North Yorkshire, where some may have been at liberty for more than eight years (Wilson 2005). Martin Goulding (personal communication 2008) received reports of repeated sightings of Wild Boar from Daventry (Long Buckby and Badbury), Cheltenham (Windrush), Evesham (Cleeve Prior), Moreton-in-Marsh (Blockley), Bishops Stortford (Quendon), High Wycombe (Great Missenden/Speen), Ringwood (Alderholt), Thetford (Ickburgh and Thetford Forest) and York (Long Marsten, Bilbrough, Nether Poppleton).

WILD BOAR

By 1998, two viable sub-populations of Wild Boar existed in England, one in the Weald on the Kent/East Sussex border and the other in west Dorset, where breeding has occurred since 1995. In 2006, the former was estimated to number 100–200 animals spread over around 1,000 sq km, and the latter 20–30. In the late 1990s, a third small although self-sustaining sub-population of less than 20 became established in the western part of the Forest of Dean in Gloucestershire (Wilson 2003).

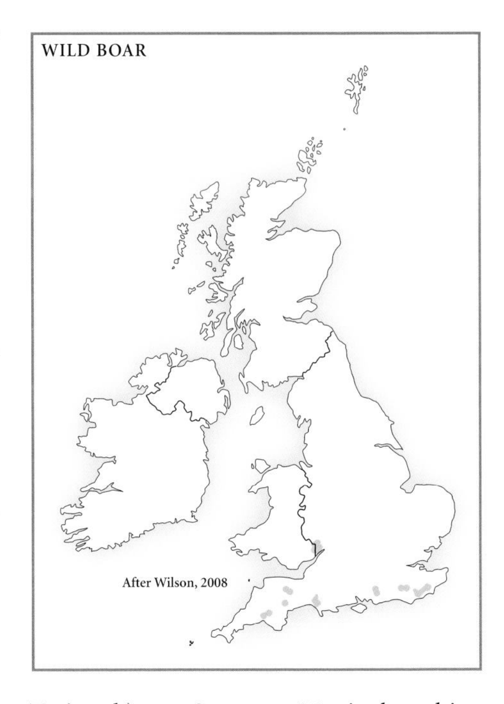

By 2006, Goulding et al. (2008) had recorded 23 escapes of Wild Boar involving a total of at least 198 individuals. Recent press reports refer to the release of Wild Boar by vandals in Devon in 2006–7, and to their presence on the Sussex Wildlife Trust Reserve at Flatropers Wood at Peasmarsh near Rye on the coast of East Sussex; in Devon (Dartmoor, Exmoor, East and West Anstey, Buckland Monachorum near Tavistock); east Somerset; Nottinghamshire; Leicestershire and Gloucestershire (near Coleford, east of Monmouth). There have been numerous reported (and doubtless unreported) escapes and sightings elsewhere.

There are currently several established, or apparently established, sub-populations of Wild Boar naturalized in England and Wales.[1] The largest, numbering around 200 individuals in 2004, is the one present on the Weald. During the last five years, Martin Goulding (personal communication 2008) received numerous reports of Wild Boar and piglets in the south-east from around Crawley (Crawley Down), East Grinstead (Forest Row) and Tunbridge Wells (Tonbridge), which he believed originated from a separate source to the population further south on the Kent/East Sussex border. The west Dorset sub-population, which derives from an escape of less than 10 animals from a farm near Bridport in 1994–5, numbers c 30. The sub-population in the western part of the Forest of Dean was joined in 2004 by a boar (possibly a hybrid) that escaped from an abattoir 10 km south of Ross-on-Wye in Herefordshire, and by at least 25–30 animals illegally released near Staunton, 10 km to the south-west. From here, some have crossed the River Wye into eastern Monmouthshire in Wales. Elsewhere in Wales, boar have been reported in the wild in Usk, Monmouthshire and Chepstow, Powys (Roger Trout, personal communication 2008). The Ross-on-Wye/Forest of Dean sub-population is, like the one in the Weald, almost certainly increasing. In the winter of 2005–6, a substantial number of Wild Boar were apparently deliberately released in Devon and on the Devon/Somerset border where, even after some were recaptured or shot, more than 50 remained at large, and breeding is known to have occurred. These sub-populations were augmented by further releases in north-west Devon in November 2006 and 2007. In West Sussex, south of Haslemere, Wild Boar may be establishing a breeding population following a recent escape (Wilson 2005 and personal communication 2008).

Most, if not all these sub-populations are increasing in numbers, and would be considerably higher were it not for culling.

Scotland

Although no Wild Boar have been recently reported in Scotland, Foggo (2008) reported the despicable practice of 'canned hunting', when 51 Wild Boar were shot on a private 80-ha fenced property in Dumfries and Galloway.

Prospects

The feral population of Wild Boar in Britain, which in 2005 was estimated to number less than 500, seems likely to continue to increase, given the species' fecundity, the continued probable recruitment to the population from further releases and/or escapes, and the absence of natural predators other than humans. At present the Wild Boar population could probably be eradicated by shooting and/or live-trapping.

Characteristics

Wild Boar are primarily animals of deciduous woodland, from which they emerge – mainly at night – to feed. Although largely herbivorous they are also omnivorous and at times scavengers.

Impact

Were Wild Boar to become widely established, they could prove an agricultural pest by their ready consumption of such crops as sugar beet, potatoes, wheat and maize, and by their destruction of fences. In woodlands, rootling by Wild Boar can have an adverse effect on Common Bluebells, Wild Daffodils and other species, and on grassland can disturb leys. Ground-nesting birds, including game-birds, could suffer from predation of both their eggs and their young, and small mammals and possibly even neonatal lambs could be predated. Wild Boar could also become a reservoir for various diseases affecting domestic livestock.

On the other hand, at moderate densities, as in Brede High Woods in East Sussex, Wild Boar could control the Common Bracken (whose rhizomes are an important winter food) and other woodland weeds, and also harmful invertebrates and small rodents. Rootling by Wild Boar could improve soil aeration, fertility and structure, and create favourable germination sites for tree seedlings. The return of a species forming an integral component of woodland ecology contributes to the biodiversity of the greatly impoverished British mammalian fauna, and provides a valuable new sporting quarry (Wilson 2005; Sims 2005).

References

Anon 2006; de Bruxelles 2005; Elliott 2005b, 2008a; Engeman et al. 2002; Fitter 1959; Foggo 2008; Goulding 2003; Goulding & Roper 2002; Goulding et al. 1998, 2003; Goulding & Roper 2002; Goulding et al., in Harris & Yalden 2008; Howells & Edwards-Jones 1977; Leaper et al. 1999; Lusher 2006; Moore 2004; Moore & Wilson 2005; Sims 2005; Tyrer 1997; Wilson 1999, 2003a, 2003b, 2004, 2005; Woodman 1978; Yalden 1999, 2001.

Note

1. I am grateful to Charles Wilson of Natural England for providing me with a map showing the distribution of Wild Boar in England and Wales, and to Martin Goulding for his cooperation.

BIRDS

ANATIDAE (DUCKS, GEESE & SWANS)

GREYLAG GOOSE *Anser anser*

Natural Distribution The Palaearctic region from Western Europe eastwards to northern China.
Reintroduced Distribution England; Wales; Ireland.

GREYLAG GOOSE, NORFOLK

England; Wales; Ireland

In the late 18th century, the Greylag Goose[1] was widely distributed in England, being especially common in East Anglia from which its disappearance by the early 19th century was due to the drainage and subsequent cultivation of the fens. Elsewhere in England, e.g. in the carrs[2] of east Yorkshire, Greylag Geese died out much earlier than in East Anglia. In Scotland, where native Greylag Geese survive only in the Western Isles and in the north, the species disappeared in the 19th century as a result of over-shooting. In Ireland, Greylag Geese bred in Co. Dublin and Co. Down until at least the 1770s, but died out sometime thereafter.

The Greylag Goose, the ancestor of the farmyard goose, has been domesticated in Britain for hundreds of years. After its disappearance from the wild, large numbers of eggs and goslings were reintroduced to East Anglia, where inevitably some birds escaped from captivity to form feral populations. Subsequently, Greylag Geese were also released elsewhere for sporting purposes by landowners and wildfowlers. Their natural spread has been aided by the formation of reservoirs and flooded gravel-pits, whose vegetated islands they favour for breeding.

The most successful early reintroductions were in the Lake District of Cumbria, Kent, Norfolk and south-western Scotland. Although birds spread from these localities, later reintroductions in the eastern Midlands resulted in an even more dramatic expansion of range from Bedfordshire east to Lincolnshire and north to Yorkshire and Humberside. Greylag Geese are, however, scarce in the south-western Midlands and south-central and south-western England.

In Wales, Greylag Geese are rare except on the Isle of Anglesey.

In Ireland, the large population established around Strangford Lough in Co. Down has spread very little, but breeding has occurred in the north-west and south-west following several further reintroductions.

Impact

Because they are non-migratory, and are thus present when young plants are growing in the spring and when cereals are ripening in late summer, reintroduced Greylag Geese are a cause of some concern to farmers, especially those cultivating low-yielding Hebridean machair.[3] During the time of the mid-summer moult, when they are flightless, Greylag Geese and their goslings take refuge from potential predators on lakes, grazing on surrounding waterside pastures and crops or emergent vegetation, where they can have a noticeable impact on small reed beds.

References

Holloway 1996; Prater, in Gibbons et al. 1994.

Notes

1. The name 'Greylag' is said to derive from the species' habit of 'lagging' behind when other geese leave Britain in the spring.
2. Wet woodland, usually dominated by the Common Alder or willows.
3. Low-lying coastal land formed of sand and shell fragments deposited by the wind.

PHASIANIDAE (TURKEYS, GROUSE, PHEASANTS & PARTRIDGES)

WESTERN CAPERCAILLIE *Tetrao urogallus*

Natural Distribution Much of the Palaearctic region apart from E Asia, parts of W Europe and the Mediterranean region. Originally indigenous in England, Scotland, Wales and Ireland.
Reintroduced Distribution Scotland.

WESTERN CAPERCAILLIE MALE DISPLAYING IN FOREST

According to Witherby et al. (V: 213), the Western Capercaillie[1] 'Became extinct in Scotland and Ireland about 1760 and England perhaps a century previously'. Pennie (1950–1), however, quotes from Angus (1886) the inscription on a drawing of two birds shot in Aberdeenshire in 1785, and Holloway (1996: 138) claims that they 'must have become extinct in England much earlier than the mid 17th century ...', but does not provide a reference. According to Harrison (1988: 91), the Western Capercaillie 'was still present in Ireland in 1790, but [was] not recorded later'. Holloway quotes Thomas Pennant as referring, in *A Tour in Wales* (1778), to the 'Ceiliog coed, or Cock of the Forest', which from Pennant's description was almost certainly the Western Capercaillie.

The species' demise was caused partially by over-shooting, but mainly by widespread deforestation in the late 17th/early 18th centuries. The reafforestation of the late 18th century came too late to prevent its ultimate extinction.

In the 19th century, several unsuccessful attempts were made to re-establish the species, including in Norfolk around 1823, Aberdeenshire (1827–31), Perthshire (late 1830s and 1845), Buckinghamshire and Lancashire (1842), Northumberland (1872–7), Yorkshire (1877), and Co. Cork and Co. Sligo in Ireland (1879). In the early 1970s, unsuccessful attempts were made to establish it in Grizedale Forest in Lancashire.

Scotland

The first successful introductions of Western Capercaille were made in 1837–8 at Taymouth Castle near Aberfeldy in Perthshire, by the Marquess of Breadalbane, who imported 48 birds from Sweden. By the autumn of 1839, the population at Taymouth was estimated at 60–70, and by 1842 the birds were deemed to be well established. Twenty years later there were believed to be 1,000–2,000 Western Capercaillies at Taymouth, from where they later 'spread over all the more wooded parts of the Highlands as far as Aberdeen'. Other lesser reintroductions were made successfully in Angus (1862–5), Kincardineshire (1870), Morayshire (1883), Inverness-shire (1895) and elsewhere. Before the turn of the century, the species had become established also in the Isle of Arran (1843), Angus (1856), Stirlingshire (1863), Fife and Kinross (1864), Dunbartonshire (1867), Argyll (1875), Kincardineshire (1878), Morayshire (1886), Aberdeenshire (1888), Ross-shire (1890), Banffshire (1890s), probably Nairn (1890s) and Inverness-shire (1892) (Bannerman 1953–63).

The expansion of the birds in Scotland has been documented by Harvie-Brown (1879, 1880, 1898, 1911) and Pennie (1950–1). The maximum spread was achieved by around 1914, when breeding was occurring from Sutherland to Stirling and from Argyll to Aberdeen, but thereafter declined due to extensive timber felling, especially during the two world wars. The general direction of advance from Perthshire until about 1914 was north-eastwards and south-westwards along the river valleys, with the hens usually preceding the cocks by several years and hybridizing with Black Grouse and, less commonly, Common Pheasants. One reason for the Western Capercaillie's failure to expand its range as mature woodlands with a dry understorey declined was its tendency to colonize younger plantations, where damp brushwood and shortage of food were responsible for the death of many fledglings.

The formation of the Forestry Commission in 1919 helped to ensure the survival of the species (and of other forest-dwelling species) and the maturation of plantations encouraged it to start re-expanding its range. By the late 1930s, Western Capercaillies were spread over the Tay, Dee and Morayshire regions north to the Dornoch Firth in Sutherland, west into southern Argyll, and south to the Firth of Forth, Stirling, Dunbarton and parts of Lanarkshire.

By the 1980s, the species was confined almost entirely to the Highlands of east-central Scotland, from the Moray Firth to the upper reaches of the River Forth in the south, being especially abundant in the valleys of the Spey, Don, Dee and Tay. Small, isolated populations became established elsewhere, but Western Capercaillies seem unable to tolerate the damper maritime climate of the western Highlands. Parslow (1973) estimated the population to number c 1,000–10,000 pairs.

A decade later, the distribution had declined (Moss, in Lack 1986), and a decade later still a further severe decline had occurred (Moss, in Gibbons et al. 1993; Holloway 1996). Catt et al. (1998) estimated the population to number 2,200 individuals from the Forth/Clyde industrial belt northwards to the Dornoch Firth, and eastwards from the central Highlands. The principal concentrations were in eastern and central Scotland (Deeside, Speyside and Perthshire). From a nadir of 1,073 birds in 1999 (Wilkinson et al. 2002) the population had recovered to a still-precarious 1,980 in 2003–4 (Eaton et al. 2007), the majority of which occurred in Badenoch and Strathspey, where the birds may be increasing.

In 2003, 2004 and 2005, Holling & the RBBP (2007, 2008) visited 75, 80 and 85 leks[2] recording 184, 222 and 234 males respectively, suggesting a small increase in the population. The leks were distributed among the Clyde (1); Moray and Nairn (11); north-east (20); Perth and Kinross (6); and Highland (47). Even if the population is not actually increasing, the decline in recent years does seem to be, albeit perhaps only temporarily, levelling out.

Characteristics

The typical habitat of the Western Capercaillie in Scotland is ancient Caledonian forests of Scots Pine, with an understorey of Common Heather, Cowberry, Bilberry and small trees such as Juniper, interspersed with open glades, and nowadays to a lesser extent 20–30-year-old larch and spruce plantations, especially where mixed with Scots Pine. The food of the Western Capercaillie is mainly various parts, according to season, of coniferous trees, with some Common Heather, berries and invertebrates.

The causes of the species' decline in the 20th century include:

- Loss of habitat due to timber felling and habitat fragmentation.
- Wetter June weather (chicks are particularly susceptible to damp).
- A reduction in berries due to changed silvicultural practices.

- Increased numbers of Red Foxes and Carrion Crows.
- Excessive shooting.
- The erection of unmarked deer fences, into which the birds fly.
- Overgrazing by sheep may have reduced the number of invertebrates available.

Recent research by the RSPB in Abernethy Forest Reserve in Perth and Kinross revealed that of 20 Western Capercaillie nests, 9 (33 per cent) were predated by Pine Martens (Summers et al. 2008).

The magnificent Western Capercaillie is one of the species most likely to suffer from global warming and wetter summers, since the chick survival rate is likely to decline. All birds hatch their eggs at a time when there is a peak abundance of food for the nestlings. If that pattern is broken, as it probably will be if global warming continues, then it will be even less likely that the young will survive. As Scotland warms up, the species will have nowhere to disperse to, and may eventually die out.

Impact

The Western Capercaillie's consumption of various parts of a variety of coniferous trees has sometimes brought it into conflict with forestry interests. Because the trees affected are also damaged by other animals, such as Red Deer, Sika and Red Squirrels, the harm attributable to the birds is hard to determine. Damage is most frequent in spring, and recovery depends largely on the vigour of the individual tree. Normally, however, the damage is not extensive, and the Western Capercaille is generally tolerated by foresters.

References

Angus 1886; Bannerman 1953–63; Bartholomew 1929; Bonar 1907, 1910; Campbell 1906; Carter et al. 2008; Catt et al. 1998; Davidson 1907; Eaton et al. 2007; Gladstone 1906, 1921a; Grant & Cubby 1972–3; Harvie-Brown 1879, 1880, 1898, 1911; Holling & RBBP 2007, 2008; Holloway 1996; Kay 1904; Lack 1986; Leckie 1897; Lever 1977, 1987, in press; Lister-Kaye 2005; Lloyd 1867; Mackeith 1916; Mackenzie 1900; Marshall 1907; Maxwell 1907; Menzies 1907; Moss, in Lack 1986, in Gibbons et al. 1993; Pennie 1950–1; Reid 1930; Ross 1897; Summers et al. 2008; Taylor 1948; Wilkinson et al. 2002; Witherby et al. 1938–41.

Notes

1. The name Capercaillie (which has a variety of spellings) probably derives either from the Gaelic *capull* = 'great horse' and *coille* = 'the wood', or from *cabher coille* = old man (of the wood).
2. Land used for the communal display in the breeding season by the males of certain birds and mammals. Believed to derive from the Swedish *leka* = 'to play'.

ACCIPITRIDAE (OSPREYS, KITES, HAWKS & EAGLES)

OSPREY *Pandion haliaetus*

Natural Distribution Widely distributed throughout much of the Holarctic, Oriental, Australasian and parts of N Ethiopian regions.
Reintroduced Distribution England.

The Osprey died out as a breeding species in Britain (Scotland) in 1916, and recolonized that country naturally in 1954. Last known breeding dates in England, recorded by Harrison (1988), were 1570 (Solent, Hampshire), 1678 (Lake District, Cumbria), 1757 (south Devon), 1838 (Lundy Island, Bristol Channel) and 1847 in Somerset.

England

In 1996–2005, a total of 75 nestlings from Scotland were 'hacked back' into the wild on Rutland Water Nature Reserve near Oakham. Sixty-four were released in 1996–2001, followed in 2005 by a further 11, mainly females, to try to restore a balanced sex ratio. First successful breeding occurred in 2001, and in 2004 two Rutland Water Ospreys bred successfully with birds in Wales. In 2006, a pair at Rutland Water fledged three young. The organizations involved in this project included Anglian Water and the Leicester and Rutland Wildlife Trust.

Characteristics

In the breeding season Ospreys in Britain are largely confined to inland waters in the neighbourhood of woods. They are entirely piscivorous.

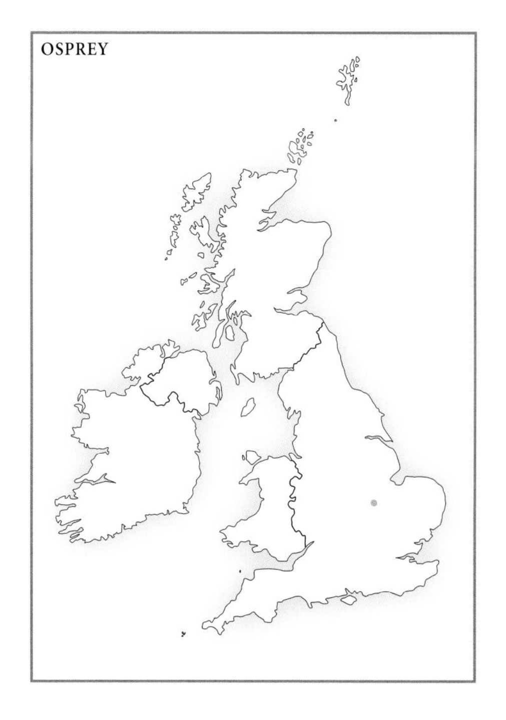

References

Carter 2007; Dennis 1996, in Wernham et al. 2002; Dennis & Dixon 1999; Witherby et al. 1939. Vol. III.

OSPREY FISHING

ACCIPITRIDAE (OSPREYS, KITES, HAWKS & EAGLES)

RED KITE *Milvus milvus*

Natural Distribution A W Palaearctic species, ranging N to 61° in Sweden and S to NW Africa, E to N Iran. Formerly abundant in Britain and Ireland,[1] but by c 1890 confined to Wales and possibly Northern Ireland.
Reintroduced Distribution England; Scotland; Ireland.

RED KITE

In the Middle Ages the Red Kite[2] was a familiar sight in both the countryside and medieval towns, including London, where it was invaluable as a scavenger of animal carcasses and waste,[3] and by the 15th century visitors were remarking on the tameness and abundance of the species in the capital's streets. Despite their depredation on domestic poultry, the birds were valued as useful scavengers until the advent, in the 19th century, of game preservation and a significant decrease in the amount of carrion available as a result of improving hygiene. Since they were easy to shoot, trap and poison, large numbers were killed. They died out in England by about 1870 and ceased to breed in Scotland around 1890 (Harrison 1988). In Northern Ireland there is evidence of possible breeding around Belfast as late as 1891. By the end of the 19th century, the population in central Wales had been reduced to only a few pairs, but thereafter it made a slow but satisfactorily steady recovery to 500–600 breeding pairs by 2006 (Carter 2007).

England; Scotland; Ireland

Although during the 20th century Red Kites became increasingly regular visitors to eastern England and, to a lesser extent, Scotland (probably originating in the burgeoning populations in central Wales and northern Europe), the only cases of confirmed breeding were in Suffolk in 1996 and 1997, and despite confirmed breeding in Herefordshire and Shropshire in 2004 and 2005, and possibly in Devon in 2006 (all almost certainly dispersers from Wales), it was clear that the species was unlikely to become re-established naturally as a widespread breeding bird (Carter 1998, 2007).

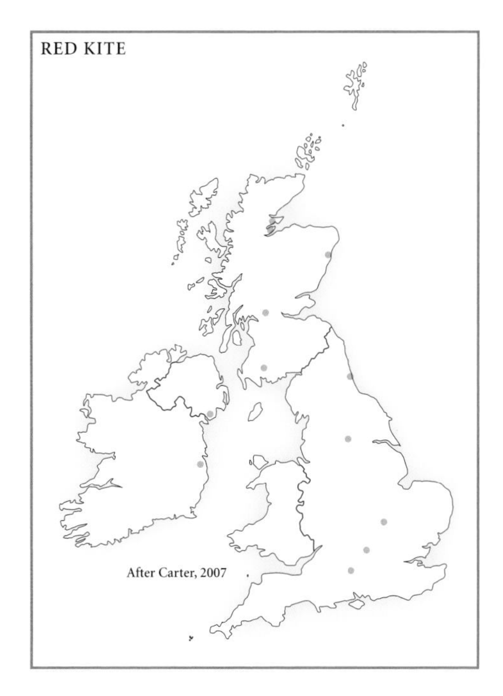

It was against this background that in 1989 the then Nature Conservancy Council and the RSPB started a programme of reintroductions to England and Scotland. In 2007, the Golden Eagle Trust began a similar project to reintroduce Red Kites to Ireland.

In 1989–94, at least 76 per cent of young birds in the Chiltern Hills and 51 per cent in the Black Isle survived their first year (Evans et al. 1999), with survival rates for older birds and ones that did not disperse considerably higher. Although the total population has increased fairly rapidly, the birds have been slow to spread far from their release sites. This is due to several factors:

- An inherited tendency, common to many raptors, to natal philopatry.
- The amount of suitable habitat available at both sites, which negates any incentive to disperse.
- The habit of local inhabitants, especially in the Chilterns, of artificially feeding 'their' Red Kites with unsuitable butchers' meat (personal observation).

The Red Kite introduction project in England and Scotland has been a major success in restoring this beautiful and graceful bird to its former haunts. By 2007, the population of 1,262–1,432 breeding pairs (>500 in England; 92 in Scotland; 670–840 in Wales) had been recorded as fledgling young (outside Wales) in Herefordshire, Shropshire, Berkshire, Buckinghamshire, Hampshire, Hertfordshire, Oxfordshire, Sussex, Wiltshire, Cambridgeshire, Lincolnshire, Leicestershire and Rutland, Northamptonshire, Co. Durham, Yorkshire, Highland, Perth and Kinross, Upper Forth and Dumfries and Galloway (Holling and RBBP 2008).

At least one bird of prey centre, near Andover in Hampshire, has released a small number of captive-bred young and rehabilitated Red Kites into the wild. Although this is a well-intentioned policy, it carries the risk of birds imprinted on humans (which is rigorously avoided in mainstream releases) losing their natural wariness and even soliciting humans for food (Carter 2007).

Although about 160 km apart, the two release sites in Ireland are separated by apparently suitable habitat, and it is hoped that the two sub-populations will eventually coalesce to form one large continuous population. Red Kites from Co. Wicklow have already

Table 25 Reintroductions of Red Kites *Milvus milvus* to England, Scotland and Ireland since 1989

Release Area	Years	Numbers	Origin	First Successful Breeding	Breeding Pairs in 2006	Remarks
ENGLAND						
*Wormsley, near Watlington, Chiltern Hills (Oxfordshire)	1989–94	93	Aragón and Navarra, N Spain (82); Wales (7) Skåne, S Sweden (4)	1992	300+	Most successful reintroduction
Brigstock & Welden, Rockingham Forest (Northhamptonshire)	1995–8	70	Segovia, Salamanca and Valladolid, Spain (35 in 1997–8); Chiltern Hills (33 in 1995–6); Wales (2)	1997	74	First use of birds from the Chilterns; 73 breeding pairs in 2007 (67 successful); 95 in 2008 (82)
Harewood House, Leeds, Yorkshire	1993–2003	69	Chiltern Hills	2000	40	Most successful release/breeding pairs ratio
Lower Derwent Valley (near Gateshead), Tyne and Wear	2004–8	94 20 (2004), 41 (2005), 33 (2006)	Chiltern Hills	2006	5	First release near conurbation. Kites now commonly seen over Newcastle suburbs. In 2007, 8 pairs reared 11 young.
SCOTLAND						
Black Isle (Highland)	1989–93	93	Skåne, S Sweden	1992	40	Much persecution on grouse moors south of release site
Argaty, Stirlingshire	1996–2001	103	Sachsen-Anhalt, Sachsen, Germany	1998	28	Slow increase due to persecution
Near Loch Ken (Dumfries and Galloway)	2001–5	104	Black Isle (54) Chilterns (41) Sachsen, Germany (7)	2003	17	Slow start but now thriving
Balnagown, Easter Ross	2005–	24 (2005–6)	Black Isle	-	-	Small release of radio-tagged birds to reinforce Black Isle release

Aberdeen ('outskirts')	2007–9	92	Black Isle and Chilterns (90); Stirlingshire (2)	-	-	Second conurbation release; 22 birds in roost spring 2008
REPUBLIC OF IRELAND						
Co. Wicklow (Redcross)	2007–11	100–120	Wales	-	-	30 released 2007 and 26 in 2008; roost mainly on Shalton and Ballyarthur estates in Avoca Valley
NORTHERN IRELAND						
Co. Down (Castle Wellan, Tollymore Forest Park)	2008–11	Up to 100	Wales	-	-	27 released 2008

Source Carter 2007, et al. *The original site selected is said to have been Windsor Great Park in Berkshire. This proposal was allegedly vetoed by the head gamekeeper, who said that the Crown Estate could have either him or Red Kites but not both!

Table 26 Breeding records for Red Kites *Milvus milvus* in Scotland, 2006–8

Area	Year	Territorial Pairs	Clutches Laid	Successful Pairs	Young Fledged
Black Isle	2007	40	39	28	65
Black Isle	2008	49	46	39	82
Stirlingshire	2007	45	34	26	59
Stirlingshire	2008	55	45	33	75–6
Dumfries & Galloway	2007	22	21	19	38
Dumfries & Galloway	2008	32	30	25	53
Aberdeen	2008	1	1	0	0
All	2006	97	85	70	154
All	2007	107	94	73	162
All	2008	137	122	97	21

Sources George Christie and Duncan Orr-Ewing, personal communications 2008.

been seen in the Irish Republic as far afield as western Co. Kerry, Co. Louth, Co. Leitrim, Co. Kildare (Naas) and Co. Dublin (Dublin city centre, Marlay Park and Killiney), and in Northern Ireland in Co. Down (Glenarm).

Characteristics

The preferred habitat of the Red Kite is hilly country with wooded valleys and cultivated districts with scattered woodland. Although primarily a scavenger, it has an exceptionally wide food spectrum, and will occasionally prey seasonally on a wide variety of small mammals and birds, and even on frogs, invertebrates and occasionally fish.

Impact

Although Witherby et al. (1939: 86. Vol. III) state that in former times 'chickens and ducklings from farmyards' formed part of the Red Kite's diet, no such predation (or on game-bird chicks or garden songbirds) has yet been reported among the British and Irish reintroduced populations.

References

Aberdeen Red Kites Newsletters 1–3 2007–8; Carter 1998, 2003, 2007; Carter & Grice 2000, 2002; Carter & Newbery 2004; Carter & Whitlow 2005; Carter et al. 1999, 2003, 2005; Craven 2007; D'Arcy 1999; Davis, in Lack 1986, in Gibbons et al. 1993; Dixon 2001; English Nature 2001; Evans et al. 1997, 1998, 1999, in Wernham et al. 2002; Golden Eagle Trust 2006, 2007, 2008; Harrison 1988; Holling & RBBP 2007, 2008; Holloway 1996; Kalaher 2004; Kalaher et al. 2008; Lovegrove et al. 1990; McGrady et al. 1994; Minns & Gilbert 2001; Ogilvie & RBBP 2003; Orr-Ewing et al. 2006; Scott 2002; Sharrock 1976; Snell et al. 2002; Witherby et al. 1938; Wildman et al. 1998; Wotton et al. 2002.

Notes

1. Although it was formerly believed that the Red Kite was never indigenous in Ireland, recent archaeological, linguistic and historical evidence suggests that it was widespread and abundant until well into the 19th century (D'Arcy 1999).
2. The suggestion that it was the closely related Black Kite that scavenged in British towns is erroneous.
3. From the 15th century, Red Kites (and Common Ravens) were given legal protection (the first non-game species to be so treated) because of their value as urban scavengers (Carter 2007).

ACCIPITRIDAE (OSPREYS, KITES, HAWKS & EAGLES)

NORTHERN GOSHAWK *Accipiter gentilis*

Natural Distribution Much of the Holarctic region, but declined in Western Europe in the 1960s/1970s, where it is now recovering.
Reintroduced Distribution England; Scotland; Wales; [Ireland].

The Northern Goshawk was probably as common in the extensive forests that covered much of Britain in the Middle Ages as it is in continental Europe. Its slow decline can be attributed to:

- Widespread deforestation.
- The introduction of firearms, which replaced birds of prey in falconry.

NORTHERN GOSHAWK, NORTH WALES

- The increasing practice of rearing game-birds artificially, which brought raptors into conflict with landowners and gamekeepers.

Petty (1996), from whom much of the following account is derived, has traced the history of the Northern Goshawk in Britain.

By the first half of the 19th century, the birds were scarce and heavily persecuted. Baxter and Rintoul (1953) and Bannerman and Lodge (1956) gathered a few breeding records for Scotland after 1850, mainly in the valleys of the Spey, in the woods of Castle Grant and Dulnan (Dulnain) (More 1865), and Findhorn, and a single instance in 1871 at Balmacara in Wester Ross. The last breeding record for Scotland may have been at Rohallion in Perthshire in the early 1880s (Harvie-Brown 1906). The last breeding records in England (and according to Harrison 1988, in Ireland) date from around the same period. A pair attempted to breed at Normanby Park in Lincolnshire in 1864

(Cordeaux 1899) and in Westerdale, Cleveland in Yorkshire in 1893, but both falcons[1] were shot (Nelson et al. 1907). The many records in eastern England in the late-19th century (almost certainly of birds from western or northern Europe) may have led to sporadic but undocumented breeding.

Thereafter there were no further breeding records until 1950, when three pairs bred in Sussex, where breeding may have occurred sporadically since 1921 or 1938; this small population was eradicated by man (Meinertzhagen 1950). Petty (1996) believed that these birds may have been natural colonists from Europe.

England; Scotland; Wales; [Ireland]

Northern Goshawks were re-established in Britain as a result of deliberate releases by falconers. One of the earliest breeding records, however, in 1951 in Shropshire, involved a female that had escaped from a local falconer (Parslow 1973). In the late 1960s and 70s, large numbers were imported to Britain from central Europe by falconers, in the expectation that the eyasses[2] could be removed from nests for falconry, and breeding became fairly widespread. In 1970–80, it was estimated that some 20 falconry birds a year escaped into the wild, with a further 30–40 deliberately released in small groups (Kenward et al. 1981). Most of the birds imported in the 1970s came from Fennoscandia (Marquiss 1981), and the British population is thus entirely composed of the nominate race *A. g. gentilis*. By the early 1980s, Harrison (1988) estimated the population to number around 100 pairs, with a post-breeding total of about 300 birds. The success or failure of these new populations varied considerably, depending on the amount of human persecution, most of which was (and still is) caused by egg collectors and gamekeepers.

Since the species is known to have been declining in mainland Europe due to contamination by the widespread use of organochlorine pesticides, natural recolonization of Britain is highly unlikely. Furthermore, many newly established colonies were in areas far removed from the English Channel, and with a known history of releases.

By the mid-1990s, discrete sub-populations were starting to amalgamate. This dispersal was due largely to the increasing availability of maturing coniferous forests. By about 1996, there were estimated to be c 120 pairs in England, 80 in Scotland and perhaps 200 in Wales, most of which were in Forestry Commission forests with little or no persecution and a robust breeding performance.

After a slow rate of increase up to the mid-1990s, followed by a period of relative stability, the number of birds reported in 2005 increased again to a new peak, although the latest estimate of 249–424 pairs is considered by Holling and RBBP (2008) to be an underestimate. Nevertheless, the higher figure exceeds that of 410 pairs given by Baker et al. (2006) and the total of about 400 pairs quoted by Carter et al. (2008). Holling and RBBP received reports as follows:

- England (120 pairs confirmed breeding plus 92 other pairs in 22 counties).
- Scotland (47 pairs confirmed breeding plus 35 other pairs).
- Wales (82 pairs confirmed breeding plus 44 other pairs in 13 counties).
- Ireland (4 pairs reported but as yet no confirmed breeding reported).

The Northern Goshawk's principal breeding range today is Wales, the Forest of Dean (Herefordshire) and the southern Pennines (northern England), the Borders and north-eastern Scotland. It is estimated that there is enough unoccupied suitable habitat available in Britain to support, if unmolested, a population of several thousand pairs.

Characteristics

Petty (1996) has also examined the ecology of British Northern Goshawks. Although principally a forest species, the Northern Goshawk is non-specific in its habitat requirements, and in Europe (though not in North America) is not inhibited by forest fragmentation. The birds breed in a wide range of woodland habitats ranging from semi-natural lowland deciduous woods to upland conifer plantations, the only essential requirement being a tree with sufficiently strong branches to bear the substantial nest and good flight access. Northern Goshawks are equally catholic in their choice of prey, feeding on whatever is locally most available. In Britain, Common Woodpigeons and corvids comprise about 68 per cent of their diet during the breeding season. There is a near absence of other predators for the Northern Goshawk's preferred size of prey, which should help it to become more widely established. There is some overlap with Eurasian Sparrowhawks and Peregrine Falcons, but the former concentrate on smaller items and the latter hunt in open habitats, while Northern Goshawks prefer to do so in a wooded environment.

Impact

Because Northern Goshawks concentrate on whatever prey animals are most abundant and vulnerable, such as Common Woodpigeons, feral pigeons and European Rabbits, they are likely to have a greater impact on common than on scarce species. Since the abundance and vulnerability of the prey animals varies according to season, Northern Goshawks tend to concentrate on one or two species until not many are left and then to switch to the next most available species. On the other hand, they are opportunistic predators and prey on rarer species whenever the opportunity to do so arises. However, healthy wildlife populations produce more offspring annually than are needed to replace adult losses, and predation will only have a negative impact on prey species if their populations are already in decline due to other factors.

Concern has also been expressed by landowners about the effect of Northern Goshawks on the poults of Common Pheasants. Petty (1996) suggests the construction of safer release pens and the use of taste-aversion chemicals and acoustic devices to deter the raptors. The likelihood of landowners and gamekeepers acting on this suggestion, however, must at best be slim.

'Goshawks are once again part of the British avifauna,' Petty (1996: 101) concludes, 'and have the potential to become far more abundant and widespread in the future. While this will be welcomed by many, there will be concern too, particularly about the impact of goshawks on prey species, although this is likely to be limited.'

References

Baker et al. 2006; Baxter & Rintoul 1953; Carter et al. 2008; Cordeaux 1899; Fitter 1959; Harrison 1988; Harvie-Brown 1906; Holling & RBBP 2008; Hollom 1957; Holloway 1996; Kenward 1981, 1983, 2006; Kenward et al. 1981; Lever 1987; Marquiss 1981, in Lack 1986, in Gibbons et al. 1993, in Wernham et al. 2002; Marquiss & Newton 1982; Meinertzhagen 1950; More 1865; Nelson et al. 1907; Newton 1993; Opdam et al. 1977; Parslow 1973; Petty 1989, in Holmes & Simons 1996; Sharrock 1976.

Notes

1. A female hawk, especially a Peregrine Falcon or Northern Goshawk. The male is known as a tiercel (or tercel), from the Latin *tertius* (third), perhaps from the belief that the third egg from a clutch produced a male.
2. An unfledged nestling hawk.

ACCIPITRIDAE (OSPREYS, KITES, HAWKS & EAGLES)

WHITE-TAILED EAGLE *Haliaeetus albicilla*

(SEA EAGLE)

Natural Distribution A widely but thinly and discontinuously distributed Palaearctic species occurring from Iceland E to Kamchatka, and from c 70°N in Scandinavia and Russia S to c 30°N in Europe and Asia. A separate form, *H. a. groenlandicus*, occurs in SW Greenland. Formerly widespread in Scotland and Ireland and in parts of England, mainly the Lake District and Isle of Man.
Reintroduced Distribution Scotland; ? England; Ireland.

WHITE-TAILED EAGLE FISHING

Since the Anglo-Saxon period, when extensive tracts of woodland began to be felled and fenlands drained for agriculture, the habitat of the magnificent White-tailed Eagle has been steadily eroded. Viable numbers might, however, have managed to survive on the more remote coasts of north-west Scotland and Ireland had not the human population in the 19th century increased and partially dispersed to the coast, where changing land usage brought man and eagle into conflict, and many eagles were killed.

As a result of human persecution, the last native White-tailed Eagle's nest on the English mainland was recorded shortly before 1800 (a single pair survived on the Isle of Man until 1818), and in Ireland, where there were at one time at least 50 eyries, in Co. Mayo, in 1898. Although on the Scottish mainland only a few pairs bred into the very early years of the 20th century (e.g. on Ardnamurchan Point, Highland), nesting did persist later on some of the offshore islands such as the Shiants (off Lewis), Rum, Yell (in the Shetlands) in 1910, and on the remote headlands of Skye, where the last pair bred at Dunvegan Head in 1916. In Wales, White-tailed Eagles are believed to have disappeared in the early 19th century.

Scotland

The earliest documented attempt to reintroduce White-tailed Eagles was made in 1959, when Pat Sandeman unsuccessfully released an adult and two juveniles from Norway in Glen Etive, Argyllshire. A second attempt by the RSPB in 1968, when two pairs from Norway were freed on Fair Isle, south of Shetland, similarly failed. Both these attempts were unsuccessful almost certainly because neither release was reinforced by further liberations in succeeding years.

It was against this background that in 1975 the then Nature Conservancy Council began a concerted effort to re-establish White-tailed Eagles in Scotland. The attempt was considered both morally and scientifically justified on several grounds:

- The birds were unlikely to recolonize Scotland naturally.
- Vast areas of the western Highlands and coast consist of suitable habitat.
- Deliberate persecution (though it still occurs) has been greatly reduced as a result of a more tolerant attitude to birds of prey.
- The dissemination of a species' population (the White-tailed Eagle was then classed as Vulnerable by the IUCN) is a recognized means of spreading the risk to a threatened species.

The 10,820-ha Island of Rum[1] (a National Nature Reserve in the Inner Hebrides) was selected as the release site for a number of reasons:

- It lies at the heart of the species' former range and is surrounded by islands with similar suitable habitat.
- Potential prey, in the form of seabirds and fish, is in abundance, and Red Deer and feral goats provide a ready source of carrion.
- White-tailed Eagles are notorious pirates, and some of the species from which they steal occur in plenty.
- Northern Fulmars, which can cause the deaths of White-tailed Eagles by clogging their feathers with regurgitated stomach oil, are relatively few.
- On other neighbouring islands and on the mainland European Rabbits, Mountain Hares and sheep carcasses, which are key constituents of the birds' diet, abound.

Norway was chosen as the source of supply because it had the most flourishing population in western Europe, which is, moreover, genetically close to the former British and Irish population.

Love (1983a, 2007) and Bainbridge et al. (2003), from whom the following account is derived, have summarized the reintroduction of the White-tailed Eagle to Scotland.

The first 4 birds (1 male and 3 females) arrived on Rum in June 1975, and by 1985 a total of 85 eaglets – mostly aged about 8 weeks – had been imported, of which 82 (3 having died while in quarantine) were ringed or wing-tagged and 'hacked back'[2] into the wild. Although seven were subsequently found dead, and most of the survivors stayed within an 80-km radius of Rum, within a few years sightings of others were reported 350 km north in Shetland and over 200 km south in Co. Antrim in Northern Ireland. The young birds' survival rate in the wild (possibly as high as 60–70 per cent) was attributed to an absence of competing adults, the tolerance of the more aggressive Golden Eagles and perhaps a reduction in the pollution levels of the sea around Rum.

The first British White-tailed Eagle clutch of eggs for 67 years was laid in 1983. Although this and other clutches in that year and in 1984 failed to hatch, in 1985 four pairs laid eggs on the Isle of Mull, of which at least 2 clutches hatched and 2 birds fledged successfully – the first White-tailed eaglets to be reared in Britain for 69 years. Successful breeding has occurred annually since 1985.

In 1981–7, the breeding population increased from 1 to 9 pairs, although the success rate was relatively low (34 per cent) with no more than 4 successful breeding pairs in any year before 1992. A total of 29 young fledged during this period.

It was judged that the small size of the population made it vulnerable to stochastic effects, such as the deaths of adults and the failure of pair formation, so it was decided that a second phase of releases was required. The site selected was at Letterewe in Wester Ross on the Scottish mainland, which provided a wider range of habitats than on Rum, including deciduous woodland and landlocked lochs. Here, in 1993–8, a further 58 young birds were released.

In 1993–2000, the breeding population increased from 8 to 22 territorial pairs, 19 of which bred in 2000. Wild-bred birds first nested in 1996, rearing young successfully two years later. In 1993–2000, there were 48 cases of successful breeding (compared with 18 for 1985–92). A total of 71 young fledged during that period.

The first breeding by birds released during the second phase occurred in 1998, when two pairs reared young successfully. The 2000 post-breeding population was estimated to be at least 80–90 individuals. By 2002, a cumulative total of 31 territories had been occupied, and the breeding population had increased to 26 territorial pairs, 24 of which bred that year. In 2003, a total of 26 young fledged from 25 nests of 31 territorial pairs. In 2004, when a total of 168 birds had fledged, there were 32 territorial pairs, and 28 nests produced 19 fledged young. Individuals had been observed in many parts of Scotland, including the Orkneys and Shetlands, and also once in England (Bowland, Lancashire) and Northern Ireland.

By 2004, the total population had increased to 250–300 individuals, and two years later the 200th eaglet since the start of the reintroduction project fledged successfully. In 2005, 28 pairs out of 33 territorial pairs fledged 24 young, and in 2007 there was a total of 42 occupied territories.

In an attempt to establish a new sub-population, 15 young White-tailed Eagles from Norway were released in 2007 in northern Fife, followed by a similar number in 2008, and up to a further 65 will be freed in eastern Scotland during the next 5 years. So far, birds from this release have been observed as far afield as Aberdeenshire, Angus, Stirling and Berwick-on-Tweed in England.

?England

Natural England is leading a consortium that is assessing the feasibility of releasing White-tailed Eagles between King's Lynn and Hunstanton in north-west Norfolk. Initial studies suggested the Suffolk coast as a suitable release site, but concern was expressed about the possibility of predation on rare Great Bitterns at the RSPB reserve at Minsmere and elsewhere along the coast – the stronghold for the species in Britain. While there is no evidence of White-tailed Eagles affecting Great Bitterns where they occur together in continental Europe, it was felt that, when equally suitable release sites exist elsewhere, it would be wrong to place them at even theoretical risk. Within a few years Great Bitterns should be more widely distributed, and the Suffolk coast sub-population will be less critical to the species' survival.

If the result of the study is positive and the application for a licence successful, the founder stock of White-tailed Eagles will be sourced from Poland (which has a healthy population of some 800 pairs that fledge about the same number of birds annually), because Polish birds nest mostly in trees (as they would have to do in East Anglia) rather than on cliffs as they do in Norway and Scotland and Ireland (Andrew Smith and Richard Saunders, personal communications 2008).

Even if the application for a licence is unsuccessful, natural recolonization of East Anglia, where at present White-tailed Eagles are only irregular winter visitors, from the expanding population in lowland Europe is probable (Andy Clements, personal communication 2009).

Ireland

In late 2007, 15 birds from Norway were released by the Golden Eagle Trust on the Iveragh Peninsula in Killarney National Park in Co. Kerry in south-west Ireland, at the start of a project to re-establish the species in the Irish Republic. These were followed by a further 20 birds in 2008. The aim is to release a total of 90–100 birds by 2011. It is considered that the coastal area within a 50-km radius of the National Park (the breeding range of White-tailed Eagles in Scotland during the first phase of reintroductions), from Kerry Head in the north to Cape Clear in the south, could support a breeding population of up to 69 pairs (Halley et al. 2006). Most sightings have so far been in the vicinity of Lough Leane, though individuals have been recorded from the Dingle Peninsula, the Skellig Islands, Ballybunion and elsewhere, including in Co. Cork (Allan Mee, personal communication 2008).

Characteristics

White-tailed Eagles frequent cliffs and rocky coasts and islands, and inland lakes in wooded districts. Although in Scotland the majority of nests have occurred on cliff ledges, some have been built in trees. The species' principal food in Scotland is Mountain Hares, European Rabbits, seabirds, fish and carrion.

Impact

Marquiss et al. (1999, 2003) examined the impact of White-tailed Eagle predation on lambs on the island of Mull. They found that some birds do eat considerable numbers of lambs, and that although most are scavenged some are killed by the birds. The majority of lambs are scavenged in May, when they are newly born and natural mortality is high. At least some of those that are killed are sickly and would probably have otherwise died of natural causes. The heaviest lamb losses tend to occur in years when European Rabbits are in short supply. Although the loss of lambs to the birds is relatively small and the loss to the island's farming economy is more than offset by the £1.4 million of income from tourism directly attributable to the presence of White-tailed Eagles, this does not preclude the possibility of a significant impact on individual affected farmers.

So far, few effects have been observed of competition between reintroduced White-tailed Eagles and Golden Eagles for either food or nesting sites. The potential impact *on* (rather than *of*) White-tailed Eagles is considerable, both through shooting, trapping and poisoning by farmers and gamekeepers, and through theft by oologists, and several have been lost through human persecution.

References

Bainbridge et al. 2003; Carter et al. 2008; Collier 1986; Dennis 1968–9; Evans et al. 2001; Green et al. 1996; Grice 2005; Halley 1998; Halley et al. 2006; Harrison 1988; Holling & RBBP 2007, 2008; Irish Raptor Study Group 2007; Jourdain 1912; Lever 1984c; Lister 2007; Love 1978, 1980a, b, c, d, 1983a, b, c, d, 1983–4, 1988, in Gibbons et al. 1993, 2003, 2007; Love & Ball 1979; Marquiss et al. 1999, 2003; Nature Conservancy Council 1985; Ogilvie & RBBP 2003; Sandeman 1965; Ward 2008; Watson et al. 1992; White-tailed Eagle Project Steering Group 2006; Witherby et al. 1939 (Vol. III); www.naturalengland.org.uk/regions/east/white-tailed-eagle.htm; Yalden 2007.

Notes

1. The correct ancient Norse or Gaelic spelling, preferred to the modern spurious Rhum.
2. The release of captive-reared birds of prey into the wild.

ACCIPITRIDAE (OSPREYS, KITES, HAWKS & EAGLES)

GOLDEN EAGLE *Aquila chrysaetos*

Natural Distribution A Holarctic species, ranging in Europe from 70°N in Norway S to S Iberia and N Africa, E to the Himalayas. In N America occurs from Alaska S to California. Still breeds in uplands of Scotland, and rarely in England.
Reintroduced Distribution Ireland.

The magnificent Golden Eagle was formerly widely distributed in Ireland, but last bred in Glenveagh, Co. Donegal (where the last one was trapped in 1926) in 1910, and perhaps in Co. Mayo in 1911 or 1912; a pair bred on Fair Head, Co. Antrim in 1956–60. Its demise in Ireland was due to persecution by gamekeepers and farmers, and to loss of habitat.

Ireland

In 2001–8, a total of 53 young Golden Eagles from Scotland (Glen Affric, Inverness-shire; Lude, Tayside; Glenfinnan, Lochaber; Isle of Skye; Assynt, Sutherland; Cowal; Island of Uist; Badenoch) were released in the 16,000-ha Glenveagh National Park in the Derryveagh Mountains of Co. Donegal in the Irish Republic, where the release of a further 15–20 is planned and where it is hoped that by 2010, 6–8 pairs will have become established. Nesting took place in 2005, and in 2007 a chick fledged successfully in Glenveagh, the first to do so in Co. Donegal in 97 years. It is hoped that eventually the Golden Eagle will become re-established in at least part of its former range along the west coast of the Irish Republic and perhaps also along the northern coast of Ulster. At least 23 potential Golden Eagle home ranges have been identified in the Irish Republic from Co. Donegal in north-western Ireland south to Co. Galway, and it is believed that in the south-west Co. Kerry and Co. Cork could support up to 20 pairs. Several other counties,

GOLDEN EAGLE

including Wicklow, could support up to 5 pairs each, and Northern Ireland perhaps 4–6 pairs. In total the island of Ireland should be able to support at least 50 pairs of Golden Eagles (Lorcan O'Toole, personal communication 2008).

GOLDEN EAGLE

Impact

Although the reintroduced Golden Eagles in Glenveagh may have killed two lambs in the last seven years, they also take a toll of Red Foxes and Hooded Crows.

References

Golden Eagle Trust 2006, 2007, 2008; Harrison 1988; Holloway 1996; Irish Raptor Study Group and Curlew Trust 2006; O'Toole et al. 2002; Watson 1997; Witherby et al. 1939; www.goldeneagle.ie

OTIDAE (BUSTARDS)

GREAT BUSTARD *Otis tarda*

Natural Distribution From SW and C Europe and NW Africa discontinuously eastwards through Kazakhstan to Mongolia, the Russian Far East and NE China.
Reintroduced Distribution England.

GREAT BUSTARD MALE DISPLAYING

'Formerly bred many parts England and in ancient times in SE Scotland. Last survivor Yorks 1832 or 1833, last bred Suffolk 1832 and Norfolk 1830 … Possibly some indigenous hens until 1845 in Norfolk' (Witherby et al. 1938–41. Vol IV: 439).

Although the precise reasons for the Great Bustard's extinction in England are not known, probable causes include a combination of:

- Field enclosure in the late 18th/early 19th centuries, which reduced the amount of desirable 'prairie'-like habitat.
- Agricultural mechanization and changes in farming practices, resulting in the destruction of many nests.
- Over-hunting for food and as trophies, aided by the advent of efficient breech-loading shotguns.
- Climate change, with increasingly cold winters and cool springs that would have increased winter mortality and reduced breeding.

England

Indigenous Great Bustards frequented mainly the open downs and chalk uplands of Wessex and the grassy heaths and brecks of East Anglia. The apparently most suitable surviving habitat in England is on Salisbury Plain, a large area of calcareous grassland in southern Wiltshire. Over half of Salisbury Plain is an EU designated Special Protection Area, and parts are classified as a Special Area of Conservation. Around 38,000 ha of the plain are used by the military as a training area, and the region is covered by a wildlife-management plan that ensures that areas of conservation importance are managed sympathetically by the army and others.

It was here in the 1970s that the Great Bustard Trust (GBT), founded by the Hon. Aylmer Tryon, introduced Great Bustards to a 4-ha pen on Porton Down. Although one chick was hatched at Porton Down it did not survive to adulthood, and none were ever released into the wild. In 1989, the surviving birds were transferred to the Whipsnade Animal Park in Bedfordshire, and in 1997 the GBT was wound up.

In the following year a successor, the Great Bustard Group (GBG) was formed, and in 2002 an application was made to the Department of Environment, Food and Rural Affairs for a licence to reintroduce to England up to 40 Great Bustards a year for 10 years; this was approved in 2003.

Since breeding Great Bustards in captivity has seldom been achieved successfully (which was the reason for the failed attempt by the GBT), the GBG decided that any reintroduction must be made with young reared from eggs gathered in the wild. It seems likely that the original British population was part of the mega-population of eastern and central Europe, where only Russia, and especially the Saratov region south-east of Moscow, has a large population. It was considered that sourcing eggs (many of which would otherwise be destroyed through agricultural practices) for England from Saratov could be done without detriment to the donor population. It was judged that 75 eggs per annum from threatened nests would need to be collected.

On 2 August 2004, 28 eight-week-old Great Bustard chicks from Saratov arrived at the GBG's facility on Salisbury Plain, where they were placed in quarantine and where a week later four died. In September the survivors were transferred to a larger soft release pen, where two males were injured and a decision was made to release the remaining 22 birds into a large, roofless enclosure (which they were free to leave at will) rather than risk further injuries. Of those that left this enclosure before the end of the year, two were taken by Red Foxes and a further two were killed by flying into barbed-wire fences.

In July 2005, 37 chicks arrived from Russia, followed a year later by a further 9. In the spring of 2007, a female Great Bustard nested on Salisbury Plain – the first to do so for 175 years; 2 eggs were laid but were deserted by the female and were later found to be infertile. In August six young birds arrived in Wiltshire from Russia, and having been fitted with GPS satellite transmitters were released into the wild. In recent years the project has been hampered by the small number of nests being found in Saratov; it is believed that exceptionally hot temperatures in Russia may have contributed to lower breeding success and also caused agricultural operations, which destroy nests and eggs, to be advanced by several weeks. In 2008, a further 19 chicks from Saratov were released, and another nest with infertile eggs was found on Salisbury Plain; this infertility may be because the males are as yet still too young to breed.

Although the anthropogenic and other factors that led to the Great Bustard's demise may have been removed, new and as yet unknown ones could materialize. Osborne (2005: 23) claims that 'the illegal killing of bustards is unlikely to pose a serious threat' which, given the birds' culinary reputation (which in fact is said to be misplaced) and value as a trophy may prove somewhat optimistic. The birds' high mortality rate of up to 80 per cent in their first year, their intolerance of disturbance when nesting, their habit of dispersing in autumn and winter, and their undoubted vulnerability to poaching must make the establishment of a viable population problematic.

References
Anon 2005b; Collar & Goriup 1980; Dawes 2006; de Bruxelles 2008; Harrison 1988; Osborne 2002, 2005; Thomas 2000; Waters 2001; Witherby et al. 1938–41; Woolcock 2007; www.greatbustard.com

STRIGIDAE (OWLS)

EURASIAN EAGLE OWL *Bubo bubo*

Natural Distribution A Palaearctic species, occurring in Europe N to the Arctic Circle in Scandinavia and S to the Iberian Peninsula, North Africa and Italy, E to Manchuria, China and Japan.
Reintroduced Distribution England; Scotland.

There is fossil evidence for the presence of Eurasian Eagle Owls in Britain throughout most of the Ice Age to around 9,000–10,000 BP, and perhaps into the Holocene when, according to Harrison (1988), they may have been exterminated by man. There is no evidence for a natural recolonization of Britain, and all the following records are almost certainly of birds that escaped from captivity or were deliberately released.

England; Scotland

Holling and RBBP (2007), from whom the following account is derived, have reviewed the records of Eurasian Eagle Owls in Britain.

The first confirmed record of attempted breeding came from a quarry in Morayshire in 1984; this attempt failed but in the following year the same pair nested again in a neighbouring quarry, where it successfully reared a single chick. The male was killed on a nearby road later in the same year, but the female remained in the area for at least a further decade, laying infertile eggs in at least seven years.

EURASIAN EAGLE OWL

In 1993, a nest containing four eggs was discovered in the Pennines; although this nest was later deserted, sightings of Eurasian Eagle Owls in the area were made later in the year and again in 2000 and 2001.

Breeding in Yorkshire was first confirmed in 1997, although a pair was known to have been present in the area since at least the previous year. When first seen the female had jesses trailing from her tarsi, indicating a captive origin. This pair continued to breed annually on remote moorland until 2005, raising a total of 23 young, all of which were ringed before fledging; 2 of these, both from the 2004 brood, have been recovered dead, one under power lines in Shropshire in 2005 and the other in the Borders in 2006. Despite surveillance this nest was vandalized on at least three occasions, once when the eggs were on the point of hatching. The female was found dead in December 2005; pellets found in her sternum showed that she had been shot but not killed, and had subsequently starved to death because the injury made her unable to eat.

Early in 2006 another female was reported in the same area, and although she was seen with the male the pair did not breed. A probable male found in Birmingham was ringed and released in the same locality in May, but is believed to have been driven away by the resident male, which remained at the site until 2007.

Elsewhere in northern England, another pair laid a clutch of infertile eggs in 2003, but only one bird was observed in 2004 and there have been no subsequent records of nesting from this site.

In 1996–2002, further individuals were reported from four separate counties, including the same area in the Pennines where breeding was attempted in 1993. In Highland in 1996, a bird was seen close to an aviary containing captive Eurasian Eagle Owls, whose owner claimed that the visiting bird had not belonged to him. Elsewhere in Highland, a male held territory in 1997–8, where calling was heard in January–March in the former year, and a nest scrape was found in January of the latter year. In 1999–2002, a male held territory in Warwickshire, and in the latter year individuals were reported in Norfolk and also in Highland.

One or two birds have been reported in parts of Lancashire since 2004, and in 2006 two eggs were laid but subsequently abandoned. In the following year what may have been the same pair nested close to a public footpath at Dunsop Bridge near Clitheroe in the same county, where with three young in the nest the adults took to mobbing pedestrians and their dogs, forcing the county council to close the right of way. At another site in northern England in 2007, the behaviour of a pair suggested young near by. In 2006, there were reports of the presence of at least 10 other birds from a minimum of 7 localities in central or northern England and southern Scotland, including 2 pairs.

This evidence suggests, as Holling and RBBP point out, that Eurasian Eagle Owls are maintaining a small presence in at least parts of central and northern England and possibly southern Scotland, with a maximum to date of 3 pairs breeding in any season and a maximum, in 2006, of around 13 other individuals. Although it is apparent that the number of records has increased since 2004, it is not yet clear whether this is a consequence of growing publicity or an actual increase in the wild population.

Characteristics

In continental Europe, Eurasian Eagle Owls live in extensive forests or the vicinity of cliffs, crags, ravines and rocky ground in both wooded and wild open country, including mountains and barren fells. No nest is constructed, the 2–3 eggs being laid in a simple scrape on the ground on ledges in gullies, on steep hillsides or in rocky crevices. The species' principal food is mainly small mammals such as European Rabbits, Common Rats, Water Voles and European Hedgehogs, but also deer calves and Red Foxes. Most frequently eaten birds include Carrion Crows and Common Woodpigeons.

Impact

Concern has been expressed that the magnificent Eurasian Eagle Owl could prey, as it does in other parts of its range, on such species as Common Buzzards, Peregrine Falcons and Ospreys. The birds have also been known to take trout from fish farms, and are probably predators of game-bird chicks. Their ability to survive in Britain depends on both the amount of suitable habitat available and the tolerance of humans.

References

Carter et al. 2008; Dennis 2005; Elliott 2005; Harrison 1988; Holling & RBBP 2007; Ogilvie & RBBP 2004; Stewart 2007; Turk 2004.

PROPOSED REINTRODUCTIONS

If several of the species that have already been introduced to Britain and Ireland have provoked objections in some quarters, the proposed reintroduction of such erstwhile native species as the Grey Wolf *Canis lupus*, Brown Bear *Ursus arctos* and Eurasian Lynx *Lynx lynx* – which survived into the Mesolithic – has caused widespread concern among farmers and landowners. Although we are unlikely ever to see such predators living truly in the wild (except perhaps on offshore islands containing a stock of Red Deer as prey), one Scottish landowner is currently proposing their release on a large fenced estate in Sutherland. This project has been widely trumpeted as the 'rewilding' of part of the Scottish Highlands, whereas at best it will amount to little more than a glorified 'safari park'. In any case, the ethical and legal aspects of having both predators and prey contained within a fenced enclosure could provide a fertile ground for the legal profession and those involved in animal welfare.

Grey Wolf

Yalden (1986) made a seriously reasoned argument for reintroducing Grey Wolves (which became extinct in England probably in the late 13th century, Scotland in 1680–1700 and Ireland in 1692–1786) to the Hebridean island of Rum. The island supports a population of some 1,600 Red Deer and the surplus annual production should be sufficient to support up to 20 Grey Wolves. Although they might increase to such an extent that excessive predation on Red Deer calves might occur in summer, the population has plenty of spare reproductive capacity. More calves would survive, and more hinds would give birth annually, were the population to be reduced so that there was more available food. Most adult Red Deer killed would be sick or injured animals. Carrion left by the Grey Wolves would benefit both Common Ravens and reintroduced White-tailed Eagles.

The controlled release of Grey Wolves in the Scottish Highlands would reduce the excessive Red Deer population (thus both relieving estates of the annual cost of culling hinds and reducing the number of ticks that cause Lyme's disease in humans), improve bird and plant diversity, aid forest regeneration, increase tourist revenue and provide jobs in a region of declining agriculture (Nilson et al. 2007). Although Grey Wolves pose an undoubted threat to sheep, where large ungulates occur with sheep they tend to prey preferentially on the former. The danger that they pose to humans is almost entirely imaginary. The most serious threat from them is as vectors of rabies, and strict quarantine precautions would be essential before any release (Yalden 1999).

When salmon are migrating upriver in British Columbia they can comprise up to 70 per cent of Grey Wolves' diet. Fish are a safer and less exhausting prey to catch than deer, and their flesh is of much higher calorific value than venison. Dr Christopher Darimont of the University of Victoria believes that Grey Wolves may well also have once similarly fished in British rivers: were they to do so again they would be likely to come into conflict with riparian owners.

Eurasian Lynx

Radiocarbon dating of bones and references in ancient literature suggests that the Eurasian Lynx, which was previously thought to have died out in Britain due to climate change some 4,000–10,000 years ago, may have survived until medieval times, when it probably succumbed to deforestation, declining numbers of Red Deer and human persecution. Computer analysis suggests that reafforestation and recovering Red Deer populations (with the expansion of Sika) provide sufficient habitat and prey in parts of Scotland and northern England to support a fragmented population of around 450 Eurasian Lynx (Hetherington et al. 2006, 2008).

Common Crane

The Common Crane *Grus grus* is believed to have bred in Ireland until perhaps the 14th century, and certainly nested in the ferns and swamps of England until it was eradicated by hunting around 1600.

Since 1981, a very small number of natural migrants from Europe have occasionally nested in East Anglia where, however, the birds have at best only a tenuous foothold. In 2009, funding from Viridor Credits Environmental Company through the Landfill Communities Fund enabled a consortium, led by the Wildfowl and Wetlands Trust, to start work on the construction of a rearing facility – based on a pilot model created in 2007 – at Slimbridge in Gloucestershire. Common Crane eggs from Germany will be imported in spring 2010 and incubated, hatched and hand reared at Slimbridge. At the age of eight weeks, the young birds will be transferred to a large holding pen at the release site in the Somerset Levels, where in autumn 2010 they will be fitted with satellite transmiters and released into the wild. The Somerset Levels were historically the stronghold of the Common Crane, and still provide the most suitable habitat for the species in the UK (Great Crane Project Press Release 26 February 2009, per Andrew Parker, WWT).

Burbot

The last native British Burbot *Lota lota*, a north Holarctic species, was recorded in the Old West River in Cambridgeshire in 1969 or in the Great Ouse system in 1972. Possible reasons for the species' extinction include being on the edge of its natural range; its preference for spawning in water under ice; increasing water temperatures in rivers as a result of global warming and, e.g. in the Trent, because of heated effluent discharged from factories; and pollution and land drainage resulting in habitat fragmentation.

In 2004 an application was made for a licence under the Import of Live Fish Act 1980 to import Burbot from Denmark to Brooksby Melton College in Leicestershire. The application was initially refused on the grounds that the monogenean parasite *Gyrodactylus salaris* was endemic in Denmark. Subsequently, since Burbot are not susceptible to *G. salaris*, and because an undertaking was given that the fish would never be released from the secure quarantine facility at Brooksby, a licence was granted and in February 2006, 42 Danish Burbot arrived at Brooksby. Here the fish are being studied to: try to determine the reasons for the species' extinction; assess whether such causes would prevent re-establishment; review the Burbot's ecological requirements; calculate which, if any, rivers within the species' historic range might be suitable for a reintroduction project; establish a spawning protocol to determine whether Burbot eggs will hatch in normal English winter temperatures (Wellby et al. 2006).

References

de Bruxelles 2007; Elliott 2008; Evans, in Davies et al. 2004; Hall, in Harris & Yalden 2008; Kitchener & Yalden, in Harris & Yalden 2008; Maitland 2004; Marlborough 1970; Nilson et al. 2007; Wellby et al. 2006; Yalden 1986, 1999; Yalden & Kitchener, in Harris & Yalden 2008.

EPHEMERAL SPECIES

EPHEMERAL SPECIES ARE, BY DEFINITION, those that only established breeding populations in the wild for a limited period. Some, such as the Coypu and Muskrat, became established in large numbers, over a wide area and for a considerable length of time. Others, like the Oriental Short-clawed Otter and Red-winged Laughingthrush, only established tenuous breeding populations for a short period in small numbers and in a limited area. No specific time period has been selected for a species' survival in the wild to qualify it for inclusion as 'ephemeral' – the Coypu, which survived in the wild between 1944 and 1989 was, by a considerable extent, the longest-surviving ephemeral species.

The variety of exotic species that have the potential to occur in the wild in Britain and Ireland is considerable. S.J. Baker (in Harris and Yalden 2008) found that in 2006 there were 67 zoos and aquaria registered with the British and Irish Association of Zoos and Aquariums, in addition to numerous private collections of exotic animals that together held many hundreds of species from around the world. These collections must be licensed and adhere to legislation stipulated in the Zoo Licensing Act (Amendment, England and Wales) Regulations 2002.

Some further animals are farmed for their meat, including various species of deer, cattle, Wild Boar and Water Buffalo. (Farming for fur, which led to the establishment in the wild of Coypu, Muskrats and American Mink, was prohibited from 1 January 2003 by the Fur Farming (Prohibition) Act 2000.) Although the Dangerous Wild Animals Act 1976 imposed stricter legislation on the keeping of potentially dangerous species, numerous unusual exotic pets continue to be kept by private individuals, and a variety of small animals (mostly mammals and amphibians) occurs in school and research laboratories. Some small species are unwittingly imported from abroad in freight or in private vehicles.

Non-native species established in the wild in Britain have died out for a variety of reasons. Some, such as the Coypu and Muskrat, were deliberately eradicated because of the ecological and/or economic threat they posed. Others, such as the various parrot species, probably found climatic conditions verging on marginal. The Guppy and Redbelly Tilapia originally thrived in the heated water discharged by an electricity-generating power station and a glassworks, but were unable to survive when the discharge of heated water ceased.

MAMMALS

BLACK-TAILED PRAIRIE DOGS GREETING

SCIURIDAE (SQUIRRELS, MARMOTS & ALLIES)

BLACK-TAILED PRAIRIE DOG *Cynomys ludovicianus*

Natural Distribution Grasslands of the USA from Texas to the Canadian border.
Naturalized Distribution England.

England

Several wildlife parks have free-ranging populations of Black-tailed Prairie Dogs, in three of which (in Cornwall and Cambridgeshire, and on the Isle of Wight) colonies have been reported outside the park perimeter. In 1976, a small population became established in the wild in Cornwall 6 km from a wildlife park, and at least six animals are known to have been captured or killed outside the park boundary. On the Isle of Wight, burrows dug by free-living Black-tailed Prairie dogs were noted for several years on farmland near a wildlife park, and breeding was reported in the wild. Individuals have been recorded at considerable distances from their presumed source in both Norfolk and Staffordshire.

Impact

In North America, Black-tailed Prairie Dogs have destroyed crops of wheat, alfalfa, hay, maize, sorghum, beans, potatoes and canteloupe melons, and have ring-barked newly planted orchard trees. Their burrows are a potential hazard for domestic stock, with whom they also compete for grazing.

References

Baker & Hills, in Harris & Yalden 2008; Long 2003.

PROCYONIDAE (RACCOONS & ALLIES)

COMMON RACCOON *Procyon lotor*

Natural Distribution From S Canada (Quebec, the prairie provinces and British Columbia) S through the United States to Panama.
Naturalized Distribution England; Scotland; Wales.

England

Raccoons were recorded in the wild at Haywards Heath in West Sussex in 1978, and in the 1980s in Norfolk. In 1984, a pregnant female escaped from captivity near Sheffield in Yorkshire, where she subsequently gave birth to two cubs. Raccoons were captured in several different localities in Somerset in 1985. There are also said to have been a number of sightings in Leicestershire (Smith 2006).

The inclusion of Common Raccoons under the Dangerous Wild Animals Act 1984, and the necessity in future of buying a licence to keep them, is believed to have led many owners to release their animals into the wild.

COMMON RACCOON YOUNG HUNTING FOR CRABS

Scotland; Wales

Wild-living Common Raccoons were reported in Strathclyde, Scotland, in 1981, and in Brecon, Powys, Wales in 1977.

Characteristics

The Common Raccoon has a catholic diet and a range of habitat requirements (tending to favour woodland near water), and although there is as yet no evidence of an established wild population in Britain, the species is clearly well able to survive here, and has been recorded outside captivity on at least 32 occasions. Individuals are known to have survived in the wild for up to four years.

The species' failure to become established in England, as it has in continental Europe, may be due to the fact that escapes have usually involved only single individuals. As Baker (in Harris and Yalden 2008: 784) says: 'It is important to distinguish such chance events preventing establishment from ecological ones … If there are ecological constraints on a species becoming established, it may never happen; if it is a matter of chance, then it may only be a matter of time.'

Impact

In the United States and in parts of mainland Europe, the species has caused problems in game-management localities and to growing and stored corn.

References

Baker 1987, 1990; Baker & Hills, in Harris & Yalden 2008; Hills, in Corbet & Harris 1991; Long 2003.

CASTORIDAE (BEAVERS)

CANADIAN BEAVER *Castor canadensis* (NORTH AMERICAN BEAVER)

Natural Distribution Forested regions of temperate North America.
Naturalized Distribution England; Scotland.

England

In the late 19th century, several colonies of Canadian Beavers became established on private estates in England. In about 1870, a number escaped or were released at Sotterley Park near Beccles in Suffolk, where in the two years that they were at liberty they bred successfully, their offspring dispersing some 7 km to Benacre Broad. In 1890, some were introduced by Sir Edmund Loder, Bt to Leonardslea near Horsham in West Sussex, where a further pair was released in 1917 and where the colony survived until at least the outbreak of the Second World War.

More recently, escaped beavers (believed to be Canadian Beavers) have been recorded in Essex in 1984, in Surrey (1990 and 1998) and in Somerset. The individual in Somerset was one of two that escaped from a wildlife park in 1977 (the other was recaptured) and took up residence on the River Axe, where fresh beaver signs were present in 1990.

CANADIAN BEAVER EATING WILLOW LEAVES

Scotland

In 1874, the Marquess of Bute released a pair of Canadian Beavers (and a pair of the European species) in the woods above Kilchattan Bay on the Isle of Bute, Strathclyde, where they survived and bred until 1890.

Impact

In North America, the animals damage trees by bark stripping, and also destroy commercial timber by waterlogging the roots with their dams. In some localities, agricultural damage has been reported.

References

Baker & Hills, in Harris & Yalden 2008; Barrett-Hamilton & Hinton 1921; Harting 1880; Hills, in Corbet & Harris 1991; Lever 1985; Loder 1898.

MYOCASTORIDAE (CAPROMIDAE) (HUTIAS & COYPU)

COYPU *Myocastor coypus*

Natural Distribution Argentina, Uruguay, Paraguay, S Brazil, SE Bolivia, Peru, Chile, the Chonos Archipelago, Chilöé Island, S to the Straits of Magellan.
Naturalized Distribution England; Scotland; Wales.

COYPU

England

Coypus, which produce a fur known in the trade as nutria, were first introduced to England in 1929 to stock fur farms, mainly in Surrey, Sussex, Hampshire, Norfolk and Devon, but also in Yorkshire and Cumberland (Cumbria). At least 49 farms were established, from 50 per cent of which escapes were reported. A fall in the value of nutria and the outbreak of the Second World War caused the closure of most of them by 1939.

The first escapes were recorded at Horsham in Sussex in 1932, and by the 1940s colonies of wild-living Coypus were established in Norfolk and Suffolk, and at Dorney near Slough in Buckinghamshire, where they survived until the mid-1950s. In 1936–44, individual Coypus were recorded in the wild in Bedfordshire, Cheshire, Essex, Gloucestershire, Somerset and Staffordshire.

From about 1940, wild Coypu colonies largely disappeared from southern England, surviving only in Norfolk and Suffolk. In the former county Coypus first escaped from fur farms in the valley of the River Yare in 1937. During the war they spread up and down the Yare and its tributaries, and along the Rivers Wensum and Tass, and by the end of hostilities they had colonized some 65 km of those waterways. At about the same time, Coypus became increasingly numerous in the marshland around Surlingham, Wheatfen and Rockland Broads, where they found an ideal habitat. Within a short time Coypus had spread east to Cantley, Reedham and the Langley marshes, and south-west to Northwold on the River Wissey and to Wroxham.

Coypus were first seen on Hickling Broad in 1948 and on Hornsey Broad in the following year. In the early 1950s, they appeared in many localities in Norfolk away from the main river system, where they formed new and discrete sub-populations on isolated ponds and flooded gravel-pits, having followed little streams from the Rivers Waveney, Yare, Wensum and Bure. By the middle of the decade, Coypus had reached the River Glaven in north Norfolk, and colonization of Suffolk was well under way. By the late 1950s, they had reached Oulton Broad and the Rivers Ant and Thurne; east and west Suffolk and Cambridgeshire as far south and east as Holbrook near Ipswich; south and west to Mildenhall on the River Lark; Wisbech and King's Lynn; and as far north as Melton Constable.

Despite the harsh winter of 1946–7, Coypus steadily increased in both range and numbers in eastern England in 1945–62, reaching a peak population in the latter year of some 200,000. A government-sponsored trapping scheme resulted in the capture of 97,000 animals in the two years up to August 1962 without, however, any apparent diminution of the species' distribution. Indeed, the animals' range actually expanded during this cull, and Coypus began increasingly to be reported from Lincolnshire, Peterborough (now Cambridgeshire), the Isle of Ely, Huntingdonshire, north-eastern Bedfordshire, south-west Hertfordshire, and various parts of Essex and Suffolk. Isolated sightings came from as far afield as Sheffield (south Yorkshire), Leicestershire, Northamptonshire and parts of Derbyshire.

In 1962–5, an intensified trapping campaign accounted for a further 40,461 Coypus. During the exceptionally severe winter of 1962–3, an estimated 80–90 per cent of the population succumbed, and by the middle of the latter year the survivors were confined to some 6,000 sq km of Norfolk and east Suffolk. Indeed, harsh winter weather was the species' principal controlling factor in England. Colder than average winters and continued trapping pressure kept the population low and declining to 1969. In 1970–5, when the winters were mild or average, the population recovered sharply from around 1,600 to some 12,000. During the late 1970s, the numbers fluctuated between about 6,000 and

12,000. Less than 10 animals were found outside East Anglia, including isolated individuals in Grimsby, Humberside and Castleton, Derbyshire.

In April 1981, an even more intensive campaign was started, which was designed to eradicate Coypus by around 1990 at an estimated cost of £1.7 m. In fact, by the time the last animal had been accounted for in 1989, the entire eradication campaign had cost the taxpayer £2.5 m.

Scotland

Nutria farms were established at Auchinroath, near Rothes on the River Spey in Morayshire, from which Coypus escaped in 1934, and on the River Tay in Perthshire.

Wales

In 1936, some Coypus escaped from a nutria farm situated near the confluence of the Rivers Vyrnwy and Severn in Montgomeryshire.

Characteristics

The Coypu is an aquatic animal that burrows into the banks of the rivers and lakes it frequents. Its burrows may extend for a distance of 6 m or more, and end in a chamber in which the young are born. Where banks are not high enough for burrowing, a platform-like nest is constructed among reeds. Although the Coypu is an excellent swimmer and diver, it is somewhat awkward and ungainly on land. It normally lives in still or only slow-flowing water, where it subsists mainly on aquatic plants.

Impact

Coypus in England were a serious, albeit local pest. They raided gardens for brassicas and agricultural land for cereals and root crops – swedes, mangolds, potatoes and, especially, sugar beet. Their burrows undermined the banks of rivers and dykes, which then collapsed, causing the inundation of valuable farmland. They trampled down and crushed marsh vegetation and ate the tender young shoots of Common Reeds, often clearing entire beds, thus converting them to open water. They also consumed the submerged parts of Reed Mace and Lesser Reed Mace, leaving the remainder floating, and ate a wide variety of other aquatic plants. Limited damage to Common Osiers and larch saplings was also reported. Some wetland plants, such as the Flowering Rush, were all but eradicated, and the destruction of much aquatic vegetation had a negative impact on the complex and diverse invertebrate community. Although the removal of large areas of reed beds did to an extent help to keep waterways clear and prevent silting, it also threatened the habitat and survival of such threatened marshland birds as Great Bitterns, Bearded Tits and Marsh Harriers, and reduced the supply of Norfolk reed for thatching.

References

Davis 1963; Davis & Jenson 1960; Fitter 1959; Gosling 1974, in Corbet & Southern 1977, 1981, 2003; Gosling & Baker, in Corbet & Harris 1991, in Harris & Yalden 2008; Laurie 1946; Lever 1977, 1985; Norris 1967; Long 2003; Smith 1995; Yalden 1999.

CRICETIDAE (VOLES, LEMMINGS, HAMSTERS & ALLIES)

GOLDEN HAMSTER *Mesocricetus auratus*

Natural Distribution *M. a. auratus*, from which all domesticated hamsters are derived, is known only from in and around Aleppo in N Syria.
Naturalized Distribution England.

GOLDEN HAMSTER AT ENTRANCE TO NEST

England

The first live specimens of *M. a. auratus* were imported from Aleppo in 1880 by a retired British Consul, J.H. Skere, and brought to Edinburgh, where a small population survived in captivity until around 1910.

All extant domesticated Golden Hamsters are the descendants of a single adult female and her 12 young captured in Aleppo in 1930 and taken to Jerusalem University, where within a year they had increased to around 150.

Golden Hamsters are today one of the most popular children's pets, and are also kept in schools for educational purposes and in laboratories for experimental research. Numerous animals have doubtless escaped and continue to do so, but there have only been a limited number of records of breeding in the wild, all in semi-urban habitats. These include:

- Six that escaped into the unheated basement of a pet shop in Bath, Avon, in 1957, where in the following year 52 were caught.
- Barrow-in-Furness, Cumbria (1961).
- Finchley, north London (4 escaped in 1960 and 25 caught in 1962).
- Bootle, Lancashire (17 trapped in a florist's shop in 1962).
- Bury St Edmunds, Suffolk (230 caught in shop basements in 1964–5).
- Burnt Oak, Barnet, north London, where in 1980–1, 150 were caught around houses, sheds and allotment gardens on a council housing estate.

Impact

Since Golden Hamsters hibernate they seem well able to survive an English winter, and would be helped if global warming were to continue. In the Netherlands and Germany, the related European Hamster has caused damage to fruit crops and stored produce, and the Golden Hamster might well become a pest if it were ever to become widely established in the wild in Britain.

Poisoning seems an ineffective means of control for two reasons: firstly, the animals hoard much more food than they require and are thus unlikely to ingest lethal concentrations; and secondly, their habit of storing food in their cheek pouches gives the poison time to react, be detected and expelled.

References

Baker 1986; Baker & Hills, in Harris & Yalden 2008; Hills, in Corbet & Harris 1991; Lever 1977, 1983, 1985; Long 2003; Rowe 1960, 1968; Yalden 1999.

CRICETIDAE (VOLES, LEMMINGS, HAMSTERS & ALLIES)

MUSKRAT *Ondatra zibethicus*

Natural Distribution North America, from Alaska and Canada (except the far N) S to South Carolina, Texas, Arizona and extreme N Baja California, Mexico.
Naturalized Distribution England; Scotland; Wales; Ireland.

England

Muskrats, whose fur is known in the trade as musquash, were introduced to England in the 1920s, and at least 87 fur farms are known to have been established by the turn of the decade. In 1929–31, Muskrats had escaped or, in the case of failed enterprises, had been deliberately released, from a total of 14 farms, and within 2–3 years they had been recorded

MUSKRATS NEXT TO LODGE

in the wild from the same number of counties. They were concentrated mainly in Shrawardine, 10 km west of Shrewsbury in Shropshire; here 120 Muskrats had been introduced from North Rice Lake in Ontario. In 1930, following an escape, some made their way to the River Severn, and by late the following year they had become established in 30 km of the river (whose banks in places they undermined) above Shrewsbury, eventually spreading over a distance of almost 60 km, with odd individuals being reported from as far distant as Welshpool upriver and downstream to Bewdley. Other localities in England where Muskrats became established included the River Wey near Farnham in Surrey, and the dykes and marshes of the River Arun in West Sussex.

In March 1932, the Destructive Imported Animals Act made it illegal in future to keep Muskrats in Britain without a licence, and in the following year their farming was banned altogether. Similar legislation was also enacted in Ireland. In June 1932, an intensive trapping programme was begun, which resulted by the end of 1934 in the capture of 4,382 Muskrats – including 2,720 in the Severn area of Shropshire, 162 in West Sussex and 55 in Surrey. Trapping by private landowners in England (and Scotland) accounted for around a further 350 animals. By the end of 1934, Muskrats must have been nearly eradicated in Britain, as no young were caught in 1935, and before the outbreak of the Second World War natural wastage had accounted for any survivors.

Scotland

Muskrats were first introduced to Scotland in about 1920, when a number turned down near Oban in Argyllshire[1] failed to become established. However, of 6 pairs placed in a fenced enclosure at Feddal, near Braco in Perthshire, several soon escaped, and within 2 years they and their progeny had constructed 16 mud and stick lodges on a marsh 3 km away. Within a few years, some 90 km of the Rivers Forth, Teith and (especially) the Earn and of the Allan Water in the Tay basin had been overrun. In 1929–31, Muskrats had escaped or been released from 5 Scottish fur farms, and within 2–3 years they occurred in 9 Scottish counties. The 1932–5 trapping campaign accounted for 958 Muskrats in Scotland. Unfortunately, a large number of non-target species were also caught, including 2,305 Water Voles and 2,178 Common Moorhens. Corresponding figures for England and Ireland were never published.

Wales

Although no musquash farms are known to have been stocked with Muskrats in Wales, animals from Shropshire apparently dispersed to Montgomeryshire (Powys).

Ireland

A pair of Muskrats imported from Malachi, Ontario to Annaghbeg near Nenagh in Co. Tipperary in 1929 promptly escaped, and within five years their offspring had colonized an area of almost 130 sq km between Lough Derg and Nenagh, where between September 1933 and May 1934, 487 Muskrats were trapped, and by 1935 the population had been eradicated.

Impact

In continental Europe, where Muskrats were first introduced to fur farms in 1905 and where they are now widely established (see Lever 1985), they are generally regarded as a pest. Their burrows undermine the banks of rivers, lakes, dykes and drainage ditches, which then collapse, resulting in erosion and flooding. In some countries damage to commercial agricultural crops has been reported, and Muskrats have been found to be vectors of *Leptospira*, which can cause Weil's disease in humans.

In Eurasia, Muskrats destroy large areas of aquatic vegetation, and in some places have displaced Russian Desmans and Water Voles through competition for food and habitat. Other species affected by Muskrats include shellfish, insects, fish and ground-breeding birds. The damage caused by Muskrats far outweighs the profit from the sale of pelts.

As mentioned, in their short time in the wild in Britain, Muskrats in places undermined the banks of the River Severn, and there seems no reason not to suppose that, if they had not been eradicated so promptly, the damage could have eventually been on a par with that experienced in mainland Europe.

References

Baker & Hills, in Harris & Yalden 2008; Fairley 1982; Fitter 1959; Gosling & Baker 1989; Hills, in Corbet & Harris 1991; Hinton 1932; Lever 1985; Long 2003; Sheail 1988; Warwick 1934, 1941; Yalden 1999.

Note

1. Long (2003: 169) places Oban, Argyllshire, in Ireland!

MURIDAE (RATS, MICE, GERBILS & ALLIES)

MONGOLIAN GERBIL *Meriones unguiculatus*

Natural Distribution Most of Mongolia and China, from Sinkiang through Inner Mongolia to Manchuria and neighbouring parts of the former USSR.
Naturalized Distribution England.

MONGOLIAN GERBIL

Mongolian Gerbils were first bred in captivity in Japan, from where a foundation stock of 11 pairs was exported to the United States in 1954. They first arrived in Britain in the 1960s and have since, like the Golden Hamster, become popular as children's pets, for educational purposes in schools and for laboratory research.

England

In 1973, at the conclusion of the filming of *Tales of the River Bank* on the Isle of Wight for BBC Television, some white rats, Golden Hamsters, Guinea Pigs and Mongolian Gerbils either escaped or, more likely, were deliberately released. The rats and Guinea Pigs soon disappeared, but the Mongolian Gerbils became established in and around a group of houses, a barn, a wood yard and waste ground at Coathy Butts near Fishbourne, 4 km west of Ryde. Within three years this population had multiplied to more than a hundred.

There are a number of records of wild-living colonies of Mongolian Gerbils in other locations, in particular from Yorkshire, including:

- Thorne, Goole and Swinfleet Moors in 1971 – almost certainly deliberate releases since the moors are far removed from human habitation.
- Armthorpe near Doncaster, where animals that had escaped from a school science laboratory were found living beneath some sheds in 1972–3, where they were joined by more from the same source in 1975.

- Bradford, where three individuals were discovered in a burrow underneath some tree-roots in woodland near a housing estate in 1975.
- Harrogate (1977).
- Doncaster (1981).
- In March 1987, Mongolian Gerbil remains were recovered from the fresh pellet of a Long-eared Owl collected at a roost near Mexborough.
- Melanics are common in the Netherlands, which was probably the source of one found in a florist's shop in Bolton, Lancashire (possibly imported with a consignment of cut flowers) in 1996.

Impact

In their natural range in the wild, Mongolian Gerbils consume vast quantities of seeds, and are known to be potential vectors of bubonic plague (caused by *Yersinia pestis*, a parasite of rodents and transmitted to humans by fleas), bilharzia or schistosomiasis (a disease caused by infestation by Trematoda blood flukes), and rabies (caused by a rhabidovirus). 'The use as pets of species such as the Mongolian Gerbil,' wrote Gulotta (1971: 3), 'if released carries a grave risk of damage to the environment and to human agriculture and health.'

References

Aistrop 1968; Baker & Hills, in Harris & Yalden 2008; Gulotta 1971; Hills, in Corbet & Harris 1991; Howes 1973, 1983, 1984; Lever 1977, 1985; Long 2003; Yalden 1999.

HYSTRICIDAE (OLD WORLD PORCUPINES)

CRESTED PORCUPINE *Hystrix cristata*

Natural Distribution E and W Africa.
Naturalized Distribution England.

England

In the summer of 1972, a pair of Crested Porcupines escaped from the botanical gardens at Alton Towers, east of Stoke-on-Trent in Staffordshire, where they took up residence at Cote Farm on the Farley Hall estate about 1.5 km north-west of Alton Towers. (One of these animals was later recaptured and returned to Alton Towers, from which it rapidly re-escaped!) The hilly terrain on the Farley Hall estate was plentifully supplied with conifer plantations intersected by small tributaries of the River Churnet, which provided an ideal porcupine habitat not dissimilar to that colonized by Himalayan Porcupines in Devon. The range of reported sightings in Staffordshire covered an area of some 7 sq km between Ramshorn, Farley and Alton Towers, but not south of the Churnet. These animals were no longer seen after the late 1970s.

CRESTED PORCUPINE MOTHER WITH OFFSPRING

Individual Crested Porcupines are from time to time recorded in the wild elsewhere, e.g. near Market Lavington in Wiltshire (three sightings in 1974); at Rye in East Sussex (1979–81); in County Durham, where a male escaped in 1983 and was at large for some 18 months until recaptured in a suburban garden in Durham; and in Norfolk, where two escaped from a wildlife collection in 1977 but were soon recaptured.

Impact

In Staffordshire at least one larch plantation was damaged by Crested Porcupines, which, however, instead of attacking young trees as Himalayan Porcupines had done in Devon, gnawed on mature trees, thus directly affecting the final crop. Because the Crested Porcupine is a larger and heavier species than the Himalayan Porcupine, it is able to cause more extensive damage, and several trees in Staffordshire were destroyed by ring-barking. On Cote Farm the animals were said to have eaten stored grain.

References

Baker 1987; Baker & Hills, in Harris & Yalden 2008; Hills, in Corbet & Harris 1991; Lever 1977, 1985; Long 2003.

HYSTRICIDAE (OLD WORLD PORCUPINES)

HIMALAYAN PORCUPINE *Hystrix brachyura*

Natural Distribution C and E Himalayas, NE India, S China and parts of Malaysia; also Borneo, Sumatra, S Thailand and formerly Singapore.
Naturalized Distribution England.

England

In 1969, a pair of Himalayan Porcupines escaped from the Pine Valley Wildlife Park near Okehampton in Devon. Unfortunately, the escape was not reported at the time, and it was not until 1971, when some quills were discovered and damage to conifers was reported, that their presence in the wild was recorded. Later in the same year an individual was killed on Forestry Commission property at Bogtown near Northlew, 10 km north-west of Okehampton, which remained the most westerly record in Devon. In November 1973 the corpse of a sub-adult, which had almost certainly been born in the wild, was found at Risdon, north of Okehampton. Another was seen between Whiddon Down and Throwleigh south-east of Okehampton, within the borders of the Dartmoor National Park, which remained the most southerly and easterly record. The most northerly evidence of Himalayan Porcupines in Devon was provided by some quills found at Lock's Hill, south of Dolton, 16 km north of Okehampton.

Thus the main range of the species in Devon extended over an area of some 5 sq km in the shape of a letter 'h', from Oaklands in the south through Hook, Abbeyford, Springett's and Parsonage Woods, past Risdon Folly Gate and Inwardleigh, as far north as Hayes Barton. They seemed to be largely confined to conifer plantations near the River Okement: at Risdon, they were apparently restricted to a plantation of Norway Spruce.

Because of the potential damage the animals might cause if permitted to remain at liberty, in June 1973 the then Ministry of Agriculture, Fisheries and Food started a live-trapping campaign. It was not, however, until January of the following year that the first animal was captured, and by 1980 only five adults and one sub-adult had been removed; thereafter no further individuals were observed in the wild.

Impact

A survey carried out in June 1973 revealed that around 15 per cent (valued at about £300) of a 4-ha Norway Spruce plantation at Folly Gate had been damaged by the animals, and the Forestry Commission estimated the cost of damage in their nearby Springett's Wood plantation at £500–£1,000. In the Risdon plantation it was found that they were feeding extensively on young Norway Spruce, the main objective apparently being the inner cambium layer, and again some 15 per cent of trees were damaged. Elsewhere, other species of conifer were attacked, and on agricultural land crops of potatoes and swedes were affected. It was considered that Himalayan Porcupines also posed a threat to native Eurasian Badgers, some of whom they had ejected from age-old setts.

References

Baker 1974; Baker & Hills, in Harris & Yalden 2008; Butcher 1974; Comfort 1974; Gosling 1980; Gosling & Wright 1975; Hills, in Corbet & Harris 1991; Lever 1977, 1985; Long 2003; Milburn 1974.

MUSTELIDAE (WEASELS & ALLIES)

ORIENTAL SHORT-CLAWED OTTER *Aonyx cinerea*

(ASIAN SHORT-CLAWED OTTER)

Natural Distribution S India, Bengal, Assam, S China, Myanmar, Thailand, Malay Peninsula, Vietnam, Sumatra, Java, Borneo. Distribution is discontinuous.
Naturalized Distribution England.

England

In 1983–93, there were six records of sightings of live Oriental Short-clawed Otters or the discovery of road-killed carcasses – all in the Oxford area: near Draycot on the River Thame in 1983; in the Bayswater Brook near Headington (1986); in the Oxford Canal/River Thames (1991); and in the Rivers Glyme, Dorn and Cherwell (all 1993). The range of sightings increased from 17 km in 1986 to 25 km in 1993.

Away from the Oxford area, individual Oriental Short-clawed Otters were reported in Kent in July 1981; in Gloucestershire (at Gloucester in November 1985); and at Bath, Avon (January 1987).

Prospects

The Oriental Short-clawed Otter can clearly tolerate the climate of southern and central England. All the areas colonized contained few if any native Eurasian Otters, and although both species live sympatrically in parts of their native range, evidence suggests that the native species might dominate the alien were the two to come into contact in

ORIENTAL SHORT-CLAWED OTTERS

England. Indeed, the absence of records of the latter in the Oxford area since 1999, when 17 Eurasian Otters were released by the Otter Trust at three sites on the upper Thames, may be no coincidence. Eurasian Otters are known to have eradicated American Mink (see page 79) in this area by around 2001, and having spread to the surrounding tributaries by 2001–2 they could also have been responsible for eliminating the few much smaller Oriental Short-clawed Otters in the area.

Impact

Even if Oriental Short-clawed Otters were to become established in the wild in Britain, they would be unlikely to prove a problem to any native species and there would be no risk of hybridization with Eurasian Otters, since none is known to occur where the two species live together in the Far East.

References

Baker & Hills, in Harris & Yalden 2008; Jefferies 1990, 1992, in Harris & Yalden 2008; Strachan & Jefferies 1996; Sykes 1995; Yalden 1999.

CERVIDAE (DEER)

PÈRE DAVID'S DEER *Elaphurus davidianus*

Natural Distribution Formerly marshy lowlands of NE China; probably extinct in the wild for over 2,000 years.[1]
Naturalized Distribution England; Scotland.

Since their extinction in the wild, Père David's Deer were preserved by successive Chinese emperors in the Imperial Hunting Park of Nan-Hai-Tze south of Peking (Beijing), where they were found by the French missionary Armand David in 1865.[2] When the park became flooded in 1894, the enclosing walls were breached and many of the deer died; those that survived were killed and eaten during the Boxer Rebellion of 1898–1900.

England

Shortly before the loss of the Nan-Hai-Tze herd, a number of Père David's Deer were acquired by various European zoos, and during the 1890s the Duke of Bedford obtained some from Paris, Berlin and Cologne as the founder stock of a breeding herd at Woburn Abbey in Bedfordshire, where, and at the neighbouring Whipsnade Zoo and elsewhere, their descendants survive.

Over the years, a number of individuals have escaped into the wild. One, seen near Aston Abbotts in Buckinghamshire in 1963–4, probably came from Woburn. A small group of about a dozen wandered away from a farm near Swindon in 1981, and several were at large for around two years.

A herd of 25–30 Père David's Deer, based on an unfenced estate at Ashton Wold near Peterborough in Northamptonshire, has been frequently observed up to 16 km from the estate. In 1977, two yearling males were reported in Monks Wood, 13 km away, and a

PÈRE DAVID'S DEER STAG WITH ANTLERS IN VELVET

further two stags were shot 14 km from the estate at Wooley. Young calves seen well away from the estate have almost certainly been born in the wild.

Standing around 1.3 m at the shoulder, this is a large deer that is easily shot, so its widespread establishment in the wild is unlikely.

Scotland

In the late 1940s, a number of Père David's Deer escaped from a zoo near Strathblane in Stirlingshire, one of which lived at Balmaha near Loch Lomond in 1952–3.

References

Baker 1990; Baker & Hills, in Harris & Yalden 2008; Beck & Wemmer 1983; Cowdy 1965; Fletcher, in Mason 1984; Long 2003; Loudon & Fletcher 1983; Mitchell 1983; Strachan & Jefferies 1996.

Notes

1. Love (2003: 440) misquotes 'Louden' [sic] and Fletcher (1983) as saying 'there probably have not been any truly wild [Père David's Deer] in China for 200–300 years'. In fact, Loudon and Fletcher say (page 91) that 'the animal does not appear to have existed in the wild for two or three thousand years'.
2. Before its 'discovery' by David, the species was known in China as the 'Milu'.

BIRDS

NUMIDAE (GUINEAFOWL)

HELMETED GUINEAFOWL *Numida meleagris*

Natural Distribution Trans-Saharan Africa.
Naturalized Distribution England.

England

In 2002, this species was recorded at six localities in three counties, but no breeding was reported. In Devon a flock of five was observed in August, and in Somerset an unpinioned but non-breeding flock appears to be well established at Ash Priors. In Norfolk, where Helmeted Guineafowl had bred successfully in the wild in 2001, no reports were received from the breeding site in 2002, although in August two were seen about 12 km away. Two males and a female were observed in another locality in April, and single pairs at two nearby sites in June and December. In 2006–7, a flock of 5–6 was established in a small copse on the Royal County of Berkshire polo ground in Winkfield, Berkshire (personal observation)

Reference

Holling & RBBP 2007; Ogilvie & RBBP 2004.

HELMETED GUINEAFOWL

PHASIANIDAE (TURKEYS, GROUSE, PHEASANTS & PARTRIDGES)

INDIAN PEAFOWL *Pavo cristatus*
(COMMON PEAFOWL)

Natural Distribution From NE Pakistan E through India and Nepal to Assam and S to Sri Lanka.
Naturalized Distribution England.

England

In 1998 and 1999, Indian Peafowl bred in the wild in Northamptonshire, and two were seen at Winfarthing in Norfolk in 2001. In 2002, the birds were observed in five places in Norfolk, but no breeding was recorded. In May 2005, an adult with young was seen in Markwell's Wood in Sussex.

It is likely that this species is greatly under-recorded in the wild, and that other successful breeding has taken place but is so far unrecorded.

INDIAN PEACOCK DISPLAYING

References

Holling & RBBP 2007; Ogilvie & RBBP 2004.

ARDEIDAE (HERONS, BITTERNS & EGRETS)

BLACK-CROWNED NIGHT HERON *Nycticorax nycticorax*
(NIGHT HERON)

Natural Distribution The race *N. n. hoactli* (which is the one that occurred in Scotland) breeds from SE Canada to Argentina and in the Hawaiian Islands. The nominate form (which is present in England) ranges from C and S Europe E to Kazakhstan through S and SE Asia to E China, Japan, Taiwan, the Sunda Islands and the Philippines, and N Africa. It winters in C Africa and SE Asia.
Naturalized Distribution England; Scotland.

England

Lord Lilford made what appears to be the earliest documented attempt to introduce this species to England when he 'turned out two young Night Herons at Lilford in the summer of 1887, but though they were seen on several occasions after their liberation, I have no evidence to prove that they remained in the neighbourhood for any length of time'.

In or before 1996, a free-flying colony of up to 30 Black-crowned Night Herons (of the nominate European race) became established in Great Witchingham Wildlife Park in

BLACK-CROWNED NIGHT HERON

Norfolk, where they nest colonially with Grey Herons. Eight pairs were present in the park in 2003 and breeding was recorded, but no breeding took place in 2004 or 2005.

Scotland

In 1936, the Royal Zoological Society of Scotland obtained three pairs of the North American race of the Black-crowned Night Heron, *N. n. hoactli*, from the National Parks Bureau in Canada. These were kept, unpinioned, in a covered aviary in the society's gardens at Murrayfield, Edinburgh, where the first pair bred successfully two years later, and where nesting took place annually until at least the late 1990s. In 1946, a further pair from the same source reinforced this colony.

In December 1950, a number of birds escaped through a hole in the aviary roof. In May of the following year the entire roof was removed, and the whole colony of around 18 birds became established in the zoo's grounds. The population initially increased slowly, probably due to a lack of fecundity coupled with predation by Grey Squirrels, Red Foxes, Common Rats, Carrion Crows, Eurasian Jackdaws, Black-billed Magpies and semi-feral domestic cats, together with some adult emigration. During the decade in which the birds escaped and were released, they were reported outside the society's grounds at:

- Union Canal, Sighthill, 3 km to the south-west in 1952.
- North Tyne River at Haddington, East Lothian, 30 km east in 1954.
- Water of Leith, Colinton, 4 km south,
- River Almond at Cramond Brig, 4 km north-west, both in 1954.
- In April 1955, a group of 5–6 birds was observed to be commuting every evening to feed in the Gogar Burn, 5 km to the south-west, returning to the zoo's grounds the following morning. Although the birds were fairly tame when in the security of the zoo's grounds, they were observed to be extremely wary elsewhere.

The population at Edinburgh varied between about 30 and 60 (in 1968). More recently, the presence of at least 30 birds was reported in the zoo grounds in 1997, including 5–10 breeding pairs, and about 35 individuals in 1998. Following a policy by the zoo to reduce the population, it was estimated in 2000 to number only 5–10 breeding birds, and shortly thereafter the colony died out (R. Thomas, personal communication 2008).

In both Europe and North America, Black-crowned Night Herons breed between April and June. In Edinburgh, however, they were known to have nested throughout the year except in August and September, when the adults moult. The reasons for this greatly extended breeding season are believed to have been twofold. Firstly, the birds had a regular and abundant supply of food within the zoo grounds without the necessity for constant external foraging. They stole food put out for other species and killed and ate the young of various birds in neighbouring enclosures. Secondly, the urge to migrate, which has an endocrinological origin connected to the reproductive instinct, had been lost (possibly due to supplementary feeding), which might also have helped more continuous breeding. There is no evidence that the birds ever nested away from the zoo grounds. There was a high (perhaps 50 per cent or more) rate of fledgling mortality, possibly due to a shortage of food suitable for young birds.

Although it has been suggested that the survival of Black-crowned Night Herons in Edinburgh was due to artificial feeding in winter, this may not necessarily have been so. In the north of its natural range, the species is migratory and seldom winters north of around 53°N. Edinburgh is situated at 56°N, and were supplementary winter feeding to have been abandoned the birds would probably have migrated south to winter elsewhere, where the natural food supply was more abundant.

References

Blair et al. 2000; Dorward 1957; Fitter 1959; Holling & RBBP 2007; Lack 1986; Lever 1977, 1987, 2005; Mead 2000; Ogilvie & RBBP 1999, 2000; Royal Zoological Society of Scotland, Annual Reports from 1936; Sharrock 1976; Welch et al. 2001; Young & Duffy 1984.

PSITTACIDAE (COCKATOOS & PARROTS)

BUDGERIGAR *Melopsittacus undulatus*

Natural Distribution Much of inland Australia, and some offshore islands, but not including Tasmania.
Naturalized Distribution England.

BUDGERIGAR AT NESTHOLE

England

The artist John Gould is credited with having introduced the first pair of Budgerigars[1] to England in about 1840, and in 1861 the animal dealer Carl Jamrach is said to have acquired no fewer than 6,000, which had been consigned to London from Port Adelaide in South Australia, and which he offered for sale to London Zoo.

In the 1930s, partly as a consequence of an outbreak of psittacosis, Budgerigars were frequently to be seen at large in the home counties. Before the First World War, Philip Gosse released over a dozen in Beaulieu in Hampshire, where at least one pair nested successfully in an English Elm. The numbers increased to around 20, but the birds subsequently disappeared. At around the same time, a pair lived for a number of years in a London square, but did not breed.

In 1969, during a visit to Tresco Abbey in the Scilly Isles, Her Majesty Queen Elizabeth the Queen Mother suggested to Lieutenant-Commander and Mrs Dorrien-Smith that the gardens might be a suitable place in which to establish a colony of free-flying Budgerigars, such as she had at Royal Lodge in Windsor Great Park in Berkshire. Accordingly, an aviary was constructed in the abbey gardens, into which were introduced in November of the same year four pairs from Royal Lodge. In April 1970, eight nesting boxes were placed in the aviary, and breeding began almost immediately. After the young had hatched, the entry/exit hole to the aviary was opened, but at first only the males chose to leave, always returning in the evening to roost with their families. Later in the summer the cocks were joined in the garden by the hens and young birds, but all continued regularly to return to the aviary to roost. In the autumn of 1970, the colony was reinforced with a further six pairs from the same source. In the spring of 1971, the number of nesting boxes in the aviary was doubled, and most were occupied. The birds spent much of the time within the aviary, only leaving for brief flights around the garden.

In 1972, however, the birds began nesting outside the aviary in holes in the trunks of Australian Cordylines and palms in the gardens. As soon as the fledglings were able to fly they were introduced to the aviary by their parents.

In 1973, when the population numbered about 60, the birds' habits began perceptively to change. They became more independent, spending more time outside the aviary, to which they only returned to feed or to shelter in stormy weather. Nesting occurred throughout the year in natural holes in a variety of trees in the gardens, and nest boxes in the aviary and on trees were ignored. By 1975, when the population numbered in excess of 100, the birds were sometimes to be seen in small flocks on the neighbouring islands of Bryher, St Martin's, St Mary's and St Agnes, but there was no evidence of breeding away from Tresco. Thereafter the population began to decline, and it died out during the hard winter of 1976–7.

On the English mainland, many people have from time to time kept free-flying colonies of 'homing' Budgerigars, which returned to their aviary to roost. Over the years some failed to 'home', and became temporarily established in the wild – for example at Margaretting in Essex, in Windsor Great Park in Berkshire and in the New Forest in Hampshire. Sharrock (1976) listed three recent Budgerigar breeding sites:

- At Wigginton, south of Tring in Hertfordshire, where a pair nested in 1971 and 1972 over 800 m from the aviary from which the birds were believed to have escaped.
- Near Downham Market in Norfolk, where a flock of 30 became established in 1970–1, and where several pairs reared young in 1971 and 1972 before the colony died out a few years later.
- At Fenstanton south of St Ives in Huntingdonshire, where a pair bred at Honey Hill and where a flock of 12–15 had built up by 1974 before disappearing.

Gibbons et al. (1993) list the presence in the wild (but not breeding) of free-flying Budgerigars in 1988–91 in 10-km squares TL27 and TL37 (near Huntingdon).

Since in the long-term Budgerigars were unable to tolerate the relatively mild climate of the Scilly Isles, they would be unlikely to survive permanently in the wild (despite artificial winter feeding) on even the southern English mainland.

References

British Trust for Ornithology 1978; Fitter 1959; Gibbons et al. 1993; Lever 1977, in Mason 1984, 1987, 1992; Long 1981; Sharrock 1976.

Note

1. In the Aboriginal Eora (Port Jackson) dialect *budgeri* = 'good' (to eat), while *gar* and *kar* occur as elements in the names of various parrot species in the dialects of other tribes in eastern New South Wales. In Australia, there is a variety of other spellings of Budgerigar.

PSITTACIDAE (COCKATOOS & PARROTS)

ALEXANDRINE PARAKEET *Psittacula eupatria*

Natural Distribution From E Afghanistan to Bangladesh, E Assam, N Burma, N and W Thailand and Indochina: also Sri Lanka and Andaman Islands.
Naturalized Distribution England.

England

Alexandrine Parakeets bred at Fazackerley, Merseyside, between at least 1997 and 1999. In 1998 two pairs were reported to have fledged five and three young respectively. Subsequently some birds were shot, although a single pair survived and bred successfully again in 1999.

In 2001, a pair of Alexandrine x Rose-ringed Parakeet hybrids is believed to have bred in a park at Sidcup in Kent, where a third bird (possibly a juvenile from the previous season) was seen from time to time. In 2002, two hybrid parakeet nests were found in the park. These three birds, and three apparently pure Alexandrine Parakeets, have been observed at the Rose-ringed Parakeet roost at Lewisham. There has been no apparent breeding since 2002.

References

Butler 2002; Lever 2005; Ogilvie & RBBP 1998–2004.

PSITTACIDAE (COCKATOOS & PARROTS)

ROSY-FACED LOVEBIRD *Agapornis roseicollis*

Natural Distribution Angola, Namibia, South Africa.
Naturalized Distribution Scotland.

Scotland

In 2002, a pair of Rosy-faced Lovebirds bred successfully at Dunbar in East Lothian, where up to five adults and juveniles were observed between January and April, and another juvenile was seen in August. This is the first documented record of the species breeding in the wild in Britain.

ROSY-FACED LOVEBIRDS

Reference

Ogilvie & RBBP 2004.

PSITTACIDAE (COCKATOOS & PARROTS)

BLUE-CROWNED PARAKEET *Aratinga acuticaudata*

Natural Distribution S America from Venezuela S to N Argentina and W Uruguay.
Naturalized Distribution England.

England

In 1997, a pair of Blue-crowned Parakeets was observed at a garden birdfeeder in Bromley in Kent, where by 1999 the number had increased to 8 (including 3 juveniles), and a flock of 15 was seen in neighbouring Beckenham.

In April 2001, a nest containing four eggs was found in Bromley; it had, however, failed, either through desertion or predation of the adults by Grey Squirrels. No breeding has since been reported.

References

Butler 2002; Butler et al. 2002; Williams 1999.

TIMALIIDAE (BABBLERS & PARROTBILLS)

RED-WINGED LAUGHINGTHRUSH *Garrulax formosus*

Natural Distribution SW Sichuan, NE Yunnan, NW Vietnam (Fansipan Mountains).
Naturalized Distribution England.

England

Between at least 1996 and 2002, Red-winged Laughingthrushes that had escaped from a wildlife park bred successfully in the wild on the Isle of Man, where since 2000 sightings have been intermittent near Ballaugh on the north-west coast, where Red-necked and Parma Wallabies are established. In 2003, 1–2 birds were seen at a single site, while in the following year the same number was reported at one site and a single male was recorded at another at least 8 km further east than any previous record; in 2005, there were reports of a single bird only. No breeding has been observed since 2002.

References

Holling & RBBP 2007; Ogilvie & RBBP 2004; Thorpe & Sharpe 2004.

FISH

POECILIIDAE (MOLLIES, GUPPIES & SWORDTAILS)
GUPPY *Poecilia reticulata*
(MILLIONS FISH)

Natural Distribution NE S America north to Trinidad, Barbados, St Thomas and Antigua.
Naturalized Distribution England.

England

In the 1960s, a population of Guppies[1] became established on an 800-m stretch of the River Lee (or Lea) in Hackney, north-east London, where the water temperature was permanently raised by the cooling water discharge from the Central Electricity Generating Board's Hackney power station. The poor quality of this stretch of water provided an effective barrier to many potential competing fish species other than Three-spined Sticklebacks. In the early 1970s, the quality of the water in the Lee improved, and native species began to recolonize this stretch of water. Following the closure in the 1970s of the generating station with a consequent reduction in the temperature of the water, and a further improvement of water quality, this thriving colony of Guppies died out.

In about 1963, another colony of Guppies became established in the 350-m Church Street stretch of the St Helens Canal in Lancashire, where the water was warmed by the discharge from the neighbouring Pilkington Brothers' glass factory. This colony, which was a result of the release of stock from a defunct tropical-fish dealer, was apparently still thriving in 1975, but has since died out.

Characteristics

The tiny Guppy is a resilient species, well able to survive for short periods in temperatures as low as 15°C, and to live in brackish and poorly oxygenated water.

GUPPY

Impact

In other parts of the world the Guppy has been widely introduced for the control of mosquito larvae. Unfortunately, it also eats the eggs of other fish, and has been blamed for the decline of a number of native species.

References

Lee Conservancy Catchment Board 1967; Lever 1977, 1996; Maitland 2004; Meadows 1968; Wheeler 1974; Wheeler & Maitland 1973.

Note

1. Named after the American naturalist and geologist Robert J.L. Guppy.

CENTRARCHIDAE (SUNFISH)

ROCK BASS *Ambloplites rupestris*

Natural Distribution S Canada, from Lake Winnipeg in Manitoba E to the coast, and in the USA from the Great Lakes E to Vermont and S to the Gulf coast.
Naturalized Distribution England.

England

From at least the 1930s until the 1970s or later, a colony of Rock Bass was established in a flooded 1.2-ha gravel-pit near Carfax on the northern outskirts of Oxford. How the fish came to be in this water is unknown, but an angler caught the first individuals in July 1937. In 1956, D.F. Leney reported that the pit was 'teeming' with Rock Bass, 'probably but few exceeding 6 or 7 in [15–18 cm]'. In 1969, R.L. Manuel informed Wheeler and Maitland (1973) that the colony had still been there 'a few years ago'. It has since apparently died out.

Characteristics

In North America, the Rock Bass, which reaches a length of 20–30 cm and a weight of 230 g, occurs mainly among stones and boulders in cool, weedy lakes and in the lower reaches of rocky streams, where it feeds principally on small aquatic invertebrates.

References

Fitter 1959; Lever 1977, 1996; Perring 1967; Wheeler 1974; Wheeler & Maitland 1973.

CENTRARCHIDAE (SUNFISH)

LARGEMOUTH BASS *Micropterus salmoides*
(BLACK BASS)

Natural Distribution From NE Mexico to Florida, USA, N to Ontario and Quebec, Canada, and the Atlantic slope N to SC South Carolina.
Naturalized Distribution England; [Scotland].

England

Patterson (1909: 372) saw a specimen of a Largemouth Bass in the Wherry Hotel at Oulton Broad, near Lowestoft in Suffolk, 'exhibited as the only survivor(!) captured out of a consignment from Austria that had been deposited in local waters; the others, it is believed, were all devoured by the Oulton Pike ... [this is] an introduced species, which did not flourish; had it done so, I think anglers would have very soon desired the extirpation of so voracious a fish'. Fitter (1959: 296) records that:

> about thirty years ago renewed attempts to establish black-bass fishing in Britain were definitely made with the large-mouth species ... a German fish breeder had established a stock which was used for the second, more successful, wave of black-bass introductions. In 1927 the Norwich Angling Club put 250 large-mouthed yearlings into a three-acre [1.2-ha] lake, but they disappeared in the following year. At about the same time fry from imported eggs were put into the River Ouse at Earith Bridge, but ... some at least of these may actually have been pike-perch [*Sander lucioperca*].

In 1934 or 1935, D.F. Leney obtained some 10–20 cm Largemouth Bass from an established population in the River Loire near Bourges in central France, for the Surrey Trout Farm and United Fisheries at Shottermill near Haslemere in Surrey. These fish were subsequently introduced to a gravel-pit at Send near Woking in Surrey, and to two disused clay-pits near Wareham in Dorset and Frome in Somerset. At Send, several young fish

LARGEMOUTH BASS

were netted in 1937 or 1938, but the survivors are believed to have died out by at least 1969. Largemouth Bass from Shottermill were also introduced to ponds in Cornwall, the Midlands, Manchester and north-west Kent, and to a lake in south Devon, where they grew 'to a fair size'. In 1935, a number were released in two reservoirs on the Lancashire moors, where they seem to have soon disappeared. A hundred yearlings placed in a lake at Coombe Abbey in Warwickshire before the Second World War similarly failed to become established.

By the early 1970s, the populations near Wareham and Frome were probably the only ones then extant. Since then:

> It is believed that all British populations of large-mouth bass have now died out. ... introductions, mostly unsuccessful, were made in England during the mid-1920s and the 1930s, including sites in Cornwall, Surrey, Kent, the Midlands and Manchester. By the mid-1970s it was suspected that the only surviving population was in a disused clay-pit near Wareham, Dorset. (Hickley, in Davies et al. 2004: 129.)

[**Scotland**]

According to Harvie-Brown and Buckley (1892: 271), Largemouth Bass were: 'imported by the Marquis of Exeter in 1879 – "So far as I recollect, I imported through my pisciculturist, Mr Silk, the Bass which were turned into Loch Baa by the Duke of Argyll in 1881 or 1882. These were, I think, mostly small-mouthed Bass [*Micropterus dolomieu*], but there were a good many of the big-mouthed among them".' Nothing more appears to have been heard of these fish.

Characteristics

In North America, where it has been widely translocated outside its natural range, the Largemouth Bass, which can weigh up to 2 kg, is a popular sporting fish. The success of this early-spawning species in northern latitudes is limited by its temperature requirement of at least 16°C for successful breeding. It favours clear, vegetated still waters and river backwaters with quiet water and overgrown banks. Those waters in which it became naturalized in England were mostly well sheltered and spring-fed, and were devoid of predatory Northern Pike but had plenty of prey species such as Roach and Perch.

Impact

Juvenile Largemouth Bass feed on aquatic invertebrates, tadpoles, insect larvae and small fish, while adults prey on larger fish, crayfish and frogs. The species has been widely introduced around the world, and in many places has been blamed for the decline or local extinction of native fish species. Had it become widely naturalized in England, it is probable that Patterson's fears would have proved well founded.

Maitland and Price (1969) reported the discovery of a North American monogenetic trematode, *Urocleidus principalis*, on the gills of Largemouth Bass from Wareham. They suggested that the species, new to Britain, was probably introduced with the original stock from the Loire.

References

Fitter 1959; Hickley, in Davies et al. 2004; Lever 1977, 1996; Maitland & Price 1969; Patterson 1909; Perring 1967; Wheeler 1974; Wheeler & Maitland 1973; Wintle 2001.

CICHLIDAE (CICHLIDS)

REDBELLY TILAPIA *Tilapia zillii*

(ZILL'S CICHLID; STRIPED TILAPIA; JORDAN ST PETER'S FISH)

Natural Distribution From W Africa E through the Chad and Nile River basins to Lakes George and Albert, N into Israel and the Jordan Valley.
Naturalized Distribution England; Wales.

England

In 1963, following the closure of a local tropical-fish shop, its stock, including Redbelly Tilapia and Guppies, was released in the Church Street stretch of the St Helens Canal in Lancashire, where the fish established breeding populations in water heated by effluent discharged from the glassworks of Pilkington Brothers. Although the water in the canal is said still to be artificially warmed by factory emissions, the Redbelly Tilapia have apparently died out.

Wales

'Redbelly tilapia have also been recorded from Llyn Trawsfynydd [in the Snowdon National Park, Merionethshire] in Wales into which warm water was discharged from the Trawsfynydd Power Station, which is now decommissioned' (Hickley, in Davies et al. 2004: 139). A survey of the lake in 1985–7 failed to reveal the presence of Redbelly Tilapia, and even if any had been present they would certainly have died out when heated water ceased to be discharged into the lake in 1991 (Linda van Gucci and Robin Ward, personal communications 2008).

Characteristics

In their native range, Redbelly Tilapia live in shallow and well-vegetated waters, where they feed principally on aquatic plants. Tilapia species in general are extremely palatable, and large numbers are being increasingly imported from Africa to British supermarkets.

References

Hickley, in Davies et al. 2004; Lever 1977, 1996; Maitland 2004; Wheeler 1974; Wheeler & Maitland 1973.

APPENDIX I

Chronology of the earliest arrival in Britain and Ireland of subsequently naturalized species

Date	Species
Neolithic	Orkney and Guernsey Voles *Microtus arvalis*
Neolithic	House Mouse *Mus domesticus*
Neolithic	Feral Cattle *Bos taurus*
Neolithic	Feral Goat *Capra hircus*
Neolithic	Feral Sheep *Ovis aries*
Bronze Age	Feral Horse *Equus caballus*
Bronze/Iron Age	Brown Hare *Lepus europaeus*
Bronze/Iron Age	Lesser White-toothed Shrew *Crocidura suaveolens*
Bronze/Iron Age	Greater White-toothed Shrew *Crocidura russula*
Iron Age	Feral Cat *Felis catus*
Saxon (9th/10th century)	Ship Rat *Rattus rattus*
Pre-Norman (? c 1050)	Common Pheasant *Phasianus colchicus*
Normans (11th/12th century)	Feral Ferret *Mustela furo*
Normans (11th/12th century)	Fallow Deer *Dama dama*
Early Plantagenet (mid- to late 12th century)	European Rabbit *Oryctolagus cuniculus*
1496–1530	Common Carp *Cyprinus carpio*
Before 1665	Greater Canada Goose *Branta canadensis*
1665	Goldfish *Carassius auratus*
c 1665	Egyptian Goose *Alopochen aegyptiacus*
1673	Red-legged Partridge *Alectoris rufa*
1725	Golden Pheasant *Chrysolophus pictus*
1728–9	Common Rat *Rattus norvegicus*
1738	Reindeer *Rangifer tarandus*
Before 1745	Mandarin Duck *Aix galericulata*
1768	Common Wall Lizard *Podarcis muralis*
1791	Black Swan *Cygnus atratus*
1828	Lady Amherst's Pheasant *Chrysolophus amherstiae*
1831	Reeves's Pheasant *Syrmaticus reevesii*
1837	Water Frogs *Rana* spp.
1838	Reeves's Muntjac *Muntiacus reevesi*
1840	Green Pheasant *Phasianus versicolor*
1841	Western Green Lizard *Lacerta bilineata*
1842	Little Owl *Athene noctua*
1860	Sika *Cervus nippon*
1840s	European Green Tree Frog *Hyla arborea*
1853 (1863)	Wels *Silurus glanis*
1855	Rose-ringed Parakeet *Psittacula krameri*
1865	Red-necked Wallaby *Macropus rufogriseus*

Date	Species
1868	American Brook Trout *Salvelinus fontinalis*
1873	Chinese Water Deer *Hydropotes inermis*
1870s	Wood Duck *Aix sponsa*
1876	Grey Squirrel *Sciurus carolinensis*
1874	Orfe *Leuciscus idus*
1878	Zander *Sander lucioperca*
1878 or 1898	Midwife Toad *Alytes obstetricans*
1884	Rainbow Trout *Oncorhynchus mykiss*
1885	Black Bullhead *Ameiurus melas*
1890–1	European Pond Terrapin *Emys orbicularis*
1899	Green Lizard *Lacerta viridis*
1902	Edible Dormouse *Glis glis*
1903	Alpine Newt *Triturus alpestris*
After 1903	Italian Crested Newt *Triturus carnifex*
c 1905	American Bullfrog *Rana catesbeiana*
Early 20th century	Bitterling *Rhodeus sericeus*
1914	Pumpkinseed *Lepomis gibbosus*
1920s	Bank Vole *Myodes glareolus*
1929	American Mink *Mustela vison*
1920s	Common Wall Lizard *Podarcis muralis*
Late 1920s	Chukar Partridge *Alectoris chukar*
1934–5	Marsh Frog *Rana ridibunda*
1930s	Ruddy Duck *Oxyura jamaicensis*
1937	Red-crested Pochard *Netta rufina*
Since 1950s	Ruddy Shelduck *Tadorna ferruginea*
1954	Yellow-bellied Toad *Bombina variegata*
Mid-1960s	Aesculapian Snake *Zamensis longissimus*
1967	African Clawed Toad *Xenopus laevis*
1960s	Chinese Grass Carp *Ctenopharyngodon idella*
1968	Channel Catfish *Ictalurus punctatus*
1986–7	Sunbleak *Leucaspius delineatus*
1980s	Red-eared Slider *Trachemys scripta*
Late 1980s	Topmouth Gudgeon *Pseudorasbora parva*
c 1987	Monk Parakeet *Myiopsitta monachus*
20th century	Bar-headed Goose *Anser indicus*
20th century	Snow Goose *Anser caerulescens*
20th century	Emperor Goose *Anser canagicus*
20th century	Fire-bellied Toad *Bombina bombina*
20th century	Sturgeon *Acipenser* spp.
Late 20th century	Alexandrine Parakeet *Psittacula eupatria*
Late 20th century	Blue-crowned Parakeet *Aratinga acuticaudata*
Late 20th century	Rosy-faced Lovebird *Agapornis roseicollis*
Late 20th century	Parma Wallaby *Macropus parma*
?	Crucian Carp *Carassius carassius*
?	Muscovy Duck *Cairina moschata*

APPENDIX II

Natural distribution and British and Irish distribution of naturalized and feral domestic species

	Natural Distribution						British and Irish Distribution			
NATURALIZED SPECIES	**EU**	**AS**	**AF**	**NA**	**SA**	**AU**	**E**	**S**	**W**	**I**
Mammals										
Red-necked Wallaby *Macropus rufogriseus*						X	X	X		
Parma Wallaby *Macropus parma*						X	X			
Grey Squirrel *Sciurus carolinensis*				X			X	X	X	X
Edible Dormouse *Glis glis*	X	X					X			
Bank Vole *Myodes glareolus*	X	X								X
Orkney and Guernsey Voles *Microtus arvalis*	X	X					X	X		
House Mouse *Mus domesticus*	X	X	X				X	X	X	X
Common Rat *Rattus norvegicus*		X					X	X	X	X
Ship Rat *Rattus rattus*		X					X	X	X	X
European Rabbit *Oryctolagus cuniculus*	X						X	X	X	X
Brown Hare *Lepus europaeus*	X						X	X	X	X
Lesser White-toothed Shrew *Crocidura suaveolens*	X	X					X			
Greater White-toothed Shrew *Crocidura russula*	X		X				X			X
American Mink *Mustela vison*				X			X	X	X	X
Sika *Cervus nippon*		X					X	X		X
Reeves's Muntjac *Muntiacus reevesi*		X					X		X	
Fallow Deer *Dama dama*	X	X					X	X	X	X
Chinese Water Deer *Hydropotes inermis*		X					X			
Birds										
Bar-headed Goose *Anser indicus*		X					X			
Snow Goose *Anser caerulescens*		X		X			X	X		
Emperor Goose *Anser canagicus*		X		X			X			
Greater Canada Goose *Branta canadensis*				X			X	X	X	X
Black Swan *Cygnus atratus*						X	X	X		
Egyptian Goose *Alopochen aegyptiacus*			X				X			
Ruddy Shelduck *Tadorna ferruginea*	X	X	X				X			X
Wood Duck *Aix sponsa*				X			X			
Mandarin Duck *Aix galericulata*		X					X	X		X
Red-crested Pochard *Netta rufina*	X	X					X			
Ruddy Duck *Oxyura jamaicensis*				X			X	X	X	X

Key EU = Europe; AS = Asia; AF = Africa; NA = North America; SA = South America; AU = Australia
E = England; S = Scotland; W = Wales; I = Ireland

Natural distribution and British and Irish distribution of naturalized and feral domestic species

	Natural Distribution						British and Irish Distribution			
NATURALIZED SPECIES *continued*	EU	AS	AF	NA	SA	AU	E	S	W	I
Birds *continued*										
Chukar Partridge *Alectoris chukar*	X	X					X	X		
Red-legged Partridge *Alectoris rufa*	X						X	X	X	X
Common Pheasant *Phasianus colchicus*	X	X					X	X	X	X
Green Pheasant *Phasianus versicolor*		X					X	X		
Golden Pheasant *Chrysolophus pictus*		X					X	X	X	
Lady Amherst's Pheasant *Chrysolophus amherstiae*		X					X	X	X	
Reeves's Pheasant *Syrmaticus reevesii*		X					X	X		
Rose-ringed Parakeet *Psittacula krameri*		X	X				X	X	X	
Monk Parakeet *Myiopsitta monachus*					X		X			
Little Owl *Athene noctua*	X						X	X	X	
Reptiles										
European Pond Terrapin *Emys orbicularis*	X	X	X				X			
Red-eared Slider *Trachemys scripta*				X			X		X	
Green Lizard *Lacerta viridis*	X						X			
Common Wall Lizard *Podarcis muralis*	X	X					X			
Aesculapian Snake *Zamensis longissimus*	X						X		X	
Amphibians										
Midwife Toad *Alytes obstetricans*	X						X			
Yellow-bellied Toad *Bombina variegata*	X						X			
Fire-bellied Toad *Bombina bombina*	X						X			
African Clawed Toad *Xenopus laevis*			X				X		X	
European Green Tree Frog *Hyla arborea*	X						X			
American Bullfrog *Rana catesbeiana*				X			X			
Marsh Frog *Rana ridibunda*	X						X			
Edible Frog *Rana* kl. *esculenta*	X						X			
Pool Frog *Rana lessonae*	X						X			
Southern Marsh Frog (Iberian Water Frog) *Rana perezi*	X						X			
Italian Pool Frog *Rana bergeri*	X						X			
Alpine Newt *Triturus alpestris*	X						X	X		
Italian Crested Newt *Triturus carnifex*	X						X			

Key EU = Europe; AS = Asia; AF = Africa; NA = North America; SA = South America; AU = Australia
E = England; S = Scotland; W = Wales; I = Ireland

Natural distribution and British and Irish distribution of naturalized and feral domestic species

	Natural Distribution						British and Irish Distribution			
NATURALIZED SPECIES *continued*	EU	AS	AF	NA	SA	AU	E	S	W	I
Fish										
Siberian Sturgeon *Acipenser baerii*		X					X			
Russian Sturgeon *Acipenser gueldenstaedtii*	X	X					X			
Sterlet *Acipenser ruthenus*	X	X					X			
Goldfish *Carassius auratus*	X	X					X	X	X	
Crucian Carp *Carassius carassius*	X	X					X	X	X	
Common Carp *Cyprinus carpio*		X					X	X	X	X
Chinese Grass Carp *Ctenopharyngodon idella*		X					X		X	
Sunbleak *Leucaspius delineatus*	X						X			
Orfe *Leuciscus idus*	X						X	X	X	
Topmouth Gudgeon *Pseudorasbora parva*		X					X	X		
Bitterling *Rhodeus sericeus*	X	X					X			
Black Bullhead *Ameiurus melas*				X			X			
Channel Catfish *Ictalurus punctatus*				X			X			
Wels *Silurus glanis*	X	X					X	X		
Rainbow Trout *Oncorhynchus mykiss*				X			X	X	X	X
American Brook Trout *Salvelinus fontinalis*				X			X	X	X	
Pumpkinseed *Lepomis gibbosus*				X			X			
Zander *Sander lucioperca*	X	X					X			

FERAL DOMESTIC SPECIES	EU	AS	AF	NA	SA	AU	E	S	W	I
Mammals										
Feral Cat *Felis catus*		X	X				X	X	X	X
Feral Ferret *Mustela furo*	X						X	X	X	X
Feral Horse *Equus caballus*		X					X	X	X	X
Reindeer *Rangifer tarandus*	X	X		X				X		
Feral Cattle *Bos taurus*	X	X					X	X	X	
Feral Goat *Capra hircus*	X	X					X	X	X	X
Feral Sheep *Ovis aries*		X					X	X	X	
Birds										
Muscovy Duck *Cairina moschata*				X	X		X			

Key EU = Europe; AS = Asia; AF = Africa; NA = North America; SA = South America; AU = Australia
E = England; S = Scotland; W = Wales; I = Ireland

APPENDIX III

Scientific names of species of fauna and flora mentioned in the text (other than main species)

FAUNA

Mammals

Aurochs *Bos primigenius*

Badger, Eurasian *Meles meles*
Banteng *Bos javanicus*
Buffalo, Water *Bubalus bubalis*

Chinchilla *Chinchilla brevicaudata laniger*

Deer, Red *Cervus elaphus*
Deer, European Roe *Capreolus capreolus*
Desman, Russian *Desmana moschata*
Dormouse, Hazel *Muscardinus avellanarius*

Eland *Taurotragus oryx*

Fox, Arctic *Alopex lagopus*
Fox, Red *Vulpes vulpes*

Guinea Pig *Cavia porcellus*

Hamster, European *Cricetus cricetus*
Hare, Irish *Lepus timidus hibernicus*
Hare, Mountain *Lepus timidus*
Hedgehog, European *Erinaceus europaeus*
Horse, Przewalski's *Equus przewalskii*
Horse, Wild *Equus ferus*

Marten, Pine *Martes martes*
Mole, European *Talpa europaea*
Mongoose, Small Indian *Herpestes javanicus/auropunctatus*
Mouflon, Asiatic *Ovis orientalis*
Mouse, Long-tailed Field *Apodemus sylvaticus*
Mouse, St Kilda House *Mus domesticus muralis*
Mouse, Wood **see** Mouse, Long-tailed Field

Otter, Eurasian *Lutra lutra*

Polecat *Mustela putorius*

Rabbit, Forest *Sylvilagus brasiliensis*

Sambar *Cervus unicolor*
Shrew, Common *Sorex araneus*
Stoat *Mustela erminea*

Tarpan **see** Horse, Wild

Urial **see** Mouflon, Asiatic

Vole, Field **see** Vole, Short-tailed
Vole, Short-tailed *Microtus agrestis*
Vole, Water *Arvicola terrestris*

Wapiti *Cervus elaphus canadensis*
Weasel *Mustela nivalis*
Wildcat *Felis silvestris*

Birds

Bittern, Great *Botaurus stellaris*
Blackbird, Common *Turdus merula*
Bullfinch, Common *Pyrrhula pyrrhula*
Buzzard, Common *Buteo buteo*

Chough, Red-billed *Pyrrhocorax pyrrhocorax*
Coot, Common *Fulica atra*
Corncrake *Crex crex*
Crow, Carrion *Corvus corone*
Crow, Hooded *Corvus cornix*

Dabchick **see** Grebe, Little
Dipper, White-throated *Cinclus cinclus*
Dove, Collared *Streptopelia decaocto*
Duck, Maccoa *Oxyura maccoa*
Duck, Tufted *Aythya fuligula*
Duck, White-headed *Oxyura leucocephala*
Dunnock *Prunella modularis*

Eider, Common *Somateria mollisima*

Falcon, Peregrine *Falco peregrinus*
Fulmar *Fulmarus glacialis*

Gadwall *Anas strepera*
Godwit, Black-tailed *Limosa limosa*
Goose, Barnacle *Branta leucopsis*
Goose, Greater White-fronted *Anser albifrons*
Goose, Lesser White-fronted *Anser erythropus*
Goose, Spur-winged *Plectropterus gambensis*
Grebe, Black-necked *Podiceps nigricollis*
Grebe, Little *Tachybaptus ruficollis*
Grebe, Slavonian *Podiceps auritus*
Grouse, Black *Tetrao tetrix*
Guillemot, Common *Uria aalge*
Gull, Black-headed *Larus ridibundus*
Gull, Common **see** Gull, Mew
Gull, Herring *Larus argentatus*
Gull, Lesser Black-backed *Larus fuscus*
Gull, Mew *Larus canus*

Harrier, Hen *Circus cyaneus*
Harrier, Eurasian Marsh *Circus aeruginosus*
Heron, Grey *Ardea cinerea*

Jackdaw, Eurasian *Corvus monedula*

Kestrel, Common *Falco tinnunculus*
Kite, Black *Milvus migrans*
Kittiwake, Black-legged *Rissa tridactyla*

Lapwing, Northern *Vanellus vanellus*
Linnet, Common *Acanthis cannabina*

Magpie, Black-billed *Pica pica*
Mallard *Anas platyrhynchos*

Merganser, Red-breasted
Mergus serrator
Moorhen, Common
Gallinula chloropus

Nightingale, Common
Luscinia megahynchos
Nuthatch, Wood *Sitta europaea*

Owl, Barn *Tyto alba*
Owl, Long-eared *Asio otus*
Owl, Short-eared *Asio flammeus*
Owl, Tawny *Strix aluco*
Oystercatcher, Eurasian
Haematopus ostralegus

Partridge, Grey *Perdix perdix*
Pheasant, Silver
Lophura nycthemera
Pigeon, Feral *Columba livia*
Pipit, Meadow *Anthus pratensis*
Pochard, Common *Aythya ferina*
Ptarmigan *Lagopus mutus*
Puffin, Atlantic *Fratercula arctica*

Quail, Common *Coturnix coturnix*

Raven, Common *Corvus corax*
Razorbill *Alca torda*

Shearwater, Manx
Puffinus puffinus
Shelduck, Common
Tadorna tadorna
Shoveler, Northern *Anas clypeata*
Skua, Great *Stercorarius skua*
Skylark *Alauda arvensis*
Sparrow, Eurasian Tree
Passer montanus
Sparrowhawk, Eurasian
Accipiter nisus
Starling, Common
Sturnus vulgaris
Stone-Curlew
Burhinus oedicnemus
Swan, Mute *Cygnus olor*

Teal, Eurasian *Anas crecca*
Thrush, Mistle *Turdus viscivorus*
Thrush, Song *Turdus philomelos*
Tern, Arctic *Sterna paradisaea*
Tern, Common *Sterna hirundo*
Tit, Bearded *Panurus biarmicus*
Tit, Great *Parus major*
Turkey, Wild *Meleagris gallopavo*

Wheatear, Northern
Oenanthe oenanthe
Whitethroat, Common
Sylvia communis
Wigeon, Eurasian *Anas penelope*
Woodlark *Lullula arborea*
Woodpecker, Green *Picus viridis*
Woodpecker, Great Spotted
Dendrocopos major
Woodpigeon, Common
Columba palumbus

Yellowhammer *Emberiza citrinella*

Reptiles

Alligator, American
Alligator mississippiensis

Lizard, Sand *Lacerta agilis*
Lizard, Viviparous or Common
Zootoca vivipara

Snake, Grass *Natrix natrix*
Snake, Western Whip
Coluber viridiflavus

Amphibians

Frog, Common *Rana temporaria*

Newt, Great Crested
Triturus cristatus
Newt, Smooth *Triturus vulgaris*
Newt, Warty **see** Newt, Great Crested

Toad, Cane *Bufo marinus*
Toad, Common *Bufo bufo*

Fish

Bleak *Alburnus alburnus*
Blenny *Lipophrys pholis*
Bream, Common *Abramis brama*
Bullhead *Cottus gobio*
Bullhead, Brown *Ictalurus (Ameiurus) nebulosus*

Charr, Arctic *Salvelinus alpinus*

Eel *Anguilla anguilla*

Grayling, European
Thymallus thymallus
Gudgeon *Gobio gobio*

Miller's Thumb **see** Bullhead

Perch *Perca fluviatilis*
Pike, Northern *Esox lucius*

Roach *Rutilus rutilus*
Rudd *Scardinius erythrophthalmus*
Salmon, Atlantic *Salmo salar*
Smelt *Osmerus operlanus*
Stickleback, Three-spined
Gasterosteus aculeatus
Sturgeon, Common
Acipenser sturio

Tench *Tinca tinca*
Trout, Brown (and Sea)
Salmo trutta

Whitefish *Coregonus lavaretus*

Insects

Adonis Blue
Polyommatus bellargus

Bee, Honey *Apis mellifera*

Duke of Burgundy Fritillary
Hamearis lucina

Flea, European Rabbit
Spilopsyllus cuniculi

Heath Fritillary *Melitaea athelia*
High Brown Fritillary
Argynnis adippe

Lulworth Skipper
Thymlicus actaeon
Large Blue *Maculinea arion*

Marbled White *Melanargia galathea*
Mosquito (Culicinae)

Pearl-bordered Fritillary
Boloria euphrosyne

Six-spot Burnet
Zygaena filipendulae
Small Pearl-bordered Fritillary
Boloria selene

White Admiral *Limenitis camilla*
Wood White *Leptidea sinapis*

Crustaceans

Crab, Shore *Carcinus maenas*
Crayfish, Freshwater or White-clawed *Astacus (Austropotamobius) pallipes*
Crayfish, Signal
Pacifastacus leniusculus

Molluscs

Mussel, Swan *Anodonta cygnea*

FLORA

Alder, Common *Alnus glutinosa/incana*
Anemone, Wood *Anemone nemorasa*
Apple, Crab *Malus sylvestris*
Apple (cultivated) *Malus domestica*
Ash, Common *Fraxinus excelsior*
Aspen *Populus tremula*

Beech, Common *Fagus sylvatica*
Beet, Sugar *Beta vulgaris*
Bilberry *Vaccinium myrtillus*
Birch, Downy *Betula pubescens*
Birch, Silver *Betula pendula*
Blackberry **see** Bramble
Blackthorn *Prunus spinosa*
Blaeberry **see** Bilberry
Bluebell, Common *Hyacinthoides non-scripta*
Bracken, Common *Pteridium aquilinum*
Bramble *Rubus fruiticosus*
Brome, False *Brachypodium sylvaticum*
Bulrush **see** Reed Mace
Bulrush, Lesser **see** Reed Mace, Lesser

Carrot *Daucus carota*
Cherry, Wild *Prunus avium*
Chestnut, Horse *Aesculus hippocastanum*
Chestnut, Sweet *Castanea sativa*
Cordyline, Australian *Cordyline australis*
Cowberry *Vaccinium vitis-idaea*
Cowslip *Primula veris*
Cuckoo Flower *Cardamine pratensis*
Cypress, Monterey *Cupressus macrocarpa*

Daffodil, Wild *Narcissus pseudonarcissus*

Elm, English *Ulmus glabra*
Elder *Sambucus nigra*

Fir, Douglas *Pseudotsuga menziesii*

Gorse, Common *Ulex europaeus*
Grass, Mat *Nardus stricta*
Grass, Tor *Brachypodium pinnatum*

Hawthorn, Common *Crataegus monogyna*
Hazel, Common *Corylus avellana*
Heather, Common *Calluna vulgaris*
Holly, European *Ilex aquifolia*
Honeysuckle *Lonicera periclymenum*
Hornbeam, European *Carpinus betulus*

Ivy *Hedera helix*

Juniper, Common *Juniperus communis*

Lady's Smock **see** Cuckoo Flower
Larch, European *Larix decidua*
Larch, Japanese *Larix leptolepis/kaempferi*
Ling **see** Heather, Common
Lords and Ladies *Arum maculatum*

Maize *Zea mays*
Melon, Canteloupe *Cucumis melo*
Mercury, Dog's *Mercurialis perennis*
Moor-grass, Purple *Molinia caerulea*
Mulberry, Black *Morus nigra*
Myrtle, Bog *Myrica gale*

Oak, Holm *Quercus ilex*
Oak, Pedunculate *Quercus robur*
Oak, Sessile *Quercus petraea*
Orchid, Bee *Ophrys apifera*
Orchid, Common Spotted *Dactylorhiza fuchsii*
Orchid, Early Purple *Orchis mascula*
Orchid, Greater Butterfly *Platanthera chlorantha*
Orchid, Pyramidal *Anacamptis pyramidalis*
Osier, Common *Salix viminalis*
Oxlip *Primula elatior*

Peony, Common *Paeonia officinalis*
Pine, Corsican *Pinus nigra*
Pine, Lodgepole *Pinus contorta*
Pine, Scots *Pinus sylvaticus*
Primrose *Primula vulgaris*

Reed, Common *Phragmites communis*
Reed Mace *Typha latifolia*
Reed Mace, Lesser *Typha angustifolia*
Rhododendron, Common *Rhododendron ponticum*
Rowan *Sorbus aucuparia*
Rush, Flowering *Butomus umbellatus*
Rush, Soft *Juncus effusus*

Sorghum *Sorghum bicolor*
Spruce, Norway *Picea abies*
Spruce, Sitka *Picea sitchensis*
Stonecrop, Australian Swamp *Crassula helmsii*
Sunflower *Helianthus annus*
Sycamore *Acer pseudoplatanus*

Tare *Vicia sativa*
Traveller's Joy *Clematis vitalba*
Thyme, Wild *Thymus polytrichus*

Water Lily, White *Nymphaea alba*
Water Lily, Yellow *Nuphar lutea*
Wheat *Triticum aestivum*
Whitebeam, Swedish *Sorbus intermedia*
Willow *Salix* spp.
Wrack, Serrated (Toothed) *Fucus serratus*

Yew *Taxus baccata*

REFERENCES

Aars, J., Lambin, X., Denny, A. & Griffin, A.C. 2001. 'Water vole in the Scottish uplands: distribution patterns of disturbed and pristine populations ahead of and behind the American mink invasion front'. *Animal Conservation* 4: 187–94.

Aberdeen Red Kites Newsletters 1–3. 2007–8. RSPB, Scotland.

Abernethy, K. 1994. 'The establishment of a hybrid zone between red and sika deer (genus *Cervus*)'. *Molecular Ecology* 3: 551–62.

Abernethy, K. 1998. *Sika Deer in Scotland.* Deer Commission for Scotland and the Stationery Office: Inverness and Edinburgh.

Adams, C.E. & Maitland, P.S. 2001. 'Invasion and establishment of freshwater fish populations in Scotland – the experience of the past and lessons for the future'. *Glasgow Naturalist* 23 (Supplement): 35–43.

Adams, C.E. & Mitchell, J. 1992. 'Introduction of another non-native fish species to Loch Lomond: crucian carp (*Carassius carassius* (L.))'. *Glasgow Naturalist* 22: 165–8.

Ainslie, D. 1907. 'The little owl in Bedfordshire'. *The Zoologist* 11: 353.

Aistrop, J.B. 1968. *The Mongolian Gerbil.* Dennis Dobson: London.

Alcamo, J. 2003. 'Ecosystems and human well-being'. In: *The Millennium Ecosystem Assessment.* Island Press: Washington, DC [www.milleniumassessment.org].

Alderson, L. 1997. *A Breed of Distinction: White Park Cattle Ancient and Modern.* Countrywide Livestock Ltd.: Shrewsbury, Shropshire.

Aldridge, D.C. 1999. 'Development of European bitterling in the gills of freshwater mussels'. *Journal of Fish Biology* 54: 138–51.

Allard, P. 1999. 'Category D species and a selected list of category F species'. In: Taylor, M., Seago, M., Allard, P. & Dorling, D. *The Birds of Norfolk*: 522–30. Pica Press: Robertsbridge.

Allen, Y., Kirby, S., Copp, G.H. & Brazier, M. 2006. 'Toxicity of rotenone to topmouth gudgeon *Pseudorasbora parva* for the species' eradication from a tarn in Cumbria'. *Fisheries Management and Ecology* 13: 337–40.

Amori, G. & Clout, M. 2002. 'Rodents on islands: a conservation challenge'. *Australian Centre for International Agricultural Research. Monograph Series 96*: 63–8.

Anderson, D. 1986. 'Distribution of Bedfordshire mammal species 1971–1985'. *Bedfordshire Naturalist* 40: 13–20.

Anderson, D. & Cham, S.A. 1987. 'Muntjac deer (*Muntiacus reevesi*) – the early years'. *Bedfordshire Naturalist* 42: 14–18.

Anderson, R. & Hughes, D. 1995. 'Recent spread of the Grey Squirrel, *Sciurus carolinensis* Gmelin, into Cos. Down and Londonderry'. *Irish Naturalists' Journal* 25: 118.

Andrews, C.H., Thompson, H.V. & Mansi, W. 1959. 'Myxomatosis'. *Nature, London* 184: 1179–80.

Angler's Mail 2005. 'Our sturgeon: unwanted pet'. (19 July). 5pp.

Angus, S. & Hopkins, P.G. 1996. 'Ship rat *Rattus rattus* confirmed on the Shiant Islands'. *Hebridean Naturalist* 13: 18–22.

Angus, W.C. 1886. 'The capercaillie'. *Proceedings of the Natural History Society of Glasgow* New Series 1: 380.

Anon. 1925. 'A new Lakeland fishery'. *The Field* (6 August).

Anon. 1965. 'Britain's wild wallabies'. *Animals* 7: 524.

Anon. 1997. 'Exotic reptiles and amphibians in the wild: Information and advice on the problems of non-native species in Britain and Ireland'. *Froglife* 1997: 1–8.

Anon. 2000a. 'England called to bullfrog alert'. *English Nature* 52: 4.

Anon 2000b. 'Bullfrog alert'. *Froglife News* 1: 3.

Anon. 2001. *Interim Summary Report on American Bullfrog Containment at Land at Cowden, Sussex/Kent border. Froglife,* Halesworth, Suffolk. 20 page mimeo.

Anon. 2002. 'News special: Zander – a problem?' *Angler's Mail* (14 December): 6–7.

Anon. 2005a. '£3.3 million for Ruddy Duck eradication'. *Aliens* 22: 21.

Anon. 2005b. 'Bustards back in business'. *Geographical* 77: 7.

Anon. 2006. 'Are wild boar to be welcomed?' *Country Life* (editorial): 2 February.

Anon. 2007. *UK Ruddy Duck Eradication Programme: Information Bulletin Detailing Progress in the First Two Years.* Department for Environment, Food and Rural Affairs.

Armistead, J.J. 1895. *An Angler's Paradise, and How To Obtain It.* Author: London.

Armitage, P.L. 1985. 'Small mammal faunas in later mediaeval towns'. *Biologist* 32: 65–71.

Armitage, P.L., West, B. & Steedman, K. 1984. 'New evidence of black rat in Roman London'. *The London Archaeologist* 4: 375–83.

Armour, C.J. & Thompson, H.V. 1955. 'Spread of myxomatosis in the first outbreak in Great Britain'. *Annals of Applied Biology* 43: 511–85.

Armour-Chelu, M. 1991. 'The faunal remains'. In: Sharples, N.M. (ed.). *Maiden Castle: Excavations and Field Survey, 1985–6. English Heritage Archaeological Report 19*: London.

Armstrong, P. 1982. 'Rabbits (*Oryctolagus cuniculus*) on islands: a case-study of successful colonization'. *Journal of Biogeography* 9: 353–62.

Arnold, H.R. 1973. *Provisional Atlas of the Amphibians and Reptiles of the British Isles.* Nature Conservancy: Peterborough.

Arnold, N. & Ovenden, D. 2004. *Collins Field Guide to Reptiles and Amphibians.* Collins: London.

Ashton, R. 2005. 'Comments from an old naturalist about exotic species and a new herpetocultural ethic'. *Iguana* 12: 48–9.

Atherton, D. 1994. 'Anglers declare war on the greedy minnow'. *Sunday Telegraph*, 4 September.

Atkinson-Willes, G.L. (ed.). 1963. *Wildfowl in Great Britain.* Nature Conservancy Monograph. Her Majesty's Stationery Office: London.

Austin, G.E., Rehfisch, M.M., Allan, J.R. & Holloway, S.J. 2007. 'Population size and differential population growth of introduced Greater Canada Geese *Branta canadensis* and re-established Graylag Geese *Anser anser* across habitats in Great Britain in the year 2000'. *Bird Study* 54: 343–52.

Avery, M. 2000. 'Ruddy Ducks and other aliens'. *British Birds* 93: 500.

Baatsen, R.G. 1990. 'Red-crested Pochard *Netta rufina* in the Cotswold Water Park'. *Hobby* 16: 64–7. Wiltshire Ornithological Society.

Backhouse, J. 1989. *The Luttrell Psalter.* British Library: London.

Backhouse, K.M. & Thompson, H.V. 1955. 'Myxomatosis'. *Nature, London* 176: 1155–6.

Bailey, J. R. 2002. 'Wild rabbits – a novel vector for vero cytotoxigenic *Escherichia coli* (VTEC) 0157'. *Communicable Disease and Public Health* 5: 74–5.

Bainbridge, I.P., Evans, R.J., Broad, C.H., Crooke, K., Duffy, R.E., Love, J.A. & Mudge, G.P. 2003. 'Re-introduction of White-tailed Eagles to Scotland'. In: Thompson, D.B.A., Redpath, S.M., Fielding, A.H., Marquiss, M. & Galbraith, C.A. (eds). *Birds of Prey in a Changing Environment*: 395–406. Scottish Natural Heritage: Edinburgh.

Baker, H., Stroud, D.A., Aebischer, N.J., Cranswick, P.A., Gregory, R.D., McSorley, C.A., Noble, D.G. & Rehfisch, M.M. 2006. 'Population estimates of birds in Great Britain and the United Kingdom'. *British Birds* 99: 25–44.

Baker, J. 1974. 'New menace of forests – bark-eating Himalayan porcupines'. *Eastern Daily Press* (24 October).

Baker, S.J. 1986. 'Free-living golden hamsters (*Mesocricetus auratus*) in London'. *Journal of Zoology* 209: 285–96.

Baker, S.J. 1987. 'Irresponsible introductions and reintroductions of animals into Europe with particular reference to Britain'. *International Zoo Yearbook* 24/25: 200–5.

Baker, S.J. 1990. 'Escaped exotic mammals in Britain'. *Mammal Review* 20: 75–96.

Balharry, D. & Daniels, M. 1998. 'Wild living cats in Scotland'. *Scottish Natural Heritage Research, Survey and Monitoring Report No. 23.*

Balon, E.K. 1969. 'Studies on the wild carp: new opinions concerning the origins of the carp'. *Práce Laboratória Rybarstva* 2: 99–120.

Banks, B. 1989. 'Alpine newts in north-east England'. *Bulletin of the British Herpetological Society* 30: 4–5.
Banks, B. & Laverick, G. 1986. 'Garden ponds as amphibian breeding sites in a conurbation in the north-east of England (Sunderland, Tyne and Wear)'. *The Herpetological Journal* 1: 44–50.
Banks, B., Foster, J., Langton, T. & Morgan, K. 'British Bullfrogs?' *British Wildlife* 11: 327–30.
Bannerman, D.A. 1953–63. *The Birds of the British Islands*. Oliver & Boyd: Edinburgh.
Barnard, S. 1997. 'A review of consented trout stockings into riverine environments'. *Environment Agency Research and Development Technical Report W74*: 1–80.
Barnett, S.A. 1951. 'Damage to wheat by enclosed populations of *Rattus norvegicus*'. *Journal of Hygiene* 49: 22–5.
Barr, D. 1976. 'In pursuit of zander'. *Country Life* (26 February).
Barrett-Hamilton, G.E.H. 1898. 'Notes on the introduction of the brown hare into Ireland'. *Irish Naturalists' Journal* 7: 69–76.
Barrett-Hamilton, G.E.H. & Hinton, M.A.C. 1910–21. *A History of British Mammals*. Gurney & Jackson: London.
Barrington, R.M. 1880. 'On the introduction of the squirrel into Ireland'. *Scientific Proceedings of the Royal Dublin Society* 2: 615–31.
Bartholomew, J. 1929. 'Capercaillie in West Stirling'. *Scottish Naturalist*: 61.
Baxter, E.V. & Rintoul, L.J. 1953. *The Birds of Scotland*. Vol. I. Oliver & Boyd: Edinburgh.
Bayliss, J.H. 1980. 'The extinction of bubonic plague in Britain'. *Endeavour* 4: 58–66.
Beck, B.B. & Wemmer, C.M. (eds). 1983. *The Biology and Management of an Extinct Species, Père David's Deer*. Noyes: Park Ridge, New Jersey.
Beebee, T.J.C. 1981. 'Habitats of British amphibians (3): river valley marshes'. *Biological Conservation* 18: 281–7.
Beebee, T.J.C. 1995. 'Even earlier breeding by alpine newts (*Triturus alpestris*) in Britain'. *Bulletin of the British Herpetological Society* 51: 5–6.
Beebee, T.J.C. & Griffiths, R.A. 2000. *Amphibians and reptiles: a Natural History of the British Herpetofauna*. Harper Collins: London.
Beebee, T.J.C., Buckley, J., Evans, I., Foster, J.P., Gent, A.H., Gleed-Owen, C.P., Kelly, G., Rowe, G., Snell, C., Wycherley, J.T. & Zeisset, I. 2005. 'Neglected native or undesirable alien? Resolution of a conservation dilemma concerning the pool frog *Rana lessonae*'. *Biodiversity and Conservation* 14: 1607–26.
Beirne, B.P. 1947. 'The history of the British land mammals'. *Annals of the Magazine of Natural History*, Series 11, 14: 501–14.
Beirne, B.P. 1952. *The Origin and History of the British Fauna*. Methuen: London.
Bell, A.P. 1978. 'An English colony of the alpine newt'. *British Journal of Herpetology* 5: 748.
Bell, D.P. & Bell, A.P. 1995. 'Distribution of the introduced alpine newt *Triturus alpestris* and of native *Triturus* species in north Shropshire, England'. *Australian Journal of Ecology* 20: 367–75.
Bell, T. 1859. 'The edible frog, long a native of Foulmire Fen'. *Zoologist* 17: 6565.
Bellrose, F.C. & Holm, D.J. 1994. *The Ecology and Management of the Wood Duck*. Stackpole Books: Mechanicsburg, Pennsylvania.
Benham, E.M. 1953. 'The distribution of squirrels in Dorset'. *Proceedings of the Dorset Natural History and Archaeological Society* 74: 121–32.
Bentley, E.W. 1959. 'The distribution and status of *Rattus rattus* L. in the United Kingdom in 1951 and 1956'. *Journal of Animal Ecology* 28: 299–308.
Bentley, E.W. 1964. 'A further loss of ground by *Rattus rattus* L. in the United Kingdom during 1956–61'. *Journal of Animal Ecology* 33: 371–3.
Berry, J. 1939. 'The status and distribution of wild geese and wild duck in Scotland'. *Wildfowl Inquiry Commission Report No. 2*: Cambridge.
Berry, J. 1985. *The Natural History of Orkney*. Collins: London.
Berry, R.J. 1970. 'The natural history of the house mouse'. *Field Study* 3: 212–62.
Berry, R.J. & Rose, F.E.N. 1975. 'Islands and the evolution of *Microtus arvalis* (Microtinae)'. *Journal of Zoology* 177: 395–409.
Beyer, K. 2004. 'Escapes of potentially invasive fishes from an ornamental aquaculture facility: the case of the topmouth gudgeon *Pseudorasbora parva*'. *Journal of Fish Biology* 65 (Supplement A): 326–7.
Bieber, C. 1998. 'Population dynamics, sexual activity, and reproduction failure in the fat dormouse (*Myoxus glis*)'. *Journal of Zoology* 244: 223–9.
Bignal, E. 1978. 'Mink predation of shelduck and other wildfowl at Loch Lomond'. *Western Naturalist* 7: 47–53.
Birks, J. 1986. *Mink*. Anthony Nelson: Oswestry, Shropshire.
Birks, J. 1990. 'Feral mink and nature conservation'. *British Wildlife* 1: 313–23.
Birks, J. 1999. 'Polecats are on their way back'. *Mammal News* 119: 10–11.
Birks, J.D.S. & Kitchener, A.C. 1999. *The Distribution and Status of the Polecat* Mustela furo *in Britain in the 1990s*. Vincent Wildlife Trust: London.
Bishop, I.R. & Delany, M.J. 1963. 'The ecological distribution of small mammals in the Channel Islands'. *Mammalia* 27: 99–110.
Blackburn, T.M. & Duncan, R.P. 2001. 'Establishment patterns of exotic birds are constrained by non-random patterns in introduction'. *Journal of Biogeography* 28: 927–39.
Blackwell, K. 1985. 'The midwife toad, *Alytes obstetricans*, in Britain'. *Bulletin of the British Herpetological Society* 14: 13.
Blair, M.J., McKay, H., Musgrove, A.J. & Rehfisch, M.M. 2000. 'Review of the status of introduced non-native waterbird species in the agreement area of the African-Eurasian Waterbird Agreement Research Contract CR0219'. *BTO Research Report* 229: 1–129.
Blathwayt, F.L. 1902; 1904. 'The little owl in Lincolnshire'. *The Zoologist* 11: 112; 74.
Blurton-Jones, N.G. 1956. 'Census of breeding Canada geese, 1953'. *Bird Study* 3: 153–70.
Bonar, H.N. 1907. 'Capercaillie in Midlothian'. *Annals of Scottish Natural History*: 51–2.
Bonar, H.N. 1910. 'Capercaillie in East Lothian'. *Annals of Scottish Natural History*: 120.
Bond, F. 1843. 'Note on the occurrence of the edible frog in Cambridgeshire'. *Zoologist* 2 (393): 677.
Bonesi, L. & Macdonald, D.W. 2004a. 'Impact of released Eurasian otters on a population of American mink: a test using an experimental approach'. *Oikos* 106: 9–18.
Bonesi, L. & Macdonald, D.W. 2004b. 'Differential habitat use promotes sustainable coexistence between the specialist otter and the generalist mink'. *Oikos* 106: 509–19.
Booth, C. & Booth, J. 1994. *The Mammals of Orkney*. The Orcadian: Kirkwall, Orkney.
Boulenger, G.A. 1884. 'On the origin of the edible frog in England'. *Zoologist* 42: 265–9.
Boulenger, G.A. 1897–98. *The Tailless Batrachians of Europe*. Two volumes. The Ray Society: London.
Boulenger, G.A. 1918. 'On the races and variation of the edible frog'. *Annals of the Magazine of Natural History* 2: 241–57.
Boyd, J.M. 1953. 'The sheep population of Hirta, St Kilda, 1952'. *Scottish Naturalist* 65: 25–28; 68: 10–13.
Boyd, J.M. 1964a. 'The Soay sheep on the island of Hirta, St Kilda'. *Proceedings of the Zoological Society of London* 142: 129–63.
Boyd, J.M. 1964b. *St Kilda National Nature Reserve Management Plan*. Nature Conservancy: Peterborough.
Boyd, J.M. 1981. 'The Boreray sheep of St Kilda, Outer Hebrides, Scotland: the natural history of a feral population'. *Biological Conservation* 20: 215–27.
Boyd, J.M. & Jewell, P.A. 1974. 'The Soay sheep and their environment: a synthesis'. In Jewell, P.A. (ed.). *Island Survivors: the Ecology of the Soay Sheep of St Kilda*: 360–73. Blackwell Science: Oxford.
Bradshaw, C.J. 1901. 'A little owl at Henley'. *The Zoologist* 9: 476.
Bramwell, D. 1990. 'Ossom's Eyrie Cave: an archaeological contribution to the recent history of vertebrates in Britain'. *Zoological Journal of the Linnean Society* 98: 1–25.
Brazil, M. 1991. *Birds of Japan*. Christopher Helm: London.
Brede, E.G., Thorpe, R.S., Arntzen, J.W. & Langton, T.E.S. 2000. 'A morphometric study of a hybrid newt population (*Triturus cristatus/T. carnifex*: Beam Brook Nurseries, Surrey, U.K)'. *Biological Journal of the Linnean Society* 70: 685–95.

Bremner, A. & Park, K. 2007. 'Public attitudes to the management of invasive non-native species in Scotland'. *Biological Conservation* 139: 306–14.

British Ornithologists' Union Records Committee Report. 1970. *Ibis* 113: 420–3.

British Trust for Ornithology. 1978. Records Committee Report. *Ibis* 120: 411.

Britton, J.R. & Brazier, M. 2006. 'Eradicating the invasive topmouth gudgeon, *Pseudorasbora parva*, from a recreational fishery in northern England'. *Fisheries Management and Ecology* 13: 329–35.

Britton, J.R. & Davies, G.D. 2006. 'Ornamental species of the genus *Acipenser*: new additions to the ichthyofauna of the UK'. *Fisheries Management and Ecology* 13: 207–10.

Britton, R., Davies, G., Page, R. & Pegg, J. 2006. 'Stocking of non-native fish into recreational fisheries: what makes a successful introduction'. In: Hickley, P. & Axford, S. (eds). *Fisheries and Conservation: Success and Failure. Proceedings of the Annual Conference of the Institute of Fisheries Management*: 90–101. Minehead, Somerset.

Browne, A.M. & O'Halloran, J. 1998. 'Canada geese in Ireland'. *Irish Birds* 6: 233–6.

Bryce, J. 1997. 'Changes in the distributions of red and grey squirrels in Scotland'. *Mammal Review* 27: 171–6.

Buchanan Smith, A.D. 1927. 'Goats for the island of Eigg'. *British Goat Society Yearbook*: 57–8.

Buchanan Smith, A.D. 1932. 'Wild goats in Scotland'. *British Goat Society Yearbook*: 120–5.

Buckhurst, E.A.J., Copland, W.O. & Roberts, E.A. 1964. 'The status of sika deer in the Poole basin'. *Proceedings of the Dorset Natural History & Archaeological Society* 86: 96–101.

Buckland, F.T. 1861. 'On the acclimatisation of animals'. *Journal of the Society of Arts* 9: 19–34. (See also 46–8; 92.)

Buckland, F. 1881. *Natural History of British Freshwater Fishes.* SPCK: London.

Buckley, J. 1986. 'Water frogs in Norfolk'. *Transactions of the Norfolk and Norwich Naturalists' Society* 27: 199–211.

Bullock, D.J. 1983. 'Borerays, the other rare breed on St Kilda'. *Ark* 10: 274–8.

Bullock, D.J. 1995. 'The feral goat – conservation and management'. *British Wildlife* 6: 152–9.

Bullock, D.J. & Kinnear, P.K. 1988. 'The use of goats to control birch in sand dunes: an experimental study'. *Aspects of Applied Biology* 16: 163–8.

Bunting, W. 1957. 'Animals and plants introduced to the Thorne district of Yorkshire'. *British Journal of Herpetology* 1: 70.

Burton, J. 1973. 'The laughing frogs of Romney Marsh'. *Country Life* (20 December).

Butcher, A.J. 1974. 'Porcupine threat in Devon'. *Western Morning News* (3 April).

Butler, C., Hazlehurst, G. & Butler, K. 2002. 'First nesting by Blue-crowned Parakeet in Britain'. *British Birds* 95: 17–20.

Buxton, P.A. 1907. 'The spread of the little owl in Hertfordshire'. *The Zoologist* 11: 430.

Caffell, G. 1995. 'Pen up your kids!' *Current Archaeology* 12: 385–6.

Calado, R. & Chapman, P.M. 2006. 'Aquarium species. Deadly invaders'. *Marine Pollution Bulletin* 52: 599–601.

Campbell, J.B. 1977. *The Upper Palaeolithic of Britain.* Clarendon Press: Oxford.

Campbell, J.M. 1906. 'Capercaillies in Ayrshire'. *Annals of Scottish Natural History*: 186.

Campbell, R.N. 1974. 'St Kilda and its sheep'. In: Jewell, P.A. (ed.). *Island Survivors: the Ecology of the Soay Sheep of St Kilda*: 8–35. Athlone Press: London.

Cannings, P. 1999. 'The Lady Amherst's pheasant'. *Bedfordshire Naturalist* 53: 68–72.

Carlton, J. & Ruiz, G. 2000. 'The vectors of invasions by alien species'. In: Preston, G., Brown, G. & van Wyk, E. (eds). *Best Management Practices for Preventing and Controlling Invasive Alien Species*: 82–9. Symposium Proceedings. The Working for Water Programme, Cape Town.

Carrington, J.T. 1876. 'Fish culture for the Thames'. *Zoologist* 2: 5110–13.

Carrington, R.P. 1950. 'The edible dormouse'. *Journal of the Association of School Natural History Societies* 3: 27–9.

Carter, I. 1998. 'The changing fortunes of the Red Kite in Suffolk'. *Suffolk Birds* 46: 6–10.

Carter, I. 2003. 'The return of the Red Kite'. *Biologist* 50: 217–21.

Carter, I. 2005. 'The benefits and dangers of species reintroductions: lessons from the Red Kite reintroduction programme and other bird reintroductions'. In: Rooney, P., Nolan, P. & Hill, D. (eds). *Restoration, Reintroduction and Translocation*: 117–22. Proceedings of the 20th Conference of the Institute of Ecology and Environmental Management: Southport.

Carter, I. 2007. *The Red Kite.* (Second, revised, edition.) Arlequin Press: Shrewsbury.

Carter, I. & Grice, P. 2000. 'Studies of re-established Red Kites in England'. *British Birds* 93: 304–22.

Carter, I. & Grice, P. 2002. *The Red Kite Reintroduction Programme in England. English Nature Research Report No. 451.* English Nature: Peterborough.

Carter, I. & Newbery, P. 2004. 'Reintroduction as a tool for population recovery of farmland birds'. *Ibis* 146 (Supplement 2): 221–9.

Carter, I. & Whitlow, G. 2005. *Red Kites in the Chilterns.* (Second edition). English Nature and Chilterns Conservation Board: Princes Risborough.

Carter, I., Evans, I. & Crockford, N. 1995. 'The Red Kite re-introduction project in Britain – progress so far and future plans'. *British Wildlife* 7: 18–25.

Carter, I., McQuaid, M., Snell, N. & Stevens, P. 1999. 'The Red Kite *Milvus milvus* reintroduction project: modelling the impact of translocating Red Kite young within England'. *Journal of Raptor Research* 33: 251–4.

Carter, I., Newbery, P., Grice, P. & Hughes, J. 2008. 'The role of reintroductions in conserving British birds'. *British Birds* 101: 2–25.

Carter, I., Cross, A.V., Douse, A., Duffy, K., Etheridge, B., Grice, P.V., Newbery, P., Orr-Ewing, D.C., O'Toole, L., Simpson, D. & Snell, H. 2005. 'Re-introduction and conservation of the Red Kite *Milvus milvus* in Britain: current threats and prospects for future range expansion'. In: Thompson, D.B.A., Redpath, S.M., Fielding, A.H., Marquiss, M. & Galbraith, C.A. (eds). *Birds of Prey in a Changing Environment*: 407–16. Scottish Natural Heritage: Edinburgh.

Carter, N.A. 1984. 'Bole scoring by sika deer (*Cervus nippon*) in England'. *Deer* 6: 77–8.

Caseby, J.A. 1936. 'The Welsh goat'. *British Goat Society Yearbook*: 120–3.

Caseldine, A. 1990. *Environmental Archaeology in Wales.* Department of Archaeology, St David's University College: Lampeter.

Cassey, P. 2001. 'Are there body size implications for the success of globally introduced land birds?' *Ecography* 24: 413–20.

Cassey, P. 2002. 'Life history and ecology influences establishment success of introduced land birds'. *Biological Journal of the Linnean Society of London.* 76: 465–80.

Cassey, P., Blackburn, T.M., Duncan, R.P. & Lockwood, J.L. 2005. 'Lessons from the establishment of exotic species: a meta-analytical case study using birds'. *Journal of Animal Ecology* 74: 250–8.

Catt, D.C., Baines, D., Picozzi, N., Moss, R. & Summers, R.W. 1998. 'Abundance and distribution of Capercaillie *Tetrao urogallus* in Scotland 1992–1994'. *Biological Conservation* 85: 257–67.

Cawkwell, C. & McAngus, J. 1976. 'Spread of the zander'. *Angler's Mail* (3 March): 12–13.

Chadwick, A.H. 1996. 'Sika deer in Scotland: density, population size, habitat use and fertility – some comparisons with red deer'. *Scottish Forestry* 50: 8–16.

Chandler, R. 2003. 'Rose-ringed parakeets – how long have they been around?' *British Birds* 96: 407–8.

Chapman, D.I. & Chaplin, R.E. 1967. 'Research on the biology of fallow deer'. *Deer* 1: 90–1.

Chapman, D.I. & Chapman, N. 1969. 'Observations on the biology of fallow deer in Epping Forest, England'. *Biological Conservation* 2: 55–62.

Chapman, D.I. & Chapman, N. 1970. *Fallow Deer.* No. 1. British Deer Society: Southampton.

Chapman, D.I. & Chapman, N. 1972. 'Muntjac deer hybrids'. *Deer* 2: 803–4.

Chapman, D.I. & Chapman, N.G. 1975. *Fallow Deer: their History, Distribution and Biology.* Terence Dalton: Lavenham, Suffolk.

Chapman, D.I. & Chapman, N.G. 1980. 'The distribution of Fallow deer: a worldwide review'. *Mammal Review* 10: 61–138.

Chapman, J. 2000. 'Beware of the frog'. *Daily Mail* (29 March).

Chapman, N. 1995. 'Our neglected species'. *Deer* 9: 360–2.

Chapman, N. 1996. 'Are deer a problem?' *Journal of Practical Ecology and Conservation, Special Publication No. 1*: 4–10.

Chapman, N. 2005. 'Muntjac in Breckland'. *Deer* 13 (6): 14–16.

Chapman, D.I. & Chapman, N. 1972. 'Muntjac deer hybrids'. *Deer* 2: 803–4.

Chapman, N. & Harris, S. 1993. 'Distribution of muntjac in Britain'. *Quarterly Journal of Forestry* 87: 116–21.

Chapman, N., Harris, S. & Stanford, A. 1994. 'Reeves' muntjac *Muntiacus reevesi* in Britain: their history, spread, habitat selection, and the role of human intervention in accelerating their dispersal'. *Mammal Review* 24: 113–60.

Chapman, N.G., Claydon, K., Claydon, M., Forde, P.G. & Harris, S. 1993. 'Sympatric populations of muntjac (*Muntiacus reevesi*) and roe deer (*Capreolus capreolus*): a comparative analysis of their ranging behaviour, social organization and activity'. *Journal of Zoology* 229: 623–40.

Chapman, N., Claydon, K., Claydon, M. & Harris, S. 1995. 'Muntjac in Britain: is there a need for a management strategy?' *Deer* 9: 226–36.

Chitty, D. & Shorten, M. 1946. 'Techniques for the study of the Norway rat'. *Journal of Mammalogy* 27: 63–78.

Chitty, D. & Southern, H.N. 1954. *Control of Rats and Mice.* Volumes 1–3. Clarendon Press: Oxford.

Clark, F.N. 1885. 'Report of operations at the Northville and Alpena Stations during the season 1884–85'. *Report of the United States Fish Commissioner of Fish and Fisheries* 12: 156–7.

Clark, M. 1971. 'Notes on muntjac in south Hertfordshire'. *Deer* 2: 725–8.

Clark, M. 1987. 'The survey of mammals, reptiles and amphibia in Hertfordshire'. *Transactions of the Hertfordshire Natural History Society* 29: 322–89.

Clinging, V. & Whiteley, D. 1980. 'Mammals of the Sheffield Area'. *Sorby Record, Special Series* 3: 1–48.

Clutton-Brock, J. 1981. *Domesticated Animals from Early Times.* William Heinemann/British Museum (Natural History): London.

Clutton-Brock, J. & Burleigh, R. 1983. 'Some archaeological applications of the dating of animal bone by radiocarbon with particular reference to post-pleistocene extinctions'. In: Mook. W.G. & Waterbolk, H.T. (eds). 'Proceedings of the 1st International Symposium on C-14 and archaeology'. *PACT Journal* 8: 409–19. Council of Europe, Strasbourg.

Clutton-Brock, J. & MacGregor, A. 1988. 'An end to medieval reindeer in Scotland'. *Proceedings of the Society of Antiquaries of Scotland* 118: 23–35.

Clutton-Brock, T.H. & Pemberton, J.M. 2004. *Soay Sheep: Dynamics and Selection in an Island Population.* Cambridge University Press: Cambridge.

Clutton-Brock, T.H., Price, O.F., Albon, S.D. & Jewell, P.A. 1991. 'Persistent instability and regulation in Soay sheep'. *Journal of Animal Ecology* 60: 593–608.

Colclough, S. 2006. Convention on the Conservation of European Wildlife and Natural Habitats: Report of the Working Group on the Elaboration of an Action Plan for the Conservation and Restoration of the European Sturgeon (*Acipenser sturio*), Bordeaux, 3–4 July 2006.

Coleridge, W.L. 1974. 'Laughing frogs'. *Country Life* (4 April): 798.

Coles, B. 2006. *Beavers in Britain's Past.* Oxbow Books: Oxford.

Collar, N.J. & Goriup, P.D. 1980. 'Problems and progress in the captive breeding of great bustards *Otis tarda* in quasi-natural conditions'. *Avicultural Magazine* 86: 131–40.

Collier, R. 1986. 'Return of the Sea Eagle'. *Bird Watching*: 30.

Collinge, W.E. 1921–2. 'The food and feeding habits of the Little Owl'. *Journal of the Ministry of Agriculture* 28: 1022–31.

Colquhoun, M.K. 1942. 'The habitat distribution of the grey squirrel in Savernake Forest'. *Journal of Animal Ecology* 11: 127–30.

Comfort, N. 1974. 'Farmers facing threat from pest porcupines'. *Daily Telegraph* (25 October).

Conroy, J.W.H., Kitchener, A.C. & Gibson, J.A. 1998. 'The history of the beaver in Scotland and its future reintroduction'. In: Lambert, R.A. (ed.). *Species History in Scotland: Introductions and Extinctions Since the Ice Age*: 107–28. Scottish Cultural Press: Edinburgh.

Cooke, A.S. 1994. 'Colonisation by muntjac *Muntiacus reevesi* and their impact on vegetation'. In: Massey, M.E. & Welch, R.C. (eds). *Monks Wood National Nature Reserve – the Experience of 40 Years 1953–93*: 45–61. English Nature: Peterborough.

Cooke, A.S. 1997. 'Effects of grazing by muntjac (*Muntiacus reevesi*) on bluebells *Hyacinthoides nonscripta* and a field technique for assessing field activity'. *Journal of Zoology* 242: 365–9.

Cooke, A.S. 2004. 'Muntjac and conservation'. In: Quine, C.P. (ed.). *Managing woodlands and their mammals*: 65–9. Proceedings of a Joint Mammal Society/Forestry Commission Symposium. Forestry Commission: Edinburgh.

Cooke, A.S. 2005. 'Muntjac deer *Muntiacus reevesi* in Monks Wood NNR: their management and changing impact'. In: Gardiner, C. & Sparks, T. (eds). *Ten Years of Change: Woodland Research at Monks Wood NNR 1993–2003. Research Report* 613: 65–74. English Nature: Peterborough.

Cooke, A. 2006a. 'There are muntjac at the bottom of my garden'. *Deer* 14 (1): 34–7.

Cooke, A.S. 2006b. 'Monitoring muntjac deer *Muntiacus reevesi* and their impacts in Monks Wood National Nature Reserve'. *Research Report* 681: 1–174. English Nature: Peterborough.

Cooke, A. 2006–07. 'The muntjac of Monks Wood'. *Deer* 14 (2): 10–13.

Cooke, A.S. & Farrell, L. 1981. *The Ecology of Chinese Water Deer (*Hydropotes inermis*) on Woodwalton Fen National Nature Reserve.* Nature Conservancy Council: Huntingdon.

Cooke, A.S. & Farrell, L. 1987. 'The utilisation of neighbouring farmland by Chinese water deer (*Hydropotes inermis*) at Woodwalton Fen National Nature Reserve'. *Huntingdonshire Fauna & Flora Society Report, 1986* 39: 28–38.

Cooke, A. & Farrell, L. 1998. *Chinese Water Deer.* The Mammal Society and the British Deer Society: London and Fordingbridge.

Cooke, A.S. & Farrell, L. 1995. 'Establishment and impact of muntjac (*Muntiacus reevesi*) on two national nature reserves'. In: Mayle, B.A. (ed.). *Muntjac Deer: their Biology, Impact and Management in Britain. Proceedings of a Conference at New Hall, Cambridge, 9 October 1993*: 48–62.

Cooke, A.S. & Farrell, L. 2001. 'Impact of muntjac deer (*Muntiacus reevesi*) at Monks Wood National Nature Reserve, Cambridgeshire, eastern England'. *Forestry* 74: 241–50.

Coote, J. 1998. 'IATA & CITES, changes for reptiles and amphibians plus future biodiversity'. *Herptile* 23: 95–101. *Reptilia* (GB) 4: 10–15.

Copp, G.H., Templeton, M. & Gozlan, R.E. 2007. 'Propagule pressure and the invasion risks of non-native freshwater fishes: a case study in England'. *Journal of Fish Biology* 71 (sd.): 148–59.

Copp, G.H., Wesley, K.J., Kovac, V. Ives, M.J. & Carter, M.G. 2003. 'Introduction and establishment of the pikeperch *Stizostedion lucioperca* (L.) in Stanborough Lake (Hertfordshire) and its dispersal in the Thames catchment'. *London Naturalist* 82: 139–53.

Copp, G.H., Wesley, K.J., Verreycken, H. & Russell, I.C. 2007. 'When an "invasive" fish species fails to invade! Example of the topmouth gudgeon *Pseudorasbora parva*'. *Aquatic Invasions* 2: 107–12.

Copp, G.H., Fox, M.G., Przybylski, M., Godinho, F.N. & Vila-Gispert, A. 2004. 'Life-time growth patterns of pumpkinseed *Lepomis gibbosus* introduced to Europe, relative to native North American populations'. *Folia Zoologica* 53: 237–54.

Copping, J. 2007. 'Russian sturgeon invading British waters'. *The Times*, 25 February.

Corbet, G.B. 1961. 'Origin of the British insular races of small mammals and of the Lusitanian fauna'. *Nature, London* 191: 1037–41.

Corbet, G.B. 1969. 'The geological significance of the present distribution of mammals in Britain'. *Bulletin of the Mammal Society* 31: 14–16.

Corbet, G.B. 1974. *The Distribution of Mammals in Historic Times.* Systematics Association Special Volume No. 6. Academic Press: London.

Corbet, G.B. 1983. 'Temporal and spatial variation of dental patterns in the voles *Microtus arvalis*, of the Orkney Islands'. *Journal of Zoology* (A) 208: 395–402.
Corbet, G.B. 1986. 'The relationships and origins of the European lagomorphs'. *Mammal Review* 16: 105–10.
Corbet, G.B. & Harris, S. (eds). 1991. *The Handbook of British Mammals*. Third edition. Blackwell Science: Oxford.
Corbet, G.B. & Southern. H.N. (eds). 1977. *The Handbook of British Mammals*. Second edition. Blackwell Scientific Publications: Oxford.
Corbett, L. 1978. 'Current research on wild cats; why have they increased'. *Scottish Wildlife* 14: 17–21.
Cordeaux, J. 1899. *A List of the British Birds Belonging To the Humber District*. R.H. Porter: London.
Countryside Council for Wales 2008. 'Position statement on Beaver reintroductions in Wales'. 1 page mimeograph.
Coward, T.A. 1920, 1926. *The Birds of the British Isles*. 3 vols. Frederick Warne & Co. Ltd: London.
Cowdy, S. 1966. 'Mammal Report 1958–65'. *Middle Thames Naturalist* 1965: 1–6.
Coy, J. 1982. 'The animal bones'. In: Gingell, C. (ed.). 'Excavations of an Iron Age enclosure at Groundswell Farm, Blunsdon St Andrews, 1976–7: 68–73'. *Wiltshire Archaeological and Natural History Society Magazine* 76: 33–75.
Coy, J. 1984. 'The small mammals and amphibia'. In: Cunliffe, B. (ed.). *Danebury: an Iron Age Hillfort in Hampshire*: 526–7. Vol. 2. CBA Research Report 52: London.
Craik, C. 1997. 'Long-term effects of North American Mink *Mustela vison* on seabirds in western Scotland'. *Bird Study* 44: 303–9.
Craik, J.C.A. 1998. 'Recent mink related declines of gulls and terns in west Scotland and the beneficial effects of mink control'. *Argyll Bird Report* 14: 98–110.
Craven, J. 2007. 'Red Kites'. *BBC Country File* (November): 16.
Crichton, M. 1974. *Provisional Distribution Maps of Amphibians, Reptiles and Mammals in Ireland*. Folens/Foras Forbatha: Dublin.
Crook, I. 1969. 'Feral goats in north Wales'. *Animals* 12: 13–15.
Crowcroft, P. 1966. *Mice All Over*. G.T. Foulis: London.
Cunningham, A.A. & Langton, T.E.S. 1997. 'Disease risks associated with translocation of amphibians into, out of and within Europe – a UK perspective'. *Journal of the British Veterinary Zoological Society* 3: 37–41.
Cunningham, A.A., Garner, T.W.J., Aguilar-Sanchez, V., Banks, B., Foster, J., Sainsbury, A.W., Perkins, M., Walker, S.F., Hyatt, A.D. & Fisher, M.C. 2005. 'Emergence of amphibian chytridiomycosis in Britain'. *Veterinary Record* 157: 386–7.
Cunningham, P. 1987. 'Mink in the Outer Hebrides'. *Scottish Bird News* 8: 4.
Cuthbert, J.H. 1973. 'The origin and distribution of feral mink in Scotland'. *Mammal Review* 3: 97–103.

Dalton, R.F. 1950. 'Distribution of the Dorset amphibia and reptilia'. *Proceedings of the Dorset Natural History and Archaeological Society* 72: 135–43.
Dandy, J.E. 1947. 'A list of the vertebrates of Hertfordshire'. *Transactions of the Hertfordshire Natural History Society Field Club* 22: 168.
Dansie, O. 1966. 'Reeves [sic] muntjac (*Muntiacus reevesi*) in the Welwyn area'. *Deer News* 1: 37–41.
Dansie, O. 1969a. 'A policy for muntjac?' *Deer* 1: 420.
Dansie, O. 1969b. 'Are muntjac deer?' *Deer* 1: 373.
Dansie, O. 1970. *Muntjac*. No 2. British Deer Society.
Dansie, O. 1971. 'The spread of the remarkable muntjac'. *Deer* 2: 616–20.
Dansie, O. 1983. 'Muntjac'. In: *Muntjac and Chinese Water Deer*: 3–24. British Deer Society: Warminster, Wiltshire.
D'Arcy, G. 1999. *Ireland's Lost Birds*. Four Court's Press: Dublin.
D'Arcy, G. & Haywood, J. 1992. *The Natural History of the Burren*. Immell Publishing: London.
Davidson, J. 1907. 'Capercaillies in Moray'. *Annals of Scottish Natural History*: 52.
Davies, P.W. & Davis, P.E. 1973. 'The ecology and conservation of the red kite in Wales'. *British Birds* 66: 183–270.
Davies, A. 1985. 'A place for Mandarins'. *Birds* 10: 12–14.
Davies, A. 1988. 'The distribution and status of the Mandarin Duck *Aix galericulata* in Britain'. *Bird Study* 35: 203–8.
Davies, A.K. & Baggot, G.K. 1989. 'Clutch size and nesting sites of the Mandarin Duck *Aix galericulata*'. *Bird Study* 36: 32–6.
Davies, C., Shelley, J., Harding, P., McLean, I., Gardiner, R. & Peirson, G. 2004. *Freshwater Fishes in Britain: the Species and their Distribution*. Harley Books: Colchester, Essex.
Davis, P.E. & Davis, J.E. 1981. 'The food of the Red Kite in Wales'. *Bird Study* 28: 33–40.
Davis, R.A. 1963. 'Feral coypus in Britain'. *Annals of Applied Biology* 5: 345–8.
Davis, R.A. & Jenson, A.G. 1960. 'A note on the distribution of the coypu (*Myocastor coypus*) in Great Britain'. *Journal of Animal Ecology* 29: 397.
Davis, S.J.M. 1987. *The Archaeology of Animals*. Batsford: London.
Davison, A., Birks, J.D.S., Griffiths, H.I., Kitchener, A.C., Biggins, D. & Butlin, R.K. 1998. 'Hybridization and the phylogenetic relationship between polecats and domestic ferrets in Britain'. *Biological Conservation* 87: 155–62.
Dawes, A. 2006. [Letter in the *Daily Mail* (11 December)].
Dawson, L.R. 1974. 'Rearing the North American ruddy duck'. *Avicultural Magazine* 80: 327.
Day, F. 1887. *British and Irish Salmonidae*. Williams & Norgate: London.
d'Ayala, R. 2003. 'Weeds and aliens'. *Reading Naturalist* 55: 10–16.
Deane, C.D. 1952. 'The black rat in the North of Ireland'. *Irish Naturalists' Journal* 10: 296–8.
Deane, C.D. 1964. 'Introduced mammals in Ireland'. *Bulletin of the Mammal Society of the British Isles* 21: 2.
Deane, C.D. 1972. 'Deer in Ireland'. *Deer* 2: 920–6.
Deane, C.D. & O'Gorman, F. 1969. 'The spread of feral mink in Ireland'. *Irish Naturalists' Journal* 16: 198–202.
de Bruxelles, S. 2005. 'Return of a native brings danger to royal forest'. *The Times* (26 January).
de Bruxelles, S. 2007. 'Scientists give "dodo of the rivers" second chance at life'. *The Times* (2 May).
de Bruxelles, S. 2008. 'Great Bustard colony is still fighting to get off the ground'. *The Times* (26 September).
Deedler, C.J. & Willemsen, J. 1964. 'Synopsis of biological data on the pike-perch (*Lucioperca lucioperca* L.)'. *FAO Fisheries Synopsis* 28: Rome.
Delany, S. 1992. *Survey of Introduced Geese in Britain, Summer 1991: Provisional Results*. Wildfowl & Wetlands Trust: Slimbridge, Gloucestershire.
Delany, S. 1993. 'Introduced and escaped geese in Britain in summer 1991'. *British Birds* 86: 591–9.
Delany, S. 1995. 'The 1991 survey of introduced geese'. In: Carter, S. (ed.). *Britain's Birds in 1991–92, the Conservation and Monitoring Review*: 107–13. British Trust for Ornithology: Thetford.
Delap, P. 1936–7. 'Deer in Co. Wicklow'. *Irish Naturalists' Journal* 6: 82–8.
Delap, P. 1967. 'Hybridization of red and sika deer in rural western England'. *Deer* 1: 131–3.
Delap, P. 1968. 'Observations on deer in north-west England'. *Journal of Zoology* 156: 531–3.
De Nahlik, A.J. 1959. *Wild Deer*. Faber & Faber: London.
De Nahlik, A.J. 1974. *Deer Management*. David & Charles: Newton Abbot, Devon.
Dennis, R. 1964. 'Capture of moulting Canada geese on the Beauly Firth'. *Wildfowl Trust Annual Report* 15: 71–4.
Dennis, R. 1968–9. 'Sea Eagles'. *Fair Isle Bird Observatory Report* 21: 17–21; 22: 23–9. (See also: *Scottish Birds* 9: 173–235 (1976)).
Dennis, R. 1996. *A Proposal to Translocate Ospreys to Rutland Water*. Anglian Water and Leicestershire and Rutland Trust for Nature Conservation.
Dennis, R. 2005. 'The eagle owl has landed'. *BBC Wildlife* 23 (13): 24–9.
Dennis, R. & Dixon, H. 1999. *Report of Translocation of Young Ospreys from Scotland to Rutland Water Nature Reserve*. Anglian Water.
Department for Environment Food and Rural Affairs (DEFRA). 2003. *Review of Non-native Species Policy: Report of the Working Group*. DEFRA: London.
Department for Environment Food and Rural Affairs (DEFRA) 2007. *The Invasive Non-native Species Framework Strategy for Great Britain: Protecting our Natural Heritage From Invasive Species*. DEFRA: London.

Devoy, R.J. 1986. 'Possible landbridges between Ireland and Britain: a geological appraisal'. *Occasional Publications of the Irish Biogepgraphical Society* 1: 15–26.
Diaz, A. 2006. 'A genetic study of sika (*Cervus nippon*) in the New Forest and in the Isle of Purbeck, Southern England: is there evidence of recent or past hybridization with red deer?' *Journal of Zoology* 270: 227–35.
Diaz, A., Hughes, S., Putman, R., Mogg, R. & Bond, J. 2007. 'Sika in the New Forest and the Isle of Purbeck: how hybridised with red deer are they?' *Deer* 14 (3): 19–23.
Dixon, W.J.B. 2001. 'A study of the reintroduced population of Red Kite *Milvus milvus* in southern England'. Unpublished PhD thesis, Oxford University.
Dobney, K. & Harwood, J. 1998. 'Here to stay? Archaeological evidence for the introduction of commensal and economically important mammals in the north of England'. In: Benecke, N. (ed.). 'The Holocene history of the European vertebrate fauna. Modern aspects of research': 373–87. Workshop, 6–9 April 1998, Berlin. Verlag Marie Leidorf, Rahden/Westf.
Domaniewski, J. & Wheeler, A. 1996. 'The topmouth gudgeon has arrived'. *Fish* 43: 40.
Doney, J. & Parker, J. 1998. 'An assessment of the impact of deer on agriculture'. In: Goldspink, C.R. (ed.). *Population Ecology, Management and Welfare of British Deer*: 38–43. British Deer Society/Universities Federation for Animal Welfare/ Manchester Metropolitan University.
Dorward, D.F. 1957. 'The night-heron colony in the Edinburgh zoo'. *Scottish Naturalist* 69: 32–6.
Dubbledam, A. 2008. 'Thoughts on wallabies'. *Manx Nature*. (Spring issue.)
Dudley, S.P., Gee, M., Kehoe, C., Melling, T.M. & BOURC. 2006. 'The British list: a checklist of Birds of Britain (seventh edition)'. *Ibis* 148: 526–63.
Dunstone, N. 1986. 'Exploited animals: the mink'. *Biologist, Institute of Biology* 33: 69–75.
Dunstone, N. 1993. *The Mink*. T. & A.D. Poyser: London.

Eastcott, B. 1971. 'Deer in Essex'. *Essex Naturalist* 32: 313–16.
Easteal, S. 1981. 'The history of introductions of *Bufo marinus* (Amphibia: *Anura*); a natural experiment in evolution'. *Biological Journal of the Linnean Society* 16: 93–113.
Easterbee, N. 1988. 'The wild cat *Felis sylvestris* in Scotland'. *Lutra* 31: 29–43.
Eaton, M.A., Marshall, K.B. & Gregory, R.D. 2007. 'Status of Capercaillie *Tetrao urogallus* in Scotland during winter 2003/04'. *Bird Study* 54: 145–53.
Edlin, H.L. 1952. *Changing Wildlife of Britain*. B.T. Batsford: London.
Edwards, K.C. 1962. *The Peak District*. Collins: London.
Elliott, V. 2005a. 'Eagle owl has landed'. *The Times* (14 November).
Elliott, V. 2005b. 'Lust for life may be the death of wild boar'. *The Times* (2 September).
Elliott, V. 2008a. 'Wild boar under fire and on the menu'. *The Times* (20 February).
Elliott, V. 2008b. 'The fish that died simply waiting to be allowed back into Britain'. *The Times* (12 May).
Ellison, A. 1907. 'Little owls breeding in Hertfordshire'. *The Zoologist* 11: 430.
Ellison, N.F. 1959. 'A new Cheshire fish'. *Cheshire Life* (November).
Ellison, N.F. & Chubb, J.C. 1963. 'The marine and freshwater fishes'. *Fauna of Lancashire & Cheshire* 44: 1–8.
Elwes, H.J. 1912. 'Notes on the primitive breeds of sheep in Scotland'. *Scottish Naturalist* 1: 1–7, 25–32, 49–52.
Engeman, R.M., Pipas, M.J., Gruver, K.S., Bourassa, J. & Allen, L. 2002. 'Monitoring changes in feral swine abundance and spatial distribution'. *Environmental Conservation* 28: 235–40.
England, M.D. 1970. 'Escapes'. *Avicultural Magazine* 76: 150–2.
England, M.D. 1974a. 'A further review of the problem of escapes'. *British Birds* 67: 177–97.
England, M.D. 1974b. 'Feral populations of parakeets'. *British Birds* 67: 393–4.
English Nature. 2001. *Return of the Red Kite*. English Nature: Peterborough.
English, S. 1998. 'The return of a long-lost friend'. *The Times* (20 March).
Evans, I.M., Dennis, R.H., Orr-Ewing, D.C., Kjellén, N., Andersson, P.O., Sylvén, M., Senosiain, A. & Carbo, F.C. 1997. 'The re-establishment of Red Kite breeding populations in Scotland and England'. *British Birds* 90: 123–38.
Evans, I.M., Cordero, P.J. & Parkin, D.T. 1998. 'Successful breeding at one year of age by Red Kites *Milvus milvus* in southern England'. *Ibis* 140: 53–7.
Evans, I.M., Summers, R.W., O'Toole, L., Orr-Ewing, D.C., Evans, R., Snell, N. & Smith J. 1999. 'Evaluating the success of translocating Red Kites *Milvus milvus* to the UK'. *Bird Study* 46: 129–44.
Evans, K. 1996. 'Beavers, fish and fishermen'. *British Wildlife* 7: 279–80.
Evans, R.J., Broad, R.A., Duffy, K., MacLennan, A.M., Bainbridge, I.P. & Mudge, G.P. 2001. 'Re-establishment of a breeding population of white-tailed eagles *Haliaeetus albicilla* in Scotland'. *Proceedings of the Sea Eagle 2000 International Conference*, Björkö, Sweden.
Ewart, J.C. 1913. 'Domestic sheep and their wild ancestors'. *Transactions of the Highland & Agricultural Society of Scotland* 5: 160–91.

Fairley, J.S. 1971. 'A critical re-appraisal of the status in Ireland of the eastern house mouse'. *Irish Naturalists' Journal* 17: 2–5.
Fairley, J.S. 1980. 'Observations on a collection of feral Irish mink *Mustela vison* Schreber'. *Proceedings of the Royal Irish Academy* B 80: 79–90.
Fairley, J.S. 1982. 'The muskrat in Ireland'. *Irish Naturalists' Journal* 20: 405–11.
Fairley, J.S. 1983. 'Export of wild mammal skins from Ireland in the eighteenth century'. *Irish Naturalists' Journal* 21: 75–9.
Fairley, J.S. 1984. *An Irish Beast Book: a Natural History of Ireland's Furred Wildlife*. The Blackstaff Press: Belfast.
Fairley, J. 2001. *A Basket of Weasels*. Privately published.
Fargher, S.E. 1977. 'The distribution of the brown hare (*Lepus capensis*) and the mountain hare (*Lepus timidus*) in the Isle of Man'. *Journal of Zoology* 182: 164–7.
Farr-Cox, L. Leonard, S. & Wheeler, A. 1996. 'The status of the recently introduced fish *Leucaspius delineatus* (Cyprinidae) in Great Britain'. *Fisheries Management and Ecology* 3: 193–9.
Fenner, F. & Ratcliffe, F.N. 1965. *Myxomatosis*. Cambridge University Press: Cambridge.
Ferguson-Lees, I.J. 1958. 'The identification of white-headed and ruddy ducks'. *British Birds* 51: 239–41.
Fickling, N.J. 1985. 'The feeding habits of the zander'. *Fourth British Freshwater Fisheries Conference, Liverpool University*: 193–202.
Fickling, N.J. & Lee, R.L.G. 1985. 'A study of the movements of the zander *Lucioperca lucioperca* L. population of two lowland fisheries'. *Aquaculture and Fisheries Management* 16: 377–93.
Fisher, J. 1948. 'St Kilda, a natural experiment'. In: *New Naturalist Journal*: 91–108. Collins: London.
Fisher, J. 1950. 'The black rat in London'. *London Naturalist* 29: 136.
Fishwick, J.L. 1904. *On the Reptiles and Batrachians*. Victoria County Histories: Bedfordshire.
Fitter, R.S.R. 1959. *The Ark in Our Midst*. Collins: London.
Flower, F.S. 1929a. 'Exhibition of skin and skull of female fat dormouse'. *Proceedings of the Zoological Society of London*: 769.
Flower, F.S. 1929b. *List of the Vertebrated Animals Exhibited in the Gardens of the Zoological Society of London, 1828–1927. Vol. 1. Mammals*. The Zoological Society of London: London.
Foggo, D. 2008. 'The boar gets it as big guns return'. *The Sunday Times* (27 January).
Forestry Commission. Leaflet No. 31. 1953. *The Grey Squirrel : a Woodland Pest*. Her Majesty's Stationery Office: London.
Frazer, D. 1983. *Reptiles and Amphibians in Britain*. Collins: London.
Fraser, D. & Adams, C.E. 1997. 'A crucian carp *Carassius carassius* (L.) in Loch Rannoch, Scotland: further evidence of the threat posed to unique fish communities by introduction of alien fish species'. *Aquatic Conservation* 7: 323–6.
Fraser, F.C. 1968. 'Animal bones from Hod Hill. I. Sites within the Roman Fort: 127'. In: Richmond, D.I. (ed.). *Hod Hill*. Vol. 2. British Museum: London.
Frazer, J.F.D. 1949. 'The reptiles and amphibians of the Channel Islands'. *British Journal of Herpetology* 1: 51–3.

Frazer, J.F.D. 1964. 'Introduced species of amphibians and reptiles in mainland Britain'. *British Journal of Herpetology* 3: 145–50.
Freethy, R. 1983. *Man and Beast: the Natural and Unnatural History of British Mammals.* Blandford Press: Poole, Dorset.
Frost, W.E. 1940. 'Rainbows of a peat lough on Arranmore'. *Salmon & Trout Magazine* 100: 234–40.
Frost, W.E. 1974. *A Survey of the Rainbow Trout in Britain and Ireland.* Salmon & Trout Association.
Frost, W.E. & Brown, M.E. 1967. *The Trout.* Collins: London.

Gadow, H. 1904. 'Reptilia and amphibia of Cambridgeshire'. In: Marr, J.E. & Shipley, A.E. (eds). *Handbook of the Natural History of Cambridgeshire.* Cambridge University Press: Cambridge.
Garner, T. 2007. '*Batrachochytrium dendrobatidis* in Europe and the UK: an introduced pathogen causing species to decline. III'. *British Herpetological Society/Herpetological Conservation Trust Scientific Meeting, Bournemouth. BHS Newsletter* 155: 3 (Abstract).
Gates, S. 1979. 'A study of the home ranges of free-ranging Exmoor ponies'. *Mammal Review* 9: 3–18.
Gates, S. 1981. 'The Exmoor pony – a wild animal?' *Nature in Devon* 2: 7–30.
Gatins, J. 1980. 'Feral mink'. In: Kernan, R.P., Mooney, O.V. & Went, A.E.J. (eds). *The Introduction of Exotic Species: Advantages and Problems*: 107–11.
Genovesi, P. & Shine, C. 2003. 'The European strategy on invasive alien species'. *Nature & Environment* 137: 1–67.
Gent, T. 2003. 'Herpetological introductions'. In: Poland Bowen (ed.). *Conference Proceedings 2001–2002. The Return of the Native – the Reintroduction of Native Species Back into their Natural Habitat*: 11–17. People's Trust for Endangered Species & Mammals Trust UK: London.
Gibbons, D.W. & Avery, M.I. 2001. 'Birds'. In: Hawksworth, D.L. (ed.). *The Changing Wildlife of Great Britain and Ireland. Systematics Association Special Volume Series* 62: 367–98.
Gibbons, D.W., Reid, J.B. & Chapman, R.A. 1993. *The New Atlas of Breeding Birds in Britain and Ireland 1988–91.* T. & A.D. Poyser: London.
Gibson, I. 2001. 'Controlling ruddy ducks – the case against'. *Glasgow Naturalist* 23 (Supplement): 99–102.
Gibson, J. 1970. [Mammals on Bute and Arran]. *Transactions of the Buteshire Natural History Society* 18: 5–20; 45–7.
Gibson, M. 1994. 'Introduction of exotic fish. Fish 2'. *English Nature, Species Conservation Handbook*: 1–7. English Nature: Peterborough.
Gill, E.L. 1994. *Ponies in the Wild.* Whittet Books: London.
Gill, R.M.A. 1992. 'A review of damage by mammals in north temperate forests. 1. Deer'. *Forestry* 65: 145–69. 3. 'Impact on trees and forests'. 65: 363–88.
Gillett, L. 1988. 'Beam Brook aquatic nurseries: an update'. *Bulletin of the British Herpetological Society* 26: 31.
Gillett, L. 1991. 'Introduced amphibians: Part 1'. *Junior Herpetologists' Newsletter* 33: 6–9.
Gladstone, H.S. 1906. 'Capercaillie in Ayrshire'. *Annals of Scottish Natural History*: 116.
Gladstone, H.S. 1921a. 'The last of the indigenous Scottish capercaillies'. *Scottish Naturalist*: 169–77.
Gladstone, H.S. 1921b. 'A sixteenth-century portrait of the pheasant'. *British Birds* 15: 67–9.
Gladstone, H.S. 1923–4. 'The introduction of the ring-necked pheasant to Great Britain'. *British Birds* 17: 36–7; 18: 84.
Gladstone, H.S. 1926. 'Birds mentioned in the Acts of Parliament of Scotland, 1124–1707'. *Transactions & Journal of the Proceedings of the Dumfries & Galloway Natural History and Antiquarian Society Series 3*: 16–46.
Glegg, W.E. 1951. 'Introduction of wels or catfish into Hertfordshire'. *Transactions of the Hertfordshire Natural History Society Field Club* 23: 131.
Golden Eagle Trust. 2006. 'The Reintroduction of the Red Kite to the Republic of Ireland and Northern Ireland. Licence Application' (16 pages).
Golden Eagle Trust. 2008. *Irish Eagle Newsletter* (8 pages).
Gordon, I.J. 1987. 'Ponies, cattle and goats'. In: Clutton-Brock, T.H. & Ball, M.E. (eds). *Rhum: the Natural History of an Island*: 110–25. Edinburgh University Press: Edinburgh.
Gordon, I.J. 1988. 'Facilitation of red deer grazing by cattle and its impact on performance'. *Journal of Applied Ecology* 25: 1–10.
Gordon, I.J. 1989. 'Vegetation community selection by ungulates on the Isle of Rhum. II. Vegetation community selection'. *Journal of Applied Ecology* 26: 53–64.
Gordon, I. & Illius, A.W. 1989. 'Resource partitioning by ungulates on the Isle of Rhum'. *Oecologia* 79: 383–90.
Gordon, S. 1937. *Edward Grey and His Birds.* Country Life: London.
Gosling, L.M. 1974. 'The coypu in East Anglia'. *Transactions of the Norfolk and Norwich Naturalists' Society* 23: 49–59.
Gosling, L.M. 1980. 'Reproduction of the Himalayan porcupine in captivity'. *Journal of Zoology* 192: 546–9.
Gosling, L.M. 1981. 'Continuous retrospective census of the East Anglia coypu population between 1970 and 1979'. *Journal of Animal Ecology* 50: 885–901.
Gosling, L.M. 2003. 'Coypus: successful removal of an introduced rodent from a threatened wetland habitat'. In: Poland Bowen, C. (ed.). *Conference Proceedings 2001–2002. Mammalians – the Problems Caused by Non-native British Mammals*: 53–6. People's Trust for Endangered Species & Mammals Trust UK: London.
Gosling, L.M. & Baker, S.J. 1989. 'The eradication of muskrats and coypus from Britain'. *Biological Journal of the Linnean Society* 38: 39–51.
Gosling, L.M. & Wright, M. 1975. 'Feral porcupines'. *Report of the Pest Infestation Control Laboratory 1971–73*: 160–1.
Goulding, M. 2003. *Wild Boar in Britain.* Whittet Books: Stowmarket, Suffolk.
Goulding, M.J. & Roper, T.J. 2002. 'Press responses to the presence of free-living Wild Boar (*Sus scrofa*) in southern England'. *Mammal Review* 32: 272–82.
Goulding, M.J., Smith, G. & Baker, S.J. 1998. *Current Status and Potential Impact of Wild Boar* (Sus scrofa) *in the English Countryside: a Risk Assessment.* Central Science Laboratory Report to the Ministry of Agriculture, Fisheries and Food.
Goulding, M.J., Roper, T.J., Smith, G.C. & Baker, S.J. 2003. 'Presence of free-living wild boar in southern England'. *Wildlife Biology* 9: (Supplement 1): 15–20.
Gozlan, R.E., Pinder, A.C. & Shelley, J. 2002. 'Occurrence of the Asiatic cyprinid *Pseudorasbora parva* in England'. *Journal of Fish Biology* 61: 298–300.
Gozlan, R.E., Pinder, A.C., Durand, S. & Bass, J. 2003. 'Could the small size of sunbleak, *Leucaspius delineatus* (Pisces: Cyprinidae) be an ecological advantage in invading British waterbodies?' *Folia Zoologica* 52: 99–108.
Gozlan, R.E., Flower, C.J. & Pinder, A.C. 2003. 'Reproductive success in male sunbleak, a recent invasive fish species in the UK'. *Journal of Fish Biology* 63: 1–13.
Gozlan, R.E., St-Hilaire, S., Feist, S.W., Martin, P. & Kent, M.L. 2005. 'Disease threat to European fish'. *Nature* 435: 1046.
Grant, A. 1976. 'Faunal remains'. In: Cunliffe, B. (ed.). *Excavations at Portchester Castle II. Saxon.* Society of Antiquaries: London.
Grant, W. & Cubby, J. 1972–73. 'The capercaillie reintroduction experiment at Grizedale'. *WAGBI Report & Yearbook*: 96–8.
Gray, R. 1882. 'On the introduction of Reeves's Pheasant into Scottish game preserves'. *Proceedings of the Royal Physical Society of Edinburgh* 70.
Green, A.J. & Anstey, S. 1992. 'The status of the white-headed duck *Oxyura leucocephala*'. *Bird Conservation International* 2: 185–200.
Green, A.J. & Hughes, B. 1995. 'Action plan for the white-headed duck *Oxyura leucocephala*'. Birdlife International, Cambridge.
Green, R.E., Pienkowski, M.W. & Love, J.A. 1996. 'Long-term viability of the re-introduced population of the white-tailed eagle *Haliaeetus albicilla* in Scotland'. *Journal of Applied Ecology* 33: 357–68.
Greig, J.C. 1969. 'The ecology of feral goats in Scotland'. MSc thesis, University of Edinburgh.
Grey of Fallodon, Viscount. 1927. *The Charm of Birds.* Hodder & Stoughton: London.
Grice, P. 2005. *Reintroducing the Sea Eagle to England.* Paper presented to Council of English Nature. 8 pages.
Grubb, P. & Jewell, P.A. 1966. 'Social grouping and home range in feral soay sheep'. *Symposium of the Zoological Society of London* 18: 179–210.
Gulotta, E.J. 1971. 'Mammalian species, No. 3. *Meriones unguiculatus*'. *American Society of Mammalogists* (19 January): 1–5.

Gunther, A. 1894. 'Mammals, reptiles and fishes of Essex'. *The Field* (8 September).

Gupta, S., Collier, J.S., Palmer-Felgate, A. & Potter, G. 2007. 'Catastrophic flooding origin of shelf valley systems in the English Channel'. *Nature, London* 448: 342–5.

Gurnell, J. 1996. 'The grey squirrel in Britain: problems for management and lessons for Europe'. In: da Luz Mathias, M. et al. (eds) *European Mammals. Proceedings of the 1st European Congress of Mammalogy, Museu Bocage, Lisboa*. 1–314.

Gurnell, J. 1999. 'Grey squirrels in woodlands; managing grey squirrels to prevent woodland damage'. *Enact* 7: 10–14.

Gurnell, J. & Mayle, B.A. 2003. 'Ecological impacts of the alien grey squirrel (*Sciurus carolinensis*) in Britain'. In: Poland Bowen, C. (ed.). *Conference Proceedings 2001–2002. Mammalians – Problems Caused by Non-Native British Mammals*: 40–5. People's Trust for Endangered Species & Mammals Trust UK: London.

Gurnell, J. & Pepper, H. 1993. 'A critical look at conserving the British Red Squirrel'. *Mammal Review* 23: 127–37.

Gurnell, J. & Steele, J. 2002. 'Grey squirrel control for red squirrel conservation. A study in Thetford Forest'. *English Nature Research Reports No. 453*: 1–49.

Gurnell, J., Gurnell, A.M., Demeritt, D., Lurz, P.W.W., Shirley, M.D.F., Rushton, S.P., Faulkes, C.G., Nobert, S. & Hare, E.J. 2009. 'The feasibility and acceptability of reintroducing the European Beaver to England'. *Natural England Commissioned Report NECR002 for Natural England and People's Trust for Endangered Species*. (VII + 106 pages.)

Guthrie, D. 1903. 'Canada geese in the Outer Hebrides'. *Annals of Scottish Natural History* 2: 119.

Hachisuka, M.U. 1926. 'Nomenclature and abberation of pheasant (*tenebrosus*)'. *Bulletin of the British Ornithologists' Club* 46:101–2; 47: 50–2.

Hagemeijer, W.J.M. & Blair, M.J. (eds). 1997. *The EBCC Atlas of European Breeding Birds: their Distribution and Abundance*. T. & A.D. Poyser: London.

Hall, C., Cranswick, P., Trinder, M. & Hughes, B. 2008. 'Monitoring of the UK Ruddy Duck population during ongoing control operations: survey results winter 2007/8'. *WWT Report to Central Science Laboratory*.

Hall, S.J.G. 1975. 'Some recent observations on Orkney sheep'. *Mammal Review* 5: 59–64.

Hall, S.J.G. 1985. 'The Chillingham white cattle'. *British Cattle Breeders' Club Digest* 40: 24–8.

Hall, S.J. G.1989. 'Chillingham cattle: social and maintenance behaviour in an ungulate which breeds all year round'. *Animal Behaviour* 38: 215–25.

Hall, S.J.G. & Clutton-Brock, J. 1988. *Two Hundred Years of British Farm Livestock*. British Museum (Natural History): London.

Hall, S.J.G. & Moore, G.F. 1986. 'Feral cattle on Swona, Orkney Islands'. *Mammal Review* 16: 89–96.

Hall, S., Fletcher, J., Gidlow, J., Ingham, B., Shepherd, A., Smith, A. & Widdows, A. 2005. 'Management of the Chillingham Wild White Cattle'. *Government Veterinary Journal* 15: 4–11.

Halley, D.J. 1998. 'Golden and white-tailed eagles in Scotland and Norway: coexistence, competition and environmental degradation'. *British Birds* 91: 171–9.

Halley, D.J., Nygård, T. & Folkestad, A.O. 2006. 'An evaluation of the proposed Sea Eagle *Haliaeetus albicilla* reintroduction area in Ireland'. Unpublished report.

Halliday, T. 2001. 'Endangered reptiles and amphibians'. In: Levin, S.A. (ed.). *Encyclopedia of Biodiversity*. Volume 2: D-Fl.: 479–86. Academic Press: San Diego, California.

Halliwell, E.C. & Macdonald, D.W. 1996. 'American mink *Mustela vison* in the Upper Thames catchment: relationship with selected prey species and den availability'. *Biological Conservation* 76: 51–6.

Hamilton-Dyer, S. 1993. 'The animal bones'. In: Zienkiewicz, J.D. (ed.). 'Excavations at the *Scamnum Tribunorum* at Caerleon. The Legionnary Museum Site 1983–5': 132–6. *Britannia* 24: 27–140.

Hänfling, B. 2007. 'Understanding the establishment of non-indigenous fishes: lessons from population genetics'. *Journal of Fish Biology* 71 (Supplement D): 115–35.

Harby, A. 2008. 'An assessment of the Ballaugh Curragh wallaby population'. MSc thesis for University College, London. 56 pp.

Harcourt, R. 1979. 'The animal bones'. In: Wainwright, R. (ed.). *Gussage All Saints: an Iron Age Settlement in Dorset*: 150–60. HMSO: London.

Hardy, E. 1954. 'The bitterling in Lancashire'. *Salmon & Trout Magazine* 142: 548–53.

Harrington, R. 1973. 'Hybridization among deer and its implications for conservation'. *Irish Forestry Journal* 30: 64–78.

Harrington, R. 1980. 'Exotic deer in Ireland'. In: Kernan, R.P., Mooney, O.V. & Went, O.E.J. (eds). 1980. *The Introduction of Exotic Species: Advantages and Problems*: 73–81. Royal Irish Academy: Dublin.

Harrington, R. 1982. 'The hybridization of red deer (*Cervus elaphus* L. 1758) and Japanese sika deer (*C. nippon* Temmink 1838)'. *Transactions of the International Congress of Game Biologists* 14: 559–71.

Harris, D.R. 1962. 'The distribution and ancestry of the domestic goat'. *Proceedings of the Linnean Society of London* 173: 79–81.

Harris, R.A. & Duff, K.R. 1970. *Wild Deer in Britain*. David & Charles: Newton Abbot, Devon.

Harris, S. 1979. 'History, distribution, status and habitat requirements of the harvest mouse (*Micromys minutus*) in Britain'. *Mammal Review* 9: 159–71.

Harris, S. & Yalden, D.W (eds). 2008. *Mammals of the British Isles Handbook*. Fourth edition. Mammal Society: Southampton.

Harris, S., Morris, P., Wray, S. & Yalden, D. 1995. *A Review of British Mammals: Population Estimates and Conservation Status of British Mammals Other Than Cetaceans*. JNCC: Peterborough.

Harrison, C. 1988. *A History of the Birds of Britain*. Collins: London.

Harting, J.E. 1880. *British Mammals Extinct within Historic Times*. Trübner: London.

Harting, J.E. 1883. 'The local distribution of the red-legged partridge'. *The Field* 61: 130–1. [Reprinted in: *The Recreations of a Naturalist* (1906)].

Harvie-Brown, J.A. 1879. *The Capercailzie in Scotland*. David Douglas: Edinburgh.

Harvie-Brown, J.A. 1880. 'The capercaillie in Scotland'. *Scottish Naturalist*: 289–94.

Harvie-Brown, J.A. 1880–1. 'Squirrels in Great Britain'. *Proceedings of the Royal Physiological Society of Edinburgh* 5: 343–8; 6: 31–63, 115–83.

Harvie-Brown, J.A. 1898. 'Capercaillie in S.E. Lanarkshire'. *Annals of Scottish Natural History*: 118.

Harvie-Brown, J.A. 1906. *A Fauna of the Tay Basin and Strathmore*. David Douglas: Edinburgh.

Harvie-Brown, J.A. 1911. 'Extension of the capercaillie in Moray'. *Annals of Scottish Natural History*: 184.

Harvie-Brown, J.A. & Buckley, T.E. 1892. *A Fauna of Argyll and the Inner Hebrides*. David Douglas: Edinburgh.

Harvey Pough, F. Andrews, R.M., Cadle, J.E., Crump, M.L., Savitzky, A.H. & Wells, K.D. 2001. *Herpetology*. Second edition. Prentice Hall: Upper Saddle River, New Jersey.

Hawkes, B. 1976. 'The invasion of the aliens'. *Surrey Life* 5: 30.

Hayden, T. & Harrington, R. 2000. *Exploring Irish Mammals*. Dychas – The Heritage Service: Dublin.

Haynes, S., Jaarola, M. & Searle, J.B. 2003. 'Phylogeography of the common vole (*Microtus arvalis*) with particular emphasis on the colonization of the Orkney archipelago'. *Molecular Ecology* 12: 951–6.

Henderson, A. 1997. 'From coney to rabbit: the story of a managed coloniser'. *Naturalist* 122: 101–21.

Henderson, J.A. 1947. 'Grey squirrels in Co. Fermanagh'. *Irish Naturalists' Journal* 9: 97.

Henderson, M. 2005. 'Cannibal fish may be threat to natives'. *The Times* (4 February).

Henderson, M. 2007. 'How flood turned Britain into an island'. *The Times* (19 July).

Hennig, A.S. 2004. '*Trachmys scripta elegans* – pet, food and inhabitant of temples'. *Radiata* 13: 13–28.

Hervey, G.F. & Hems, J. 1968. *The Goldfish*. Faber & Faber: London.

Hetherington, D.A., Lord, T.C. & Jacoby, M. 2006. 'New evidence for the occurrence of Eurasian lynx (*Lynx lynx*) in medieval Britain'. *Journal of Quaternary Science* 21: 3–8.

Hetherington, D.A., Miller, D.R., Macleod, C.D. & Gorman, M.L. 2008. 'A potential habitat network for the Eurasian lynx *Lynx lynx* in Scotland. *Mammal Review* 38: 285–303.

Hewson, R. 1951. 'Some observations on the Orkney vole *Microtus o. orcadensis* (Millais)'. *Northwestern Naturalist* 23: 7–10.

Hibbert-Ware, A. 1937. 'Report of the little owl food inquiry, 1936–1937'. *British Birds* 31: 162–87, 205–29, 249–64.

Hickley, P. 1986. 'Invasion by zander and the management of fish stocks'. *Philosophical Transactions of the Royal Society* (B) 314: 571–82.

Hickley, P. & Chare, S. 2004. 'Fisheries for non-native species in England and Wales: angling or the environment'. *Fisheries Management and Ecology* 11: 203–12.

Hickley, P. & North, E. 1983. 'Zander threaten the Severn: true or false?' *Third British Freshwater Fisheries Conference, Liverpool University*: 106–14.

Hill, M., Baker, R., Broad, G., Chandler, P.J., Copp, G.H., Ellis, J., Jones, D., Hoyland, C., Laing, I., Longshaw, M., Moore, N., Parrott, D., Pearman, D., Preston, C., Smith, R.M. & Waters, R. 2005. *English Nature Research Report No. 662. Audit of Non-native Species in England.* Centre for Ecology & Hydrology, National Environment Research Council.

Hills, D. 1991. 'Ephemeral introductions and escapes'. In: Corbet, G.B. & Harris, S. (eds). *The Handbook of British Mammals*: 576–80. Blackwell Science: Oxford.

Hinton, M.A.C. 1924. 'On a new species of *Crocidura* from Scilly'. *Annals of the Magazine of Natural History Series 9*, 14: 509–10.

Hinton, M.A.C. 1931. *Rats and Mice as Enemies of Mankind.* British Museum (Natural History): London.

Hinton, M.A.C. 1932. 'The muskrat menace'. *Natural History Magazine* 3: 177–84.

Holdgate, M.W. 1986. 'Summary and conclusions: characteristics and consequences of biological invasions'. *Philosophical Transactions of the Royal Society.* B 314: 733–42.

Holling, M. & the Rare Breeding Birds Panel 2007a. 'Non-native breeding birds in the United Kingdom in 2003, 2004 and 2005'. *British Birds* 100: 638–49.

Holling, M. & the Rare Breeding Birds Panel 2007b, 2008. 'Rare breeding birds in the United Kingdom in 2003, 2004 and 2005'. *British Birds* 100: 321–67; 101: 276–316.

Hollom, P.A.D. 1957. 'The rarer birds of prey: their present status in the British Isles. Goshawk'. *British Birds* 50: 135–6.

Holloway, S. 1996. *The Historical Atlas of Breeding Birds in Britain and Ireland 1875–1900.* T. & A.D. Poyser: London.

Holmes, J.S. & Galbraith, C.A. 1994. 'The UK ruddy duck *Oxyura jamaicensis* working group'. *Oxyura* 7: 61–6.

Holmes, J.S. & Simons, J.R. 1996. *The Introduction and Naturalisation of Birds.* Stationery Office: London.

Hoodless, A. & Morris, P.A. 1993. 'An estimate of population density of the fat dormouse (*Glis glis*)'. *Journal of Zoology* 230: 337–40.

Hope, W.H. St John & Fox, G.E. 1899 & 1905. 'Excavations on the site of the Roman city of Silchester, Hampshire'. *Archaeologica* 1: 87–112 (1899): 60: 149–68 (1905).

Horwood, M.T. 1973. 'The world of Wareham sika'. *Deer* 2: 978–84.

Horwood, M.T. & Masters, E.H. 1970. *Sika Deer.* No. 3. British Deer Society: Southampton.

Howells, O. & Edwards-Jones, G. 1977. 'A feasibility study of reintroducing wild boar (*Sus scrofa*) to Scotland: are existing woodlands large enough to support a minimum viable population?' *Biological Conservation* 81: 77–89.

Howes, C.A. 1973. *Annual Report, Yorkshire Naturalists' Union 1972*: 407.

Howes, C.A. 1983. 'An atlas of Yorkshire mammals'. *The Naturalist* 108: 41–82.

Howes, C.A. 1984. 'Free range gerbils'. *Bulletin of the Yorkshire Naturalists' Union* 1: 10.

Hudson, R. 1974a. 'Feral parakeets near London'. *British Birds* 67: 33.

Hudson, R. 1974b. 'Parakeets in the London area'. *British Birds* 67: 174.

Hudson, R. 1976a. 'Ruddy ducks in Britain'. *British Birds* 69: 132–43.

Hudson, R. 1976b. 'Ruddy immigrant'. *Country Life* 59: 1351.

Hughes, B. 1992. 'The ecology and behaviour of the ruddy duck *Oxyura jamaicensis jamaicensis* in Great Britain'. PhD thesis, University of Bristol.

Hughes, B. 1996. *The Feasibility of Control Measures for North American Ruddy Ducks* Oxyura jamaicensis *in the United Kingdom. Report to the Department of the Environment.* Wildfowl & Wetlands Trust: Slimbridge.

Hughes, B. 2002. 'The ruddy duck *Oxyura jamaicensis* in Europe and the threat to the white-headed duck *Oxyura leucocephala*'. *Proceedings of Anatidae 2000 Conference, Strasbourg, France. Gibier Faune Sauvage* 12: 17–21.

Hughes, B. 2003. 'Ruddy Duck control in Europe and Asia.' *Aliens* 18: 20.

Hughes, B. & Grussu, M. 1994. 'The ruddy duck *Oxyura jamaicensis* in the United Kingdom: distribution, monitoring, current research and implications for European colonization'. *Oxyura* 7: 29–47.

Hughes, B. & Grussu, M. 1995. 'The ruddy duck in Europe and the threat to the white-headed duck'. In: Carter, S.P. (ed.). *Britain's Birds in 1992–1993: the Conservation and Monitoring Review*: 17–19. BTO/JNCC: Thetford.

Hughes, B., Criado, J., Delany, S., Gallo-Orsi, U., Green, A., Grussu, M., Perennou, C. & Torres, J. 2000. 'The status of the North American ruddy duck *Oxyura jamaicensis* in the Western Palaearctic: towards an action plan for eradication'. *TWSG News* 12: 26–33.

Hunt, P.C. 1972. 'A brief assessment of the rainbow trout in Britain'. *Fisheries Management* 2: 52–5.

Hutchinson, C.D. 1989. *Birds in Ireland.* T. & A.D. Poyser: Calton.

Idle, E.T. & Mitchell, J. 1968. 'The fallow deer of Loch Lomondside'. *Deer* 1: 263–5.

Imms, A.D. 1938. *A Scientific Survey of the Cambridge District.* Report of the British Association for the Advancement of Science.

Ingham, B. 2003. 'The Chillingham wild white cattle'. *Veterinary History* NS 12: 11–22.

Irish Raptor Study Group. 2006. 'The reintroduction of the Golden Eagle to the Republic of Ireland'. Unpublished proposal document (18 pp.).

Irish Raptor Study Group. 2007. 'Birds of Prey and Owls in Ireland: restoration'. (9 pages.)

Jackson, J.E. 1994. 'The edible or fat dormouse (*Glis glis*) in Britain'. *Quarterly Journal of Forestry* 88: 119–25.

Jackson, J.S. 1961. 'Two records of grey squirrel, *Sciurus carolinensis* Gmelin, shot in counties Armagh and Monaghan'. *Irish Naturalists' Journal* 13: 215.

Jarvis, P.J. 1980. 'Beavers and bustards: attitudes toward reintroductions'. *Ecos* 1: 24–6.

Jefferies, D.J. 1990. 'Short-clawed Otter *Aonyx conerea* (Illiger) living wild in Britain'. *Journal of the Otter Trust* 2: 21–5.

Jefferies, D.J. 1992. 'Another record of an Asian short-clawed otter living free in the Oxford area of England; with notes on its implications'. *Journal of the Otter Trust* 11: 9–12.

Jefferies, D.J. 2003. *The Water Vole and Mink Survey of Britain 1996–1998 with a History of the Long Term Changes in the Status of Both Species and their Causes.* Vincent Wildlife Trust: Ledbury.

Jewell, P.A. (ed.). 1974. *Island Survivors: the Ecology of the Soay Sheep of St Kilda.* Athlone Press: University of London.

Jewell, P. 1980. 'The Soay sheep – part 1'. *The Ark* 7: 51–6; Part 2, 7: 87–93.

Jewell, P.A. 1984. 'The golden fleece or a summer visit to St Kilda'. *The Ark* 11: 47–50.

Johnson, E. & Hornby, J. 1980. 'Age and seasonal coat changes in long haired and normal fallow deer (*Dama dama*)'. *Journal of Zoology* 19: 501–9.

Johnston, F.J. 1937. 'The grey squirrel in Epping Forest'. *London Naturalist* 1937: 94–9.

Jones, R.T., Sly, J. & Hocking, L. 1987. 'The vertebrate remains'. In: Olivier, A.C.H. (ed.). 'Excavations of a Bronze Age Funerary Cairn at Manor Farm, near Borwick, North Lancashire': 163–70. *Proceedings of the Prehistoric Society* 53: 129–86.

Jones, J.L. 1976. 'From fur to fugitive: the spread of wild mink'. *Country Life* 159: 420–2.

Jones, R.T. & Ruben, I. 1987. 'Animal bones, with some notes on the effects of differential sampling'. In: Beresford, G. (ed.). *Goltho: the Development of an Early Medieval Manor, c. 850–1150.* Historic Buildings and Monuments Commission for England: London.

Jones-Walters, L.M. & Corbet, G.B. 1991. 'Fat Dormouse'. In: Corbet, G.B. & Harris, S. (eds). *The Handbook of British Mammals*: 264–7. Blackwell Science: Oxford.

Jope, M. & Grigson, C. 1965. 'Faunal remains: frequencies and ages of species'. In: Smith, I.F. (ed.). *Windmill and Avebury Excavations by Alexander Keiller 1925–1939*: 142–5. Clarendon Press: Oxford.

Jourdain, F.C.R. 1912. 'Extermination of the Sea Eagle in Ireland'. *British Birds* 5: 138–9.

Kalaher, M. 2004. 'Breeding Red Kites'. *Sussex Ornithological Society Annual Report*, 2004.

Kalaher, M., Collins, B. & Law, C. 2008. 'The West Sussex Red Kite study area, 2004–2007'. *Sussex Ornithological Society Annual Report*, 2008.

Kay, F.C. 1904. 'Capercaillie in Argyll'. *Annals of Scottish Natural History*: 189.

Kear, J. 1965. 'The assessment of goose damage by grazing trials'. *Transactions of the 6th International Union of Game Biologists*: 333–9.

Kear, J. 1990. *Man and Wildfowl*. T. & A.D. Poyser: London.

Kell, L.T. 1985. 'The impact of an alien piscivore, the zander (*Stizostedion lucioperca* (L.)) on a freshwater community'. PhD theses, University of Liverpool. 420pp. (Unpublished).

Kelly, G. 2004. 'Literature/archive search for information relating to pool frogs *Rana lessonae* in East Anglia'. *English Nature Research Reports No. 480*: 1–35.

Kenward, R.E. 1981. 'Goshawk re-establishment in Britain – causes and implications'. *The Falconer* 7: 304–10.

Kenward, R.E. 1983. 'The causes of damage by red and grey squirrels'. *Mammal Review* 13: 159–66.

Kenward, R. 1989. 'Bark stripping by grey squirrels in Britain and North America'. In: Putman, R.J. (ed.). *Mammals as Pests*: 144–54. Christopher Helm: London.

Kenward, R.E. 2006. *The Goshawk*. Poyser: London.

Kenward, R.E. & Hodder, K.H. 1998. 'Red squirrels (*Sciurus vulgaris*) released in conifer woodland: the effects of source habitat, predation and interactions with grey squirrels (*Sciurus carolinensis*)'. *Journal of Zoology* 244: 23–32.

Kenward, R.E. & Parish, T. 1986. 'Bark stripping by grey squirrels'. *Journal of Zoology* (A) 210: 473–81.

Kenward, R.E., Marquiss, M. & Newton, I. 1981. 'What happens to goshawks trained for falconry?' *Journal of Wildlife Management* 45: 803–6.

Kenward, R.E., Hodder, K.H., Rose, R.J., Walls, C.A., Parish, T., Holm, J.L., Morris, P.A., Walls, S.S. & Doyle, F.I. 1998. 'Comparative demography of red squirrels (*Sciurus vulgaris*) and grey squirrels (*Sciurus carolinensis*) in deciduous and conifer woodland'. *Journal of Zoology* 244: 7–21.

Key, G., Platenberg, R., Easby, A. & Mais, K. 1996. 'The potential impact of introduced commensal rodents on island flora'. *Proceedings of the Vertebrate Pest Conference* 17: 172–8.

Key, G., Fielding, A.H., Goulding, M.J., Holm, R.S. & Stevens-Woods, B. 1998. 'Ship rats *Rattus rattus* on the Shiant Islands, Hebrides, Scotland'. *Journal of Zoology* 245: 228–33.

Kiesecker, J.M. 2003. 'Invasive species as a global problem: toward understanding the worldwide decline of amphibians'. In: Semlitsch, R.D. (ed.). *Amphibian Conservation*: 113–26. Smithsonian Books: Washington and London.

King, B. 1959–60. 'Feral North-American Ruddy Ducks in Somerset'. *Wildfowl Trust Annual Report*: 167–8.

King, B. 1976. 'Association between male North-American Ruddy Ducks and stray ducklings'. *British Birds* 69: 34.

Kirby, K.J. 2001. 'The impact of deer on the ground flora of British broadleaved woodland'. *Forestry* 74: 219–30.

Kirkham, G. 1981. 'Economic diversification in a marginal economy: a case study'. In: Roebuck, P. (ed.). *Plantation to Partition: Essays in Ulster History*. Blackstaff Press: Belfast.

Kitchener, A. 2001a. *Beavers*. Whittet Books: Stowmarket, Suffolk.

Kitchener, A.C. 2001b. 'Alien mammals: wreaking havoc or missing the boat?' *Glasgow Naturalist* 23 (Supplement): 13–22.

Kitchener, A.C. 2005. 'A diagnosis for the Scottish wildcat: a tool for conservation action for a critically endangered felid'. *Animal Conservation* 8: 223–37.

Kitchener, A.C. & Conroy, J. 1996. 'The history of the beaver in Scotland and the case for its reintroduction'. *British Wildlife* 7: 156–61.

Klee, C. 1981. 'An assessment of the contribution made by zander to the decline of fisheries in the lower Great Ouse area'. *Proceedings of the 2nd British Freshwater Fisheries Conference 1981*: 80–93.

Knight, M. 1948. 'Mystery of the Marsh frog'. *Country Life* (October).

Kottelat, M. & Freyhof, J. 2007. *Handbook of European Freshwater Fishes*. Authors: Cornol, Switzerland.

Kovač, V., Copp, G.H. & Craig, J.F. (eds). 2007. 'Non-native fishes: integrated biology of establishment success and dispersal. The Fisheries Society of the British Isles Annual Symposium, Exeter, UK 2007'. *Journal of Fish Biology* 71 (Supplement D): 1–284.

Lack, P. 1986. *The Atlas of Wintering Birds in Britain and Ireland*. T. & A.D. Poyser: Calton.

Langham, S. *The Wall Lizard Project*. Surrey Amphibian and Reptile Group; 25/06/2008. http://www.surrey-arg.org.UK/SARG/02000-Activities/SurveyAndMonitoring/Wall Lizard/PMSitePicker.asp (View at The Wall Lizard Project ver 1.0).

Langley, G. 2004. 'Second nature'. *Birdwatch* (October): 31–4.

Langton, T. & Burton, J.A. 1997. *Amphibians and Reptiles: Conservation and Management of Species and Habitats. Planning and Management Series, No. 4*. Council of Europe Publishing: Strasbourg.

Lansbury, I. 1956. 'Some notes on fauna and flora in southern Hertfordshire and north-eastern Middlesex'. *Entomologist's Gazette* 7: 97–111.

Larner, J.B. 1977. 'Sika deer damage to mature woodlands in southwestern Ireland'. *Proceedings of the 13th International Congress of Game Biology*: 192–202.

Laurie, E.M.O. 1946. 'The coypu (*Myocastor coypus*) in Great Britain'. *Journal of Animal Ecology* 15: 22–34.

Leaper, R., Massei, G., Gorman, M.L. & Aspinall, R. 1999. 'The feasibility of reintroducing wild boar to Scotland'. *Mammal Review* 29: 239–59.

Leckie, N. 1897. 'Capercaillie in Linlithgowshire'. *Annals of Scottish Natural History*: 44.

Lee Conservancy Catchment Board. 1967. *Annual Report, 1965–6*.

Leeke, C.J. 1970. 'Notes on feral barking deer in the Reading area'. *Reading Naturalist* 22: 25–9.

Legg, R. 1990. *Steep Holm Wildlife*. Wincanton Press: Wincanton.

Leicester, Earl of. 1921. 'Date of the introduction into England of the red-legged partridge'. *The Field* 137: 372.

Leutscher, A. 1954. 'The edible dormouse'. *Discovery* 15: 70–1.

Leutscher, A. 1975. 'An invader's rise and fall'. *Country Life* (20 February).

Lever, C. 1957. 'The Mandarin duck in Britain'. *Country Life* 122: 829–31.

Lever, C. 1969. 'The invaders, and how they came here'. *The Field* (28 August): 394.

Lever, C. 1977. *The Naturalized Animals of the British Isles*. Hutchinson: London.

Lever, C. 1978a. 'The not so innocuous mink'. *New Scientist* 22: 812–14. [Reprinted in the *British Waterfowl Association Yearbook 1979–80*: 7–12].

Lever, C. 1978b. 'Are wild mink a threat?' *New Scientist* 30: 712.

Lever, C. 1978c. 'A new place like home'. *Bird Life* (July/August): 12–15.

Lever, C. 1978d. 'Do we really want to bring back the beaver?' *The Times* (12 August).

Lever, C. 1979. 'A question of survival'. *Country Life* (15 February).

Lever, C. 1980a. 'Naturalised animals of the British Isles'. *Animals* 3: 3–7.

Lever, C. 1980b. 'Naturalized reptiles and amphibians in Britain'. *Bulletin of the British Herpetological Society* 1: 27–30.

Lever, C. 1980c. 'Unwanted visitors: Britain's pest problem'. In: Sitwell, N. (ed.). *Wildlife '80: the World Conservation Yearbook*: 76–81. Grolier Enterprises Corp: USA.

Lever, C. 1980d. 'Plight of the wallabies'. *Daily Telegraph* (24 March).

Lever, C. 1980e. 'No to the beaver'. *Ecos* 1: 22–3.

Lever, C. 1980f. 'No beavers for Britain'. *New Scientist* 87: 471–2.
Lever, C. 1981. 'The case against the beaver'. *Illustrated London News*: 38.
Lever, C. 1983. 'The golden hamster in the London area'. *London Naturalist* 62: 111.
Lever, C. 1984a. 'In two minds about mink: on the other hand ...' *BBC Wildlife* (September): 435–7.
Lever, C. 1984b. 'The little owl in Britain'. *Hawk Trust Annual Report* 14: 12–14.
Lever, C. 1984c. [Review of Love 1983]. *Ibis* 126: 267.
Lever, C. 1987. *Naturalized Birds of the World.* Longman Scientific & Technical: Harlow, Essex.
Lever, C. 1990. *The Mandarin Duck.* Shire Publications: Princes Risborough, Buckinghamshire.
Lever, C. 1992. *They Dined on Eland: the Story of the Acclimatization Societies.* Quiller Press: London.
Lever, C. 1994. 'The proposed reintroduction of the beaver to Britain'. *Reintroduction News* 9: 14–15.
Lever, C. 1996. Introduction (vi–viii). In: Holmes, J.S. & Simons, J.R. (eds). *The Introduction and Naturalisation of Birds.* The Stationery Office: Norwich.
Lever, C. 1997. 'Wood Duck (84), Mandarin (85) and Ring-necked Parakeet (393)'. In: Hagemeijer, W.J.M. & Blair, M.J. (eds). *The EBCC Atlas of European Breeding Birds: their Distribution and Abundance.* T. & A.D. Poyser: London.
Lever, C. 2003. *Naturalized Reptiles and Amphibians of the World.* Oxford University Press: Oxford.
Lever, C. 2005a. *Naturalised Birds of the World.* T. & A.D. Poyser: London.
Lever, C. 2005b. 'The ecology of naturalized vertebrates'. *Neobiota* 6: 3–12.
Lever, C. in press. 'Vertebrate animal introductions'. In: Maclean, N. (ed.). *Silent Summer.* Cambridge University Press: Cambridge.
Levitan, B. 1990. 'The vertebrate remains'. In: Bell, M. (ed.). *Brean Down Excavations 1983–1987. English Heritage Archaeological Report* 15. London.
Lilford, Lord. 1895. *Notes on the Birds of Northamptonshire and Neighbourhood.* Porter: London.
Lilford, Lord. 1903. *Lord Lilford on Birds.* Hutchinson: London.
Limentani, J.D. 1975. 'Changes in wintering habits by Canada geese'. *Cambridgeshire Bird Report* No. 48.
Linfield, R.S.J. 1984. 'The impact of zander (*Stizostedion lucioperca* L.) in the United Kingdom and the future management of the affected fisheries in the Anglian region'. In: *EIFAC Technical Paper* 42 (Supplement 2): 353–61. FAO: Rome.
Linfield, R.S.J. & Rickards, R.B. 1979. 'The zander in perspective'. *Fisheries Management* 10: 1–16.
Linn, I.L. (ed.). 1979. *Wildlife Introductions to Great Britain. Report by the Working Group on Introductions of the UK Committee for International Nature Conservation* (32 pp.). Nature Conservancy Council: London.
Linn, I. & Chanin, P. 1978a. 'Are mink really pests in Britain?' *New Scientist* 77: 560–2.
Linn, I. & Chanin, P. 1978b. 'More on the mink "menace"'. *New Scientist* 22: 38–40.
Linn, I. & Stevenson, J.H.F. 1980. 'Feral mink in Devon'. *Nature in Devon* 1: 7–27.
Lister, D. 2007a. 'Where sea eagles dare – freed raptors get a taste for city life'. *The Times* (27 October).
Lister, D. 2007b. 'Landowner flies in elk to get a really wild show off the ground'. *The Times* (3 December).
Lister, D. & Smith, L. 2008. 'Squirrel wars: reds, greys and blacks battle for supremacy'. *The Times* (26 April).
Lister-Kaye, J. 2005. 'Out there'. *The Times* (26 February).
Livingstone, S., Senn, H. & Pemberton, J. 2006. 'The distribution and spread of Japanese sika deer in Scotland'. *Deer* 13 (8): 21–3.
Lloyd, B. 1947. 'A list of vertebrates of Hertfordshire. 4. Mammals'. *Transactions of the Hertfordshire Natural History Society* 22: 227–38.
Lloyd, H.G. 1962. 'The distribution of grey squirrels in England and Wales, 1959'. *Journal of Animal Ecology* 31: 157–66.
Lloyd, H.G. 1970. 'Post-myxomatosis in England and Wales'. *EPPO Publicity Series A 58*: 197–215.
Lloyd, H.G. 1983. 'Past and present distributions of red and grey squirrels'. *Mammal Review* 13: 69–80.
Lloyd, H.G. & Walton, K.C. 1969. 'Rabbit survey in West Wales'. *Agriculture* 75: 32–6.
Lloyd, L. 1867. *The Game Birds and Wild Fowl of Sweden and Norway.* Day & Son: London.
Lockley, R.M. 1940. 'Some experiments in rabbit control'. *Nature, London* 145: 767.
Lockley, R.M. 1960. 'Wild sheep in Wales'. *Nature in Wales* 6: 75–8.
Lockley, R.M. 1964. *The Private Life of the Rabbit.* André Deutsch: London.
Lockwood, J.L. 1999. 'Using taxonomy to predict success among introduced avifauna; relative importance of transport and establishment'. *Conservation Biology* 13: 560–7.
Loder, E. 1898. 'On the beaver pond at Leonardslee'. *Proceedings of the Zoological Society of London 1898*: 201–2.
Long, J.L. 1981. *Introduced Birds of the World.* A.H. & A.W. Reed: Sydney, NSW.
Long, J.L. 2003. *Introduced Mammals of the World.* CABI Publishing: Wallingford.
Lord, D. 2001. 'A preliminary hare distribution survey of Cornwall, with reference to introductions'. Environmental Records Centre for Cornwall and the Isles of Scilly. Occasional Paper 1: 1–6.
Loudon, A. & Fletcher, J. 1983. [Père David's Deer]. *New Scientist* 99: 88–92.
Love, J.A. 1978. 'The reintroduction of the Sea Eagle to the Isle of Rhum'. *Hawk Trust Annual Report 1977*: 16–18.
Love, J.A. 1980a. 'White-tailed Eagle re-introduction on the Isle of Rhum'. *Scottish Birds* 11: 65–73.
Love, J.A. 1980b. 'Return of the Sea Eagle'. *British Trust for Ornithology News* 108: 4–5.
Love, J.A. 1980c. 'Reintroducing Sea Eagles on Rhum: a progress report'. *Hawk Trust Annual Report 1979*: 32–3.
Love, J.A. 1980d. 'The return of the Sea Eagle to Scotland'. *Hebridean Naturalist* 4: 46–8.
Love, J.A. 1983a. *The Return of the Sea Eagle.* Cambridge University Press: Cambridge.
Love, J.A. 1983b. 'White-tailed Eagle – reintroduction experiment'. *Birds* 9: 27–8.
Love, J.A. 1983c. 'A saga of Sea Eagles'. *Scottish Wildlife* (September): 12–15.
Love, J.A. 1983d. 'First nesting attempts by reintroduced Sea Eagles'. *Hawk Trust Annual Report 1982*: 21.
Love, J.A. 1983–4. *Nature Conservancy Council White-tailed Sea Eagle Reintroduction Project Report*, 1982 and 1983.
Love, J.A. 1988. *The Reintroduction of the White-tailed Eagle to Scotland: 1975–1987. NCC Research and Survey in Nature Conservation Report No. 12.* Nature Conservancy Council: Peterborough.
Love, J. 2003. 'Sea Eagle 2000. A history of the White-tailed Sea Eagle in Scotland'. In: Helander, B., Marquiss, M. & Bowermann, W. (eds). *Proceedings of an International Conference, Bjökö, Sweden, September 2000*: 39–50.
Love, J.A. 2007. 'White-tailed Eagle *Haliaeetus albicilla*'. In: Forrester, R.W., Andrews, I.J., McInerny, C.J., Murray, R.D., McGowan, R.Y., Zonfrillo, B., Betts, M.W., Jardine, D.C. & Grundy, D. (eds.). *The Birds of Scotland*: 451–5. The Scottish Ornithologists' Club: Aberlady.
Love, J.A. & Ball, M.E. 1979. 'White-tailed Sea Eagle reintroduction to the Isle of Rhum, Scotland, 1975–77'. *Biological Conservation* 16: 23–30.
Lovegrove, R., Elliott, G. & Smith, K. 1990. 'The Red Kite in Britain'. *RSPB Conservation Review* 4: 15–21. RSPB: Sandy.
Lovegrove, R., Williams, G. & Williams, I. 1994. *Birds in Wales.* T. & A.D. Poyser: London.
Lowe, J. 1864. 'The arrival of *Silurus glanis* in England'. *The Field* (17 September).
Lowe, P.R. 1930. 'Some remarks on *Phasianus colchicus*, mut. *tenebrosus*'. *Ibis* 12: 314–20.
Lowe, P.R. 1933. 'The introduction of the pheasant into the British Isles'. *Ibis* 13: 332–43.
Lowe, V.P.W. 1993. 'The spread of the grey squirrel (*Sciurus carolinensis*) into Cumbria since 1960 and its present distribution'. *Journal of Zoology* 231: 663–7.
Lowe, V.P.W. & Gardiner, A. 1975. 'Red deer – sika deer hybridization'. *Journal of Zoology* 177: 553–66.

Luff, R. 1982. *A Zooarchaeological Study of the Roman N. W. Provinces.* BAR International Series 137. Oxford.
Lusher, A. 2006. 'Villagers under siege in the Exmoor boar war'. *Sunday Telegraph* (1 January).

MacConnochie, A.I. 1923. *The Deer and Deer Forests of Scotland.* H.F. & G. Witherby: London.
MacCrimmon, H.R. 1971. 'World distribution of rainbow trout'. *Journal of the Fisheries Research Board of Canada* 28: 663–704.
MacCrimmon, H.R. & Campbell, J.S. 1969. [with Gotts, B. 1971]. 'The world distribution of brook trout'. *Journal of the Fisheries Research Board of Canada* 26: 1699–1725. [28: 452–6].
Macdonald, A.W., Tattersall, F.H., Brown, E.D. & Balharry, D. 1995. 'Reintroducing the European Beaver to Britain: nostalgic meddling or restoring biodivesity?' *Mammal Review* 25: 161–200.
Macdonald, D. 2003. 'American mink and the ethics of extinction'. In: Poland Bowen, C. (ed.). *Conference Proceedings 2001–2002. MammAliens – the Problems Caused by Non-Native British Mammals*: 46–8. People's Trust for Endangered Species & Mammals Trust UK: London.
Macdonald, D. & Strachan, R. 1999. *The Mink and the Water Vole: Analyses for Conservation.* Wildlife Conservation Research Unit and the Environment Agency: Oxford.
Macdonald, D.W. 1999. 'The impact of American Mink, *Mustela vison*, as predators of native species in British freshwater systems'. In: Cowand, D.P. & Feare, C.J. (eds). *Advances in Vertebrate Pest Management.* Filander Verlag: Fürth.
Macdonald, D. & Burnham, D. 'The state of Britain's mammals'. *Report by People's Trust for Endangered Species and Wildlife Conservation Research Unit. Oxford University.* (23 pages.)
Macdonald, D.W. & Harrington, L.A. 2003. 'The American mink: the triumph and tragedy of adaptation out of context'. *New Zealand Journal of Zoology* 30: 421–41.
Macintyre, B. 2005. 'Invading parakeets warm to wilds of suburbia'. *The Times* (10 September).
Macintyre, B. 2007. 'What big teeth you have! Ideal for the Highlands'. *The Times* (2 February).
MacKeith, T.T. 1916. 'The capercaillie in Renfrewshire'. *Scottish Naturalist*: 270.
MacKenzie, W.D. 1900. 'Capercailzie in Strathnairn'. *Annals of Scottish Natural History*: 51.
MacKinnon, K. 1978. 'Competition between Red and Grey squirrels'. *Mammal Review* 8: 185–90 (1983).
Macklin, R.N. 1990. 'Chinese water deer at Minsmere, 1989'. *Transactions of the Suffolk Naturalists' Society* 26: 5.
MacRae, F. 2005. 'Fiend of the fish world: Japanese invader threatens salmon and trout stocks'. *Daily Mail* (1 December).
MacRae, F. 2006. 'Extinct by 2016: virus carried by grey invaders threatens to wipe out our remaining red squirrels'. *Daily Mail* (22 August).
Maitland, P.S. 1964. 'A population of common carp in the Loch Lomond district'. *Glasgow Naturalist* 18: 349–50.
Maitland, P.S. 1966. 'Rainbow trout, *Salmo irideus* Gibbons, in Loch Lomond'. *Glasgow Naturalist* 18: 421–3.
Maitland, P.S. 1969. 'The distribution of freshwater fish in the British Isles'. *Journal of Fish Biology* 1: 45–58.
Maitland, P.S. 1972. *A Key to the Freshwater Fishes of the British Isles, with Notes on their Distribution and Ecology. Scientific Publications No. 27.* Freshwater Biological Association: Windermere.
Maitland, P.S. 2004. *Keys to the Freshwater Fish of Britain and Ireland, with Notes on their Distribution and Ecology. Scientific Paper No. 62.* Freshwater Biological Association: Ambleside.
Maitland, P.S. 2006. *Scotland's Freshwater Fish: Ecology, Conservation and Folklore.* Trafford Publishing. (trafford.com/06-2823).
Maitland, P.S. & Campbell, R.N. 1992. 'Pikeperch, *Stizostedion lucioperca* (Linnaeus 1758) Zander'. In: *Freshwater Fishes. New Naturalist Series* 75: 288–92. HarperCollins: London.
Maitland, P.S. & Price, C.E. 1969. '*Urocleidus principalis* (Mizelle 1936), a North American monogenetic trematode new to the British Isles, probably introduced with the largemouth bass *Micropterus salmoides* (Lacepede 1802)'. *Journal of Fish Biology* 1: 17–18.
Mallon, D. 1970. 'Britain's wild wallabies'. *Animals* 13: 256–7.
Maltby, M. 1979. *Faunal Studies on Urban Sites. The Animal Bones from Exeter 1971–1975. Exeter Archaeological Reports*, Vol. 2. Department of Prehistory and Archaeology, University of Sheffield.
Maltby, M. 1984. 'The animal bones'. In: Fulford, M.G. (ed.). *Silchester: Excavations in the Defences, 1974–1980*: 199–212. *Britannia Monograph* 5. London.
Maltby, M. 1985. 'The animal bones'. In: Fasham, P. (ed.). *The Prehistoric Settlement at Winnal Down, Winchester.* Hampshire Field Club & Archaeological Society Monograph No. 2.
Malvern, J. 2008. 'Captured squirrels live to nibble again'. *The Times* (16 February).
Manchester, S.J. & Bullock, J.M. 2000. 'The impacts of non-native species on UK biodiversity and the effectiveness of control'. *Journal of Applied Ecology* 37: 845–64.
Mansfield, K. 1958. 'Pike-perch in England'. *Salmon & Trout Magazine* 153: 94–8.
Maples, S. 1907. 'The little owl in Hertfordshire'. *The Zoologist* 11: 353.
Marlborough, D. 1970. 'The status of the Burbot, *Lota lota* (L.) (Gadidae) in Britain'. *Journal of Fish Biology* 2: 217–22.
Marquiss, M. 1981. 'The goshawk in Britain – its provenance and current status'. In: Kenward, R.E. & Lindsay, M.I. (eds). *Understanding the Goshawk*: 43–55. International Association for Falconry: Oxford.
Marquiss, M. & Newton, I. 1982. 'The goshawk in Britain'. *British Birds* 75: 243–60.
Marquiss, M., Madders, M. & Carss, D. 2003. 'White-tailed Eagles (*Haliaeetus albicilla*) and Lambs (*Ovis aries*)'. In: Thompson, D.B.A., Redpath, S.M., Fielding, A.H., Marquiss, M. & Galbraith, C.A. 2003. *Birds of Prey in a Changing Environment.* Scottish Natural Heritage: Edinburgh.
Marquiss, M., Madders, M., Irvine, J. & Carss, D. 1999. *The Impact of White-tailed Eagles on Sheep Farming on Mull.* Institute of Terrestrial Ecology: Banchory.
Marshall, F. 1974. 'The future for fallow deer'. *Deer* 3: 210–13.
Marshall, H.B. 1907. 'Capercaillie in Peeblesshire'. *Annals of Scottish Natural History*: 244.
Mascheretti, S., Rogatcheva, M.B., Gündüz, I., Fredga, K. & Searle, J.B. 2003. 'How did pygmy shrews colonize Ireland? Clues from a phylogenetic analysis of mitochondrial *b* sequences'. *Proceedings of the Royal Society of London, Series B 270*: 1593–9.
Mason, I.L. (ed.). *Evolution of Domesticated Animals.* Longman: London.
Matheson, C. 1931. *The Brown and the Black Rat in Wales.* The National Museum of Wales: Cardiff.
Matheson, C. 1933. *Changes in the Fauna of Wales within Historic Times.* National Museum of Wales: Cardiff.
Matheson, C. 1939. 'A survey of the status of *Rattus rattus* and its sub-species in the sea-ports of Great Britain and Ireland'. *Journal of Animal Ecology* 8: 76–93.
Matheson, C. 1941. 'The rabbit and the hare in Wales'. *Antiquity* 15: 371–81.
Matheson, C. 1944. 'The domestic cat as a factor in urban ecology'. *Journal of Animal Ecology* 13: 130–3.
Matheson, C. 1956–7. 'Game records in Wales'. *Journal of the National Library of Wales* 9: 288–94; 10: 205–14.
Matheson, C. 1958. 'The black-rat at British sea-ports'. *Pest Technology* 1: 4–7.
Matheson, C. 1962. *The Brown Rat. Animals of Britain No. 16.* Sunday Times Publications: London.
Matheson, C. 1963. 'The Pheasant in Wales'. *British Birds* 44: 452–6.
Matthews, L.H. 1952. *British Mammals.* Collins: London.
Maxwell, H. 1905. 'Naturalization of the golden pheasant'. *Annals of Scottish Natural History* 5: 53–4.
Maxwell, H. 1907. 'Capercaillie in the south of Scotland'. *Annals of Scottish Natural History*: 116.
May, D. 1998. 'Eager beavers to return home'. *The Times* (22 March).
Mayle, B.A. 2003. 'The impact of introduced deer on the natural environment'. In: Poland Bowen, C. (ed.). *Conference Proceedings 2001–2002. MammAliens – the Problems Caused by Non-Native British Mammals*: 49–52. People's Trust for Endangered Species & Mammals Trust UK: London.
McCurdy, J. 1988. 'Sika deer in Ireland: a success story'. *Deer* 7: 239–40.

McDougall, P. 1975. 'Feral goats of Kielderhead Moor'. *Journal of Zoology* 176: 215–46.
McGrady, M.J., Orr-Ewing D.C. & Stowe, T.J. 1994. 'The reintroduction of the red kite *Milvus milvus* in Scotland'. In: Chancellor, R.D. & Meyburg, B.U. (eds). *Raptor Conservation Today*: 471–8. WWGBP/Pica Press.
McNeely, J.A. 2003. 'Human dimensions of invasive species of mammals'. In: Poland Bowen, C. (ed.). *Conference Proceedings 2001–2002, MammAliens – the Problems Caused by Non-native British Mammals*: 28–34. People's Trust for Endangered Species & Mammals Trust UK: London.
Mead, C. 2000. *The State of the Nation's Birds*. Whittet Books: Stowmarket, Suffolk.
Meadows, B.S. 1968. 'Guppy in the River Lee'. *Essex Naturalist* 32: 186–9.
Measey, G.J. 1998a. 'Terrestrial prey capture in *Xenopus laevis*'. *Copeia* 1998: 787–91.
Measey, G.J. 1998b. 'Diet of feral *Xenopus laevis* (Daudin) in South Wales, UK'. *Journal of Zoology* 246: 287–98.
Measey, G.J. 2001. 'Growth and ageing of feral *Xenopus laevis* (Daudin) in South Wales'. *Journal of Zoology* 254: 547–55.
Measey, G.J. & Tinsley, R.C. 1998. 'Feral *Xenopus laevis* in South Wales'. *Herpetological Journal* 8: 23–7.
Meinertzhagen, R. 1950. 'The goshawk in Great Britain'. *Bulletin of the British Ornithologists' Club* 70: 46–9.
Menzies, J.I. 1962. 'The marsh frog in England'. *British Journal of Herpetology* 11: 43–54.
Menzies, W.S. 1907. 'Capercaillie and willow grouse in Moray'. *Annals of Scottish Natural History*: 116–17.
Merne, O.J. 1970. 'The status of the Canada goose in Ireland'. *Irish Bird Report* 17: 12–17.
Middleton, A.D. 1930. 'The ecology of the American grey squirrel in the British Isles'. *Proceedings of the Zoological Society of London*: 809–43.
Middleton, A.D. 1931. *The Grey Squirrel*. Sidgwick & Jackson: London.
Middleton, A.D. 1932. 'The grey squirrel in the British Isles, 1930–32'. *Journal of Animal Ecology* 1: 166–7.
Middleton, A.D. 1935. 'The distribution of the grey squirrel in Great Britain in 1935'. *Journal of Animal Ecology* 4: 274–6.
Middleton, A.D. 1937. 'Whipsnade ecological survey, 1936–37'. *Proceedings of the Zoological Society of London*, Series A. 471–81.
Middleton, A.D. & Parson, B.T. 1937. 'The distribution of the grey squirrel in Great Britain in 1937'. *Journal of Animal Ecology* 6: 286–90.
Milburn, F.M.H. 'Porcupines defended'. *Western Morning News* (13 April).
Miller, G.S. 1933. 'The origin of the ferret'. *Scottish Naturalist* 1933: 153–4.
Mills, J. 2005. 'The Domesday goats face a death sentence'. *Daily Mail* (29 April).
Milner, C., Goodier, R. & Crook, I.G. 1968. 'Feral goats'. *Nature in Wales* 11: 3–11.
Ministry of Agriculture, Fisheries & Food 1962. *Squirrels: their Biology and Control. Bulletin No. 184.* MAFF: London.
Minns, D. & Gilbert, D. 2001. *Red Kites – Naturally Scottish.* Scottish Natural Heritage: Battleby.
Mitchell, G.F. 1941. 'The reindeer in Ireland'. *Proceedings of the Royal Irish Academy* 46 B: 183–8.
Mitchell, J. 1983. 'Strange beasts on bonny banks'. *Scottish Wildlife* 19: 20–4.
Mitchell, J. 2001. *Loch Lomondside*. Collins: London.
Mitchell, W.R. & Robinson, J. 1971. 'The Bowland sika: their history, status and distribution'. *Deer* 2: 729–32.
Mivart, St. G. 1881. *The Cat*. John Murray: London.
Moffat, C.B. 1938. 'The mammals of Ireland'. *Proceedings of the Royal Irish Academy* 44 B: 61–128.
Mole, S. 2007. 'The impact of introduced Lacertids on the native herpetofauna of Boscombe Cliffs, Bournemouth. British Hertpeological Society/Herpetological Conservation Trust Scientific Meeting, Bournemouth'. IV. *BHS Newsletter* 155: 3 (Abstract).
Mole, S. 2008. 'Investigation into the effects of the Western Green Lizard (*Lacerta bilineata*) and the Common Wall Lizard (*Podarcis muralis*) introduced onto Boscombe Cliffs, Dorset, UK'. Thesis for Sparsholt College, Hampshire. 53 pp. mimeograph.
Mooney, O.V. 1952. 'Irish deer and forest relations'. *Irish Forestry* 9: 11–27.
Moore, H.D.M. 1997. 'On developing an immunocontraceptive vaccine for the grey squirrel'. In: Gurnell, J. & Lurz, P.W.W. (eds). *The Conservation of Red Squirrels* Sciurus vulgaris *L.* People's Trust for Endangered Species: London.
Moore, N. 2004. 'The ecology and management of wild boar in southern England'. *DEFRA Final Project Report, VC0325.*
Moore, N.P. & Wilson, C.J. 2005. 'Feral wild boar in England: implications for future management options'. *Report to DEFRA European Wildlife Division.*
Moors, P.J., Atkinson, I.A.E. & Sherley, G.H. 1992. 'Reducing the rat threat to island birds'. *Bird Conservation International* 2: 93–114.
More, A.G. 1865. 'On the distribution of birds in Great Britain during the nesting season'. *Ibis* 1: 1–27.
Morgan, A. 2000. 'Beware the amphibious landing'. *The Times* (7 October).
Morris, P.A. 1986. 'An introduction to reintroductions'. *Mammal Review* 16: 49–52.
Morris, P. 1997. *The Edible Dormouse*. The Mammal Society: London.
Morris, P.A. 2003. 'The edible dormouse (*Glis glis*) in Britain'. In: Poland Bowen, C. (ed.). *Conference Proceedings 2001–2002. MammAliens – the Problems Caused by Non-Native British Mammals*: 37–9. People's Trust for Endangered Species & Mammals Trust UK: London.
Morris, P.A. & Hoodless, A. 1992. 'Movements and hibernaculum site in the fat dormouse (*Glis glis*)'. *Journal of Zoology* 228: 685–7.
Mouland, B. 2005. 'Battle of the beavers: the fur flies as rodents are freed into the wild after 500 years'. *Daily Mail* (28 October).
Muir-Howe, H. 2007. 'The history and current status of the Midwife Toad (*Alytes obstetricans*) in Britain'. V. British Herpetological Society/Herpetological Conservation Trust Scientific Meeting, Bournemouth'. *BHS Newsletter* 155: 3 (Abstract).
Mulloy, F. 1970. 'A note on the occurrence of deer in Ireland'. *Deer* 2: 502–4.

Nature Conservancy Council. 1985. *The Sea Eagle*. NCC: London.
Nature-Times News Service 1974. *Problem of Escaping Birds.* Science Report (Ornithology).
Neale, G. 1990. 'Vanishing wallabies fight to bounce back in Britain'. *Sunday Telegraph* (8 November).
Nelson, T.H., Clarke, W.E. & Boyes, F. 1907. *The Birds of Yorkshire.* Brown & Sons: London.
Nevard, T. 1991. 'Reintroducing large vertebrates – what benefits'. *Marine Ecology* (Stazione Zoologica di Napoli) 12 (3): 44–6.
Newdick, J. 1979. *The Complete Freshwater Fishes of the British Isles.* A. & C. Black: London.
Newling, D. 2006. 'The British safari'. *Daily Mail* (18 September).
Newton, A. 1859. 'The naturalization of the edible frog in England'. *Zoologist* 17: 6538–40.
Newton, I. 1993. 'Predation and limitation of bird numbers'. *Current Ornithology* 11: 143–98.
Niall, I. 1957. 'Wild goats'. *Country Life* (24 January).
Nightingale, B. 2005. 'The status of Lady Amherst's Pheasant in Britain'. *British Birds* 98: 20–5.
Ní Lamhna, E. 1979.*Provisional Distribution Atlas of Amphibians, Reptiles and Mammals in Ireland.* Second edition. An Foras Forbatha: Dublin.
Nilson, E.B. 2007. 'Wolf reintroduction to Scotland: public attitudes and consequences for red deer management'. *Philosophical Transactions of the Royal Society of London Series B. 274*: 995–1002.
Noddle, B. 1993. 'Bones of larger mammals'. In: Casey, P.J., Davies, J.L. & Evans, J. (eds). *Excavations at Segontium (Caernarfon) Roman Fort 1975–1979. CBA Research Report* 90: 97–118. London.
Norris, J.D. 1967. 'A campaign against feral coypus (*Myocastor coypus*) in Great Britain'. *Journal of Applied Ecology* 4: 191–9.
Nuttall, N. 1994. 'One man's mission to bring back animals from the past'. *The Times* (1 October).

O'Connor, T.P. 1987. 'Why bother looking at archaeological wild animal assemblages?' *Circaea* 4: 107–14.

O'Connor, T.P. 1991. 'On the lack of bones of the ship rat *Rattus rattus* from Dark Age York'. *Journal of Zoology* 224: 318–20.
O'Connor, T.P. 1992. 'Pets and pests in Roman and Medieval Britain'. *Mammal Review* 22: 107–13.
Ogilvie, M.A. 1977. 'The numbers of Canada geese in Britain'. *Wildfowl* 28: 27–34.
Ogilvie, M. & Rare Breeding Birds Panel [RBBP]. 1999–2003. 'Non-native birds breeding in the UK in 1996–2002'. *British Birds* 92: 176–82, 472–6; 93: 428–33; 94: 518–22; 95: 631–35; 96: 620–5; 97: 633–7.
O'Gorman, F. 1970. 'The development of game in Ireland'. *Proceedings of the 8th International Congress of Game Biologists*, Helsinki: 387–96.
Okubo, A., Maini, P.K., Williamson, M.H. & Murray, J.D. 1989. 'On the spatial spread of the grey squirrel in Britain'. *Proceedings of the Royal Society of London, Series B 238*: 113–25.
Opdam, P., Thissen, J., Verschuren, P. & Müskens, G. 1977. 'Feeding ecology of a population of goshawks *Accipiter gentilis*'. *Journal für Ornithologie* 118: 35–51.
Orkin, P.A. (ed.). 1957. *Freshwater Fishes*. Thames & Hudson: London.
Orr-Ewing, D. 2007. 'Red Kite'. In: Forrester, R. & Andrews, I. 2007. *Birds of Scotland* (3). Scottish Ornithologists' Club: Edinburgh.
Orr-Ewing, D.C., Duffy, K., O'Toole, L., Stubbe, M. & Schönbrodt, R. 2006. 'Final report on the translocation of Red Kites *Milvus milvus* from Germany to central Scotland'. *Populationsökologie Greifvögel-und Eulenarten* 5: 261–72.
Osborne, P.E. 2002. 'Application to the Department for Environment Food and Rural Affairs for a licence to re-introduce Great Bustards *Otis tarda* to Britain'. Great Bustard Consortium: Salisbury.
Osborne, P.E. 2005. 'Key issues in assessing the feasibility of reintroducing the great bustard *Otis tarda* to Britain'. *Oryx* 39: 22–9.
Ó Teangana, D., Reilly, S., Montgomery, W.I. & Rotchford, J. 2000. 'Distribution and status of the red squirrel (*Sciurus vulgaris*) and grey squirrel (*Sciurus carolinensis*) in Ireland'. *Mammal Review* 30: 45–56.
O'Toole, L., Fielding, A.H. & Haworth, P.F. 2002. 'Reintroduction of the Golden Eagle into the Republic of Ireland'. *Biological Conservation* 103: 303–12.

Page, F.J.T. 1954. 'The wild deer of East Anglia'. *Transactions of the Norfolk and Norwich Naturalists' Society* 17: 316–21.
Page, F.J.T. & Tittensor, A.M. 1969. *The Sussex Mammal Report*. Sussex Naturalists' Trust: Henfield.
Parslow, J. 1973. *Breeding Birds of Britain and Ireland*. Poyser: Berkhamsted.
Patterson, A.H. 1909. 'Fish and fisheries of east Suffolk'. *Zoologist* 13: 361–92; 414–21.
Pelletier, F., Clutton-Brock, T., Pemberton, J., Tuljapurkar, S. & Coulson, T. 2007. 'The evolutionary demography of ecological change: linking trait variation and population growth'. *Science* 315: 1571–4.
Pemberton, J.M. 1993. 'The genetics of fallow deer production'. In: Asher, G.W. (ed.). *Proceedings of the First World Forum on Fallow Deer Farming*: 129–35. Mudgee, Australia.
Pemberton, J. 2006. 'Hybridisation between red and sika deer in Scotland'. *Deer* 13 (9): 22–6.
Pennie, I.D. 1950–1. 'The history and distribution of the capercaillie in Scotland'. *Scottish Naturalist* 62: 65–87, 157–78; 63: 4–18, 135.
Pernetta, J. 1973. 'The ecology of *Crocidura suaveolens cassiteridum* (Hinton) in a coastal habitat'. *Mammalia* 37: 241–56.
Pernetta, J.C. & Handford, P.T. 1970. 'Mammalian and avian remains from possible Bronze Age deposits on Nornour, Isles of Scilly'. *Journal of Zoology* 162: 534–40.
Perrin, M.R. & Gurnell, J. 1971. 'Rats on Lundy'. *Annual Report of the Lundy Field Society* 22: 35–40.
Perring, F.H. 1967. 'Distribution data relating to freshwater fish'. *Proceedings of the British Coarse Fish Conference* 3: 57–61.
Perrins, C.M. & Overall, R. 2001. 'Effects of increasing numbers of deer on bird populations in Wytham Woods, central England'. *Forestry* 74: 299–309.
Petty, S.J. 1989. *Goshawks: their Status, Requirements and Management. Forestry Commission Bulletin No.81*. HMSO: London.
Phillips, W.M., Stephens, M.N. & Worden, A.N. 1952. 'Observations on the rabbit in West Wales'. *Nature, London* 169: 869–70.
Pickvance, J.J. & Chard, J.S.R. 1960. 'Feral muntjac (*Muntiacus* spp.) in the west Midlands, with special reference to Warwickshire'. *Proceedings of the Birmingham Natural History and Philosophical Society* 19: 1–8.
Pierce, A. 2005. 'Beavers are making a comeback after a 500-year absence'. *The Times* (27 October).
Pinder, A.C. 2005. 'Larval taxonomy and early identification of topmouth gudgeon *Pseudorasbora parva*: a screening tool to reduce risks of unintentional species translocation'. *Fisheries Management and Ecology* 12: 99–104.
Pinder, A.C. & Gozlan, R.E. 2003. 'Sunbleak and topmouth gudgeon – two new additions to Britain's freshwater fishes'. *British Wildlife* 15: 77–83.
Pinder, A.C., Gozlan, R.E. & Britton, J.R. 2005. 'Dispersal of the invasive topmouth gudgeon, *Pseudorasbora parva*, in the UK: a vector for an emergent infectious disease'. *Fisheries Management and Ecology* 12: 411–14.
Pithon, J.A. & Dytham, C. 1999. 'Census of the British Ring-necked Parakeet *Psittacula krameri* population by simultaneous counts at roosts'. *Bird Study* 46: 112–15.
Pithon, J.A. & Dytham, C. 2001. 'Determination of the origin of British feral Rose-ringed Parakeets'. *British Birds* 94: 74–9.
Poland Bowen, C. (ed.). 2003. *The Return of the Native – the Reintroduction of Native Species Back into their Natural Habitats. Conference Proceedings, 2001–2002*. People's Trust for Endangered Species/Mammals Trust UK: Lyndhurst.
Platt, F.B.W. & Rowe, J.J. 1964. 'Damage by the edible dormouse (*Glis glis* L.) at Wendover Forest (Chilterns)'. *Quarterly Journal of Forestry* 58: 228–33.
Pocock, R.I. 1932. 'Ferrets and polecats'. *Scottish Naturalist* 196: 97–108.
Pollard, E. & Cooke, A.S. 1994. 'Impact of muntjac deer *Muntiacus reevesi* on egg-laying sites of the white admiral butterfly *Ladoga camilla* in a Cambridgeshire wood'. *Biological Conservation* 70: 189–91.
Poore, M.E.D. & Robertson, V.C. 1949. 'The vegetation of St Kilda in 1948'. *Journal of Ecology* 37: 82–99.
Porter, V. 1991. *Cattle*. Christopher Helm: London.
Porter, V. & Brown, N. 1997. *The Complete Book of Ferrets*. D & M Publications: Bedford.
Powerscourt, Viscount. 1884. 'On the acclimatization of the Japanese deer [sika] of Powerscourt'. *Proceedings of the Zoological Society of London 1884*: 207–09.
Pratt, R.M. 1986. 'Habitat use of free-ranging cattle and ponies in the New Forest of southern England'. *Journal of Applied Ecology* 23: 539–57.
Prevot-Julliard, A-C., Gousset, E., Archinard, C., Cadi, A. & Girandot, M. 2007. 'Pets and invasion risks: is the slider turtle strictly carnivorous?' *Amphibia-Reptilia* 28: 139–43.
Pritchard, K.V. 1982. 'Muntjac deer wild in Britain'. *Reading Naturalist* 34: 13–15.
Putman, R.J. 1986. 'Competition and coexistence in a multispecies grazing community: the large herbivores of the New Forest'. *Acta Theriologica* 31: 271–91.
Putman, R.J. 1994a. 'Damage by deer in coppice woodlands: an analysis of factors affecting the severity of damage and options for management'. *Quarterly Journal of Forestry* 88: 45–54.
Putman, R.J. 1994b. 'Effects of grazing and browsing by mammals in woodland'. *British Wildlife* 5: 205–13.
Putman, R.J. 1996. *Competition and Resource Partitioning in Temperate Ungulate Assemblies*. Chapman & Hall: London.
Putman, R.J. & Hunt, E.J. 1994. 'Patterns of hybridisation and introgression between red and sika deer in different populations in the North of Scotland and Argyll'. *Deer* 9: 104–10.
Putman, R.J. & Moore, N.P. 1998. 'Impact of deer in lowland Britain on agriculture, forestry and conservation habitats'. *Mammal Review* 28: 141–64.
Putman, R.J. & Sharma, S.K. 1987. 'Long term changes in New Forest deer populations and correlated environmental change'. In: Harris, S. (ed.). *Mammal Population Studies: Symposia of the Zoological Society of London* 58: 167–79.

Putman, R.J., Pratt, A.M., Ekins, J.R. & Edwards, P.J. 1987. 'Food and feeding behaviour of cattle and ponies in the New Forest, Hampshire'. *Journal of Applied Ecology* 24: 369–80.
Pyman, G.A. 1959. 'The status of the Red-crested Pochard in the British Isles'. *British Birds* 52: 42–56.

Quayle, A. & Noble, M. 2000. 'The wall lizard in England'. *British Wildlife* 12: 99–106.

Rackham, J. 1979. '*Rattus rattus*: the introduction of the black rat into Britain'. *Antiquity* 53: 112–20.
Rackham, O. 1986. *The History of the Countryside*. J.M. Dent & Sons: London.
Rackham, O. 1990. *Trees and Woodland in the British Landscape*. Second edition. J.M. Dent: London.
Ratcliffe, P.R. 1987. 'Distribution and current status of sika deer, *Cervus nippon*, in Great Britain'. *Mammal Review* 17: 39–58.
Ratcliffe, P.R. 1989. 'The control of red and sika deer populations in commercial forests'. In: Putman, R.J. (ed.). *Mammals as Pests*: 98–115. Chapman & Hall: London.
Raven, M.J., Noble, D.G. & Baillie, S.R. 2007. *The Breeding Bird Survey 2006. BTO Research Report 471*. British Trust for Ornithology: Thetford.
Rehfisch, M.M., Austin, G.E., Holloway, S.J., Allan, J.R. & O'Connell, M. 2002. 'An approach to the assessment of change in the numbers of Canada Geese *Branta canadensis* and Greylag Geese *Anser anser* in southern Britain'. *Bird Study* 49: 50–9.
Reid, B.N. 1930. 'Spread of the capercaillie in Ross-shire'. *Scottish Naturalist*: 26.
Reid, N. & Montgomery, W.I. 2007. 'Naturalisation of the brown hare in Ireland: a threat to the endemic Irish hare?' *Biology and Environment. Proceedings of the Royal Irish Academy 107B*: 129–38.
Reynolds, J.C. 1985. 'Details of the geographic replacement of the red squirrel (*Sciurus vulgaris*) by the grey squirrel (*Sciurus carolinensis*) in eastern England'. *Journal of Animal Ecology* 54: 149–62.
Richmond, F.G. 1919. 'About rainbow trout'. *Salmon & Trout Magazine* 20: 63–73.
Richmond, F.G. 1932. 'Future of rainbow trout in England'. *The Field* (30 April).
Richmond, W.K. 1955. 'Wild goats'. *Scottish Field* (May).
Rickards, R.B. & Fickling, N. 1979. *Zander*. A. & C. Black: London.
Ridley, Viscount. 1981. 'An early reindeer introduction to Britain'. *Deer* 5: 210.
Risdon, D.H.S. 1973. 'Report on the Tropical Bird Gardens at Rode, 1972'. *Avicultural Magazine* 79: 52–5.
Ritchie, J. 1920. *The Influence of Man on Animal Life in Scotland*. Cambridge University Press: London.
Ritchie, J.N., Hudson, J.R. & Thompson, H.V. 1954. 'Myxomatosis'. *Veterinary Record* 66: 796–804.
Robinson, J. 1973. 'The world of Bowland sika'. *Deer* 2: 984–7.
Rose, H. 2008. 'Beaver reintroduction on the cards yet again'. *The Field* (14 March).
Ross, D.M. 1897. 'Capercailzie in the Mid-Deveron district'. *Annals of Scottish Natural History*: 254.
Rowe, F.B. 1960. 'Golden hamsters living free in an urban habitat'. *Proceedings of the Zoological Society of London* 134: 499–503.
Rowe F.P. 1968. 'Further records of free-living golden hamsters'. *Journal of Zoology* 156: 529–30.
Rowell, H., Ward, R., Hall, C. & Cranswick, P. 2004. *The Naturalised Goose Survey 2000*. Wildfowl & Wetlands Trust Report: Slimbridge.
Royal Society for the Protection of Birds. 2006. *The State of the UK's Birds, 2006*. RSPB: Sandy, Bedfordshire.
Royal Zoological Society of Scotland 1936–1975. *Annual Reports*. Edinburgh.
Russell, H. 1904. 'Occurrence of the edible frog (*Rana esculenta, forma typica*) in Surrey. Edible Frog (*Rana esculenta*) in Surrey'. *Zoologist* 8: 352–3, 390–1.
Russell, V. 1976. *The New Forest Ponies*. David & Charles: Newton Abbot, Devon.
Ruttledge, R.F. 1924. 'Note on the distribution of squirrels in Ireland'. *Irish Naturalist* 33: 73.
Ryan, A., Duke, E. & Fairley, J.S. 1996. 'Mitochondrial DNA in bank voles *Clethrionomys glareolus* in Ireland: evidence for a small founder population and localized founder effects'. *Acta Theriologica* 41: 45–50.
Ryder, M.L. 1968. 'The evolution of Scottish breeds of sheep'. *Scottish Studies* 12: 127–67.
Ryder, M.L. 1975. 'Development, structure and seasonal change in the fleeces of unimproved Scottish blackface sheep from the Hebrides'. *Journal of Agricultural Science, Cambridge* 85: 85–92.
Ryder, M.L. 1995. 'When did the Soay sheep reach St Kilda?' *The Ark* 22: 293–4.
Rzebik, B. 1968. '*Crocidura* Wagler and other Insectivora (Mammalia), from the Quaternary deposits of Tornewton Cave in England'. *Acta Zoologica Cracoviensia* 13: 251–63.
Rzebik-Kowalska, B. 1998. 'Fossil history of shrews in Europe'. In: Wójcik, J.M. & Wolsan, M. (eds). *Evolution of Shrews*: 223–92. Mammal Research Institute, Polish Academy of Sciences: Bialowieza.

Sachs, T.R. 1878. 'Transportation of live pike-perch'. *Land & Water* (25 May).
Sage, B. 1981. 'The beaver reintroduction project'. *Discussion Paper in Conservation No. 30*. University College: London.
Salkeld, L. 2006. '3in invader that could kill off our salmon and trout'. *Daily Mail* (29 June).
Salomonsen, F. 1968. 'The moult migration'. *Wildfowl* 19: 5–24.
Samstag, T. 1985. 'Chinese deer "thriving throughout England"'. *The Times* (8 April).
Sangster, G., Collinson, J.M., Helbig, A.J., Knox, A.G. & Parkin, D.T. 2005. 'Taxonomic recommendations for British birds: third report'. *Ibis* 147: 821–6.
Sandeman, P.W. 1965. 'Attempted reintroduction of White-tailed Eagle to Scotland'. *Scottish Birds* 3: 411–12.
Savage, C. 1952. *The Mandarin Duck*. A. & C. Black: London.
Scaife, H.R. 2006. 'Wild rabbits (*Oryctolagus cuniculus*) as potential carriers of VTEC'. *Veterinary Record* 159: 175–8.
Schwartz, E. 1935. 'On ibex and wild goat'. *Annals of the Magazine of Natural History* 16: 433–7.
Schwarz, E. & Schwarz, H.K. 1943. 'Wild and commensal stocks of house mouse'. *Journal of Mammalogy* 24: 59–72.
Scott, D. 2002. 'Attempted nesting of the Red Kite *Milvus milvus* in Northern Ireland during 2002'. *Welsh Kite Trust Newsletter* 12: 16–17.
Scott, T. & Brown, A. 1901. 'The marine and freshwater fishes'. In: *Handbook of the British Association*: 173–80. Maclehose: Edinburgh.
Scott, W.A. 1967. 'Fallow deer in the Teign Valley'. *Deer* 1: 45.
Scott Porter Research & Marketing Ltd. 1998. 'Re-introduction of the European beaver to Scotland: results of a public consultation'. *Scottish Natural Heritage Research, Survey & Monitoring Report No. 121*.
Scottish Natural Heritage 2005. [*Application ... to release European beaver ... in Knapdale, Argyll*]. Scottish Natural Heritage: Edinburgh.
Seth-Smith, D. 1932. 'The origin of the melanistic pheasant'. *Ibis* 13: 438–41.
Sharrock, J.T.R. (ed.). 1976. *The Atlas of Breeding Birds in Britain and Ireland*. British Trust for Ornithology: Tring, Hertfordshire.
Sheail, J. 1970. 'Historical material on a wild animal – the rabbit'. *Local Historian* 9: 59–64.
Sheail, J. 1971. *Rabbits and their History*. David & Charles: Newton Abbot, Devon.
Sheail, J. 1988. 'The extermination of the muskrat (*Ondatra zibethicus*) in inter-war Britain'. *Archives of Natural History* 15: 155–70.
Sheail, J. 1999. 'The grey squirrel (*Sciurus carolinensis*) – a UK historical perspective on a vertebrate pest species'. *Journal of Environmental Management* 55: 145–56.
Sheail, J. 2003. 'Government and the management of an alien pest species: a British perspective'. *Landscape Research* 28: 101–11.
Sheppard, R. 2004. 'Brown hares *Lepus europaeus* Pallas in N.W. Ireland'. *Irish Naturalists' Journal* 2712: 484–5.
Shorten, M. 1946. 'A survey of the distribution of the American grey squirrel and the British red squirrel in England and Wales in 1944–1945'. *Journal of Animal Ecology* 15: 82–92.
Shorten, M. 1948. 'Grey squirrels in Britain'. In: *The New Naturalist Journal*: 42–6. Collins: London.

Shorten, M. 1951. 'Some aspects of the biology of the grey squirrel in Great Britain'. *Proceedings of the Zoological Society of London* 121 (2): 427–59.
Shorten, M. 1953. 'Notes on the distribution of the grey squirrel and the red squirrel in England and Wales from 1945–1952'. *Journal of Animal Ecology* 22: 134–40.
Shorten, M. 1954. *Squirrels*. Collins: London.
Shorten, M. 1957. 'Squirrels in England, Wales and Scotland in 1955'. *Journal of Animal Ecology* 26: 287–94.
Shorten, M. 1962. *The Grey Squirrel. (Animals in Britain No. 5)*. Sunday Times Publications: London.
Shorten, M. 1964. 'Introduced menace'. *Natural History Magazine* 10: 43–8.
Shorten, M. & Courtier, F.A. 1955. 'A population study of the grey squirrel in May 1954'. *Annals of Applied Biology* 43: 492–510.
Shurtleff, L.L. & Savage, C. 1996. *The Wood Duck and the Mandarin*. University of California Press: Berkeley.
Simmonds, M. 1982–83. 'The amphibian invader'. *World Wildlife News* (Winter): 6–7.
Simmons, I.G. & Tooley, M.J. (eds). 1981. *The Environment in British Prehistory*. Duckworth: London.
Sims, N.K.E. 2005. 'The ecological impacts of wild boar rooting in East Sussex'. D.Phil. thesis, University of Sussex. Unpublished.
Sinel, J. 1905 & 1908. 'Notes on the lizards of the Channel Islands. The reptiles, batrachia, and mammalia of the Channel Islands'. *Report & Transactions of the Guernsey Society of Natural Science* 5: 308–17, 466–78.
Sitwell, N. 1977. 'Bring back the beaver'. *Wildlife* 19: 493–7.
Sjoberg, G. 1996. 'Genetic characteristics of introduced birds and mammals'. *Wildlife Biology* 2: 159–64.
Sleeman, P. 1997. 'Mammals and mammalogy'. In: Foster, J.W. (ed.). *Nature in Ireland*: 241–62. Lilliput Press: Dublin.
Smal, C.M. 1991. 'Population studies on feral mink *Mustela vison* in Ireland'. *Journal of Zoology* 24: 233–49.
Smal, C.M. & Fairley, J.S. 1984. 'The spread of the bank vole *Clethrionomys glareolus* in Ireland'. *Mammal Review* 14: 71–8.
Smartt, J. 2007. 'A possible genetic base for species replacement: preliminary results of interspecific hybridisation between native crucian carp *Carassius carassius* and introduced goldfish *Carassius auratus*'. *Aquatic Invasions* 2: 59–62.
Smith, A.J. 1985. 'A catastrophic origin for the paleovalley system of the eastern English Channel'. *Marine Geology* 64: 65–75.
Smith, C. 1994. 'Animal bone report'. In: Smith, B.B. (ed.). *Four Millenia of Orkney Prehistory*: 139–53. Monograph Series No. 9. Society of Antiquaries of Scotland: Edinburgh.
Smith, E.P. 1939. 'On the introduction of *Rana esculenta* in east Kent'. *Journal of Animal Ecology* 8: 168–70.
Smith, L. 2006a. 'Crocodiles, big cats and wolves: it's a jungle in the countryside'. *The Times* (18 September).
Smith, L. 2006b. 'How goldfish are turning carp into small fry'. *The Times* (30 December).
Smith, L. 2006c. 'Virus fatal to red squirrel is spread by grey invader'. *The Times* (22 August).
Smith, L. 2006d. 'Wild horses will create perfect habitat for otters and voles'. *The Times* (30 November).
Smith, L. 2007a. 'As the weather gets hotter so the sheep are getting smaller'. *The Times* (16 March).
Smith, L. 2007b. 'Population boom puts parakeet invaders on hitlist'. *The Times* (22 March).
Smith, M. 1949–50. 'The midwife toad in Britain'. *British Journal of Herpetology* 1: 55–6; 89–91.
Smith, M. 1951a. *The British Amphibians and Reptiles*. Collins: London.
Smith, M. 1951b. 'The wall lizard'. *British Journal of Herpetology* 1: 99–100.
Smith, M. 1953. 'The feeding habits of the marsh frog'. *British Journal of Herpetology* 1: 170–2.
Smith, P. & Briggs, J. 1999. 'Zander – the hidden invader'. *British Wildlife* 11: 2–8.
Smith, P.A. 1993. 'The ship rat (*Rattus rattus*) on Lundy, 1991'. *Journal of Zoology* 231: 689–95.
Smith, P.A. & Eaton, J.W. 1998. 'The colonization of a British canal by zander (*Stizostedion lucioperca*)'. *Proceedings PERCIS II Symposium*. Riistan-JA Kalantutkimus, Helsinki.
Smith, P.A., Leah, R.T. & Eaton, J.W. 1996. 'Removal of pikeperch (*Stizostedion lucioperca*) from a British Canal as a management technique to reduce impact on prey populations'. *Annales Zoologicì Fennici* 33: 537–45.
Smith, P.A., Leah, R.T. & Eaton, J.W. 1998. 'A review of the current knowledge on the introduction, ecology and management of zander, *Stizostedion lucioperca*, in the UK'. In: Cowx, I.G. (ed.). *Stocking and Introduction of Fish*: 209–24. Fishing News: Oxford.
Smith, R.H. 1980. 'The genetics of fallow deer and their implications for management'. *Deer* 5: 79–83.
Smith, S. 1995. 'The coypu in Britain'. *British Wildlife* 6: 279–85.
Smith-Jones, C. 2004–5. 'Managing an alien species [muntjac]'. *Deer* 13 (4): 30–2.
Snell, C. 1981. 'Breeding of the European tree frog, *Hyla arborea*, and a note on a wild breeding population of *Podarcis muralis* in Britain'. *Bulletin of the British Herpetological Society* 3: 48.
Snell, N., Dixon, W., Freeman, A., McQuaid, M. & Stevens, P. 2002. 'Nesting behaviour of the Red Kite in the Chilterns'. *British Wildlife* 13: 177–83.
Sol, D., Timmermans, S. & Lefebre, L. 2002. 'Behavioural flexibility and invasion success in birds'. *Animal Behaviour* 63: 495–502.
Sommerville, A. 2001. 'The introduction of the beaver'. *Glasgow Naturalist* 23 (Supplement): 103–7.
Soper, E.A. 1969. *Muntjac*. Longman: London.
Southwell, T. 1901. *On the Reptiles and Amphibians*. Victoria County Histories: Norfolk.
Speed, J.G. & Etherington, M.G. 1952 & 1953. 'The Exmoor pony – and a survey of the evolution of horses in Britain: Part I. Exmoor ponies'. *British Veterinary Journal* 108: 329–38. 'Part II. The Celtic pony'. 109: 315–20.
Spellerberg, I.F. 1975. 'Britain's reptile immigrants'. *Country Life* (20 February).
Spencer, H.E.P. 1956. 'Rabbit'. *Transactions of the Suffolk Naturalists' Society* 9: 369–70.
Springthorpe, G. 1969. 'Long-haired fallow deer at Mortimer Forest'. *Journal of Zoology* 159: 537.
Springthorpe, G. & Voysey, J. 1969. 'The fallow deer of Mortimer Forest'. *Deer* 1: 407–09.
Stadler, S.G. 1988. 'Observations on the behaviour of Chinese water deer (*Hydropotes inermis* Swinhoe 1870)'. *Deer* 7: 300–1.
Staines, B. 1986. 'The spread of grey squirrels (*Sciurus carolinensis* Gm) into north-east Scotland'. *Scottish Forestry* 40: 190–6.
Steele-Elliot, J. 1907. 'The little owl in Bedfordshire'. *The Zoologist* 11: 384.
Stelfox, A.W. 1927. 'The grey squirrel spreads to Westmeath'. *Irish Naturalists' Journal* 1: 275.
Stelfox, A.W. 1965. 'Notes on the Irish "Wild Cat"'. *Irish Naturalists' Journal* 15: 57–60.
Stephens, M.N. 1952. 'Seasonal observations on the wild rabbit in West Wales'. *Proceedings of the Zoological Society of London* 122: 417–34.
Stewart, A.E. 1933. 'Wild goats in Scotland'. *The Field* (16 December).
Stewart, J.R. 2007. 'The fossil and archaeological record of the Eagle Owl in Britain'. *British Birds* 100: 481–6.
Stokes, K., O'Neill, K. & McDonald, R. 2004. 'Invasive species in Ireland. Quercus project QU03-01'. Quercus, Queen's University: Belfast: 1–152.
Storer, Rev. J. 1877. *The Wild White Cattle of Great Britain*. Cassell, Peter & Gilpin: London.
Strachan, C. & Jefferies, D.J. 1996. 'An assessment of the diet of the feral American mink *Mustela vison* from scats collected in areas where water voles *Arvicola terrestris* occur'. *Naturalist* 121: 73–81.
Strachan, C., Jefferies, D.J., Barreto, G.R., Macdonald, D.W. & Strachan, R. 1998. 'The rapid impact of resident American mink on water voles: case studies in lowland England'. *Symposia of the Zoological Society of London* 71: 339–57.
Strachan, R. 2008. 'A short history of Mink in Britain'. 2 page mimeograph (unpublished).
Strachan, R. & Jefferies, D.J. 1996. *Otter Survey of England 1991–1994*. Vincent Wildlife Trust: London.
Street, P. 1955. 'The edible dormouse in England'. *Country Life* 188: 1484–5.
Stringer, C. 2007. *Homo Britannicus*. Allen Lane: London.

Stuart, T.A. 1967. 'Trout fishery of the Lake of Menteith'. *Report on the Fisheries of Scotland*: 155–7.
Stuart, T.A. 1969–71. 'Studies on the Lake of Menteith'. *Directorate of Fisheries Research Report for 1968*: 132–4 (1969); for 1970: 114–15 (1971).
Suchentrunk, F. Jaschke, C. & Haiden, A. 2001. 'Little allozyme variation and mtDNA variability in brown hares (*Lepus europaeus*) from New Zealand and Britain – a legacy of bottlenecks?' *Mammalian Biology* 66: 48–59.
Summers, R.W., Willi, J. & Selvidge, J. 2008. 'Capercaillie nest attendance and survival'. *RSPB Research Report No. 29*. 21 page mimeograph.
Sumption, K.J. & Flowerdew, J.R. 1985. 'The ecological effects of the decline in rabbits (*Oryctolagus cuniculus* L.) due to myxomatosis'. *Mammal Review* 15: 151–86.
Sutcliffe, A.J. & Kowalski, K. 1976. 'Pleicestocene rodents of the British Isles'. *Bulletin of the British Museum (Natural History) Geology* 27: 31–147.
Sutherland, W.J. & Allport, G. 1991. 'The distribution and ecology of the naturalized Egyptian Goose *Alopochen aegyptiacus* in Britain'. *Bird Study* 38: 128–34.
Sykes, M. 1979. 'Beavers in Wales'. *Animals* 21: 21–3.
Sykes, N. 2004. 'The introduction of fallow deer to Britain: a zooarchaeological perspective'. *Environmental Archaeology* 9: 75–83.
Sykes, N. 2005. 'Zooarchaeology of the Norman Conquest'. *Anglo-Norman Studies* 27: 185–97.
Sykes, N.J. 2007. *The Norman Conquest: a Zooarchaeological Perspective*. BAR International Series 1656: Oxford.
Sykes, T. 1995. *Upper Thames Otter Habitat Project: Final Report*. Buckinghamshire, Berkshire and Oxfordshire Wildlife Trust: Oxford.

Taylor, A.B. 1967–8. 'The Norsemen in St Kilda'. *Saga Book of the Viking Society* 17 (2): 116–44; (3): 106–14.
Taylor, A. 1994. 'Zander control'. In: *Predators of Freshwater Fishes*: 1–7. Institute of Fisheries Management Publication.
Taylor, J.C. 1966. 'Home range and agonistic behaviour in the grey squirrel'. *Symposium of the Zoological Society of London* 18: 229–35.
Taylor, R.H.R. 1948. 'The distribution of reptiles and amphibia in the British Isles'. *British Journal of Herpetology* 1: 1–25.
Taylor, R.H.R. 1963. 'The distribution of amphibians and reptiles'. *British Journal of Herpetology* 15: 95–101.
Taylor, W.L. 1939 & 1948a. 'The distribution of wild deer in England and Wales'. *Journal of Animal Ecology* 8: 6–9; 17: 151–4.
Taylor, W.L. 1948b. 'The capercaillie in Scotland'. *Journal of Animal Ecology* 17: 155–7.
Tegner, H. 1952. 'Wild goats of the Border'. *Country Life* (29 February).
Tegner, H. 1965. 'Wild goats of Britain'. *Scottish Field* (March).
Tegner, 1976. 'Wild cats: feral cats'. *Wildlife* 18: 78–9.
Temple, R. & Morris, P. 1997. 'The lesser white-toothed shrew on the Isles of Scilly'. *British Wildlife* 9: 94–9.
Tetley, H. 1939. 'On the British polecats'. *Proceedings of the Zoological Society of London 109B*: 37–9.
Tetley, H. 1945. 'Notes on British polecats and ferrets'. *Proceedings of the Zoological Society of London* 115: 212–17.
Tew, W.E. 1930. 'The rainbows of the Derbyshire Wye'. *Salmon & Trout Magazine* 61: 362–4.
Thaler, L., Bonhomme, F. & Britton-Davidian, J. 1981. 'Processes of speciation and semi-speciation in the House Mouse'. *Symposium of the Zoological Society of London* 47: 27–41.
Thomas, A.S. 1956. 'Botanical effects of myxomatosis'. *Bulletin of the Mammal Society of the British Isles* 5: 16–17.
Thomas, A.S. 1963. 'Further changes in vegetation since the advent of myxomatosis'. *Journal of Ecology* 51: 151–86.
Thomas, J.A. 1980. 'Why did the Large Blue become extinct in Britain?' *Oryx* 15: 243–7.
Thomas, J. 2000. 'The great bustard in Wiltshire: flight into extinction?' *Wiltshire Archaeological & Natural History Magazine* 93: 63–70.
Thomas, R. 2002. 'Animals, economy and status: the integration of historical and archaeological evidence in the study of a medieval castle'. PhD thesis, Birmingham University. Unpublished.
Thompson, H.V. 1953. 'The edible dormouse in England 1902–51'. *Proceedings of the Zoological Society of London* 122: 1017–24.
Thompson, H.V. 1954. 'The rabbit disease: myxomatosis'. *Annals of Applied Biology* 41: 358–66.
Thompson, H.V. 1956. 'The origin and spread of myxomatosis with particular reference to Great Britain'. *Terre et Vie* 193: 137–51.
Thompson, H.V. 1962a. 'Wild mink in Britain'. *New Scientist* 13: 130–2.
Thompson, H.V. 1962b. 'Mink at large in Britain'. *Country Life* 131: 645.
Thompson, H.V. 1964. 'Wild mink'. *Agriculture* 71: 564–7.
Thompson, H.V. 1965. 'Feral mink in Britain'. *Animals* 6: 240–4.
Thompson, H.V. 1967. 'Control of wild mink'. *Agriculture* 74: 114–16.
Thompson, H.V. 1968a. 'British wild mink'. *Annals of Applied Biology* 61: 345–9.
Thompson, H.V. 1968b. 'Mink, the new British beast of prey'. *The Field* 232: 1094–5.
Thompson, H.V. 1966. 'Animal immigrants'. *Geographical Magazine* 1966: 762–72.
Thompson, H.V. 1969. 'Wild mink in Britain'. *Country Landowner* 20: 100.
Thompson, H.V. 1971. 'British wild mink – a challenge to naturalists'. *Agriculture* 78: 421–5.
Thompson, H.V. (ed.). 1994. *The European Rabbit: the History of a Successful Colonizer*. Oxford University Press: Oxford.
Thompson, H.V. & Peace, T.R. 1962. 'The grey squirrel problem'. *Quarterly Journal of Forestry* 56 (1): 33–42.
Thompson, H.V. & Platt, F.B. 1964. 'The present status of *Glis* in England'. *Bulletin of the Mammal Society of the British Isles* 21: 5–6.
Thompson, H.V. & Worden, A.N. 1956. *The Rabbit*. Collins: London.
Thomson, A.P.D. 1951. 'A history of the ferret'. *Journal of the History of Medicine and Allied Sciences* 6: 471–80.
Thorn, V.M. 1986. *Birds in Scotland*, T. & A.D. Poyser: Calton.
Thorpe, J.P. & Sharpe, C.M. 2004. 'Occurrence of Redwinged Laughing thrushes *Garrulax formosus* on the Isle of Man'. *Peregrine* 8: 348–51.
Tibbets, G. 2007. 'Grey squirrels may be sterilised to curb numbers'. *Daily Telegraph* (3 October).
Tinsley, R.C. & Kobel, H.R. (eds). 1996. *The Biology of Xenopus*. Clarendon Press: Oxford (for the Zoological Society of London).
Tittensor, A. 1982. 'Return of the rabbit'. *Country Life* 172: 1397–8.
Tomlinson, D. 1976. 'Surrey's Chinese duck'. *Country Life* 159: 1248–9.
Tomlinson, D. 2006. 'Foreign fowl here to stay'. *Country Life* 176: 55–7.
Torbett, H.D. 1961. *The Angler's Freshwater Fishes*. Putnam: London.
Tortonese, E. 1967. 'I pesi gatto'. *Rivista italiana di piscicoltora e ittiopatologia* 11: 46–7.
Tosh, D.G., Lusby, J., Montgomery, W.I. & O'Halloran, J. 2008. 'First record of greater white-toothed shrew *Crocidura russula* in Ireland'. *Mammal Review* (in press).
Trodd, P. & Kramer, D. 1991. *Birds of Bedfordshire*. Castlemead Publications: Welwyn Garden City.
Trout, R.C. 1986. 'Recent trends in the rabbit population in Britain'. *Mammal Review* 16: 117–23.
Troy, C.S.. MacHugh, D.E., Bailey, J.F., Magee, D.A., Loftus, R.T., Cunningham, P., Chamberlain, A.T., Sykes, B.C., & Bradley, D.G. 2001. 'Genetic evidence for Near-Eastern origins of European cattle'. *Nature* 410: 1088–91.
Tubbs, C.R. 1986. *The New Forest*. Collins: London.
Turk, F.A. 1964. 'Blue and brown hares associated together in a Bronze Age fissure cave burial'. *Proceedings of the Zoological Society of London* 142: 185–8.
Turk, F.A. 1968–9. 'Notes on Cornish mammals'. *Bulletin of the Cornwall Naturalists' Trust* New Series 4: 1–4.
Turk, T. 2004. 'The eagle owl in Britain 2004. Has the native returned?' *Tyto* 9: 9–20.
Turner, D.T.L. 1965. 'A contribution to the ecology and taxonomy of the vole *Microtus arvalis* on the island of Westray, Orkney'. *Proceedings of the Zoological Society of London* 144: 143–50.

Tutton, T. 1994. 'Goats versus holm oak'. *Enact* 2: 8–9.
Twigg, G. 1975. *The Brown Rat.* David & Charles: Newton Abbot, Devon.
Twigg, G. 1984. *The Black Death: a Biological Reappraisal.* Batsford: London.
Tyrer, N. 1997. 'It's back – and this time it's hungry'. *Daily Telegraph* (22 February).

Universities' Federation for Animal Welfare 1981. *The Ecology and Control of Feral Cats.* UFAW: Potters Bar, Hertfordshire.
Usher, M.B., Crawford, T.J. & Banwell, J.L. 1992. 'An American invasion of Great Britain: the case of the native and alien squirrel (*Sciurus*) spp'. *Conservation Biology* 6: 108–15.

Veale, E.M. 1957. 'The rabbit in England'. *Agriculture History Review* 5: 85–90.
Venner, B.G. 1970. 'Fallow deer in the Forest of Dean'. *Deer* 2: 446–8.
Vesey-Fitzgerald, B. 1936. 'Welcome or unwelcome guest?' *The Field* 168: 1075.
Vesey-Fitzgerald, B. 1938. 'Squirrel-tailed dormouse'. *The Field* 171: 927.
Vevers, G.M. 1947. 'The fat dormouse and other wild mammals at Whipsnade'. *Bedfordshire Naturalist* 2: 42–4.
Vilà, C., Leonard, J., Gotherstrom, A., Marklund, S., Sandberg, K., Liden, K., Wayne, R.K. & Ellegren, H. 2001. 'Widespread origins of domestic horse lineages'. *Science* 291: 474–7.
Vinicombe, K.E. & Chandler, R.J. 1982. 'Movements of Ruddy Ducks during the hard winter of 1978/79'. *British Birds* 75: 1–11.
Vizoso, M. 1967. 'Squirrel populations and their control'. *Forestry* (Supplement): 15–20.

Wade-Evans, A.A. 1909. *Welsh Medieval Law*: 226–7. Oxford University Press: Oxford.
Walker, A.F. 1976. 'The American brook trout in Scotland'. *Rod and Line* 16: 24–6.
Walker, A.F.G. 1970. 'The moult migration of Yorkshire Canada geese'. *Wildfowl* 21: 99–104.
Walker, A.F. 2003. 'Status of rainbow trout in Scotland: the results from a questionnaire survey'. *Scottish Fisheries Information Pamphlet* 23.
Walker, A.F. 2004. 'An investigation of escaped rainbow trout in the upper River Earn and Loch Earn during 2002/03'. *Scottish Fisheries Information Pamphlet* 24.
Walker, C.E. & Petterson, C.S. 1898. *The Rainbow Trout.* Lawrence & Bullen: London.
Walsh, P.M. 1988. 'Black rats *Rattus rattus* (L.) as prey of short-eared owls *Asio flammeus* on Lambay Island, Co. Dublin'. *Irish Naturalists' Journal* 22: No. 12.
Walton, P. 2001. 'The ruddy duck and the white-headed duck: the case for ruddy duck population control in the UK'. *Glasgow Naturalist* 23 (Supplement): 91–8.
Ward, A. 2005a. 'New population estimates for British mammals'. *Deer* 13: 8.
Ward, A.I. 2005b. 'Expanding ranges of wild and feral deer in Great Britain'. *Mammal Review* 35: 165–73.
Ward, A. 2005c. 'The spread of deer across Britain from 1972 to 2002'. *Deer* 13: 20–3.
Ward, M. 2008. 'The sea eagles are back – and they are slaughtering our lambs, say crofters'. *The Times* (24 September).
Warner, L.J. 1972. 'A muntjac in Windsor Forest'. 2: 876–7.
Warwick, T. 1934. 'The distribution of the muskrat (*Fiber zibethicus*) in the British Isles'. *Journal of Animal Ecology* 3: 250–67.
Warwick, T. 1941. 'Contribution to the ecology of the muskrat in the British Isles'. *Proceedings of the Zoological Society of London* 110: 165–201.
Waters, E. 2001. *The Great Bustard.* The Great Bustard Group: Salisbury.
Watkins-Pitchford, D. 1963. 'Wild goats of the Welsh mountains'. *Country Life* (21 November).
Watola, G., Allan, J.R. & Feare, C.J. 1996. 'Problems and management of naturalised introduced Canada Geese *Branta canadensis* in Britain'. In: Holmes, J.S. & Simons, J.R. (eds). *The Introduction and Naturalisation of Birds*: 71–8. The Stationery Office: London.
Watson, J. 1997. *The Golden Eagle.* Poyser: London.
Watson, J., Leitch, A.F. & Broad, R.A. 1992. 'The diet of the Sea Eagle *Haliaeetus albicilla* and Golden Eagle *Aquila chrysaetos* in western Scotland'. *Ibis* 134: 27–31.
Watt, H.B. 1923. 'The American grey squirrel in Ireland'. *Irish Naturalist* 32: 95.
Watt, H.B. & Darling, F. 1937. 'On the wild goat in Scotland. Habits of wild goats in Scotland'. *Journal of Animal Ecology* 6: 15–22.
Webster, B. 1983. 'The beaver's major effects on ecology'. *International Herald Tribune* (20 January).
Weir, A., Mcleod, J. & Adams, C.E. 1995. 'The winter diet and parasitic fauna of a population of red-necked wallabies *Macropus rufogriseus* recently introduced to Scotland'. *Mammal Review* 25: 111–16.
Weir, B. 2008. 'Face it – tiny aliens must be eradicated'. *Anglers' Mail* (March): 8.
Welch, D., Carss, D.N., Gornall, J., Manchester, S.J., Marquiss, M., Preston, C.D., Telfer, M.G., Arnold, H. & Holbrook, J. 2001. 'An audit of alien species in Scotland'. *Scottish Natural Heritage Review No. 139.*
Welcomme, R.L. 1998. 'International introductions of inland aquatic fish species'. *FAO Fisheries Technical Paper* 294. (318 pages)
Wellby, I., Pearson, R. & Easton, K. 2006. 'The future of the burbot *Lota lota* in British freshwaters'. *Proceedings of the Annual International Fisheries Management Conference*: 29–33.
Went, A.E.J. 1950. 'Notes on the introduction of some freshwater fish into Ireland'. *Journal of the Department of Agriculture, Dublin* 47: 119–24.
Went, R. 1975. 'Rare Birds: the North American Ruddy Duck'. *Bird Life 1975*: 18–19.
Wernham, C., Toms, M., Marchant, J., Clark, J., Siriwardena, G. & Baillie, S. 2002. *The Migration Atlas: Movements of the Birds of Britain and Ireland.* T. & A.D. Poyser: London.
Wheatley, J.J. 1970. 'Status of Carolina Duck (*Aix sponsa*) in Surrey'. *Surrey Bird Club Quarterly Bulletin* 55: 15.
Wheeler, A. (ed.). 1971. *The Freshwater Fish of Britain and Europe.* Collins: London.
Wheeler, A.C. 1974. *Changes in the Freshwater Fish Fauna of Britain. Systematics Association Special Volume No. 6.* Academic Press: London.
Wheeler, A.C. 1977. 'The origin and distribution of the freshwater fishes of the British Isles'. *Journal of Biogeography* 4: 1–24.
Wheeler, A.C. 1978. '*Ictalurus melas* (Rafinesque, 1820) and *I. nebulosus* (Lesueur, 1819): the North American catfishes in Europe'. *Journal of Fish Biology* 12: 435–9.
Wheeler, A.C. 1998. 'Field key to the freshwater fishes and lampreys of the British Isles'. *Field Studies* 9: 355–94.
Wheeler, A.C. 2000. 'The status of the crucian carp, *Carassius carassius* (L), in the UK'. *Fisheries Management and Ecology* 7: 315–32.
Wheeler, A. 2001. 'Fishes'. In: Hawksworth, D.L. (ed.). *The Changing Wildlife of Great Britain and Ireland. Systematics Association Special Volume Series* 62: 410–21.
Wheeler, A. & Maitland, P.S. 1973. 'The scarcer freshwater fishes of the British Isles. 1. Introduced species'. *Journal of Fish Biology* 5: 49–68.
Wheeler, A.C., Merrett, N.R. & Quigley, D.T.G. 2004. 'Additional records and notes for Wheeler's (1992) list of the common and scientific names of fishes of the British Isles'. *Journal of Fish Biology* 65 (Supplement B): 1–40.
Whitehead, G.K. 1945. 'Wild goats of Scotland'. *The Field* (10 March).
Whitehead, G.K. 1952. 'An historic herd of "wild" goats'. *Country Life Annual.*
Whitehead, G.K. 1957. 'Wild goats of Wales'. *Country Life* (26 September).
Whitehead, G.K. 1964. *The Deer of Great Britain and Ireland.* Routledge & Kegan Paul: London.
Whitehead, G.K. 1972. *The Wild Goats of Great Britain and Ireland.* David & Charles: Devon.
White-tailed Eagle Project Steering Group 2006. 'The Reintroduction of the White-tailed Eagle to the Republic of Ireland. Licence Application for the collection of White-tailed Eagles from Norway'.

Whitfield, D.P., Evans, R.J., Broad, R.A., Fielding, A.H., Haworth, P.F., Madders, M. & McLeod, D.R.A. 2002. 'Are introduced white-tailed eagles in competition with golden eagles?' *Scottish Birds* 23: 36–45.

Wiepkema, P.R. 1961. 'An ethological analysis of the reproductive behaviour of the bitterling (*Rhodeus amarus* Bloch)'. *Archieves Neerlandaises Zoologie* 14: 103–99.

Wijngaarden-Bakker, L.H. van. 1974. 'The animal remains from the Beaver Settlement at Newgrange, Co. Meath: first report'. *Proceedings of the Irish Academy* 74C: 313–83.

Wilberly, E.J. 1979. 'The Orcadian sheep of Lihou'. *The Ark* 6: 282–7.

Wildman, L., O'Toole, L. & Summers, R.W. 1998. 'The diet and foraging behaviour of the Red Kite in Scotland'. *Scottish Birds* 19: 134–40.

Wilkinson, N.I., Langston, R.H.W., Gregory, R.D., Gibbons, D.W. & Marquiss, M. 2002. 'Capercaillie abundance and habitat use in Scotland, in winter 1998–99'. *Bird Study* 49: 177–85.

Willet, J.A. & Mulloy, F. 1970. 'Wild deer, their status and distribution'. *Deer* 2: 498–504.

Williams, M. 1999. 'Parrot puzzle'. *BBC Wildlife* 17: 42.

Williamson, K. & Boyd, J.M. 1960. *St Kilda Summer*. Hutchinson: London.

Williamson, M. 1993. 'Invaders, weeds and the risk from genetically modified organisms'. *Experientia* 49: 219–24.

Williamson, M. & Fitter, A. 1996. 'The varying success of invaders'. *Ecology*, Washington 77: 1661–6.

Wilson, C.J. 1999. 'Wild boar in south west England'. In: Thomaidis, C. & Kypridemos, N. (eds). *Proceedings of the International Union of Game Biologists XXIVth Congress 1999*: 490–2. Thessaloniki, Greece.

Wilson, C.J. 2003a. 'Distribution and status of feral wild boar *Sus scrofa* in Dorset, southern England'. *Mammal Review* 33: 302–7.

Wilson, C.J. 2003b. *Preliminary Report on Distribution and Status of Feral Wild Boar in England*. Unpublished report to DEFRA European Wildlife Division.

Wilson, C.J. 2004. 'Rooting damage to farmland in Dorset, southern England, caused by wild boar *Sus scrofa*'. *Mammal Review* 34: 331–5.

Wilson, C.J. 2005. *Feral Wild Boar in England: Status, Impact and Management*. Report to DEFRA European Wildlife Division.

Winfield, I.J. & Durie, N.C. 2004. 'Fish introductions and their management in the English Lake District'. *Fisheries Management and Ecology* 11: 195–201.

Wintle, M. 2001. 'When Leney brought black bass into the UK'. *Classic Angling* 14: 15–17.

Witherby, H.F. & Ticehurst, N.F. 1908. 'Spread of the little owl in Britain'. *British Birds* 1: 335–42.

Witherby, H.F., Jourdain, F.C.R., Ticehurst, N.F. & Tucker, B.W. 1938–41. *The Handbook of British Birds*. Five volumes. H.F. & G. Witherby: London.

Wood, J.S. 1964. 'Normal development and causes of reproductive failure in Canada geese'. *Journal of Wildlife Management* 28: 197–208.

Woodman, P.C. 1978. *The Mesolithic in Ireland*. BAR British Series 58: Oxford.

Woolcock, N. 2007. 'Great bustard egg is first since 1832'. *The Times* (24 July).

Worden, J., Cranswick, P.A., Trinder, M.N. & Hughes, B. 2006, 2007. 'Monitoring the UK ruddy duck population during ongoing control operations: results of the January 2006 and winter 2006/07 surveys'. *WWT Wetlands Advisory Service Reports to Central Science Laboratory*. (12 and 29 pages.)

Worthington, E.B. 1941. 'Rainbows: a report on attempts to acclimatize rainbow trout in Britain'. *Salmon & Trout Magazine* 100: 241–60; 101: 62–99.

Wotton, S.R., Carter, I., Cross, A.V., Etheridge, B., Snell, N., Duffy, K., Thorpe, R. & Gregory, R. D. 2002. 'Breeding status of the Red Kite *Milvus milvus* in Britain in 2000'. *Bird Study* 49: 278–86.

www.surrey-arg.org.uk ('Wall Lizards – keeping track of an introduced species').

www.greatbustard.com

Wycherley, J. 2003. 'Water frogs in Britain'. *British Wildlife* (April): 260–9.

Wycherley, J. & Anstis, R. 2001. *Amphibians and Reptiles of Surrey*. Surrey Wildlife Trust: Pirbright.

Wycherley, J., Doran, S. & Beebee, T.J.C. 2003. 'Tracing aliens: identification of introduced water frogs in Britain by male advertisement call characteristics'. *Herpetological Journal* 13: 43–50.

Yalden, D.W. 1965. 'Distribution of reptiles and amphibians in the London area'. *London Naturalist* 44: 58–69.

Yalden, D. 1977. 'Small mammals and the archaeologist'. *Bulletin of the Peakland Archaeological Society* 30: 18–25.

Yalden, D.W. 1981. 'The occurrence of the pygmy shrew *Sorex minutus* on moorland, and the implications for its presence in Ireland'. *Journal of Zoology* 195: 147–56.

Yalden, D.W. 1982. 'When did the mammal fauna of the British Isles arrive?' *Mammal Review* 12: 1–57.

Yalden. D.W. 1986. 'Opportunities for reintroducing British mammals'. *Mammal Review* 16: 53–63.

Yalden, D.W. 1988. 'Feral wallabies in the Peak District, 1971–1985'. *Journal of Zoology* 215: 369–74.

Yalden, D.W. 1992. 'Changing distribution and status of small mammals in Britain'. *Mammal Review* 22: 97–106.

Yalden, D. 1999. *The History of British Mammals*. Poyser: London.

Yalden, D. 2001. 'The return of the prodigal swine'. *Biologist* 48: 259–62.

Yalden, D.W. 2007. 'The older history of the White-tailed Eagle in Britain'. *British Birds* 100: 471–80.

Yalden, D.W. & Hosey, G.R. 1971. 'Feral wallabies in the Peak District'. *Journal of Zoology* 165: 513–20.

Yeaman, J.A. 1932. 'Polecat-Ferrets on Mull'. *Scottish Naturalist* 1932: 66.

Yeomans, W.E., Chubb, J.C., Sweeting, R.A., Smith, S. & Whitfield, P.J. 1996. 'Fish parasites as indicators of environmental degradation'. *Institute of Fisheries Management Annual Study Course Proceedings* 27: 81–119.

Yeomans, W.E., Chubb, J.C. & Sweeting, R.A. 1997. '*Khawia sinensis* (Cestoda: Caryophyllidea) – an indicator of legislative failure to protect freshwater habitats in the British Isles?' *Journal of Fish Biology* 51: 880–5.

Young, H.G. 1987–8. 'Zoology section reports for 1986 and 1987'. *Annual Bulletin de la Société Jersiaise* 24: 324–5, 474–5.

Young, H.G. & Duffy, K. 1984. 'Night-Herons in Scotland'. *Annual Report of the Royal Zoological Society of Scotland*: 40–6.

Zeisset, I. & Beebee, T.J.C. 2003. 'Population genetics of a successful invader: the marsh frog *Rana ridibunda* in Britain'. *Molecular Ecology* 12: 639–46.

Zeuner, F.E. 1963. *A History of Domesticated Animals*. Hutchinson: London.

INDEX

Common names and page numbers names written in **bold** denote main entries.

PICTURE CREDITS

All photographs NHPA/Photoshot

Front cover Ernie Janes. **Back cover** clockwise from top, left: NHPA/Photoshot, Alan Williams, LUTRA, Stephen Dalton. **Spine** Ernie Janes.
A.N.T. Photo Library 20; Alan Barnes 169; Bruce Beehler 121; Joe Blossom 38, 40, 95, 174, 178, 201 (top), 212, 213, 223, 232, 301, 367; Laurie Campbell 298; John Cancalosi 369; Bill Coster 352, 360, 362, 375, 380; Lee Dalton 181, 377; Stephen Dalton 42, 48, 78, 217; Manfred Danegger 57, 69, 327, 355; Reinhard Dirscherl 236 (top); Melvin Grey 24; Dan Griggs 363; Erlend Haarberg 323; Martin Harvey 133, 198, 378; Daniel Heuclin 73, 75, 199, 207, 228; Ernie Janes 19, 34, 44, 87, 114, 165, 331, 379, 384; John Jeffery 188; T. Kitchin & V. Hurst 219; Stephen Krasemann 80; Gerard Lacz 236 (bottom), 261, 374; Mike Lane 147 150, 170, 240, 291, 318. 337; Michael Leach & Meriel Lland 293, 371; Jean-Louis Le Moigne 210; LUTRA 235, 242, 245, 259, 265, 275, 280, 283; Trevor McDonald 249, 256; NHPA/Photoshot 201 (bottom); Pierre Petit 228 (bottom); Andy Rouse 119, 137, 321, 332; Jany Sauvanet 187; John Shaw 123, 139, 364; David Slater 2; Mirko Stelzner 196; Jason Stone 159; Karl Switak 214; Robert Thompson 343; David Tipling 346; Ann & Steve Toon 117, 304, 351; Hellio & Van Ingen 338; Dave Watts 130, 382; Alan Williams 104, 143, 161, 295, 31; Martin Zwick 227.